LC-MS in Drug Bioanalysis

Q. Alan Xu • Timothy L. Madden
Editors

LC-MS in Drug Bioanalysis

Springer

Editors
Q. Alan Xu
Department of Experimental Therapeutics
University of Texas
MD Anderson Cancer Center
Houston, TX, USA

Timothy L. Madden
Department of Experimental Therapeutics
University of Texas
MD Anderson Cancer Center
Houston, TX, USA

ISBN 978-1-4899-9089-1 ISBN 978-1-4614-3828-1 (eBook)
DOI 10.1007/978-1-4614-3828-1
Springer New York Heidelberg Dordrecht London

Printed on acid-free paper

Springer is part of Springer Science+Business Media (www.springer.com)

Preface

The analysis of drugs in complex biological matrices plays a very important role in drug development, the conduct of clinical trials, and in therapeutic drug monitoring. Numerous analytical techniques have been developed to perform bioanalysis, which include immunoassays, gas chromatography-flame ionization detection (GC-FID), gas chromatography-mass spectrometry (GC-MS), high performance liquid chromatography-ultraviolet (HPLC-UV), high performance liquid chromatography-mass spectrometry (HPLC-MS or LC-MS), and high performance liquid chromatography-tandem mass spectrometry (HPLC-MS/MS or LC-MS/MS). A huge selection of analytical columns and mobile phases provides for chromatographic analysis having extreme separation power, while the addition of tandem mass spectrometry as a detection and quantification methodology provides a high degree of selectivity and sensitivity. In addition, the combination of liquid chromatography and tandem mass spectrometry (LC-MS/MS) can provide high-throughput analysis for complex biological samples. LC-MS/MS is increasingly becoming the method of choice for the determination of drug(s) in drug development, clinical pharmacology, toxicology, and therapeutic drug monitoring.

However, LC-MS methods can be quite expensive and require highly skilled and well-trained personnel, and the LC-MS/MS technique is also prone to some pitfalls. It is our; hope that *LC-MS in Drug Bioanalysis* will help readers to improve their technical skills in LC-MS/MS method development, validation, and application to the analysis of drugs and drug metabolites. First, this book presents discussions on the application of internal standardization, method development and validation for regulated quantitative bioanalysis, and associated pitfalls of LC-MS/MS, such as ion suppression or signal enhancement from matrix. It also includes information on newer sampling techniques which are becoming increasingly popular—dried blood spots (DBS) and microflow liquid chromatography-mass spectrometry. Second, this book provides detailed information on the applications of LC-MS/MS in bioanalysis, including the analysis of antipsychotic drugs, antidepressants, illicit drugs, steroid hormones, and tropane alkaloids in human biological samples. Matrix-assisted laser desorption/ionization imaging mass spectrometry is briefly introduced.

The planar integrated micro mass spectrometer—one of the new mass spectrometers under development—is also introduced here. Such a portable handheld micro mass spectrometer would be very useful for bedside therapeutic drug monitoring.

We believe this book will be a great aid for college students and academics, clinical pharmacologists and toxicologists, and pharmaceutical scientists. We sincerely hope that readers will find this book useful.

We would like to acknowledge all of the authors who found time in their busy schedule to contribute the thoughtful chapters. Our task of compiling this book was made easy by their high-quality efforts. We would also like to thank Mr. Kenneth Howell and Christopher Balmes at Springer for their much valued assistance throughout the preparation of this book.

Houston, Texas Q. Alan Xu
Houston, Texas Timothy L. Madden

Contents

Contributors

Jochen Beyer Department of Forensic Medicine, Victorian Institute of Forensic Medicine, Monash University, Southbank, VC, Australia

Patrick K. Bennett Thermo Fisher Scientific, San Jose, CA, USA

Nadine Boudreau PharmaNet Canada Inc., Quebec, QC, Canada

Thierry Buclin Division of clinical Pharmacology, Service of Biomedicine, University Hospital Center and University of Lausanne, Lausanne, Switzerland

Ana de Castro Forensic Toxicology Service, Institute of Forensic Sciences, University of Santiago de Compostela, Santiago de Compostela, Spain

Chang Cheng Analytical Development, Albany Molecular Research, Inc., Rensselaer, NY, USA

Chantal Csajka Department of Pharmaceutical Sciences, University of Geneva, Geneva, Switzerland

Angelines Cruz Forensic Toxicology Service, Institute of Forensic Sciences, University of Santiago de Compostela, Santiago de Compostela, Spain

Elyes Dahmane Division of clinical Pharmacology, Service of Biomedicine, University Hospital Center and University of Lausanne, Lausanne, Switzerland

Department of Pharmaceutical Sciences, University of Geneva, Switzerland, Geneva, Switzerland

Laurent Decosterd Division of Clinical Pharmacology-Laboratory, Department of Medicine, Centre Hospitalier Universitaire Vaudois, University of Lausanne, Lausanne, Switzerland

Innovation and Development Unit, Service of Biomedicine, University Hospital Center and University of Lausanne, Lausanne Switzerland

Olaf H. Drummer Department of Forensic Medicine, Victorian Institute of Forensic Medicine and Monash University, Southbank, VIC, Australia

Xiaotao Duan Department of Pharmaceutical Sciences, University at Buffalo, State University of New York, Amherst, NY, USA

Tawakol A. El-Shourbagy Global Pharmaceutical Research and Development, Abbott Laboratories, Dept R46W, Abbott Park, IL, USA

Dimitri Gerostamoulos Department of Forensic Medicine, Victorian Institute of Forensic Medicine and Monash University, Southbank, VIC, Australia

Amina Haouala Innovation and Development Unit, Division of Clinical Pharmacology, Service of Biomedicine, University Hospital Center and University of Lausanne, Lausanne, Switzerland

Harald John Bundeswehr Institute of Pharmacology and Toxicology, Munich, Germany

Ann Lévesque PharmaNet Canada Inc., Québec, QC, Canada

Shaorong Liu Department of Chemistry and Biochemistry, University of Oklahoma, Norman, OK, USA

Manuel López-Rivadulla Forensic Toxicology Service, Institute of Forensic Sciences, University of Santiago de Compostela, Santiago de Compostela, Spain

Kara L. Lynch Department of Laboratory Medicine, University of California, San Francisco, CA, USA

San Francisco General Hospital, San Francisco, CA, USA

Timothy L. Madden Department of Experimental Therapeutics, University of Texas MD Anderson Cancer Center, Houston, TX, USA

Min Meng Tandem Labs, Salt Lake City, UT, USA

Jörg Müller Institute for Microsystem Technology, Hamburg University of Technology, Hamburg, Germany

Sabino Napoletano Department of Chemistry, Sapienza University of Rome, Roma, Italy

Marine Neeman Innovation and Development Unit, Division of Clinical Pharmacology, Service of Biomedicine, University Hospital Center and University of Lausanne, Lausanne, Switzerland

Jihai Pang Department of Experimental Therapeutics, UT MD Anderson Cancer Center, Houston, TX, USA

Jun Qu Department of Pharmaceutical Sciences, University at Buffalo, State University of New York, Amherst, NY, USA

Grigoriy Quiring Institute for Microsystem Technology, Hamburg University of Technology, Hamburg, Germany

Maria Reinhardt-Szyba Institute for Microsystem Technology, Hamburg University of Technology, Hamburg, Germany

Ramona Rodila Global Pharmaceutical Research and Development, Abbott Laboratories, Dept R46W, Abbott Park, IL , USA

Eva Saar Department of Forensic Medicine, Victorian Institute of Forensic Medicine and Monash University, Southbank, VIC, Australia

Christoph Seger Institute of Medical and Chemical Laboratory Diagnostics (ZIMCL), University Hospital Innsbruck, Innsbruck, Austria

Manuel Sergi Department of Food Science, University of Teramo, via C. Lerici 1, Mosciano S.A, TE, Italy

Aimin Tan BioPharma Services, Inc., Toronto, ON, Canada

Michael Vogeser Institute of Clinical Chemistry, Hospital of the University of Munich, Munich, Germany

Henning Wehrs Institute for Microsystem Technology, Hamburg University of Technology, Hamburg, Germany

Nicolas Widmer Division of clinical Pharmacology, Service of Biomedicine, University Hospital Center and University of Lausanne, Lausanne, Switzerland

Régulo Miguel Ramírez Wong Institute for Microsystem Technology, Hamburg University of Technology, Hamburg, Germany

Alan H.B. Wu Department of Laboratory Medicine, University of California, San Francisco, CA, USA

San Francisco General Hospital, San Francisco, CA, USA

Huaiqin Wu Global Pharmaceutical Research and Development, Abbott Laboratories, Dept R46W, Abbott Park, IL, USA

Q. Alan Xu Department of Experimental Therapeutics, University of Texas, MD Anderson Cancer Center, Houston, TX, USA

Haoying Yu Department of Pharmaceutical Sciences, University at Buffalo, State University of New York, Amherst, NY, USA

Zimeng Yan Research Pharmaceutical Services, Inc., Fort Washington, PA, USA

Jun Zhang Global Pharmaceutical Research and Development, Abbott Laboratories, Dept R46W, Abbott Park, IL, USA

Internal Standards for Quantitative LC-MS Bioanalysis

Aimin Tan, Nadine Boudreau, and Ann Lévesque

Abstract Internal standards play critical roles in ensuring the accuracy of reported concentrations in LC-MS bioanalysis. How do you find an appropriate internal standard so that analyte losses and experimental variations during sample preparation, chromatographic separation, and mass spectrometric detection could be corrected? How is the concentration of an internal standard determined? Should internal standard responses be monitored during the analysis of incurred samples? What are the main causes for internal standard response variations? How do they impact the quantitation? Why are stable isotope labeled internal standards preferred? And yet one should still have an open-mind in their usage for the analysis of incurred samples. All these questions are addressed in this chapter supported by theoretical considerations and practical examples.

1 Internal Standards and Analytical Calibrations

Biological samples (plasma, serum, blood, and urine) are very complex. They contain a wide variety of matrix components such as proteins, lipids, and salts. To quantify trace amount of analytes (e.g., drug and its metabolites) in complex biological samples by liquid chromatography-mass spectrometry (LC-MS), the samples should be properly treated prior to being injected onto an LC-MS instrument,

A. Tan (✉)
BioPharma Services, Inc., Toronto, ON, Canada
e-mail: Aimintan@hotmail.com

N. Boudreau • A. Lévesque
PharmaNet Canada, Inc., Québec, QC, Canada

Q. Alan Xu and Timothy L. Madden (eds.), *LC-MS in Drug Bioanalysis*,
DOI 10.1007/978-1-4614-3828-1_1, © Springer Science+Business Media, LLC 2012

which typically includes dilution, extraction, evaporation, and reconstitution. Variable losses of the analytes may occur during these sample treatment steps. In addition, there might be variations during the LC-MS analysis, such as variations in injection volume and particularly in ionization (ion suppression or enhancement caused by coeluted matrix components; refer to [1]).

To reduce the impact of analyte losses and instrumental variations on the quantitation of an analyte, an internal standard (IS), which has the same or similar physical and chemical properties as the analyte, is added in equal amount to both concentration-known (calibration standards and quality controls) and unknown samples prior to sample treatment. By using signal ratios of the analyte to its IS (instead of using analyte absolute signal) for quantitation, the losses and variations can be corrected, which improves the precision and accuracy of final analytical results for unknown samples.

Internal standards could be used in external calibration, matrix-matched external calibration, and standard addition calibration [2]. However, the use of internal standards in LC-MS quantitative methods should not be confused with internal calibration in which an internal standard is employed as a calibrant and the concentration of a unknown sample is calculated from the concentration of this internal standard and its analyte/IS signal ratio, i.e., the concentration of the unknown sample is calculated without the need for a calibration curve [3]. The use of internal standards in most LC-MS quantitative methods belongs to "signal-ratio calibration" or internal standardization [2, 4]. In fact, the majority of bioanalytical LC-MS methods use matrix-matched signal-ratio external calibration.

The analytes of interest in quantitative bioanalysis vary, from small molecules with molecular weights usually less than 1,000 Da (e.g., drugs and their metabolites) to large biopolymers, such as proteins. The focus of this chapter is on the quantitation of small molecules in biological samples by LC-MS, though some of the principles presented in this chapter are also applicable to the quantitative analysis of large molecules in biological samples.

2 Selection and Use of Internal Standards for Quantitative LC-MS Bioanalysis

2.1 Requirements on Internal Standards

An internal standard should meet the following three requirements. First, it should have the same or very similar physical–chemical properties as the analyte, particularly hydrophobicity and ionization characteristics, so that it can mimic closely the performance of the analyte in every stage of analysis, i.e., from sample preparation, chromatographic separation, to mass spectrometric detection. In this way, any losses during sample preparation or variations in the mass spectrometry detection can be corrected.

Second, an internal standard must have adequate purity. Preferably, the contribution of an internal standard to any analyte should be less than 20 % of the corresponding lower limit of quantitation (LLOQ) of the analyte. Otherwise, the significant amount of analyte from the added internal standard can bias the reported signal to noise (S/N) ratio at the LLOQ and cause larger variability at low concentrations. The interference of an internal standard to other cointernal standards in a multianalyte method is rare, but it should be also evaluated. Though there are no reported criteria for this, it should be at least less than 15 % of the concentration of a cointernal standard in a multianalyte method. In addition, an internal standard should not correspond to any in vivo metabolic products of the analyte (e.g., hydroxylated metabolite, N-dealkylation metabolite).

Lastly, an internal standard should be stable during sample processing and LC separation. It should not significantly degrade, in particular not decompose to components that can interfere with the determination of the analyte.

2.2 *Types of Internal Standards*

There are two main types of internal standards. The first ones are stable isotope labeled (SIL) internal standards. They are compounds in which several atoms in the analytes are replaced by their respective stable isotopes, such as deuterium (^{2}H, D or d), ^{13}C, ^{15}N, or ^{17}O. Labeling with the first three isotopes are most common, particularly labeling with deuterium (due to less difficulty in synthesis and therefore less expensive). For examples, raloxifene-d_4-6-glucuronide was used as the internal standard for the determination of raloxifene-6-glucuronide [5] and 1, 2, 3, 4-$^{13}C_4$ estrone ([$^{13}C_4$]E1) was used as the internal standard for estrone (E1) [6]. The usage of stable isotope labeled internal standards in quantitative LC-MS or GC-MS analysis is often termed as isotope dilution mass spectrometry (IDMS) [7].

The second ones are structural analogues with different masses or even the same mass. In the latter case, chromatographic separation between the analyte and its internal standard must be achieved when distinctive MRM (multiple reaction monitoring) transitions could not be found. It is preferable that the key structure and functionalities (e.g., –COOH, –SO_2, –NH_2, halogens, and heteroatoms) of an internal standard are the same as those of the analyte and differ only by C–H moieties (length and/or position). Modifications in functionalities would result in significant differences in ionization efficiency and extraction recovery [8].

2.3 *Selection of Internal Standards*

An internal standard is expected to track the analyte in all the three distinctive stages of LC-MS bioanalysis, i.e., sample preparation (extraction), chromatographic separation, and mass spectrometric detection. Though the emphasis should usually be placed on tracking the analyte in mass spectrometric detection due to the possibility of ion suppression or enhancement from coeluting component(s), all these three stages should be considered as a whole while choosing an internal standard. In other words, these three stages affect the selection of an internal standard interactively. For example, when a sample preparation method produces extremely clean extracts, then the emphasis should be shifted from tracking the analyte in MS detection to tracking the analyte in extraction. On the other hand, if the extracts from the sample preparation contain coeluting matrix components that cause ion suppression or enhancement, then tracking the analyte during MS detection to correct matrix effects becomes more important. Moreover, different extraction methods may have different requirements on the internal standards. Apparently, the requirement on internal standards for tracking an analyte during a simple "dilution-and-shoot" treatment would be less stringent than that for liquid–liquid extraction (LLE) or solid-phase extraction (SPE) methods.

Among the two types of internal standards, SIL internal standards, particularly those labeled with ^{13}C and/or ^{15}N, are most effective and should therefore be used whenever possible [9]. The molecular weight (MW) of an SIL internal standard should be at least 3 Da (ideally 4 or 5 Da) higher than that of the analyte, though this is not an absolute necessity. For example, norethindrone-$^{13}C_2$ has been successfully used for the determination of norethindrone over the concentration ranges of 2.5–500 pg/mL and 0.05–10 ng/mL [10]. In addition, the location of stable isotope atoms should be given consideration in synthesizing a deuterated internal standard, so that deuterium exchange would not occur during sample preparation. It is also preferable that the stable isotope atoms are included in the product ions as well to avoid possible cross-contamination during mass spectrometric detection, though it is rare in modern mass spectrometers.

Despite the very desirable performances of SIL internal standards, they are not always available or are very expensive, especially when seeking exclusively non-deuterium-labeled internal standards. Then, structural analogues can be used. In this case, coelution of an analyte and its internal standard is preferred in order to reduce matrix effect [11] or to expand linearity range [12].

There are many different ways to find structural analogues, such as literature search or key chemical structure search. Merck index is a valuable reference source. Usually, internal standards can be found from the same therapeutic class as the analyte. For example, to find a potential internal standard for penciclovir, acyclovir, ganciclovir, and valacyclovir could be considered because all of them belong to guanine analogue antiviral drug, i.e., containing a key moiety of guanine (Fig. 1). Using skeleton structure search is another useful way to find an internal standard. Alternatively, structural analogues can be synthesized by adding a nonessential group, such as an extra methylene group ($-CH_2-$).

Once potential structural analogues are found, their physical–chemical properties, such as $\log D$ (hydrophobicity) vs. pH, can be calculated and compared with those of the analyte using software (e.g., Pallas) prior to being experimentally tested. As shown in Fig. 2, both acyclovir and ganciclovir could be used as the internal standard for penciclovir, particularly the latter due to the same number of hydroxy

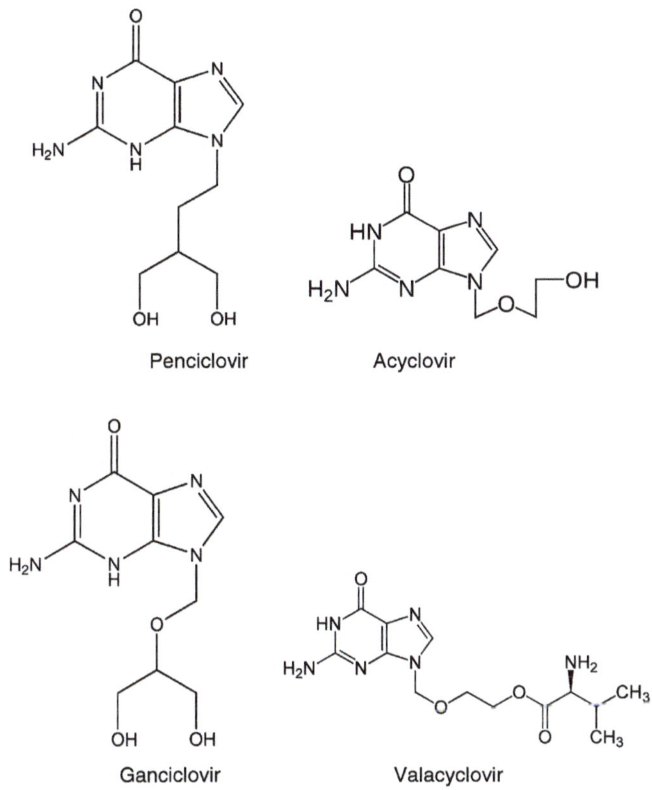

Fig. 1 Chemical structures of penciclovir, acyclovir, ganciclovir, and valacyclovir

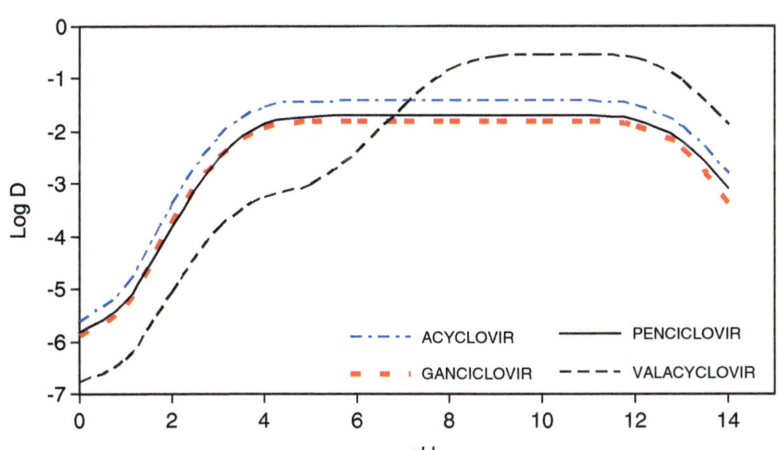

Fig. 2 Hydrophobicity (log*D*) vs. pH profiles calculated by Pallas software (version 3.1, CompuDrug International, Inc., Sedona, Arizona, USA)

moieties. On the other hand, valacyclovir is not as good because of the introduction of an extra amine moiety and the absence of the hydroxy moiety.

2.4 Concentration of Internal Standards

It is generally believed that as long as the same amount of an internal standard is added to all the samples in a batch (run), i.e., calibration standards, quality controls, and unknown samples, the concentration of an internal standard is not important. This is probably why not much information exists as how to determine an appropriate concentration for an internal standard. Some researchers proposed that the concentration of an internal standard should be approximately half of the upper limit of quantitation (ULOQ) of the analyte [13, 14] or even higher than the ULOQ [2], while others suggested a relatively lower concentration corresponding to about the first third of the calibration range, in order to minimize potential interferences with the analyte due to potential impurities from SIL internal standards [15]. Unfortunately, none of these were followed by more detailed theoretical considerations or supporting experimental data.

Based on our experience, it is difficult to set a clear-cut guideline as what IS concentration should be used. Instead, all of the following factors should be taken into consideration while determining an appropriate concentration for an internal standard with the emphasis on the first factor.

2.4.1 Purity of Reference Standards and the Targeted Calibration Range

An internal standard may contain trace amount of the analyte of interest, especially a SIL internal standard due to similar synthesizing routes. In this case, a maximum IS concentration can be obtained based on its impurity (e.g., $n\,\%$ of the analyte in the IS reference standard) and the ±20 % acceptance criterion for bias at the LLOQ level [13, 16–17]. In this regard, the concentration of an IS should be therefore kept as low as possible to reduce its contribution to analyte concentration.

$$C_{\text{IS - max}} = 20\ \text{LLOQ} / n. \tag{1}$$

On the other hand, the analyte or analyte reference standard may contribute to the response of an internal standard either due to impurity or because of natural abundance of stable isotopes (say $m\,\%$). Therefore, the concentration of the internal standard must not be too low. Otherwise, the contribution from the analyte or its reference standard would be significant, e.g., more than 5 % of IS concentration or response [16], and linearity and accuracy could be greatly affected. Accordingly, a minimum IS concentration could be obtained based on the ULOQ for the analyte. For this reason, a high IS concentration is preferred.

$$C_{\text{IS - min}} = m \text{ ULOQ} / 5. \tag{2}$$

Apparently, the $C_{\text{IS-max}}$ must be equal to or greater than the $C_{\text{IS-min}}$, i.e., $C_{\text{IS-max}} \geq C_{\text{IS-min}}$ to simultaneously satisfy the criteria of purity for both the analyte and its internal standard. Accordingly, the following formula is obtained.

$$\text{ULOQ} / \text{LLOQ} < 100 / (mn). \tag{3}$$

In other words, the purity of analyte and IS reference standards determines the maximum possible concentration span. Sometimes, it is impossible to simultaneously satisfy both (1) and (2), i.e., $C_{\text{IS-min}} > C_{\text{IS-max}}$. In this case, either calibration range needs to be adjusted (narrowed) or analyte or internal standard reference standards of higher purity must be used.

2.4.2 MS Sensitivity Towards an Analyte and Its Internal Standard

When the MS sensitivity towards an internal standard is higher than the analyte, relatively lower IS concentration can be used. Otherwise, higher IS concentration is necessary. The aim is to achieve S/N high enough for the internal standard so that noises in IS signal is negligible, i.e., no inclusion of any variations that have not been experienced by the analyte. For this reason, the IS concentration should be as high as possible. On the other hand, the IS signal should not be so high that the ratios for lower concentration standards are too low. When inappropriate regression algorithm or calculation precision is used, unreliable or even incorrect regression results could be obtained.

2.4.3 Ion Suppression or Enhancement

When matrix effect exists, it is usually preferable to coeluate the analyte and its internal standard to better reduce the impact of matrix effect on quantitation. The more their chromatographic peaks overlap, the better the correction is. Since the concentration of the analyte varies while the amount of IS added is constant, a choice must be made as to match which part of a calibration curve. Usually, the segment between 1/3 and 1/2 of the ULOQ is most important because this segment is expected to cover the average C_{max} for most drugs and metabolites. This is probably why other researchers have proposed to use IS concentrations around 1/3 or 1/2 of the ULOQ of an analyte.

In addition, mutual ion suppression or enhancement may exist when an analyte is coeluted with its internal standard. To maintain low detection limit, low IS concentration should be used. On the other hand, high IS concentration is necessary to obtain good reproducibility when an internal standard is suppressed by the analyte [14, 18]. Therefore, a balance needs to be made.

2.4.4 Solubility and Loading Capacity of Extraction Cartridges

The IS concentration should not be so high as to cause issues of solubility or exceed the loading capacity of solid-phase extraction (SPE) cartridges or similar products.

To conclude, the determination of IS concentration is not a trivial task because an inappropriate IS concentration could affect linearity, accuracy, and precision. Several different factors should be taken into consideration. Sometimes a high IS concentration is preferred, while in other times a low IS concentration is necessary. The most important is to know how to adjust the concentration of an internal standard when desired performances are not obtained. Once an appropriate concentration has been determined, it must be experimentally tested to check linearity, accuracy, and precision [18].

2.5 Introduction of Internal Standards

Internal standards may be introduced in three different ways, i.e., addition to samples prior to extraction, after the extraction but prior to LC-MS analysis, or even after chromatographic separation.

The most common way of introducing an internal standard is the addition to samples prior to any extraction procedures, e.g., before the addition of buffer and organic solvent in a liquid–liquid extraction, and as early as possible. In this way, the internal standard goes through the same preparation and extraction steps as the analyte does. This might be why an internal standard was initially termed "processed internal standard" [8, 19]. Apparently, if an appropriate internal standard is used, any variations from sample extraction to MS detection could be corrected, such as variability in dilution, organic transfer, recovery, adsorption, evaporation, injection, and ion suppression/enhancement.

However, it is sometimes very challenging to find an internal standard that can track the analyte equally well during sample extraction and LC-MS analysis. The sample extraction could be quite precise and rugged due to the properly designed procedures and/or use of automation. Accordingly, the correction of variability in MS detection becomes much more important. Moreover, the sample integrity could be sometimes compromised by the addition of an internal standard. In these cases, an internal standard can be introduced after sample extraction and prior to LC-MS analysis to address mainly instrumental errors (e.g., variation in injection volume and fluctuations in the ionization process). This approach was used in the development of a reliable method for the simultaneous determination of free and liposome-encapsulated drug forms in human serum because the addition of internal standard to study samples or even dilution of study samples with water or buffer could release the drug from liposome encapsulation, which compromises the integrity of samples. By thoughtful design, a solid-phase extraction procedure by direct loading of study samples without the addition of an internal standard or buffer was successfully developed and validated. The internal standard was introduced only after the SPE

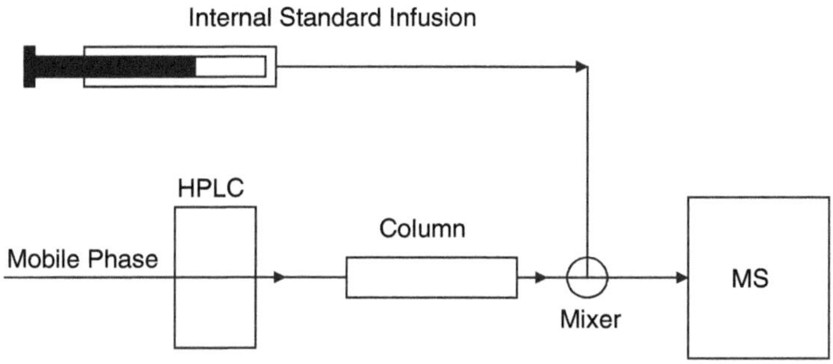

Fig. 3 Schematic representation of continuous infusion of an internal standard and postcolumn (postelution) mixing with LC eluent prior to being directed to a mass spectrometer (MS) for detection

separation of free and liposome-encapsulated drug forms. Good interday precision and accuracy were obtained, for example, % CV ranging from 4.20 to 8.26 ($n=84$) for free drug form and from 3.99 to 10.30 ($n=54$) for liposome-encapsulated drug form, which demonstrate the adequacy of this approach [20]. Similar approach was also adopted by others for this type of compound [21].

Apart from the above approaches, an internal standard can also be introduced after chromatographic separation with the aim to mainly compensate for quantitative errors attributed to signal suppression or enhancement in MS detection (Fig. 3), particularly when the sample loss during sample preparation procedure is minimal or has been evaluated. This approach is useful to avoid the usage of multiple internal standards in a multicomponent analytical method and to obtain the benefit of matrix effect correction with coelution of an analyte and the internal standard (even with a structural analogue internal standard). In addition, the quantity of the internal standard introduced may be readily controlled by adjusting the infusion flow rate and/or the concentration of IS solution. It has been demonstrated by Choi et al. [22] that one single SIL internal standard was good for both the parent drug and its metabolite. However, despite the coelution of an analyte and its internal standard, how well the matrix effect could be corrected will still depends on how close their physical–chemical properties are, especially ionization related properties. In addition, an additional pump is necessary for the introduction of the internal standard in this case.

2.6 *Alternative Approaches without an Internal Standard*

When an appropriate internal standard could not be found, it is sometimes possible to develop a method without an internal standard [19], especially for usage in early drug discovery stage where less strict criteria could be used. Nevertheless, efforts should be made to minimize the variations in sample extraction, LC separation, and

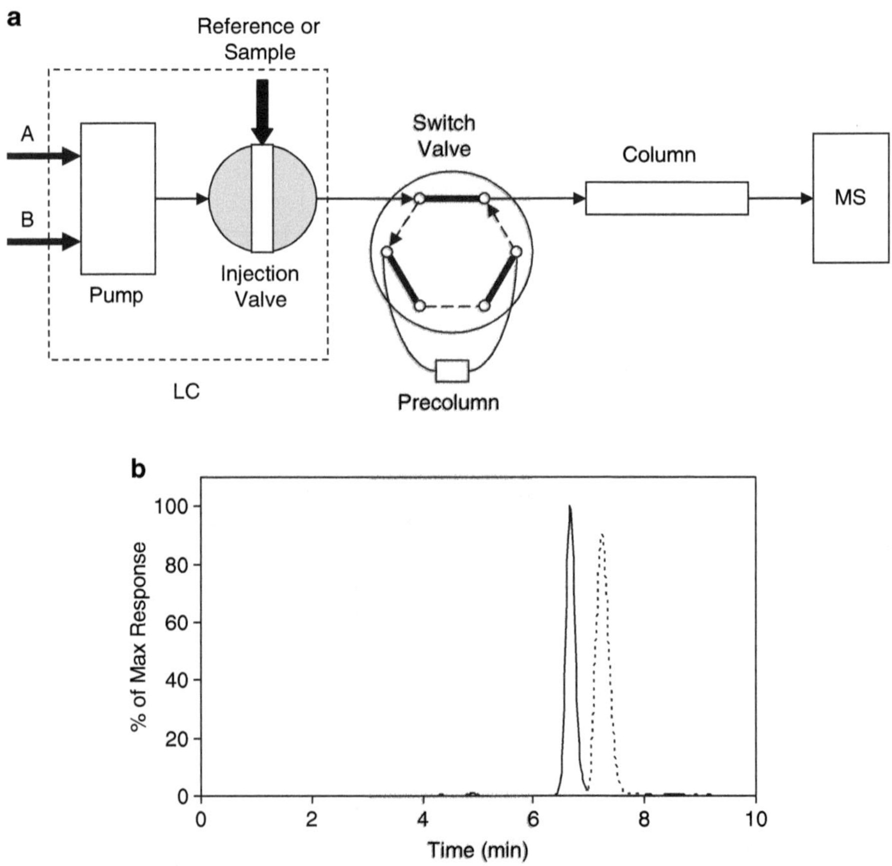

Fig. 4 Instrumental setup (**a**) and representative peaks (**b**) of echo-peak technique. An unknown sample is first injected and the mobile phase flows directly to the separation column. Within a short delay, a reference standard solution is injected and the switching valve turns to a different position to allow the mobile phase to flow through a short precolumn (as shown by *dashed arrows*) prior to passing through the same separation column. Two chromatographic peaks are therefore recorded, one for the unknown sample (*solid line*) and the other for the reference solution (*dotted line*). *A and B* mobile phases, *LC* liquid chromatograph, *MS* mass spectrometer

MS detection, such as using automated sample preparation, obtaining clean extracts, and achieving good separation between the analyte and the coextracted matrix components that could cause matrix effect.

In addition, a new technique termed ECHO peak technique could be considered [23, 24]. With this new technique, two injections are carried out in each analysis, namely within a short time period (typically 30–50 s) the unknown sample and a standard solution. As a result, the peak of the analyte from the standard elutes in close proximity to the peak of the analyte from the sample, thus forming the so-called echo peak (Fig. 4). It is expected that both peaks elute so closely that they are affected in the same manner by the coeluted matrix components, which usually have broader peak shape. Quantitation is based on the responses (e.g., peak areas) of the two peaks without the need for an internal standard. This technique has been used mainly in the analysis of pesticide residuals in agricultural produces (e.g., fruits and vegetables) by LC-MS, especially as the first screening measurement to identify samples with residues above the maximum residue limits (MRLs).

Since it is always necessary to maintain a gap between the two peaks, the ability of this technique to correct matrix effect will be impacted when there is a change in matrix effect during the gap. Moreover, like postcolumn introduction of an internal standard, any variations in sample preparation as well as in injection volume cannot be corrected.

3 Performance of Internal Standards in LC-MS Bioanalysis

3.1 Stable Isotope Labeled Internal Standards vs. Structural Analogue Internal Standards

As demonstrated by many studies, SIL internal standards outperform structural analogue internal standards in terms of precision and accuracy provided the extraction and LC separation are comparable [8, 25–30]. In addition, SIL internal standards can also extend linearity range, enable the usage of less reliable transitions (e.g., loss of water), and prolong in-processing and postpreparation stabilities [8, 28].

3.2 Unexpected Results Using Deuterated Internal Standards

Despite the overall good performance of SIL internal standards, one must not take it for granted due to the complexity of biological samples, particularly when deuterated internal standards are used. Deuteration could cause differences in hydrophobicity, reaction rates, and noncovalent interactions [31, 32]. It is usually observed that a deuterated internal standard elutes slightly earlier than the analyte does in reversed-phase LC. This is even more pronounced with extensive deuteration and long retention time. Sometimes, base-line resolution between an analyte and its deuterated internal standard could be achieved. For example, when D_{10} internal standards were used and the retention time was longer than 15 min, pibutidine metabolites were completed separated from their deuterated internal standards (Fig. 5; [33]).

Sometimes, even the slight separation between an analyte and its deuterated internal standard could cause significant quantitation errors due to differential ion suppression towards the analyte and its deuterated internal standard [34, 35]. As shown in Fig. 6, the elution of carvedilol and its internal standard (D_5-carvedilol) overlaps with the declining edge of a matrix suppression region in a matrix lot. This resulted in more pronounced ion suppression for the slightly later eluted

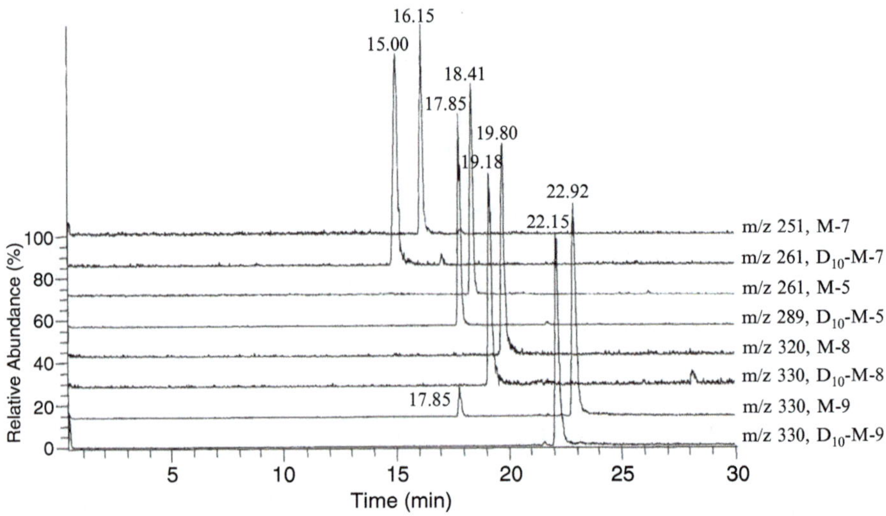

Fig. 5 Reversed-phase LC separation of pibutidine metabolites and their corresponding deuterium-labeled analogues. Reproduced from ref. [33] with permission from Elsevier

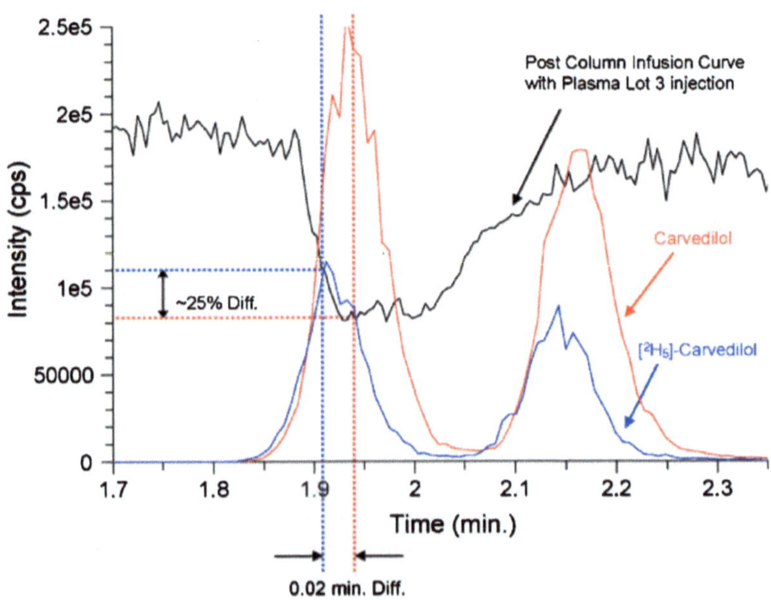

Fig. 6 Profile of postcolumn infusion of carvedilol with an injection of control plasma extract (lot 3, the problematic lot) overlaid with the LC-MS/MS chromatograms of carvedilol and its deuterated internal standard (D_5-carvediol) to demonstrate a significant difference in ion suppression (~25 %) due to even a very small difference in retention time (0.02 min) between carvediolol-S (1.93 min) and its deuterated internal standard (1.91 min). Reproduced from ref. [35] with permission from Elsevier

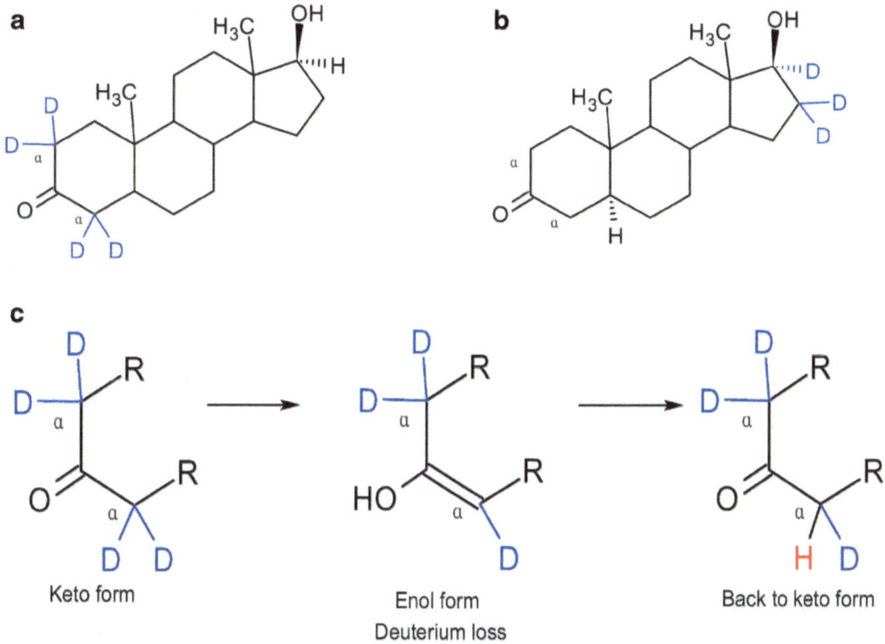

Fig. 7 Chemical structures of 5α-androstan-17β-ol-3-one-2, 2, 4, 4-D_4 (**a**) and 5α-androstan-17β-ol-3-one-16, 16, 17-D_3 (**b**) and schematic representation showing how deuterium–hydrogen exchange occurs during keto- and enol- form conversions (**c**)

analyte (carvedilol-S at the retention time of 1.93 min) than for the corresponding internal standard (at the retention time of 1.91 min) with a difference of about 25 %. This type of differential ion suppression was also believed to be the cause for the significant differences between original and reinjection concentrations observed for run reinjection within validated postpreparative stability on a different LC-MS instrument, though all calibration standards and quality control samples were accepted in both runs [36].

Another issue with deuterated internal standards is the possibility of deuterium and hydrogen exchange in sample matrix or solvent. For example, [$^{13}CD_3$]rofecoxib was shown to be isotopically unstable in plasma and water containing solvent and an efficient exchange between deuterium and hydrogen prevented its use as the internal standard in an LC-MS method. On the other hand, the isotopic integrity of [$^{13}C_7$]rofecoxib was maintained [37]. Moreover, the positions of deuterium atoms can also have an impact on the stability of a deuterated internal standard. As shown in Fig. 7, when deuterium atoms are adjacent to a keto group (5α-androstan-17β-ol-3-one-2, 2, 4, 4-D_4), a significant loss of deuterium atoms was observed at pH 6 due to potential conversions between keto- and enol- forms [38]. While the deuterium atoms are far away from the keto- group, such as 5α-androstan-17β-ol-3-one-16, 16, 17-D_3, no such deuterium–hydrogen exchange was observed.

Moreover, significant difference in extraction recovery has been observed for an analyte and its deuterated internal standard. For example, the extraction recoveries of haloperidol and d_4-haloperidol were 72 % and 44 %, respectively [19]. The large difference in extraction recovery could be due to differences in the aforementioned physicochemical properties, such as p*K*a, or hydrogen–deuterium exchange.

Since no similar issue has been reported for other SIL internal standards, such as ^{13}C and ^{15}N labeled internal standards, it is therefore necessary to distinguish deuterated internal standards from other SIL internal standards. Nevertheless, one should always have an open mind while using SIL internal standards in LC-MS bioanalysis.

3.3 Internal Standard Response Variations During Incurred Sample Analysis

3.3.1 Why Monitor Internal Standard Response Variations?

Bioanalytical chemists usually have contradictive expectations regarding internal standard response variations. In one hand, internal standard response variations in bioanalysis by LC-MS are somewhat expected due to the purpose of using internal standards, i.e., correction of variations (otherwise there is no need to use an internal standard), particularly if considering the many differences between spiked samples, e.g., calibration standards (CS) and quality control (QC) samples prepared in a pooled control blank plasma, and incurred samples (samples collected from dosed subjects in a preclinical or clinical study, also termed as study samples), such as absence of metabolites and formulation related components in spiked CS and QC samples as well as difference in number of matrix lots and freeze–thaw cycles (Table 1). On the other hand, too much variation in internal standard responses during bioanalysis could trigger doubt in the integrity of the quantitative results obtained because the same amount of an internal standard is added to all the samples. To maintain the integrity of a study and to avoid economic losses, it is therefore critical to monitor IS response variations during bioanalysis and to quickly identify the root causes if variations are observed, especially for those causes that affect an analyte and its internal standard differently.

3.3.2 How to Monitor Internal Standard Response Variations?

Though monitoring of internal standard response variations in incurred sample analysis is necessary and has been recommended in the literatures (e.g., [26]), no agreement was reached on the inclusion of internal standard criteria or on the magnitude of acceptable internal standard precision during the Third American Association of Pharmaceutical Scientists (AAPS)/Food and Drug Administration (FDA) Bioanalytical Workshop in 2007 [9]. It was only suggested that objective

Table 1 Differences between CS/QC and incurred samples

	CS/QC	Incurred Sample
Screening criteria for matrix sources	Usually loose	Usually specific and strict dependent on the objectives of a study, such as age 40–50 and nonsmoker
No. of lots/sources	Usually more than one source (pooled)	One single source
pH	Averaged due to pooling	More variable
Extra components associated with medication	None	Metabolite(s), comedication, and nonactive ingredients in formulation
Amount collected	Usually large, e.g., 200 mL per collection	Usually small, e.g., 7 mL per sampling time
No. of freeze–thaw cycles prior to being extracted	Usually two or more	Usually one
Storage tube and preuse storage	Usually stored at −20 °C and without special protection until being selected for a specific study	Could be collected under sodium light and stored at −80 °C immediately after collection
Amount of anticoagulant	May be different because of different amounts collected	

Note: *CS* Calibration standard, *QC* Quality control
Reproduced from ref. [36] with permission from Elsevier

criteria are necessary and need to be established a priori if incurred samples or analytical runs are rejected or repeated based on internal standard response variability.

In the authors' laboratory, the IS response of a sample is compared to the mean IS response of the accepted calibration standards and quality controls in the same run, i.e., those that meet the acceptance criterion of accuracy and do not show other abnormality (e.g., poor chromatography). When the IS response of a sample is outside ±50 % of the mean IS signal of calibration standards and quality controls, the sample will be repeated. Moreover, an investigation may be initiated for repeated abnormal IS signal and when there is a pattern or trend. This acceptance criterion was also recommended by others (e.g., [13]). Alternatively, though not reported, some compare the IS response of a sample to those of adjacent samples or to the mean IS response of all the samples in a batch. No matter what approach is used, it is important to be able to single out abnormal samples and to perform corrective actions to ensure that their reported concentrations are accurate.

3.3.3 Typical Causes of and Solutions to Internal Standard Response Variations

Variation or Error in the Addition of Internal Standard

Apparently, any variation or error in the amount of the internal standard added to a sample has a direct impact on IS response. Since the quantitation is based on analyte/IS response ratios, a prerequisite for good accuracy is that the same amount of

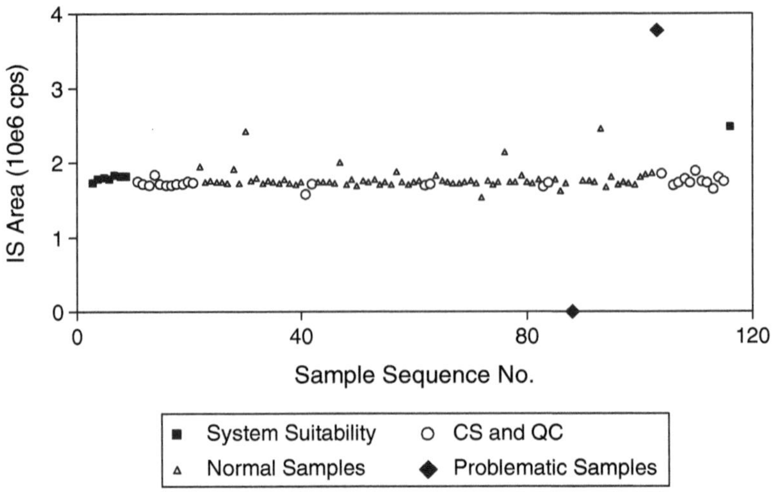

Fig. 8 Example of missed and double addition of the internal standard. Analyte: clopidogrel carboxylic acid; extraction: evaporation-free protein precipitation; sample volume: 50 μL; IS volume: 150 μL. Reproduced from ref. [36] with permission from Elsevier

an internal standard is added to all the samples including concentration-known calibration standards and quality controls as well as concentration-unknown incurred samples. Therefore, this type of variation or error directly affects the accuracy of reported concentration, and it should be minimized or eliminated whenever possible.

It is relatively easy to spot variation or an error in the addition of IS for CS and QC samples. As their concentrations are known, large variations in internal standard addition for CS and QC samples would result in the rejection of the CS or QC samples, even a whole run. However, the variation or error in IS addition for unknown incurred samples would be difficult to be ascertained unless the addition of the internal standard is doubled or missed (Fig. 8). Even when a doubled or near zero IS response is observed, there could be other reasons than the addition of the IS. For example, missed addition or incorrect amount of a derivatization reagent can produce near zero IS response as well.

Though human error is usually the cause, proper method development can make a difference in reducing this type of error or variations. First, a large volume of IS, such as 200 μL should be used when it is added by a repeater pipette because small volumes (such as 50 μL or less) are more prone to imprecision than large ones. In addition, it would be extremely difficult to visually spot missed or doubled addition for an internal standard when its volume is much smaller than that of samples and/or other reagents (e.g., buffer). Second, it would be helpful to reduce errors by adding the usually colorless IS solution first and then incurred samples, which are usually colored, such as slight yellowish for plasma samples or dark red for whole blood samples.

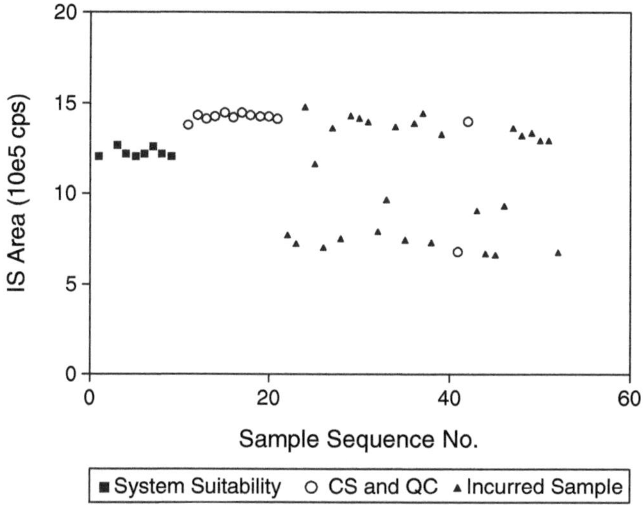

Fig. 9 Low and variable internal standard responses caused by autosampler problem occurred during the middle of run injection. Extraction: evaporation-free protein precipitation; sample volume: 50 μL; IS volume: 150 μL. Reproduced from ref. [36] with permission from Elsevier

Malfunction of Autosampler

In bioanalysis, extracted samples are usually stored in either autosampler vials or wells in a plate (such as 96-well plate) sealed with pierceable caps or covers. During injection, the autosampler needle has to pierce the caps or covers to load samples. The debris may completely or partially block the autosampler needle, which would result in no sample or variably low sample volumes injected. Accordingly, no or randomly low IS responses are observed. As most autosamplers have a built-in needle flushing mechanism, the debris in the needle might be flushed out later partially or completely. Therefore, the injected volume can be back to normal at a later time without an operator's intervention. Apparently, when a needle will be blocked and when the blocked needle will be cleared by flushing, as well as how it will be blocked (completely or partially) are difficult to predict. Hence, there would be no clear pattern for this type of IS variations. However, the affected injections normally have lowered IS responses (Fig. 9). Despite lowered IS responses, the accuracy of quantitation can usually be maintained except for situations where no or very low amount of samples are injected, resulting in responses outside the limit of linear range or unacceptable S/N.

To solve the problem, the affected run can be reinjected in a different LC-MS system or on the same system after the needle of the autosampler is cleaned.

Fouling or Malfunction of Mass Spectrometer

During routine use, a mass spectrometer might get dirty or contaminated, such as the coronal needle, orifice, or rods of a quadrupole, particularly when extracted samples are not clean enough, such as those by protein precipitation. When this happens, the IS response will normally decrease. Sometimes, the signal might be completely lost, e.g., blockage of the orifice. A special phenomenon with the commonly used triple quadrupole MS is so-called "charging," where IS responses decrease sharply at the beginning of a batch and then decrease more and more slowly, quite similar to the current vs. time curve during the charging of an electric capacitor (Fig. 10).

However, a contaminated mass spectrometer may cause various response trends toward different compounds. For example, after a triple quadrupole MS was contaminated with ascorbic acid, different response patterns were observed for levothyroxine (ANA), liothyronine (ANB), and their internal standard (thyroxine-$^{13}C_6$) during the repetitive injections of pure reference solutions containing these three components (Fig. 11). The responses of ANA were relatively stable while those of ANB and IS decreased significantly over a time period of about 1 and half hours, which resulted in increasing analyte/IS response ratios for levothyroxine and decreasing ratios for liothyronine.

To solve these problems, the contaminated components in the mass spectrometer should be cleaned. Moreover, postcolumn switching or splitting should be considered to prevent or reduce the occurrence of contamination as well as to improve signal stability [39].

Ion Suppression or Enhancement

Ion suppression or enhancement (matrix effect) by coeluting components including those from previous injections (late-eluting components) is a common phenomenon in LC-MS and it can seriously degrade the accuracy of analytical results if it is not appropriately corrected by the use of internal standards. Although it is more common with electrospray ionization (ESI), it can also exist to a less degree with APCI (atmospheric pressure chemical ionization). The compounds that cause ion suppression or enhancement are usually from coextracted matrix components, but they could also come from other sources, such as remaining salt residues from sample clean-up, additives in a mobile phase, Li-heparin anticoagulant, and polymers contained in storage devices [13, 34, 40].

To reduce matrix effect, it is important to obtain cleaner extract and use adequate separation. In the presence of late-eluting components that cause matrix effect, adjustment of retention time, run injection cycle time, and/or postinjection column washing should be used to eliminate or reduce their impact.

Inter Matrix Lot or Inter Sample Differences

Despite the wide application of matrix-matched calibration, it is very difficult or even impossible to completely match the matrix of every sample due to diversities of subjects, particularly human subjects, and the dynamic nature of in vivo metabolism.

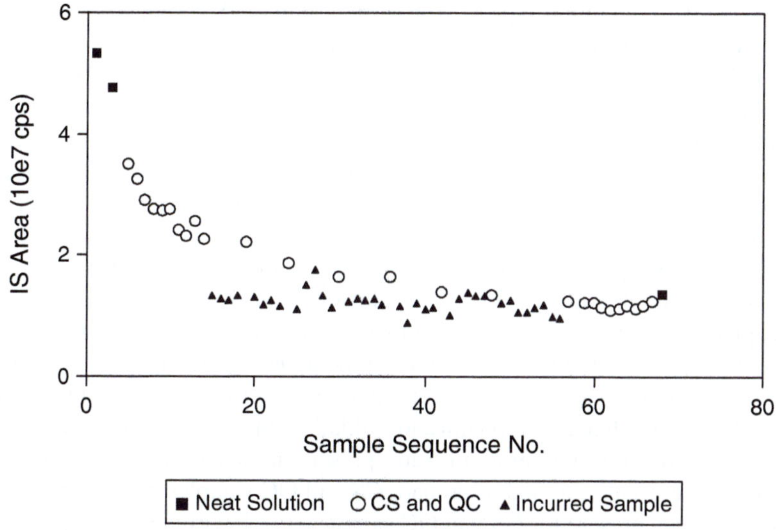

Fig. 10 Gradual decrease of internal standard response caused by the charging of the mass spectrometer. Analyte: escitalopram. Reproduced from ref. [36] with permission from Elsevier

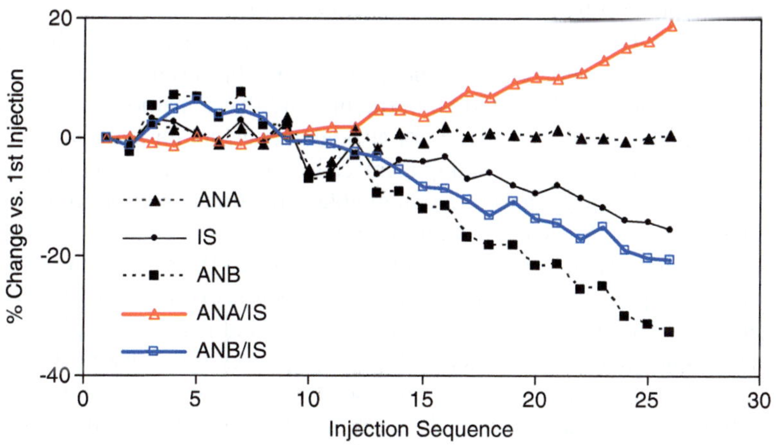

Fig. 11 An ascorbic acid-contaminated triple quadrupole mass spectrometer affected the responses of levothyroxine (ANA), liothyronine (ANB), and $^{13}C_6$-thyroxine (IS) differently when a neat solution sample was repeatedly injected, resulting in a gradual increase of analyte/IS response ratios for one analyte (ANA) while gradual decrease for the other (ANB)

Therefore, it is expected that there might be differences in matrix components for different samples because each sample is from an individual subject and collected at different sampling time. These differences could cause many different types of IS response variations.

One relatively common type is consistently higher IS response for one whole subject, including the predose sample. If a particular matrix component is absent or exists in less amount in one subject and this component happens to be one of the those that cause ion suppression, then the IS responses for all the samples from this subject will be higher than those of other subjects. For example, consistently high IS responses (higher than those of CS and QC samples) were observed for all the samples of a subject (including the predose sample) during the analysis of repaglinide in a study (Fig. 12a). Postcolumn (postelution) infusion tests demonstrated the presence of ion suppression in the pooled control human plasma used for the preparation of CS and QC samples (Fig. 12b). However, this ion suppression was absent for the predose sample from the subject that had high IS responses (Fig. 12c). Since the number of lots tested in method development and validation (e.g., 6) is usually much smaller than that of a study (e.g., 40), this interlot difference might not be observed during method development and validation. In addition, as the component(s) that can cause this ion suppression is (are) present in most of the blank sources and several lots are usually pooled to prepare CS and QC samples, the ion suppression would be always present for CS and QC samples. For incurred samples, the matrix is from individual source. Therefore, this type of IS response variations could occur to some subjects.

The interlot or intersample differences can also cause variations in the recovery for an analyte and its internal standard. For example, in a method based on liquid–liquid extraction for *p*-hydroxy-atorvastatin, the recovery of internal standard varied from 67.19 to 89.99 % (1.5-fold) for the four subjects tested despite the fact that the ratios of analyte to IS were relatively independent of subject sources, i.e., no impact on the quantitation [36]. It should be borne in mind that a method is usually optimized aiming the maximum recovery for an analyte, i.e., not for any matrix components. In case where a matrix component causes matrix effect and the optimal extraction conditions happen to be an unreliable extraction condition for the matrix component, variable IS response is very likely.

Moreover, this inter matrix lot difference can result in IS variation through other routes. For example, rosuvastatin in plasma samples was extracted by LLE with the automatic transfer of organic layers by a liquid handling system [41]. During incurred sample analysis, consistently low IS responses were observed for a few subjects in the study, including the predose samples (Fig. 13). Through careful examination of the extraction procedure, it was found that the intermediate layers between organic and aqueous layers in LLE were thicker for some subjects than those of CS and QC samples, based on which the aspirating height had been set during the method development. As this preset aspirating height was not appropriate for some subjects, a little bit of the intermediate layer between the organic and aqueous layers during LLE, which contains salts, was transferred, leading to subsequent ion suppression during MS detection. The problem was successfully solved by readjusting the aspirating height. It should be noted that this type of IS response variations can also happen to other type of LLE with decanting of organic phases, such as flash-freeze LLE or normal LLE.

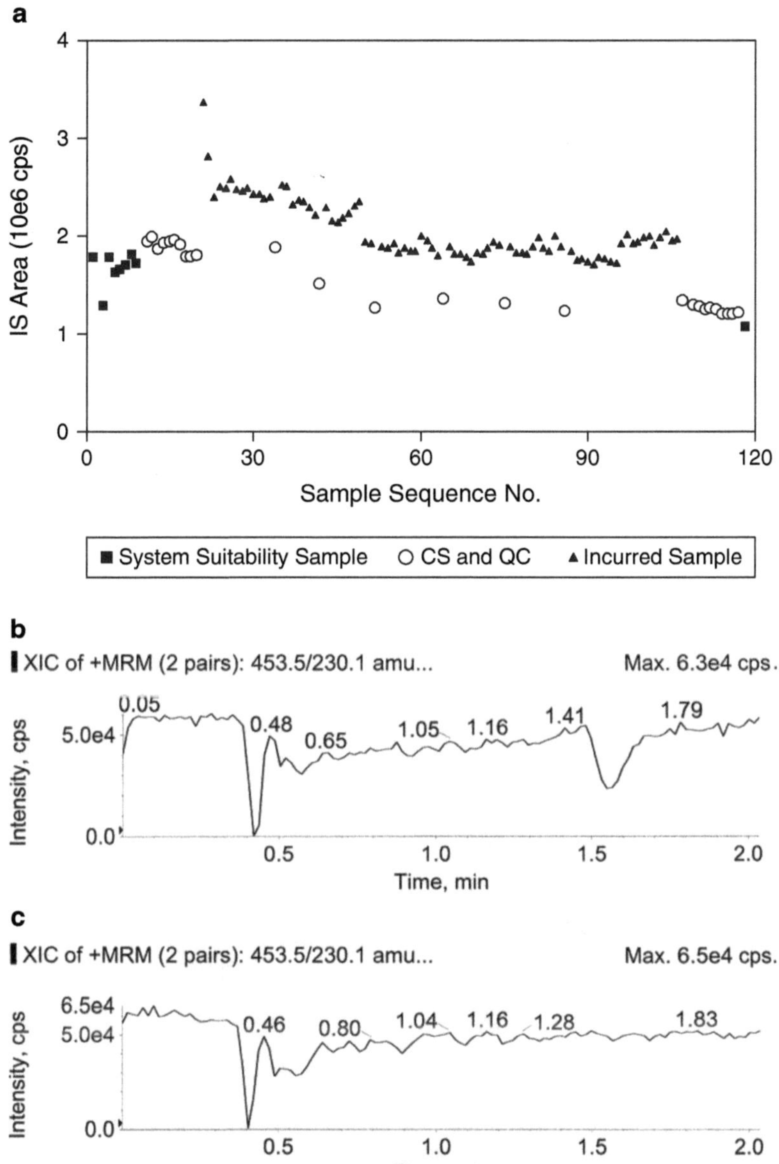

Fig. 12 (**a**, *top*): High internal standard responses were observed for incurred samples only. Analyte: repaglinide; extraction: automatic liquid–liquid extraction. (**b**, *middle*): Postelution infusion results show that ion suppression existed near the retention time of the analyte (1.57 min) from the pooled control blank used for the preparation of calibration standards and quality controls. (**c**, *bottom*): Absence of ion suppression near the retention time of the analyte in subject predose sample. Reproduced from ref. [36] with permission from Elsevier

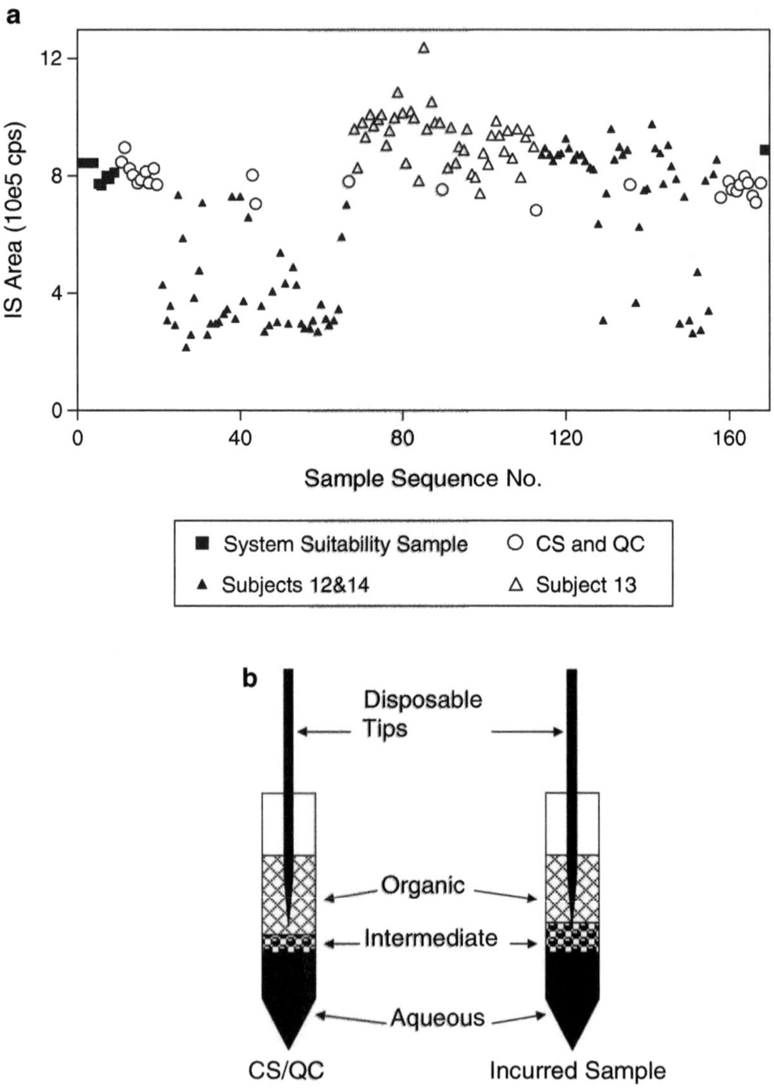

Fig. 13 (**a**, *top*): Low internal standard responses were observed for incurred samples of some subjects. (**b**, *bottom*): Due to variable thickness of intermediate layer between aqueous and organic phases in liquid–liquid extraction, an inappropriately set aspirating height could result in partial transfer of salt-containing intermediate layer, which caused ion suppression. Reproduced from ref. [36] with permission from Elsevier

Improper Use of Internal Standard

Initially, vidarabine was used as the internal standard for penciclovir in a method based on mixed-mode strong cation exchange (MCX) solid-phase extraction due to their similar properties, particularly hydrophobicity (Fig. 14a). Quite stable IS responses were obtained in a run consisted mainly of CS and QC samples with a CV of 13.02 % (Fig. 14b). However, 43 % of the CS and QC samples did not meet the acceptance criterion of accuracy. On the other hand, when penciclovir-d_4 was used as the internal standard, all the CS and QC samples met the acceptance criterion in accuracy though the IS responses were more variable (the CV in IS responses was 23.81 %, Fig. 14c).

A careful examination of the extraction procedure revealed the root cause, i.e., improper use of the internal standard. Specifically, despite the overall similarity in hydrophobicity vs. pH characteristics, there is a relatively larger difference in hydrophobicity between vidarabine and penciclovir in acidic conditions (pH 1–2), in which the mixture of a sample and the internal standards was loaded onto an MCX plate and the loaded MCX plate was washed. For vidarabine, its hydrophobicity is relatively less variable than that of penciclovir in this pH range. Because of this difference, minor change in pH caused differential recovery variation for the analyte and its internal standard, i.e., potential impact on the accuracy of quantitation.

Solubility or Stability of Internal Standard

The solubility or stability of an internal standard in IS working solution could cause IS response variations. Sometimes, an issue of solubility could appear as an issue of stability [42]. For example, high interbatch IS response variation was observed for a method of raloxifene glucuronides based on protein precipitation by acetonitrile containing deuterated raloxifene glucuronides. However, internal standard responses within a batch were relatively stable and no such variations in the responses of the corresponding analytes were observed. It was further noticed that the internal standard responses of a batch appeared to be correlated to the operation speed of the lab technician who performed the batch. The faster a lab technician's speed was, the higher the IS responses of his run were.

Initially, it was suspected that the deuterated raloxifene glucuronides might be unstable in acetonitrile. It was therefore replaced by methanol, despite the fact that acetonitrile is usually a desirable solvent for protein precipitation. This suspicion appeared to be supported by the results of a comparison test using both solvents (Fig. 15). Finally, the root-cause was identified as an issue of solubility by extensive troubleshooting, instead of stability. Specifically, raloxifene glucuronides and their deuterated internal standards are very hydrophilic and not very soluble in acetonitrile. The seeming "stability" issue disappeared when 10 % of water was added to acetonitrile. As to why the analytes (raloxifene glucuronides) did not show similar variable responses between batches as the corresponding deuterated internal

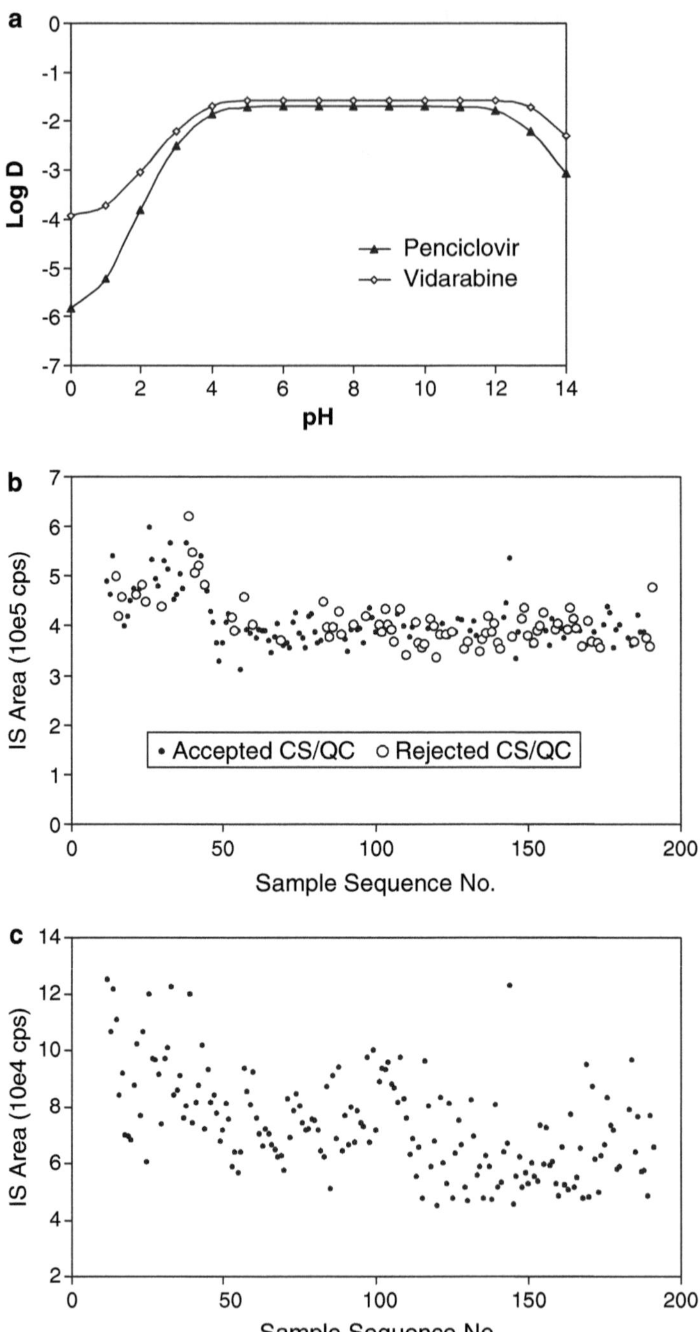

Fig. 14 (**a**, *top*): Hydrophobicity (log *D*) vs. pH curves for penciclovir and vidarabine. (**b**, *middle*): Less internal standard response variation was observed while using vidarabine as the internal standard (CV = 13.02 %) but 43 % of the calibration standards (CS) and quality controls (QC) were rejected. Extraction: MCX (mixed-mode strong cation exchange)-based solid-phase extraction. (**c**, *bottom*): More IS response variation was observed while using a deuterated internal standard, penciclovir-d_4 (CV = 23.81 %) but 100 % of the CS and QC samples were accepted. Reproduced from ref. [36] with permission from Elsevier

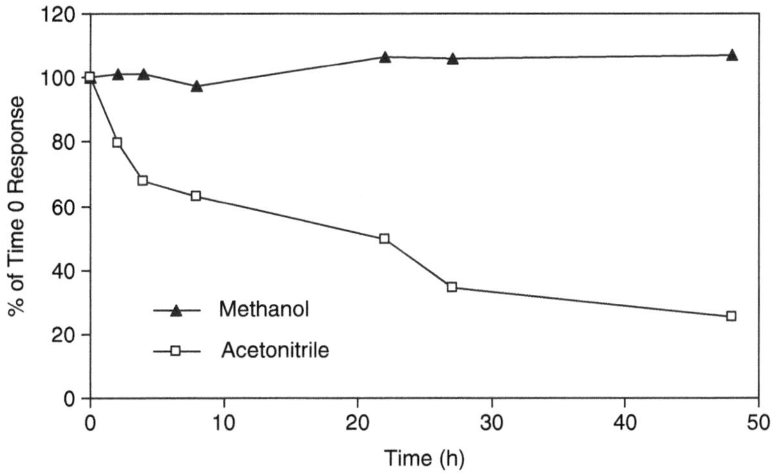

Fig. 15 Response ratio of raloxifene-D$_4$-6-glucuronide to raloxifene-6-glucuronide vs. the delay of usage after the preparation of raloxifene-D$_4$-6-glucuronide in methanol or acetonitrile

standards, it was because they were not in acetonitrile. They were in plasma matrix from the very beginning and there was enough water for their solubilization.

Furthermore, sometimes a change of solvent supplier or lot could cause issues of stability for an internal standard and therefore IS response variations. For example, Napoli reported unusual IS response variation and stability problem when a different lot of acetonitrile was used for the preparation of ascomycin working solution (as the internal standard for tacrolimus, [43]).

Lengthy Sample Treatment Procedure

When multiple sample treatment steps are used, especially derivatization is involved, larger IS response variations are usually expected. For instance, in a method based on LLE followed by derivatization and SPE, as large as fourfold difference in IS responses was observed for CS and QC samples as well as incurred samples (Fig. 16). Despite this large IS response variation, the accuracy of the concentration-known CS and QC samples was not impacted, it can be therefore concluded that the concentrations of unknown samples are also reliable.

Others

In addition to the above, there might be other less common causes for IS response variations. For example, due to mixed usage of rubber-lined caps and PTFE (polytetrafluoroethylene)-lined caps during the liquid–liquid extraction shaking (mixing) step, randomly low IS responses were observed for extracted samples, i.e.,

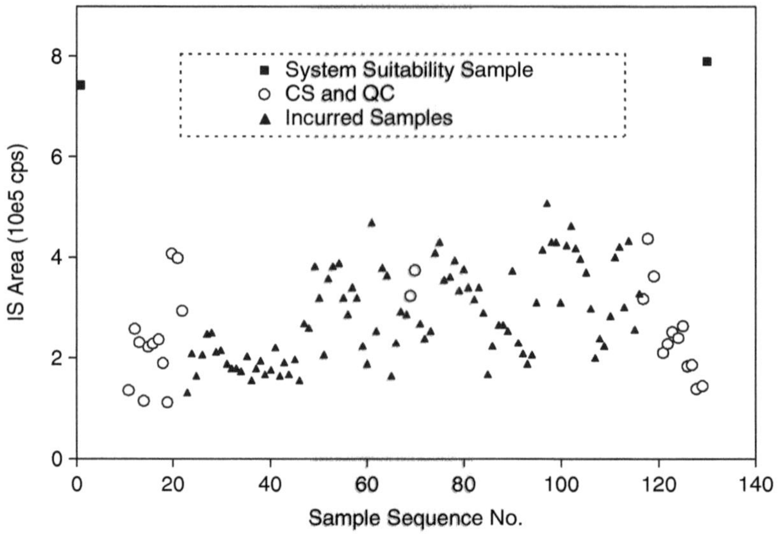

Fig. 16 Variable internal standard responses due to multiple sample processing steps (liquid–liquid extraction, derivatization, and solid-phase extraction). Analyte: gestodene; internal standard: norethindrone

incurred samples and CS/QC, because some analyte and IS in the organic solvent layer have been soaked into the rubber and were retained, which reduced recovery. On the other hand, the internal standard responses of the nonextracted system suitability samples injected at the beginning and the end of the run were high and stable (Fig. 17).

Another example is an almost linear increase of IS responses with regard to the injection sequence for all the extracted samples (incurred samples, CS and QC samples) while the IS responses of the system suitability samples injected at the beginning and the end of the run were normal (Fig. 18). The root cause was identified as the inadequate mixing of the 100 μL aliquot of supernatant transferred from protein precipitation with the 400 μL of reconstitution solution prefilled at the bottom of wells (96-well plate). As the autosampler needle loaded a sample near the bottom of a well, the loaded and later injected sample did not contain its "share" of the supernatant because homogeneous mixing had not been reached. Apparently, the mixing continued over the course of the run injection through diffusion, which resulted in the gradual increase in IS responses observed in this example.

Sometimes, the root-cause might be difficult to identify due to timeline or unavailability of incurred samples (e.g., not enough volume or no authorization from the client). For example, during the initial sample analysis, variable and randomly low IS responses were observed for incurred samples only (Fig. 19). Despite similar pattern as in the case of autosampler malfunctioning, the root cause was unlikely related to autosampler issue because no CS or QC samples were affected in this case. In addition, since all the CS and QC were accepted, there should be no

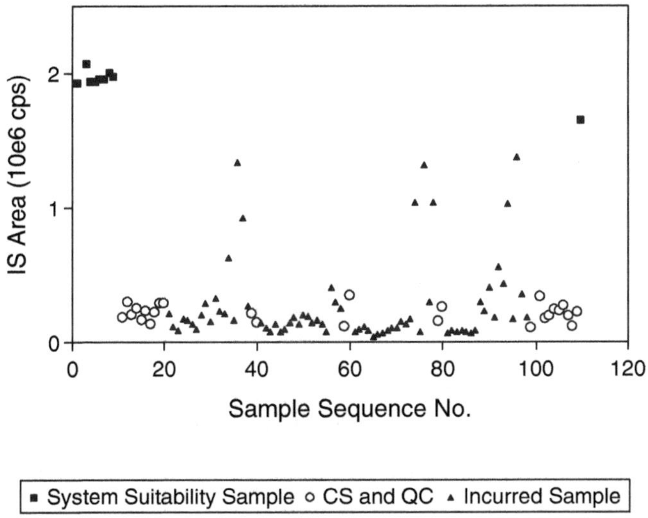

Fig. 17 Low internal standard responses caused by using rubber-line caps, instead of PTFE-lined caps during liquid–liquid extraction. Analyte: tamsulosin; internal standard: tamsulosin-d_4. Reproduced from ref. [36] with permission from Elsevier

issue with the LC-MS/MS system or reagents used. Furthermore, when those incurred samples with unaccepted low internal standard responses were reanalyzed, their internal standard responses were back to normal. The reassay results matched those of the first analyses if low internal standard responses were not considered.

Based on the above, it was deemed necessary to use the incurred samples for further troubleshooting. Unfortunately, this was not granted by the client, which left the root cause unidentified. The speculation is that the randomly low IS responses were related to the ascorbic acid added to the incurred samples at the clinic for the stabilization of the analyte. Although ascorbic acid was also added in the pooled control plasma used for the preparation of the CS and QC samples, there exist differences as outlined in Table 1. For example, the incurred samples had gone through only one thawing during the first analysis while the plasma matrix for CS and QC samples had gone through thawing twice when they were extracted (the first thawing occurred during the preparation of the CS and QC samples, which were usually aliquoted and stored in a freezer after the preparation, while the second thawing took place when the aliquoted CS and QC samples were retrieved from the freezer prior to being used for sample analysis).

3.3.4 The Impact of Internal Standard Response Variations on Quantitation

The main purpose of troubleshooting IS response variations is to make sure that the quantitation of unknown samples has not been impacted despite the variable or abnormal IS responses observed. Unfortunately, there is no clear-cut "yes or no" answer to

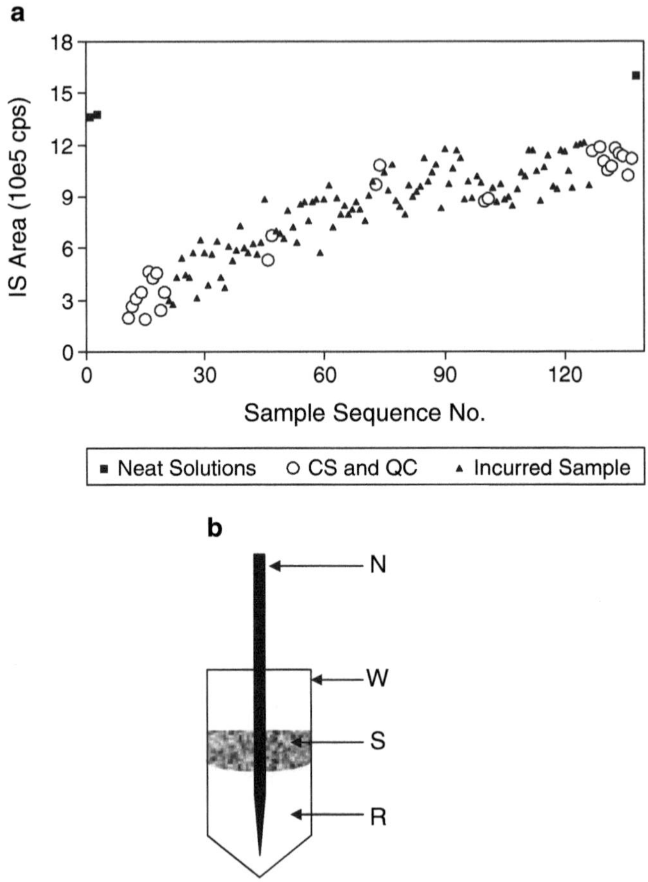

Fig. 18 (**a**, *top*): Gradually increasing IS responses due to increased mixing of the transferred supernatant (100 μL) with reconstitution solution (400 μL). (**b**, *bottom*): Schematic diagram showing the autosampler needle loaded a sample near the bottom of a well (96-well plate), where was mainly reconstitution solution (deficient of the analyte and its internal standard) when homogeneous mixing had not been reached. *N* autosampler needle, *W* a well in a 96-well plate, *S* supernatant, *R* reconstitution solution. Reproduced from ref. [36] with permission from Elsevier

this question in most of the cases. The best approach would be to monitor IS response variations during the analysis of incurred samples by using some predefined acceptance criteria, such as within ±50 % of the mean IS response of the accepted CS and QC samples in the same run. Once variable or abnormal IS responses are observed, each case should be evaluated or investigated when there is a pattern or trend. Based on the outcome of the evaluation or investigation, the affected samples may be reinjected, reanalyzed or their results may be accepted together with some scientific proof. The last approach would not only be preferable, i.e., saving time and cost, but it may be also the only option in some cases. For examples, when all the samples from a subject have consistently higher or lower IS responses than those of the CS and QC

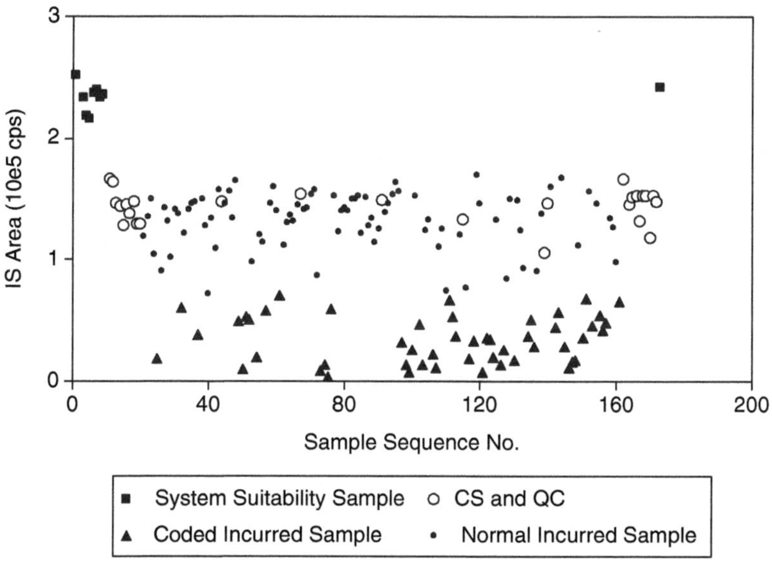

Fig. 19 Randomly scattered low internal standard (IS) responses observed for incurred samples only, whose IS responses were within normal range during repeat analyses. Analyte: olanzapine; IS: olanzapine-d_3; sample pretreatment at clinic: 25 % (w/v) L-ascorbic acid added to plasma in a ratio of 1.25:100 (v/v); extraction: MCX (mixed-mode strong cation exchange)-based solid-phase extraction. An incurred sample was coded for reassay when its IS response was outside ±50 % of the mean IS response of the accepted calibration standards and quality controls. Reproduced from ref. [36] with permission from Elsevier

samples due to interlot matrix differences, the same or similar IS responses would be repeated during reassays. Without proper investigation or evaluation, either there would be no reportable values for a whole subject (due to abnormal IS responses) or there would be uncertainty on the accuracy of the results obtained if they are to be reported. Both should be avoided during the analysis of incurred samples.

4 Conclusions

Internal standards play critical roles in ensuring the accuracy of final reported concentrations in quantitative LC-MS bioanalysis through the correction of variations during sample preparation, LC-separation, and MS detection. The physical–chemical properties of an internal standard, particularly hydrophobicity and ionization properties should be as close as possible to those of the corresponding analyte to better track the variations the analyte experiences. For this reason, stable isotope labeled internal standards should be used whenever possible. However, efforts should still be made to obtain clean extracts, adequate chromatographic separation, and optimized ionization mode and conditions.

Once an internal standard is selected, its concentration must be properly determined based on the particular situation, such as impurities of reference standards, concentration range, and detection sensitivity towards the analyte and its internal standard, and finally experimentally verified to ensure adequate linearity, accuracy, and precision.

Due to the differences between spiked samples (calibration standards and quality controls) and incurred samples as well as the potential inter sample differences, variations in internal standard response during incurred sample analysis are somewhat expected and they should be monitored with predetermined criteria to identify any potential bioanalysis abnormality. Many different factors (not necessarily only matrix effect) could cause variations in internal standard responses during incurred sample analysis. The same phenomenon, e.g., consistently higher IS responses for all the samples of a subject, can be caused by different reasons. Accordingly, each case should be dealt with individually with an open mind.

Stable isotope labeled internal standards may be the best, but they cannot always follow an analyte to compensate the variations of experimental conditions, particularly deuterated internal standards. In addition, low variation in internal standard responses may not be interpreted as good results, though it is favored. Stable internal standard response is good only when it is sure that the internal standard behaves the same way as the analyte does.

Acknowledgments We would like to thank Dr. Wen Jin in the University of Guelph for her reviewing of the draft manuscript and valuable comments. In addition, Tan would like to thank his family (Cailin and Joyce) for their support during the preparation of this book chapter.

References

1. Eeckhaut AV, Lanckmans K, Sarre S, Smolders I, Michotte Y (2009) Validation of bioanalytical LC-MS/MS assays: evaluation of matrix effects. J Chromatogr B 877:2198–2207
2. Cuadros-Rodríguez L, Bagur-González MG, Sánchez-Viñas M, González-Casado A, Gómez-Sáez AM (2007) Principles of analytical calibration/quantification for the separation sciences. J Chromatogr A 1158:33–46
3. Nilsson LB, Eklund G (2007) Direct quantification in bioanalytical LC-MS/MS using internal calibration via analyte/stable isotope ratio. J Pharm Biomed Anal 43:1094–1099
4. Doig MV (2000) Applications of mass spectrometry: quantitative mass spectrometry. In: Venn RF (ed) Principles and practice of bioanalysis. Taylor & Francis, London
5. Tan A, Hussain S, Vallée F (2009) Evaporation-free extraction and application in high-throughput bioanalysis by LC-MS/MS. LCGC North America 27:414–427
6. Yamashita K, Okuyama M, Watanabe Y, Honma S, Kobayashi S, Numazawa M (2007) Highly sensitive determination of estrone and estradiol in human serum by liquid chromatography-electrospray ionization tandem mass spectrometry. Steroids 72:819–827
7. Moore LJ, Machlan LA (1972) High accuracy determination of calcium in blood serum by isotope dilution mass spectrometry. Anal Chem 44:2291–2296
8. Stokvis E, Rosing H, Beijnen JH (2005) Stable isotopically labeled internal standards in quantitative bioanalysis using liquid chromatography/mass spectrometry: necessity or not? Rapid Commun Mass Spectrom 19:401–407

9. Viswanathan CT, Bansal S, Booth B, DeStefano AJ, Rose MJ, Sailstad J, Shah VP, Skelly JP, Swann PG, Weiner R (2007) Workshop/conference report — quantitative bioanalytical methods validation and implementation: best practices for chromatographic and ligand binding assays. AAPS J 9:E30–E42
10. Li W, Li YH, Li AC, Zhou S, Naidong W (2005) Simultaneous determination of norethindrone and ethinyl estradiol in human plasma by high performance liquid chromatography with tandem mass spectrometry – experiences on developing a highly selective method using derivatization reagent for enhancing sensitivity. J Chromatogr B 825:223–232
11. Kitamura R, Matsuoka K, Matsushima E, Kawaguchi Y (2001) Improvement in precision of the liquid chromatographic-electrospray ionization tandem mass spectrometric analysis of 3′-C-ethynylcytidine in rat plasma. J Chromatogr B 754:113–119
12. Shi G (2003) Application of co-eluting structural analog internal standards for expanded linear dynamic range in liquid chromatography/electrospray mass spectrometry. Rapid Commun Mass Spectrom 17:202–206
13. Bakhtiar R, Majumdar TK (2007) Tracking problems and possible solutions in the quantitative determination of small molecule drugs and metabolites in biological fluids using liquid chromatography-mass spectrometry. J Pharmacol Toxicol Methods 55:227–243
14. Sojo LE, Lum G, Chee P (2003) Internal standard signal suppression by co-eluting analyte in isotope dilution LC-ESI-MS. Analyst 128:51–54
15. Ansermot N, Rudaz S, Brawand-Amey M, Fleury-Souverain S, Veuthey JL, Eap CB (2009) Validation and long-term evaluation of a modified on-line chiral analytical method for therapeutic drug monitoring of (R, S)-methadone in clinical samples. J Chromatogr B 877:2301–2307
16. ANVISA (2003) Guide for validation of analytical and bioanalytical methods. Resolution RE no. 899, 23 May 2003 Brazilian Sanitary Surveillance Agency, Brazil
17. Bergeron A, Furtado M, Garofolo F (2009) Importance of using highly pure internal standards for successful liquid chromatography/tandem mass spectrometric bioanalytical assays. Rapid Commun Mass Spectrom 23:1287–1297
18. Liang HR, Foltz RL, Meng M, Bennett P (2003) Ionization enhancement in atmospheric pressure chemical ionization and suppression in electrospray ionization between target drugs and stable-isotope-labeled internal standards in quantitative liquid chromatography/tandem mass spectrometry. Rapid Commun Mass Spectrom 17:2815–2821
19. Wieling J (2002) LC-MS-MS experiences with internal standards. Chromatographia 55:S107–S113
20. Viel F, Santos N, Tan A, Laforest L, Boudreau N, Lévesque A, Musuku A, Massé R (2010) Simultaneous quantitation of free and liposomal drug forms in human serum by evaporation-free extraction. Presented at 58th ASMS conference on mass spectrometry and allied topics, Salt Lake City, Utah, USA, 23-27 May
21. Lee JW, Petersen ME, Lin P, Dressler D, Bekersky I (2001) Quantitation of free and total amphotericin B in human biologic matrices by a liquid chromatography tandem mass spectrometry method. Ther Drug Monit 23:268–276
22. Choi BK, Gusev AI, Hercules DM (1999) Postcolumn introduction of an internal standard for quantitative LC-MS analysis. Anal Chem 71:4107–4110
23. Alder L, Lüderitz S, Lindtner K, Stan HJ (2004) The ECHO technique-the more effective way of data evaluation in liquid chromatography-tandem mass spectrometry analysis. J Chromatogr A 1058:67–79
24. Zrostlíková J, Hajšlová J, Poustka J, Begany P (2002) Alternative calibration approaches to compensate the effect of co-extracted matrix components in liquid chromatography-electrospray ionization tandem mass spectrometry analysis of pesticide residues in plant materials. J Chromatogr A 973:13–26
25. Fu I, Woolf EJ, Matuszewski BK (1998) Effect of the sample matrix on the determination of indinavir in human urine by HPLC with turbo ion spray tandem mass spectrometric detection. J Pharm Biomed Anal 18:347–357
26. Jemal M, Schuster A, Whigan DB (2003) Liquid chromatography/tandem mass spectrometry methods for quantitation of mevalonic acid in human plasma and urine: method validation,

demonstration of using a surrogate analyte, and demonstration of unacceptable matrix effect in spite of use of a stable isotope analog internal standard. Rapid Commun Mass Spectrom 17:1723–1734
27. Kato K, Jingu S, Ogawa N, Higuchi S (2000) Development and validation of a liquid chromatographic-tandem mass spectrometric method for the determination of pibutidine in human urine. J Chromatogr B 740:187–193
28. Lanckmans K, Sarre S, Smolders I, Michotte Y (2007) Use of a structural analogue versus a stable isotope labeled internal standard for the quantification of angiotensin IV in rat brain dialysates using nano-liquid chromatography/tandem mass spectrometry. Rapid Commun Mass Spectrom 21:1187–1195
29. O'Halloran S, Ilett KF (2008) Evaluation of a deuterium-labeled internal standard for the measurement of sirolimus by high-throughput HPLC electrospray ionization tandem mass spectrometry. Clin Chem 54:1386–1389
30. Taylor PJ, Brown SR, Cooper DP, Salm P, Morris MR, Pillans PI, Lynch SV (2005) Evaluation of 3 internal standards for the measurement of cyclosporine by HPLC-mass spectrometry. Clin Chem 51:1890–1893
31. Iyer SS, Zhang ZP, Kellogg GE, Karnes HT (2004) Evaluation of deuterium isotope effects in normal-phase LC-MS-MS separations using a molecular modeling approach. J Chromatogr Sci 42:383–387
32. Wade D (1999) Deuterium isotope effects on noncovalent interactions between molecules. Chem Biol Interact 117:191–217
33. Kato K, Jingu S, Ogawa N, Higuchi S (2000) Determination of pibutidine metabolites in human plasma by LC-MS/MS. J Pharm Biomed Anal 24:237–249
34. Lindegardh N, Annerberg A, White NJ, Day NPJ (2008) Development and validation of a liquid chromatographic-tandem mass spectrometric method for determination of piperaquine in plasma: stable isotope labeled internal standard does not always compensate for matrix effects. J Chromatogr B 862:227–236
35. Wang S, Cyronak M, Yang E (2007) Does a stable isotopically labeled internal standard always correct analyte response? a matrix effect study on a LC/MS/MS method for the determination of carvedilol enantiomers in human plasma. J Pharm Biomed Anal 43:701–707
36. Tan A, Hussain S, Musuku A, Massé R (2009) Internal standard response variations during incurred sample analysis by LC-MS/MS: case by case trouble-shooting. J Chromatogr B 877:3201–3209
37. Chavez-Eng CM, Constanzer ML, Matuszewski BK (2002) High-performance liquid chromatographic-tandem mass spectrometric evaluation and determination of stable isotope labeled analogs of rofecoxib in human plasma samples from oral bioavailability studies. J Chromatogr B 767:117–129
38. Savard C, Pelletier N, Boudreau N, Lachance S, Lévesque A, Massé R (2010) Relative instability of deuterated internal standard under different pH conditions and according to deuterium atoms location. Presented at 58th ASMS conference on mass spectrometry and allied topics, Salt Lake City, Utah, USA, 23–27 May
39. Kloepfer A, Quintana JB, Reemtsma T (2005) Operational options to reduce matrix effects in liquid chromatography — electrospay ionization-mass spectrometry analysis of aqueous environmental samples. J Chromatogr A 1067:153–160
40. Mei H, Hsieh Y, Nardo C, Xu X, Wang S, Ng K, Korfmacher WA (2003) Investigation of matrix effects in bioanalytical high-performance liquid chromatography/tandem mass spectrometric assays: application to drug discovery. Rapid Commun Mass Spectrom 17:97–103
41. Hussain S, Patel H, Tan A (2009) Automated liquid-liquid extraction method for high-throughput analysis of rosuvastatin in human EDTA K_2 plasma by LC-MS/MS. Bioanalysis 1:529–535
42. Tan A, Montminy V, Gagné S, Musuku A, Massé R (2009) Trouble-shooting of least-expected causes in bioanalytical method development and application. Presented at 2009 AAPS annual meeting and exposition, Los Angeles, California, USA, 8–12 Nov
43. Napoli KL (2006) Organic solvents compromise performance of internal standard (ascomycin) in proficiency tesing of mass spectrometry-based assay for tacrolimus. Clin Chem 52:765–766

Method Development, Validation, and Sample Analysis for Regulated Quantitative Bioanalysis Using LC-MS/MS

Min Meng and Patrick K. Bennett

Abstract There are three primary stages of the regulated bioanalysis process using Liquid Chromatography–Tandem Mass Spectrometry (LC-MS/MS): method development, method validation, and sample analysis including incurred sample reanalysis (ISR). Robust and rugged LC-MS/MS methods are essential in support of drug discovery, toxicology studies, and clinical trials. The development of a robust bioanalytical method requires careful consideration of many critical parameters, such as accuracy and precision, linearity, matrix effect, sensitivity, selectivity, stability, throughput, and ruggedness (or reproducibility). Because bioanalytical data is critical for determining the safety and efficacy of a new drug, a bioanalytical method must be validated following the Food and Drug Administration's (FDA) guidance for the industry, the recommendations from various white papers and Standard Operating Procedures (SOPs). A complete regulated method validation in biological matrix minimally requires three interday precision and accuracy runs, various short- and long-term solution and matrix stability assessments, extraction recovery, dilution capability and linearity, extract stability, reinjection reproducibility, selectivity and specificity, assessment of matrix effects, interference from concomitant medications and prodrug/metabolites, etc. The final evaluation of any high quality bioanalytical method is not complete until it passes the ultimate test of regulated sample analysis and incurred sample reanalysis (ISR) which are also conducted following the similar rules as validation.

M. Meng, Ph.D. (✉)
Tandem Labs, 1121 East 3900 South Salt Lake City, UT 84124, USA
e-mail: MengM@LabCorp.com

P.K. Bennett
Thermo Fisher Scientific, San Jose, CA, USA
e-mail: patrick.bennett@thermofisher.com

Q. Alan Xu and Timothy L. Madden (eds.), *LC-MS in Drug Bioanalysis*,
DOI 10.1007/978-1-4614-3828-1_2, © Springer Science+Business Media, LLC 2012

1 Introduction

New drug development is a lengthy and costly process. It typically lasts 15–17 years and roughly 1 out of 5,000 compounds may come out as commercial drug. In order to ensure the safety of the drug and the well-being of the public population, the new drug development is conducted under a very constrained environment and closely monitored by regulatory agencies. Except early drug screening and discovery which is exempt from regulatory monitoring, most preclinical and clinical trials, i.e., animal toxicology studies, in vitro laboratory experiments, and clinical trials, are conducted according to FDA guidance, white paper recommendations and the SOPs [1–3]. Preclinical study is typically classified as regulated Good Laboratory Practice (GLP) study. GLP specifically refers to management controls for laboratories and institutions to ensure the uniformity, consistency, reliability, reproducibility, quality, and integrity of pharmaceutical safety and efficacy tests. Although the scope and rigidness of clinical study are similar as preclinical study, clinical study is normally classified as regulated but non-GLP study due to lack of central study director, for clinical study typically involves multiple testing facilities.

Throughout new drug development, multiple bioanalytical studies are conducted in which the amount/concentration of drug and/or its metabolites in biological matrix is measured and evaluated. For the same drug candidate, multiple methodologies may be developed to meet the specific goals at each stage. In early drug discovery, the emphasis is speed and fast turnaround. Therefore, it is imperative to use a nonspecific method suitable for wide range of analytes with diverse structures. For later stage drug development such as GLP toxicology study and clinic study, the reliability and reproducibility of the methodology are critical and essential. Therefore, it is worthwhile to develop robust and rugged method.

The most frequently used techniques in bioanalytical studies are liquid chromatography coupled with tandem mass spectrometry (LC–MS/MS) and noncompetitive hybridization enzyme-linked immunosorbent assay (ELISA). Traditionally, hybridization ELISA is very useful technique for biopharmaceuticals and biomarkers such as protein and oligonucleotides while the LC-MS/MS is more suitable for small molecules [4, 5]. This paradigm has changed in recent years. LC-MS/MS has become more popular for the quantitation of macromolecules due to the advantage of high selectivity, wide linearity range, and fast method development [6, 7]. The focus of this chapter highlights the systematic approaches for method development, validation, and sample analysis for regulated quantitative bioanalysis using LC-MS/MS technique.

2 Sample Preparation

The common biological matrixes for bioanalysis are various tissues, body fluid, whole blood, plasma, serum, and urine. More recently, dried-blood spot sample as alternative matrix has gained popularity in the industry [8]. Prior to LC-MS/MS analysis, biological samples need to be extracted from biological matrixes via various cleanup procedures. Currently the widely used extraction techniques are protein precipitation extraction (PPE), liquid-liquid extraction (LLE) and solid phase extraction (SPE). Additionally, several novel extraction techniques, i.e., supported-liquid extraction (SLE), online solid phase extraction (online-SPE), phospholipid removal SPE/PPE extraction are also available.

2.1 Protein Precipitation Extraction (PPE)

PPE is the most commonly used extraction method. In principle, the underlying mechanism of protein precipitation is to lower the solubility of the analytes by adding organic solvent such as methanol (MeOH) or acetonitrile (MeCN) or saturated buffer salt such as 10 M zinc sulfate ($ZnSO_4$) or high concentration of strong acid or base such as 5–10 % Trifluoroacetic acid (TFA) or hydroperchloric acid ($HClO_4$). In LC-MS/MS application, due to incompatibility of high salt or strong acid or base with MS source, the most common approach is to add organic solvent at minimal ratio of 1:3 (v/v) of matrix–organic solvent. If the active drug is highly bound to protein, a volatile acid or base such as formic acid (FA) or ammonium hydroxide (NH_4OH) respectively at 5–10 % is utilized to disrupt the binding thus increasing the recovery of the drug.

Historically PPE has been conducted in tubes. In recent years, PPE in 96-well format such as commercial PPE precipitation plate has gained popularity in order to increase throughput. PPE in tubes is inexpensive and reliable but requires manual transfer/pipetting, vortex mixing, and centrifugation. In PPE in 96-well format, all 96 samples can be transferred simultaneously which increases the throughput tremendously. Because the transfer volume among samples is the same and programmed in advance, it ensures consistent recovery, thus improving the overall performance of the assay.

2.2 Liquid–Liquid Extraction and Supported–Liquid Extraction

Although PPE is the most efficient and inexpensive extraction technique, it is also the most nonspecific extraction procedure which is known to be susceptible to matrix effect for LC-MS/MS assay. In contrast, LLE provides a much cleaner extract. LLE, also known as solvent extraction and partitioning, separates analytes based on their relative solubility in two different immiscible solvents, usually water and an organic solvent. The most commonly used solvents for LLE are ethyl acetate (EtOAc), methyl *tert*-butyl-ether (MtBE), methylene chloride (CH_2Cl_2), and hexane or the combination of the above solvents. In order to manipulate the polarity of the analytes, often a volatile acid or base such as FA or NH_4OH respectively at 5–10 %

is utilized. LLE for bioanalysis is typically conducted in tubes. The transfer can be achieved by either pipetting or freeze-pour. For freeze-pour approach, the tubes are merged in acetone or methanol–dry ice bath for approximately 1 min, the aqueous layer which is normally at the bottom layer is frozen, thus the top organic layer can be pour out easily. Compared to manual pipetting transfer, the freeze-pour is much easier and quicker which is quite commonly used in bioanalytical laboratory. LLE can also be conducted in 96-well plate format in which the extraction was achieved by mixing using automation robots via pipette tips for 30–40 times. This extraction approach is labor free and efficient. However, it is less forceful than by mixing using shaker or vortexer. Thus, the recovery is normally lower than tube extraction.

In recent years, supported-liquid extraction (SLE) has merged and become increasingly popular [9]. Commercial SLE plate consists of 96 individual wells each packed with proprietary and modified form of diatomaceous earth which has a high capacity for retaining aqueous samples. When the plasma sample is loaded onto the extraction well, the analytes absorb over the surface of the support in a very thin layer. Unlike manual LLE in which the emulsions could potentially cause problem, in SLE extraction, emulsion is completely eliminated for the sample and water immiscible extraction solvent are never in direct contact. It also removes all manual steps such as capping, mixing, centrifuging, and decapping et al., thus greatly reducing contamination. Because all procedural steps can be fully automated, SLE is as efficient as using PPE plate but provides much cleaner extracts. The only disadvantage for SLE is the limited sample aliquot volume (<400 μL) in order to maintain in 96-well plate format.

2.3 *Offline and Online Solid Phase Extraction (SPE)*

Another popular and selective extraction technique widely used in bioanalysis is solid phase extraction (SPE). SPE is a separation process utilizing the affinity of the analytes to a solid stationary phase. By manipulating the polarity and pH of the mobile phase, the analytes of interest or undesired impurities pass through stationary phase sequentially according to their physical and chemical properties. For a SPE procedure, a wash step refers to the elution of the unwanted impurities which are discarded and the elution step refers to the elution of the analytes of interest which are collected. While the fundamental remains the same in decades, the continuing invention and introduction of new commercial stationary phases and accessory devices have boosted the application of SPE in bioanalysis and many other fields.

SPE initially comes in the form of a packed syringe-shaped cartridge for manual extraction, and then followed by 96-well plate each of which can be mounted on its specific type of extraction manifold. The first generation of SPE stationary phases was made of silica backbone bonded to hydrocarbon chain of variable length such as C18, C8, and phenyl. The second generation of SPE stationary phases was still primarily made of silica backbone but with more modification for these functional groups.

In addition to hydrocarbon chains of variable length (for reversed phase SPE), quaternary ammonium or amino groups (for anion exchange), and sulfonic acid or carboxyl groups (for cation exchange) were introduced for mix mode extraction with increased extraction selectivity. The third generation of SPE stationary phases was made from hydrophobic, yet water-wettable polymeric sorbents. Several commonly known brand and products are silica based Isolute® SPE (Biotage, Uppsala, Sweden) and more recently polymer based SPE sorbent such as Oasis® SPE (Waters, Milford, MA, USA) and Strata™ SPE (Phenomenex, Torrance, CA, USA). These products contain generic reversed phase SPE sorbents as well as more advanced mix mode such as strong and weak ion exchange sorbent. All of the newer SPE sorbents were made of water-wettable polymers which provide much wider pH range from 0 to 14.

Another advanced and novel sample preparation technique is online solid phase extraction (online SPE). Although the extraction mechanism is as the same as traditional offline SPE, online SPE offers several advantages. Because sample preparation is carried out during analysis, it eliminates the time needed for sample preparation thus increases throughput significantly. Additionally, because the entire sample is eluted to the LC-MS/MS system, it may increase assay sensitivity. Finally, because there is no manual extraction involved, it may reduce human error, potential contamination, and inconsistent recovery.

The direct comparison of the advantage and disadvantage of offline and online SPE was evaluated in our laboratory using raltegravir as model compound. Raltegravir is a new class of drug used to treat HIV infection. Two separate methods were developed to quantify raltegravir in human plasma at linear range of 1.00–1,000 ng/mL. As shown in Table 1, the offline extraction method was developed using HPLC coupled with Sciex API 5000™ (MDS Sciex, Concord, Ontario) in negative ion mode using ESI source. Strata-X 96-well plate following multiple conditioning, equilibration, washing, and elution steps was used for sample preparation. A Waters Xbridge™ phenyl 2×50 mm column was utilized to achieve appropriate retention and separation with cycle time of 3 min. In contrast, online-SPE method uses Symbiosis™ Pharma (Spark Holland, The Netherlands) coupled with Sciex API 4000™ in negative ion mode using ESI source. The online SPE sample preparation is as easy as diluting plasma sample with internal standard and buffer solution. Online SPE is conducted under high pressure with two syringe pumps, where the HySphere™ C18HD 7 μm cartridge (Spark Holland) provides adequate sample cleanup and chromatographic separation. The diluted sample with ca. 3 μL of plasma is extracted and analyzed simultaneously by Symbiosis Pharma coupled with Sciex API 4000 with a cycle time of 100 s. The analytical column was omitted which reduces the cycle time and cost of analytical column. As shown in Figs. 1 and 2, because the sensitivity of online SPE is much higher than that of offline SPE, the mass spectrometer was changed from a Sciex API 5000 to API 4000 to achieve the equivalent sensitivity at the same Lower limit of Quantitation (LLOQ) of 1 ng/mL.

Although the advantage of online SPE is apparent, it has several limitations as well. Because online SPE method is operated under "elute and shoot" mode, the elution solvent has to be compatible with HPLC mobile phases. If the pH of the

Table 1 Comparison of offline and online SPE for the quantitation of raltegravir in human plasma using LC-MS/MS

	Offline SPE	Online SPE
LC-MS		
Mass Spec.	Sciex API 5000™ ESI (–)	Sciex API 4000™ ESI (–)
Source temperature	500 °C	500 °C
Column	Waters Xbridge® Phenyl 2×50 mm	None
Flowrate	0.600 mL/min	0.600 mL/min
Mobile phase	A: 0.1 % Formic acid in water	A: 0.1 % formic acid in water
	B: MeCN	B: MeCN
Needle Wash	1: 0.1 % Formic acid in water/MeCN	1: TEA and EDTA in water/ DMF
	2: 0.1 % Formic acid in water	2: 0.1 % formic acid in water
LC Program	Isocratic	gradient
Cycle time	3 min	100 s
SPE		
Automated SPE	N/A	Symbiosis™ Pharma
Cartridge/plate	Phenomenex Strata® X 33 μm	Spark-Holland HySphere® C18 HD 7 μm
Conditioning	MeCN	MeOH
Equilibration	5 % Formic acid in water	5 % Formic acid in water
Wash	5 % Formic acid in water	5 % Formic acid in water
Elution	5 % Formic acid in water/MeCN	Mobile phase from gradient pumps
Cartridge flush	N/A	MeCN/MeOH

elution solvent does not match the pH of the mobile phase or the composition of the elution solvent is stronger than that of mobile phases, an addition step called "peak focusing" need to be considered. Additionally, because samples injected are plasma diluted with internal standard and certain buffer, analytes that are unstable in plasma may not be suitable. Finally, it may have greater potential for carryover and is difficult to troubleshoot due to the complexity of the system. In the case of raltegravir, we found that the combination of triethylamine (TEA) and ethylenediaminetetraacetic acid (EDTA) mixture solution is the key to eliminate system carryover and SPE cartridge carryover.

Based on the newly gained knowledge of phospholipids as interferences in bioanalytical LC-MS/MS analysis, several specialty PPE stationary phases such as HybridSPE® (Sigma-Aldrich) and Ostro™ 96-well plate (Waters), Isolute® PPT+ plate (Biotage AB Corporation) have merged specifically designed for the depletion of phospholipid interferences during sample preparation. Per manufacturer, this type of specialty stationary phases can reduce ion-suppression through the complete removal of phospholipids and precipitated proteins concurrently. It can serve as generic sample cleanup procedure and minimal method development for bioanalytical applications.

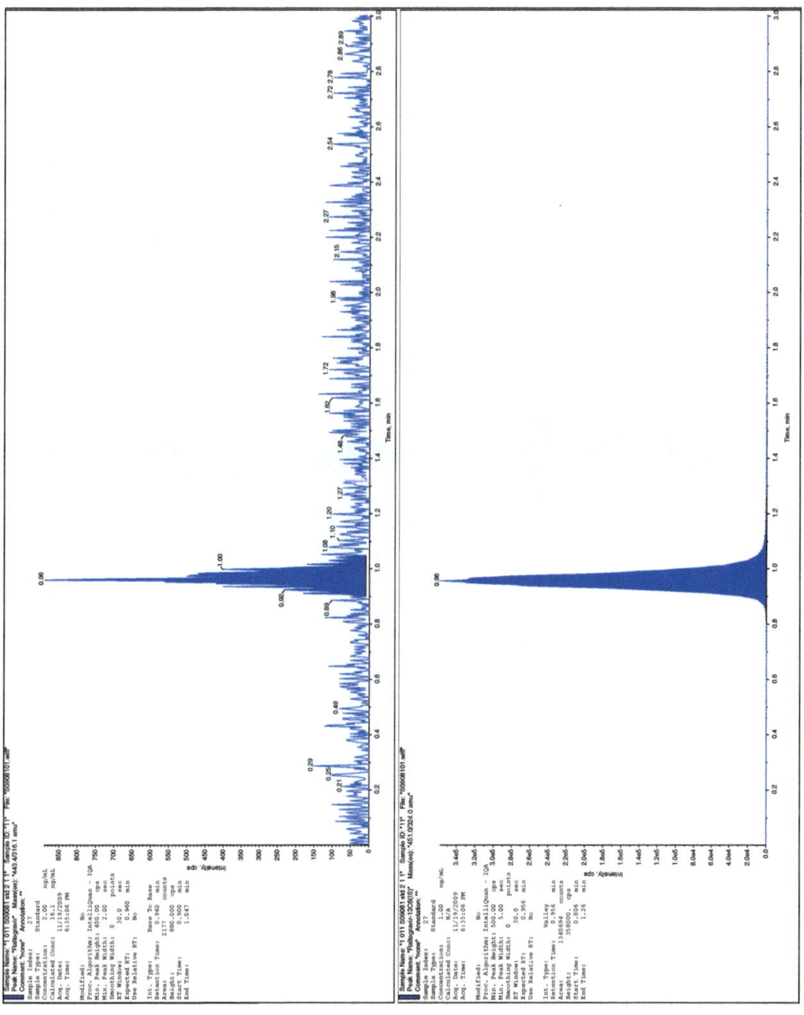

Fig. 1 Representative LLOQ of raltegravir (1 ng/mL) using offline SPE-API 5000. Strata-X 96-well plate and a Waters Xbridge™ phenyl 2×50 mm column was utilized

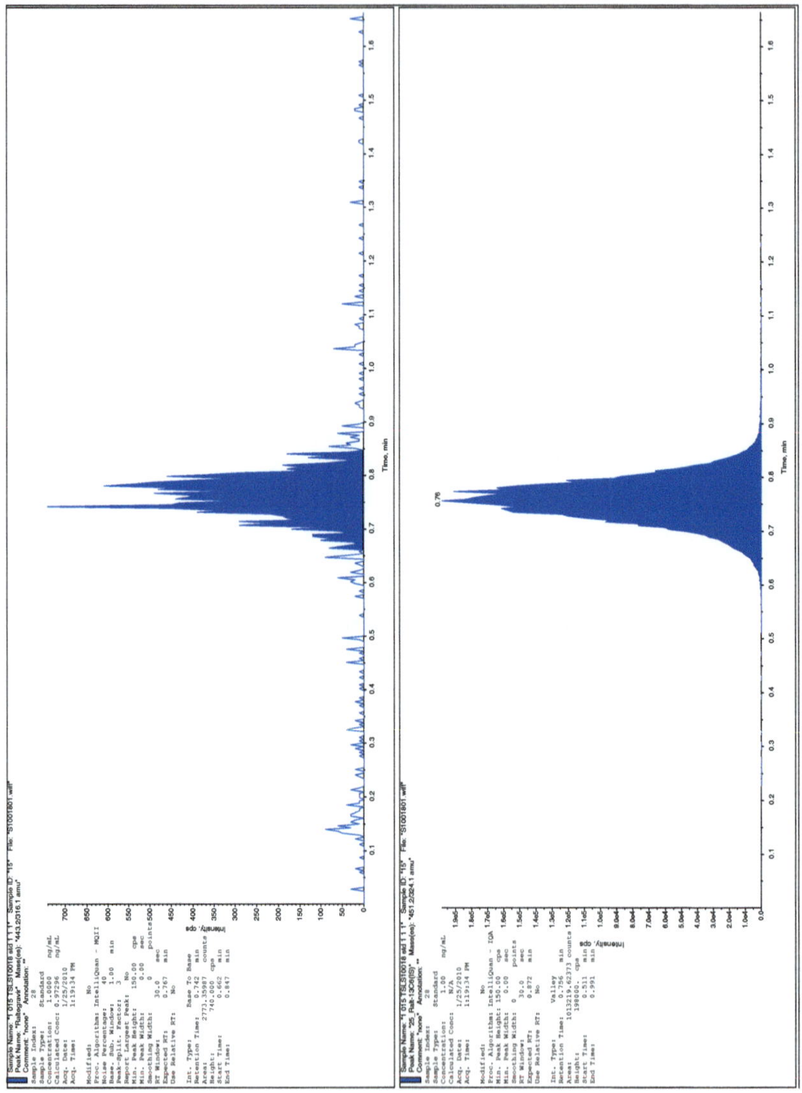

Fig. 2 Representative LLOQ of raltegravir (1 ng/mL) using online SPE-API 4000. HySphere™ C18HD 7 μm cartridge was utilized and there was no analytical column

2.4 *Selection of Sample Cleanup Procedure*

Many factors need to be taken into consideration while choosing a particular sample cleanup procedure for a LC-MS/MS assay. It is very beneficial to obtain critical information prior to initializing laboratory work, such as the scope of the project, the timeline of the project, the target Lower Limit of Quantitation (LLOQ) level, the maximal aliquot size, the polarity and stability of the analyte of interest. It is also important to recognize the benefit and pitfall of each sample cleanup technique. However, automation should be always considered in order to improve throughput and eliminate human error. In general, PPE extraction should be the first choice for an assay if the target LLOQ is at ng/mL or sub ng/mL level and the method is intended to support early drug screening and discovery study. Because PPE is the most nonselective extraction technique and tends to have severe matrix effect, it is recommended that the maximal sample aliquot volume should not be more than 100 μL. If the target LLOQ is at pg/mL or sub pg/mL level and it is used to support GLP study or clinical study for long period, LLE or SLE or SPE at 96-well plate format should be explored. LLE or SLE extraction can remove most matrix effect components and produce clean extracts. Based on the requirement of sample volume needed for desired LLOQ, automated SLE extraction can be used for sample aliquot volume less than 400 μL and larger sample aliquot volume for manual LLE extraction. While SLE/LLE is only suitable for nonpolar analytes, SPE is the best extraction option for an assay containing prodrug, active drug, and metabolite with vastly diversified polarity.

When developing sample cleanup procedure, the traditional approach is to start from the simplest extraction such as PPE, then move into cleaner extraction such as SLE or SPE. This linear and trial and error approach is time-consuming and inefficient. In our laboratory, a well-thought and systematic approach, nicknamed as "Amoeba Method Development Program," is implemented. Using Amoeba Sample Preparation Screening Protocol, three extraction techniques, i.e., PPE, SLE and SPE under various pH and solvent conditions were evaluated simultaneously. The various conditions tested in this protocol are: PPE with acidic, or neutral or basic acetonrile (MeCN); SLE with three different pH and three different organic solvents such as EtOAc, MtBE, and 1:1 hexane–EtOAc using Isolute® 200 mg SLE+plate; SPE under various combination of wash and elution solvent with acidic or neutral or basic conditions using Oasis MD SPE plate (Waters, Milford, MA, USA). With replicate of three for each condition, one test batch contains 26 different extraction conditions and a total of 81 samples including three neat control samples.

An example of the above extraction screening protocol is shown in Fig. 3 using guanfacine as model compound. In PPE and SLE groups, the letter of A or B or N represents acidic or basic or neutral condition. In SPE group, the first letter "A" in A/B represents acidic wash solvent and the second letter "B" means basic elution solvent. As shown in Fig. 3, the highest recovery was obtained using SLE and MtBE under basic condition. It was also noticed that there was relatively high recovery using various SPE conditions and the most interesting results were from MCX

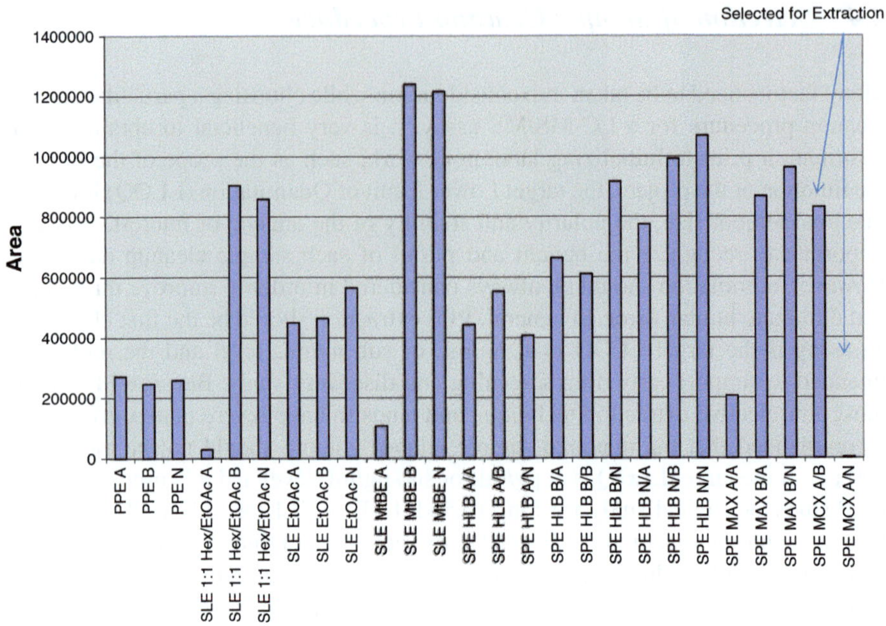

Fig. 3 Amoeba sample screening preparation of Guanfacine. Three extraction techniques (PPE, SLE, and SPE) and twenty-six different conditions were tested simultaneously

extraction. Because there is high recovery under MCX A/B condition but no recovery under MCX A/N condition, it suggested that mechanism of guanfacine using MCX plate is true ion exchange mechanism and 100 % organic solvent under neutral condition can be used for the second wash solvent to provide much cleaner extracts. Although SLE provided the highest recovery, in order to achieve the target LLOQ at 50.0 pg/mL, MCX extraction was ultimately selected due to large sample volume required.

3 Chromatography Development

Method development for mass spectrometry portion of LC-MS/MS assay is relatively simple and straightforward. For an experienced method development (MD) scientist, the optimization typically takes only hours to complete. In contrast, method development for optimal liquid chromatography conditions can be one of the most challenging tasks. Chromatography development can be very time-consuming. The task is further complicated by the nearly infinite choices in chromatography options such as vendor, sorbent, solvent selection, particle size, and column dimensions.

During last decade, liquid chromatography (LC) column technology has evolved exponentially, resulting in massive selection of commercially available LC columns. In addition to traditional reverse and normal stationary phases such as silica, phenyl, C8, or C18, a new generation of LC columns with various modifications has emerged. These new columns attempt to improve the retention of analytes, extend the lifetime of column and generally increase the throughput of the LC analysis. A few representative columns belong to this category are polar-embedded C8 or C18 columns, PFP (Pentaflourophenylpropyl) columns, HILIC (Hydrophilic Interaction Liquid Chromatography) and monolithic columns. Among these new columns, the properties of the PFP column are very unique. To certain extent, the PFP columns mimic traditional reversed phase phenyl columns in which molecules containing π–π interaction (such as aromatic ring or resonance structure) but provide much greater retention due to the presence of five strong electron-drawing atoms (i.e., pentafluorine). Because of the strong retention, PFP column typically requires much higher organic content which boosts ionization efficiency significantly. Another interesting phenomenon of PFP column is that PFP column, like CN or NH_2 column, changes to normal phase mechanism under very high organic conditions, thus provides additional and unique selectivity.

In addition to the increased variety of different stationary phases, the size of sorbent particles used for liquid chromatography has also evolved from the traditional 5 μm and 3 μm particles to smaller particle sizes such as 2.6 μm and 1.7 μm for both true Ultrahigh Performance Liquid chromatography (UHPLC) and standard HPLC that can mimic UHPLC performance. Porous silica rod or monolithic columns are also available to achieve higher throughput and/or better resolution; however, the popularity of monolithic columns is relatively limited due to increased solvent consumption, and their use has been larger superseded by the rapid growth in UHPLC technology. Typical UHPLC analyses can achieve very high sample throughput and excellent chromatographic peak resolution through the use of very fine particle sizes (1.7 μm), high pressure solvent pumps (up to 15,000 psi.), and limited void volumes throughout the system. UHPLC is very useful for multiple analyte assays or the separation of isomers which have same selective reaction monitoring (SRM) transition.

Most drug candidates are basic compounds with pKa values of around 9.5. In order to keep basic substances uncharged in and increase the retention of the analytes on the column, it is common practice to use a mobile phase pH that is up to two units higher than the pKa value of the analyte. Traditional silica based columns are only suitable for acidic and neutral condition and have limited life time at high pH. To meet the challenge and demand, several newer and modified LC column such as Gemini® (Phenomenex), Xbridge® (Waters) and Kromasil® (Agilent) have merged and can be used for mobile phases up to pH 12. While it is a well-known fact that traditional basic modifiers such as Triethylamine (TEA) suffer ion suppression due to ion pair, recently, it has been found that ammonium bicarbonate–ammonium hydroxide (NH_4HCO_3–NH_4OH) solution is an effective and great substitute buffer at high pH range [10].

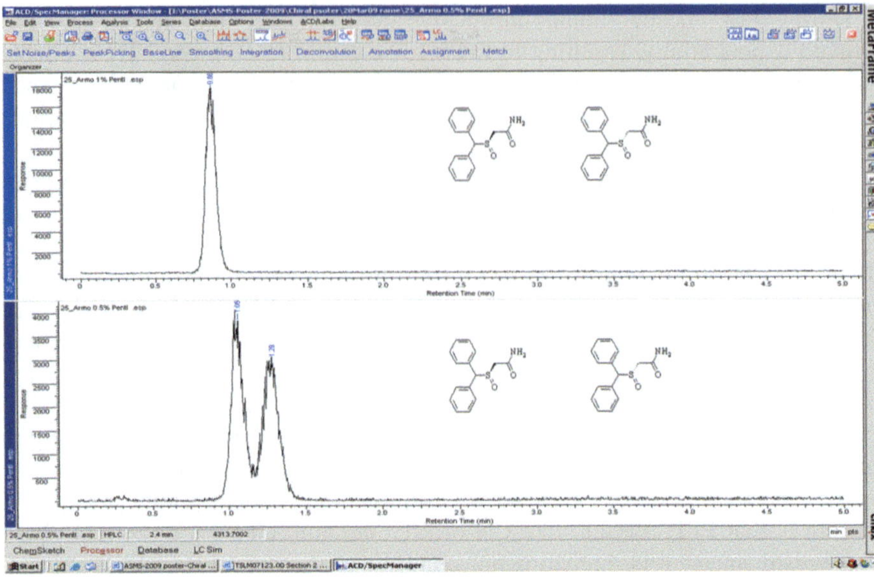

Fig. 4 Armodafinil Analyte wiff file imported to ACD SpecMananger using pentanol as mobile phase B (*top panel*: 1 % pentanol; *bottom panel*: 0.5 % pentanol, 2×50 mm AGP column)

Chiral chromatography development is time-consuming and challenging even for a very skillful and experienced chromatographer. Computer software assisted HPLC development has been reported since the late 1970s. Currently, a number of computer software programs are commercially available, and each year new papers using these are published. Among them, the most popular software programs are DryLab® (LC Resources, USA) and ChromSword® (Merck KGaA, Germany). ACD Lab® software (Advacend Chemistry Development, Canada) was introduced to the industry during last decade. In comparison, DryLab software assisted separations are optimized by simulating new runs based on two or more experimental runs. ChromSword and ACD Lab software can predict retention data based on one initial experimental chromatographic condition and molecular structural data. The following is an example of chiral chromatography development using armodafinil as model compound and ACD Lab® software. *Step 1*: Select initial conditions using a Chiral AGP® 2×50 mm column (ChromTech Ltd) and pentanol as mobile phases. A test sample containing mixture of 1:1 ratio of R- and S- armodafinil was injected at 0.5 % and 1 % pentanol. The first two injections performed had either no separation using 1 % pentanol or partial separation at 0.5 % Pentanol (Fig. 4). *Step 2*: Import these two chromatograms into the ACD/LC Simulator software. The software simulated the conditions and predicted the best separation at 0.4 % Pentanol using Chiral AGP® 2×100 mm column (Fig. 5). *Step 3*: Reproduce the condition in laboratory based on the computational predication (Fig. 6). As shown in this example, ACD/LC® Simulator can be successfully and efficiently used to assist chiral chromatographic method development. However, like any other computer

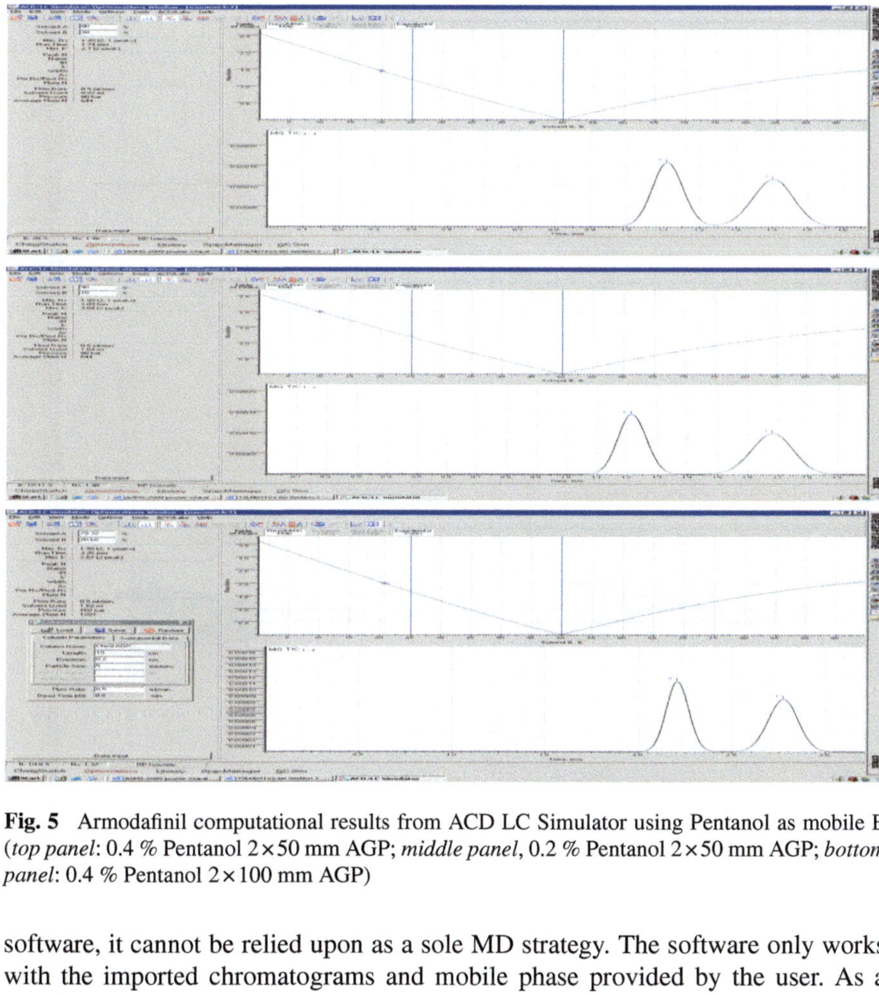

Fig. 5 Armodafinil computational results from ACD LC Simulator using Pentanol as mobile B (*top panel*: 0.4 % Pentanol 2×50 mm AGP; *middle panel*, 0.2 % Pentanol 2×50 mm AGP; *bottom panel*: 0.4 % Pentanol 2×100 mm AGP)

software, it cannot be relied upon as a sole MD strategy. The software only works with the imported chromatograms and mobile phase provided by the user. As a result, due to these limitations it can only serve as a method development aid. Sound scientific judgment and laboratory experiment is still required to select the appropriate starting condition. Using this approach, we successfully developed four chiral assays such as eszopiclone, armodafinil, ramelteon, and dexlansoprazole [11, 12].

4 Considerations While Developing LC-MS/MS Methods

4.1 Phospholipids Related Matrix Effect

The focus in early quantitative LC-MS/MS analysis was on recovery, simplicity of extraction, and fastness. Researchers gradually discovered and began to report the

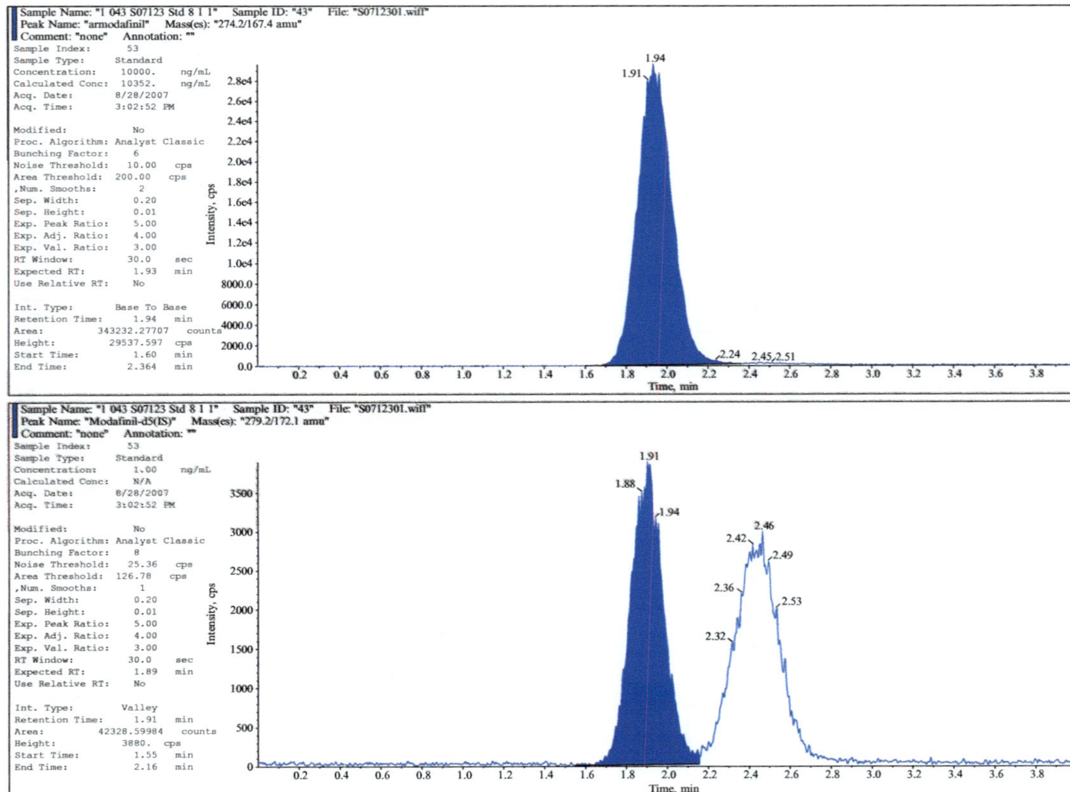

Fig. 6 Armodafinil final results from Analyte file based ACD simulator predication (0.4 % pentanol, 2×100 mm column, and column temp 25 °C, *top panel*: armodafinil; *bottom panel*: modafinil-d_5)

Table 2 Common phospholipid SRM transitions

Phospholipid	SRM transitions
Lyso-phosphatidylcholine; 16:0	496→184
Lyso-phosphatidylcholine; 18:0	524→184
Phosphatidylcholine; 30:1	704→184
Phosphatidylcholine; 34:2	758→184
Phosphatidylcholine; 38:6	806→184

phenomenon of "matrix effects," even though the sources causing the matrix effects were unknown at that time.

The term "matrix effect" can be broadly applied to any change in the behavior of a compound (analyte or internal standard) due to the presence of any other coeluting compound(s). Specifically for LC-MS/MS application, the impact could be either ionization enhancement or suppression. Coeluting compounds could be salts; endogenous compounds such as fatty acids and triglycerides; exogenous compounds such as drugs and their metabolites, including their corresponding internal standards; dosing vehicles such as carboxy-methyl-cellulose, DMSO, phosphate buffered saline, polyethylene glycol, etc.; anticoagulants such as EDTA or heparin; or constituents from the laboratory analysis process such as polymers, surfactants, and mobile phase additives. Although the reasons behind the matrix effects remained unclear for several years, the infusion experiment was well known and popularly used to showcase the matrix effect in a qualitative manner [13]. In this experiment, an extracted control blank sample is injected while the analyte of interest was infused into the HPLC effluent postcolumn using a syringe pump. Any ion suppression or enhancement would be observed as a decrease or increase of the MS signal at the time of analyte elution, without any real information as to the source of the effect. In 2003, as one of the pioneer research groups, our laboratory reported that endogenous phospholipids (PLs) were the primary source of the matrix effects observed when using ESI in the positive ion mode [14, 15]. Since then, numerous additional publications addressing matrix effects in great detail have been published as the phenomenon becomes better understood each year [16–20].

To ensure that appropriate resolution and lack of interference is obtained during method development, it is highly recommended actively monitoring PLs and aware of the retention/elution times and intensity of the PLs relative to the retention/elution times for the analytes of the interest. Table 2 summarizes several representative Selective Reaction Monitoring (SRM) transitions for PLs monitoring purpose during method development.

In general, phospholipids can be managed either via extraction or chromatographic separation. The best approach is to selectively remove the phospholipids from the sample during extraction. However, if nonselective extraction has to be used, significant amounts of phospholipids remaining in the sample will be injected onto LC column along with the analytes. In general, if the chromatography requires a higher organic composition, the PLs will either elute in the same injection or subsequent injection(s) and cause ion suppression (or enhancement) and may result in various analytical problems including, but not limited to: decreased or increased analyte or internal standard signal; divergent standard calibrators, raised baseline, imprecision and inaccuracy. To avoid these issues, the following strategies are recommended: (1) changing the LC column type and gradient, (2) changing the acidic or basic modifier, or (3) using a column switching technique to achieve two or three dimensional chromatography, such as the "heart-cut" sampling technique [21, 22]. If the chromatography requires lower organic compositions, the PLs may be retained on the LC column. This can be resolved by one of two approaches: (1) leaving the retained PLs intact on the LC column and flushing the column with a strong solvent such as 10:90 water–acetone after the completion of the entire batch (2) using column switching technique after each injection such as guard column trapping and flushing, column backflushing or forward flushing.

The above strategies are elucidated in the following two examples. In the first example, a LC-MS/MS assay was developed for the quantitation of a proprietary drug and its metabolite in human plasma with a single structural-analog internal standard using combination of PPE and Waters Oasis® HLB SPE extraction. Under the initial LC condition, the parent analyte coeluted with lysophospholipids (lyso-PLs) [lysophosphatidylcholine (16:0), 496→184 SRM transition and lysophosphatidylcholine (18:0), 524→184 SRM transition] and the full length PLs eluted after all analytes [phosphatidylcholine (38:6), 806→184 SRM transition]. The issue of late eluting PLs was resolved by adding column backflush and switching valve after all three analytes eluted. Because the lyso-PLs and analytes were intermingled, the only viable solution was to use a different separation mechanism, i.e., different LC stationary phase and/or LC gradient. The problem was resolved by switching to Metasil AQ® C18 (2×50 mm) column with a different LC gradient program (Fig. 7). Using the new condition, the lyso-PLs were still intermingled with the analytes; however, complete and consistent resolution between the analytes and the lyso-PLs was obtained, and the late elution PLs were flushed out from the column after each injection. The second example describes a proprietary assay for the simultaneous quantitation of eight analytes in human plasma using a single structural-analog internal standard and PPE extraction. As expected, the anticipated matrix effects caused by phospholipids were observed. Using Waters BEH® C18 (2×50 mm, 1.7 μm) column, 0.2 % NH_4OH in NH_4HCO_3 pH unadjusted and MeOH as mobile phases and a LC gradient program from 65 to 75 % (strong mobile phase), all nine analytes eluted before PLs. However, it was found that the PLs were eluted out of the column after several sequential injections, and sometime, they overlap with the analytes which cause imprecision. The final solution was to ramp the gradient to 100 % MeOH after all analytes are eluted out. Figure 8 showed that there is that adequately resolution between the analytes and PLs using a LC gradient program from 65 to 75 % to 100 %.

4.2 *Regression, Linearity, and Carryover*

As per the FDA guidance, the simplest regression model that adequately describes the concentration–response relationship should be used for quantitation. Linear

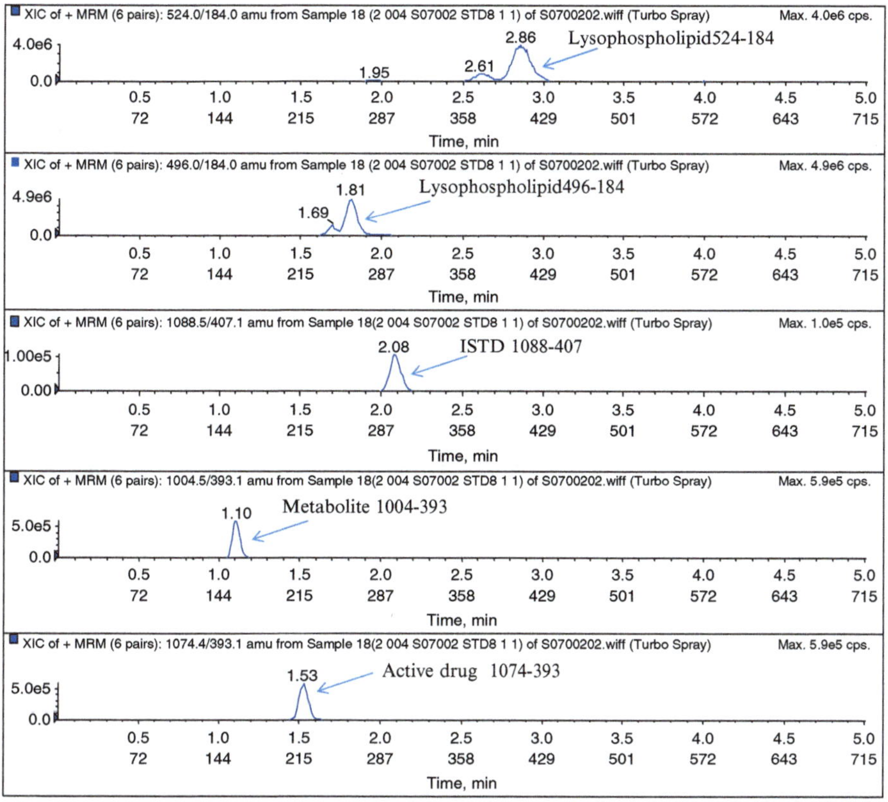

Fig. 7 Representative chromatography of a proprietary assay. The lyso-PLs were resolved from the analytes and the late elution PLs were washed out of the column via backflush and switching valve

regression is always preferred over quadratic regression. Justification for using a quadratic regression must be documented. In general, ESI-LC/MS/MS assays have tendency to demonstrate a quadratic behavior due to ion evaporation and detector capacity, especially with wide dynamic ranges (>3 orders of magnitude). Therefore, the first option to avoid quadratic regression model is to reduce linear range. Additionally, because APCI-LC/MS/MS assays do not have the same limitations as ESI techniques, they demonstrate less quadratic behavior and are less prone to be impacted by matrix effects. Therefore, APCI is preferred over ESI. If the anticipated concentrations for study samples cannot be adequately covered in a single assay with a dynamic range of ~3 orders or less, it is often better to develop two separate assays with complementary low and high concentration ranges. This eliminates excessive dilutions to bring the samples into the validated assay.

Not only will a narrow dynamic range increase the likelihood for a true linear regression, but it also provides another strategic benefit of a reducing carryover. Carryover is one of the most commonly occurring issues during LC-MS/MS method

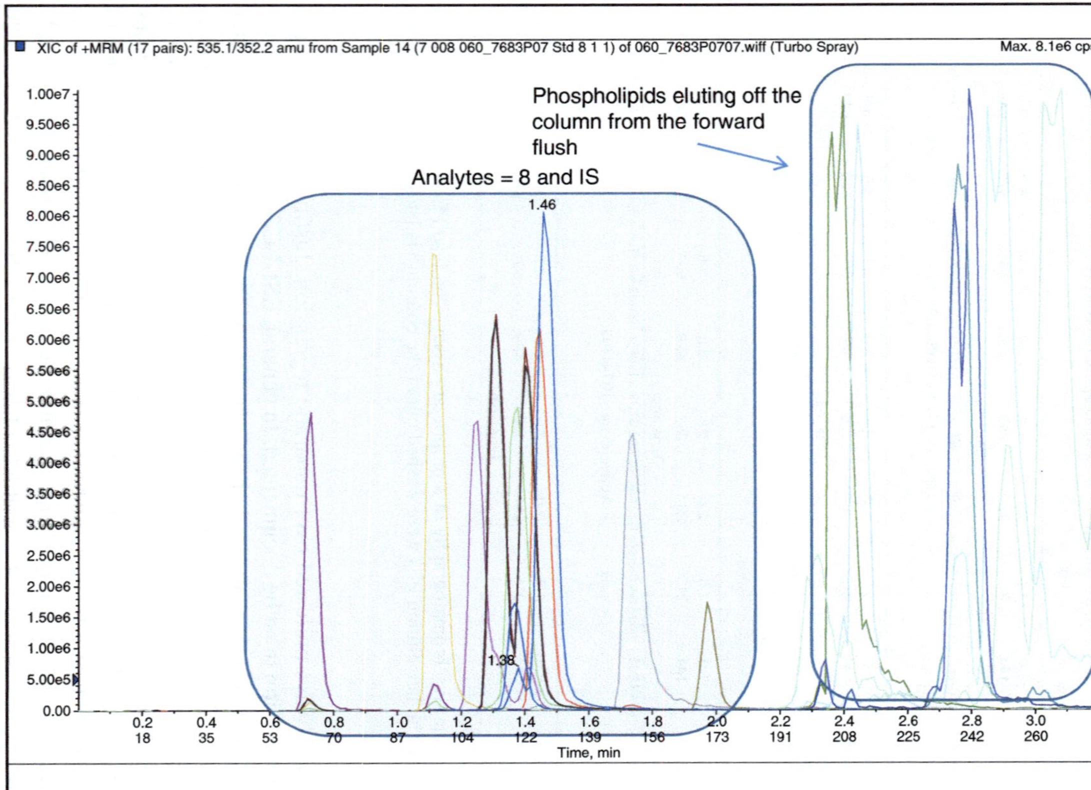

Fig. 8 Representative chromatography of a proprietary assay. The lyso-PLs and late elution PLs were eluted after the analyte via forward flush

development and can persist into routine method validation and sample analysis as well. In general, carryover is assessed by injecting one or more control blanks (double blanks) or QC0 (control blank spiked with internal standard only) immediately after the Upper Limit of the Quantitation (ULOQ) sample. If carryover is observed during method development, it is critical to determine the source of carryover and eliminate or minimize it accordingly. In general, there are two types of carryover: autosampler injector carryover and LC column carryover. The latter is caused by an ill-suited combination of column type, mobile phase solvents, and LC gradient elution program. For example, a fast and steep gradient on a retentive C18 LC column is more prone for column carryover. Fortunately, column carryover can be overcome by simply changing to isocratic conditions or using a shallower gradient or switching to a less retentive LC column type such as C4, C8, or phenyl. For autosampler injector carryover, the selection and combination of appropriate wash solvents must be evaluated. For a strong wash solvent, the strength of the organic solvent is critical. However, the addition of acid or base modifier which can facilitate the solubility of the analyte is very useful.

4.3 *Internal Standard Response*

LC-MS/MS assays typically rely on the use of an internal standard that mimics the performance of the analyte to improve the precision, reproducibility and reliability of the assay. An ideal internal standard candidate is a stable-isotope labeled ("stable labeled") form of the drug. Because synthesizing stable labeled chemicals can be expensive and time-consuming, it is very common to use a chemically similar "structural analog" of the analyte(s) as the internal standard, especially during the early phases of drug development.

From a practical point of view, internal standard in a LC-MS/MS assay serves three distinct purposes in the analytical process. The first purpose is to compensate extraction recovery inconsistencies. The second purpose is to compensate injection volume variation. The third purpose is to compensate possible "matrix effects" during the MS ionization process as has already been discussed in detail above. In 2009, Tan A. et al. reported 12 case studies from incurred sample analyses using a wide variety of bioanalytical methods for the investigation of inconsistent internal standard response [23]. For similar reasons, it has now become common for laboratory SOPs to contain specific requirements for the acceptable internal standard response of each individual sample within a sample batch during regulated bioanalysis. These requirements (e.g., 60–140 %, 50–150 % of the average internal standard area for all samples in the batch) ensure that the behavior of the internal standard, regardless of how well it tracks the analyte, is under control, and is consistent in all samples.

Although these measures and criteria were necessary during sample analysis, it is recommended to take proactive action steps to ensure appropriate IS behavior during method development. In general, inconsistent internal standard response is typically caused by either nonideal extraction conditions such as poor extraction recovery or extract insolubility or nonideal chromatography conditions such as matrix effects or an incompatibility with some combination of mobile phase, needle wash, or sample reconstitution solvent. The systematic approach to troubleshoot an inconsistent IS response is to perform intra- and intervial injection. If the inter and intravial injections are very precise, it clearly indicates that the inconsistent IS response is related to extraction. While this simple experiment does not conclusively identify the actual source of the erratic internal standard behavior, it can at least isolate it to one of two areas for further investigation.

This strategy can be demonstrated clearly in the following example. This is a proprietary assay for the quantitation of two analytes (parent analyte and its sulfoxide metabolite) in human urine using two stable labeled internal standards. Because of the extremely high linear range (5.00–100 μg/mL), a simple dilution of urine was utilized for sample preparation. The initially developed method utilized normal phase liquid chromatography [MonoChrom® Diol LC column (2.0×50 mm, 5 μm)] and a single premixed mobile phase (2 % MeOH in MeCN with 0.1 % Formic acid) under isocratic conditions. The internal standard response for the metabolite is extremely imprecise (Fig. 9). A quick intravial injection experiment outlined above was conducted to identify that the primary source of the imprecision was due to chromatographic condition. Because normal phase chromatography was used, the more polar sulfoxide metabolite eluted before the parent analyte and was the only analyte affected, suggesting that some early eluting interferences from the urine (not phospholipids in this case) might be causing an undesired matrix effect. Although an alternative and more selective extraction would likely solve this issue, it was determined that the simple dilution procedure was still the most desirable option due to the high concentration range. Thus, the final solution was to add a switching valve into the LC program in which the column was backflushed with strong organic solvent (100 % MeOH) at a high flow rate (1.0 mL/min) after the two analytes had eluted. Using the new conditions, the same sample batch was injected and a much improved internal standard response was obtained for the sulfoxide metabolite (Fig. 10).

5 Method Validation

Regulated method validation must be conducted according to Method Validation (MV) SOP which is typically written following FDA guidance and white papers for regulated bioanalysis validation. The MV SOP describes various validation tests to verify that a method is reliable and reproducible using calibration standards (STD) and quality control (QC) samples. For a formal MV, a Principle Investigator (PI) or Study Director (SD) is assigned for assuring that the validation tests meet the acceptance criteria and are adequate for the analysis of study samples. A Study Protocol (SP) has to be written and approved *a priori* to initialing validation. Minimally, the experiments, reference materials and calibration curve range(s) and statistical tests

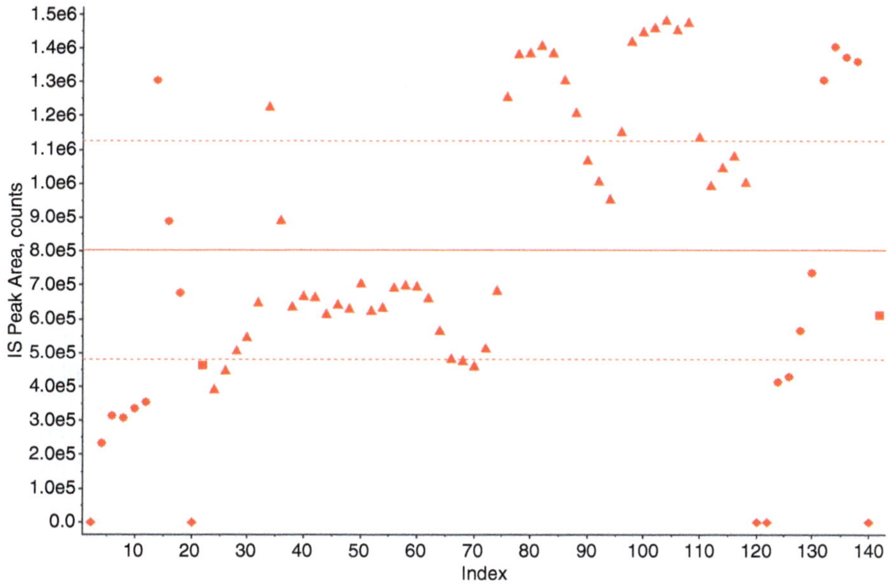

Fig. 9 Erratic IS response from a proprietary assay using dilution sample preparation and normal phase chromatography

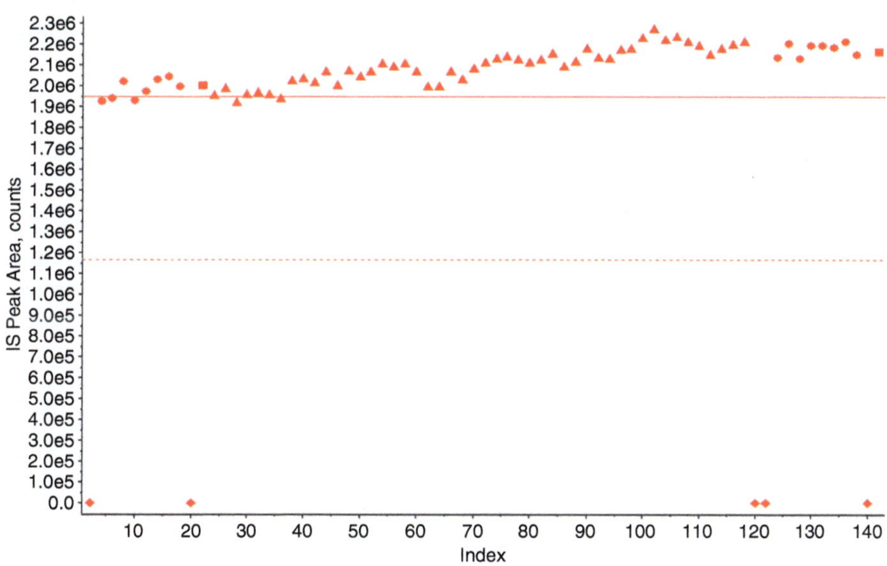

Fig. 10 Consistent IS response from the same batch shown in Fig. 9 with modified condition via column backflush and switching valve

to be utilized are defined in the protocol. Any planned deviation must be documented in protocol or amendment *a priori* and approved by PI or SD and Laboratory Director (LD).

5.1 General Guidance and Requirement

5.1.1 Matrices

An assay should be validated using the same biological matrix as the study samples. If a matrix has limited availability, a suitable substitute may be used (e.g., plasma ultrafiltrate to substitute cerebrospinal fluid) for the preparation of standard (STD) calibrators. When spiking solution is needed to prepare STDs and Quality controls (QCs) in matrix, the spiking volume should be less than 5 % of the total volume of the matrix.

5.1.2 Stock Solution

Two individually prepared and verified stock solutions are used to prepare STDs and QCs, respectively. Some laboratories require that the stability of the stock solution has to be established prior to formal validation. But some requires establishing at least one time point during validation to cover the time of solutions used during the validation.

5.1.3 Regression Model

The simplest model that adequately describes the concentration–response relationship should be used, e.g., a linear model is simpler than a quadratic model. At the completion of the validation, evaluation of different regression models must be performed. Justification for using a quadratic regression equation must be documented.

5.1.4 STD and QC Concentrations

Typically, six to eight different nonzero concentrations that cover the dynamic range of the assay should be used to define the curve. Two STDs should be at concentrations lower than the Low QC sample and two STDs should be at concentrations above the High QC sample. Five levels of analytical QC, i.e., LLOQ (Lower Limit of Quantitation), Low, geometric Medium (optional), Medium and High QC, are recommended. In general, the QC concentrations should differ from standard concentrations. The corresponding concentration distributions are: LLOQ = same concentration as lowest standard; Low = 3× the concentration of the lowest standard; Medium: ~40–60 % of the ULOQ; High: ~75–85 % of the ULOQ.

5.2 *Method Validation Experiments*

5.2.1 Intra- and Interday Precision and Accuracy Evaluation

Purpose. To evaluate assay accuracy and precision based on the overall performance of analytical QCs as well as STDs.

Experiment. Three test batches which are prepared, extracted, and analyzed in three separate days. It contains replicate of two sets of STDs and six replicates QC at LLOQ, Low, Medium, and High QC. The STDs must be prepared fresh at the day of extraction. QCs can be from previously prepared and qualified pools.

Acceptance. For STDs, the back-calculated concentration must not deviate ≤15.0 % (≤20.0 % for the LLOQ) from their individual target concentrations. At least three-fourths of the individual STDs must be acceptable. For QC, the mean of the replicates at each QC concentration and Relative Standard Deviation (RSD) must be ≤15.0 % (≤20.0 % for the LLOQ) from their target concentrations. At least two-thirds of the individual QCs at each level must be acceptable.

5.2.2 Dilution Integrity and Linearity

Purpose. to evaluate accuracy and precision of the QC with an analyte concentration originally above the ULOQ.

Experiment. Two levels of Dilution QC are required. The Dilution QC for dilution integrity is typically at 10× of the High QC. The Dilution QC for dilution linearity should be at a concentration above projected maximum concentration (C_{max}). If the projected C_{max} is not available, the recommended concentration would be 50× or 100× of High QC. To test dilution integrity or linearity, the Dilution QCs need to be diluted at dilution factor (DF) of 10, 50 or 100 to bring the final concentration within the range. It may require multiple steps of dilution. This experiment only needs to be conducted once per each validation.

Acceptance. Same as described in intra- and interday precision and accuracy evaluation.

5.2.3 Carryover of Analyte and Internal Standard

Purpose. To evaluate the impact of carryover on the quantitation of the analyte and internal standard.

Experiment. Place a control blank after the second ULOQ sample in each accuracy and precision run.

Acceptance. It is acceptable if carryover blank has analyte peaks that are ≤20.0 % of the lowest peak area of the acceptable LLOQ and carryover blank has internal standard peaks that are ≤5.00 % of the mean peak area of internal standard in the run.

5.2.4 Method Selectivity (Matrix Effects)

Purpose. To demonstrate the ability of the method to measure what it is intended to measure and not be affected by other sample components such as endogenous matrix elements. A minimum of six different matrix lots from individual donors has to be evaluated. It contains three individual experiments:

Experiment#1. Selectivity Blank Test: Extract selectivity blanks from six different lots.

Acceptance. Same as above for carryover assessment.

Experiment #2. Selectivity LLOQ Test: Extract selectivity LLOQ at replicate of one from six different lots.

Acceptance. The mean and the RSD of the selectivity LLOQ concentration must not deviate ≤20.0 % from their target concentrations. At least five out of six individual selectivity LLOQ must be acceptable.

Experiment #3. Matrix Factor (MF) Test: Extract six blanks from six different lots. Spike appropriate and same amount of analyte and IS into the postprocessed extract and neat sample, respectively. MF is the ratio of the peak area of the analyte in extract vs. neat sample. For stable labeled IS, the response can be normalized and peak area ratio (instrument response) can be used for MF calculation. For analog IS, the matrix factor must be calculated individually for analyte and IS. The recommended concentration for MF test is at medium QC level.

Acceptance. The RSD of the matrix factor for all lots must not deviate ≤15.0 %. If MF = 1, it indicates that there is no matrix effect. If MF >1, it indicates that there is ionization enhancement. If MF <1, it indicates that there is ionization suppression.

5.2.5 Interference

Purpose. To determine the ability of the method to measure what it is intended to measure and not be affected by other compounds, i.e., concomitant compounds, prodrugs, metabolites. It contains two separate experiments:

Experiment #1. For interference compounds which are not quantified within the assay, spike the interference compound at the C_{max} or minimum at ULOQ level into Low QC.

Acceptance. Same as described in intra- and interday precision and accuracy evaluation.

Experiment #2. For interference compounds which are quantified within the assay, each analyte need to be spiked individually, and the contribution of the single analyte to all other analytes are evaluated.

Acceptance. Same as above for carryover assessment.

5.2.6 Relative Extraction Recovery

Purpose. To obtain the relative extraction recovery from matrix for all analyte and internal standard. This experiment is required for stable isotope labeled IS.

Experiment. Three analyte levels generally near the low, medium, and high QC concentrations and the internal standard at the working spiking concentration were tested. The response of extracted samples (Preprocess) is compared to the response of samples representing 100 % recovery (Postprocess).

Acceptance. There is no requirement for overall recovery as long as the RSDs for replicates at each concentration (preprocess and postprocess) are ≤20.0 %

5.2.7 Short- and Long-Term Solution Stability

Purpose. To evaluate the stability of the analyte and internal standard in stock and working solvent if they differ. It contains two sets of experiments:

Experiment #1. Short-term analyte and IS bench top (i.e., room temperature or wet ice bath) solution stability. A minimum of one concentration such as ULOQ spiking solution must be tested. Some laboratories require solution stability at standard 2 (which has concentration typically twice of the LLOQ) or LLOQ spiking solution level. Test for minimum 6 h.

Experiment #2. Long-term analyte and IS storage solution stability.

Acceptance. The stability of solution is acceptable for the period of storage if the % difference between the stability of the bench top solution and control solution is ≤10.0 %.

5.2.8 Short and Long-Term Matrix Stability

Purpose. To evaluate the stability of the analyte in matrix. There are three different experiments:

Experiment #1. Freeze–thaw stability. Following an initial freezing period of at least 24 h, each freeze/thaw cycle must include thawing and frozen storage for at least 12 h. Samples should be thawed by the same procedure as is used for study samples, i.e., thawing, vortex mixing, uncapping and capping, ambient temperature or ice water bath, etc. The time interval for each cycle must be at least 30 min from the removal of the sample from the freezer to return of the sample to the freezer.

Experiment #2. Short-term matrix bench top matrix stability. Samples should be thawed by the same procedure as is used for study samples. Test for minimum 6 h.

Experiment #3. Long-term matrix storage stability. Test for the designated temperature, i.e., −20 °C or −70 °C to cover the time of sample analysis.

Acceptance. Same as described in intra- and interday precision and accuracy.

5.2.9 Extract Stability

Purpose. To determine the quantitative reproducibility of processed extract samples after storage for an extended time period under specified conditions (i.e., room temperature, 5 °C).

Experiment. Reinject previously analyzed six QC replicates of low, medium, and high QC that have been stored for a specified period under specified conditions against the freshly prepared calibration standards. The starting point is defined as the time when the extraction was completed. The end point is defined as the time when the fresh extraction was completed.

Acceptance. Same as described in intra- and interday precision and accuracy.

5.2.10 Reinjection Reproducibility

Purpose. To determine the quantitative reproducibility of autosampler after storage for an extended time period under specified conditions (i.e., room temperature, 5 °C).

Experiment. Reinject previously analyzed Accuracy/Precision batch that have been stored for a specified period under specified condition. The starting point is defined as the time when the extraction was completed. The end point is defined as the time when the last acceptable QC was injected in the reinjection run.

Acceptance. Same as described in intra- and interday precision and accuracy.

5.2.11 Whole Blood Sample Collection Stability

Purpose. To evaluate the stability of the analyte in whole blood during sample collection.

Experiment. Prepare test whole blood QC pools at medium QC level. Store at room temperature and 5 °C for 0 (control), 1 and 2 h. Centrifuge the whole blood sample and collect the plasma (test matrix) for extraction and analysis following the validation method. Analyze six (6) replicates for each group. Compare the mean instrument response of stability test samples to that of the control group.

Acceptance. The whole stability is acceptable for the period of storage and condition if the % difference between the stability group and control group is ≤15.0 % and the RSD for each group is ≤15.0 %.

5.2.12 Evaluation of Hemolysis

Purpose. To evaluate the presence of hemoglobin in plasma/serum samples for its effect on the precision and accuracy of the assay being validated.

Experiment. Prepare hemolyzed plasma or serum by spiking hemolyzed whole blood into nonhemolyzed plasma at low QC level. Hemolyzed blood for these experiments can be prepared by freezing, thawing and centrifuging whole blood. Three levels of hemolysis are used: 0 % (100 % normal plasma/serum), 0.5 % of above (99.5 % normal plasma/serum), and 2 % of above (98 % normal plasma/serum). Analyze six (6) replicates for each group. Compare the mean instrument response of the 0.5 % or 2 % hemolysis test samples to that of the 0.0 % control group.

Acceptance. Same as described in whole blood stability.

5.3 Cross Validation and Revalidation

5.3.1 Cross Validation

Cross validation is needed on a method that has been received from other institution or between sites of the same institution. This is to verify ability of the current laboratory to perform the assay. Most of the time, it requires three precision and accuracy runs. The evaluations of carryover, recovery, and the ability to dilute may be excluded if there is reason to believe these parameters will not be affected. Some or all of the stability evaluations may be excluded if the stability determinations have been adequately evaluated elsewhere and documentation of stability is available.

5.3.2 Revalidation

Revalidation is required for major changes in assay methods. The extent of the revalidation depends on those parameters that could be affected by the change that needs to be revalidated. Minor assay modifications may be made without revalidation if these changes do not alter the fundamental properties of the assay. A full revalidation is required for (1) A change in matrix/or species. i.e., dog plasma to rat plasma, (2) a change in anticoagulant, i.e., Na_2EDTA to Na-Heparin, or Na_2EDTA to K_3EDTA, (3) a change of internal standard, i.e., from structure analog to stable isotope labeled analog, or from D_4-labeled IS to D_5-labeled IS, and (4) a change of extraction procedure or LC condition or MS condition.

6 Regulated Sample Analysis

Regulated bioanalytical sample analysis must be conducted according to Sample Analysis (SA) SOP which is typically written following FDA guidance and white papers. The SA SOP describes procedures for bioanalysis performed in support of submissions to regulatory agencies. A validated method has to be used and analysis will be conducted by scientists who have been certified on the assay. Similarly as formal MV, a Principle Investigator (PI) or Study Director (SD) is assigned for assuring that the validated method is approved, that sufficient stability is established, and adequate for the analysis of study samples, and that the execution of the sample analysis follows SOP and meets FDA guidance. A Study Protocol (SP) is written and approved prior to sample analysis in which the criteria for data acceptance and criteria for reanalysis of samples are outlined. Any planned deviation must be documented in protocol or amendment *a priori* and approved by PI and Laboratory Director (LD). The following sections detail the chronicle of the study samples from sample receiving to data reporting in a GLP bioanalytical laboratory.

6.1 Sample Accession and Storage

Biological samples typically arrive in bioanalytical laboratory in secured boxes with dry ice or ice pack. Formal sample accession is the process of inspecting sample physically and registering into a LIMS (Laboratory information management system). The initial physical examinations of the samples include any missing vial, leakage, broken tubes, or thawing. The verification of documentation of the samples includes the number of the samples, subject ID, time point, period, group number, dose information et al. Any discrepancies and irregularities need to be documented and the Study Director and testing facility need to be notified immediately. After accession, samples are stored in secured environment designated by the protocol. Refrigerator or freezer for sample storage is typically monitored by a temperature monitoring software. Matrix STD and QC samples have to be stored in the same condition and, at least one set of high and low QC should be stored in the same storage unit along with study samples.

6.2 Preparation of Calibration Standards and QCs

STDs in matrix may be prepared ahead of the time and stored as long as analytes stability in matrix has been demonstrated. STD and QC pools are prepared according to the specific validated analytical methods. Spiked solutions cannot be used beyond the established stability. After qualification, the standard and QC pools should be prealiquoted out and stored under designated condition and temperature.

6.3 Prestudy Assay Evaluation (PSAE)

Prior to sample analysis, a test batch is required if (1) STD and QC pools need to be qualified, (2) there has been a significant lapse in time since any previous analysis, (3) the scientist does not have documentation with the assay or a similar assay. The test batch should contain minimally low, medium, and high QC at six replicates. Dilution QC samples may be included if dilutions are expected during sample analysis. The acceptance of PSAE is the same as described in intraassay precision and accuracy.

6.4 Sample Batch

Each sample batch contains replicates of two sets of standard calibrators which are placed at the beginning and end of the batch to bracket study samples and QC samples. The total numbers of QC samples from all levels should be >5 % of the number of unknown samples in the run and at a minimum of $N = 2$ at each level or $N = 3$ for Dilution QC. Minimum three system suitability test samples (SST) should be included in the beginning of the batch: (1) SST-LLOQ sample to evaluate the signal to noise ratio of the instrument on the day of the analysis. (2) SST-QC0 sample (a blank sample is fortified with IS) to evaluate any potential contamination of IS solution. (3) SST-Control blank (a double blank sample) to evaluate any interference or contamination of the blank matrix lot.

6.5 Sample Extraction and Analysis

Extraction scientist who is responsible for sample preparation should be certified prior to extracting real study samples. The validated extraction method has to be followed exactly. The raw data entries have to be documented promptly such as lot numbers of STD and QC, IS, extraction reagents, matrix, the IDs of automation tool and pipette, the time for study sample removal and return to storage and the completion of extraction. Instrument operator who is responsible for analysis has to perform SST test and assess sensitivity and carryover prior to initialing batch. Instrument operator has to follow the validated LC-MS/MS condition exactly and document raw data promptly such as injection volume, LC column serial number, lot number for mobile phases and needle wash solvent, and ID for autosampler, LC pumps, and mass spectrometer.

6.6 Integration and Data Review

Peak integration is conducted using peak processing software. The peak must exhibit a signal to noise (S/N) of at least 5 to 1 in order to be considered a quantifiable peak. The scientist must optimize the integration parameters to automatically integrate all peaks consistently and appropriately with respect to retention time, baseline, and peak width. The same set of integration parameters should be used for all STDs, QCs and unknown samples. If one or more peak processing parameter need to be changed to obtain a proper integration for a few samples, the scientist needs to get approval from Laboratory Director and the copies of the original and reintegrated chromatogram need to be placed in study record and reviewed by PI, LD, and QA auditors. Manual integration is highly prohibited.

6.7 Run Acceptance Criteria

For STDs, the back-calculated concentration must not deviate ≤15.0 % (≤20.0 % for the LLOQ) from their individual target concentrations. At least three-fourths of the individual STDs must be acceptable. For QC, at least two-thirds of the QC samples at all levels must be ≤15.0 % from their target concentrations. At least 50 % of the QC samples at each level (low, medium, and high) must be ≤15.0 % from their target concentrations. For dilutional QC, for each dilution factor, at least two-thirds of the QC samples at each dilution must be ≤15.0 % from their target concentrations.

7 Incurred Sample Reanalysis

Over the years, regulatory authorities noticed an alarming phenomenon in which large discrepancies between original and repeat values were observed in some submissions. In all cases the assays were formally validated and met acceptance criteria based on the performance of standard calibrator and quality control samples. It was gradually recognized that the spiked standards and QCs is not sufficient to fully assess the accuracy and reliability of a method. Although the purpose of QC samples is to mimic the performance of incurred samples, there is ample evidence to show that these entities often behave in very different manners. The topic of mandatory incurred sample reanalysis (ISR) was the primary item of interest discussed at the third Crystal City meeting in May 2006. This meeting resulted in a formal White

Paper published in September, 2007 by Rocci et al. entitled "Confirmatory Reanalysis of Incurred Bioanalytical Samples" [24].

Based on the recommendation of Rocci et al.'s publication and other regulatory documentations, the following section summarizes the main features which are essential for ISR practice:

Type of studies. Incurred sample reanalysis is applicable to bioequivalence, phase I-IV human trials and GLP preclinical studies, or any study with a unique population. For GLP preclinical studies, the ISR evaluation only needs to be performed once per method, per species.

Number of ISR samples. A minimum of 20 ISR samples for small studies (<200 study samples), at least 10 % ISR samples for medium sized studies (201–999 study samples), with an additional 5 % ISR samples for larger studies (>1,000 study samples).

Selection of ISR samples. ISR samples should be chosen from samples with reportable concentrations at approximately >3 times the LLOQ, mid-range, and near the approximate C_{max}.

Dilution scheme. Whenever possible, a sample requiring dilution to yield a reportable result should be reanalyzed using the same dilution factor as the original analysis. Samples should not be diluted to levels <3× the LLOQ.

Timing of analysis. ISRs will be run within a time frame reasonably close to the original analysis to allow real-time identification of potential problems and reduce the impact that stability may have on reassay results.

Acceptance criteria. Two-thirds (67 %) of the ISR samples repeat values must be within ±20 % of the mean of original and ISR values.

Reporting. A comparison table of results showing original result, ISR result and % bias will be presented in the Sample Analysis report, along with the overall percentage of ISR samples meeting the acceptance criteria. A conclusion statement will be included in the report text indicating if ISR met the acceptance criteria including any follow-up actions if applicable.

Investigation. If ISR acceptance criteria are not met, an investigation is conducted. Future analysis will not be conducted until management is confident that the assay is capable of producing reliable and reproducible results.

8 Conclusion

Robust and rugged LC-MS/MS methods are essential in support of drug discovery, toxicology studies, and clinical trials, for the data generated from these bioanalytical methods is used to evaluate the bioavailability, bioequivalence, toxicokinetic, and pharmacokinetic parameters of drug candidates. Thus, it is critical to invest significant thought and effort in the method development process [25–27]. Fast sample preparation and fast chromatography do not necessarily guarantee a robust and reliable bioanalytical method suitable for regulated analysis. The development of a robust bioanalytical method requires the careful consideration of many critical parameters.

A complete regulated method validation minimally requires three interday precision and accuracy runs and various stability and reproducibility experiments. It is normally executed in approximately one or two weeks and involves the laboratory, Quality control (QC) and Quality Assurance (QA) departments. Run failures during method validation are especially costly in terms of time, effort, and perception. The best practice to ensure the success of method validation is to conduct a thorough and comprehensive method development process. With this approach, the actual method validation itself should be a matter of formality, as there is an extremely high level of confidence that all validation tests will pass. The final evaluation of any high quality bioanalytical method is not complete until it passes the ultimate test of regulated sample analysis and incurred sample reanalysis. A rugged method is a prerequisite. However, regulated sample analysis and incurred sample reanalysis will not be successful without proper planning and sound execution.

Unlike basic research in academic laboratory setting in which an individual researcher works on a single project from beginning to end, regulated bioanalysis process typically involve multiple individuals, departments, or companies. Typically a principle investigator (PI) or study director (SD) is responsible for monitoring project status, reviewing data and reports, and communicating with the client. Multiple scientists will be responsible for the project at different stages. Multiple departments will also be involved to perform required activities such as Sample Management, laboratory, Quality Control (QC) and Quality Assurance (QA), and Data Management. Regulatory inspections are routinely conducted by internal QA auditors, external QA auditors, consultants, or even regulatory (FDA) auditors. Because all study records must be clearly documented and properly maintained to ensure the integrity of the study, it is therefore vital to build a system that there produces a clear hand off between each stage of laboratory activity and a centralized location to retrieve and review data. In our laboratory, the solution is to use forms (checklists) for each of three important stages during method development and validation and sample analysis. Overall, these checklists provide the required details necessary for conducting projects in a regulated environment. These checklists ensure that all required documentation is recorded and retained, eliminate the need to rely on the memory of a single individual at any time, increase the confidence of our clients based on the timely and successfully completion of contracted work, and most importantly, ensure the integrity of all studies in all circumstances.

References

1. Shah VP, Midha KK, Dighe S et al (1992) Analytical methods validation: bioavailability, bioequivalence, and pharmacokinetic studies. Pharm Res 9:588–592
2. FDA/CDER, Guidance for the industry: Bioanalytical method validation, May 2001, US Department of Health and Human Services, FDA (CDER) and (CVM), Rockville, MD, USA

3. Viswanathan CT, Bansal S, Booth B et al (2007) Quantitative bioanalytical methods validation and implementation: best practices for chromatographic and ligand binding assays. AAPS J 9:E30–E42
4. Yu RZ, Baker B, Chappell A, Geary RS, Cheung E, Levin AA (2002) Development of an ultrasensitive noncompetitive hybridization ligation enzyme-linked immunosorbent assay for the determination of phosphorothioate oligodeoxynucleotide in plasma. Anal Biochem 304(1):19–25
5. Deverre JR, Boutet V, Boquet D, Ezan E, Grassi J, Grognet JM (1997) A competitive enzyme hybridization assay for plasma determination of phosphodiester and phosphorothioate antisense oligonucleotides. Nucleic Acids Res 25(18):3584–3589
6. Deng P, Chen X, Zhang G, Zhong D (2010) Bioanalysis of an oligonucleotide and its metabolites by liquid chromatography–tandem mass spectrometry. J Pharm Biomed Anal 52:571–579
7. Zhang G, Lin J, Srinivasan K, Kavetskaia O, Duncan JN (2007) Strategies for bioanalysis of an oligonucleotide class macromolecule from rat plasma using liquid chromatography–tandem mass spectrometry. Anal Chem 79(9):3416–3424
8. Li WK, Tse FL (2010) Dried blood spot sampling in combination with LC-MS/MS for quantitative analysis of small molecules. Biomed Chromatogr 24(1):49–65
9. Pan JW, Jiang XY, Chen YL (2010) Automatic supported liquid extraction (SLE) coupled with HILIC-MS/MS: an application to method development and validation of erlotinib in human plasma. Pharmaceutics 2:105–118
10. Wang J, Aubry A et al (2010) Effect of mobile phase pH, aqueous-organic ratio, and buffer concentration on electrospray ionization tandem mass spectrometric fragmentation patterns: implications in liquid chromatography/tandem mass spectrometric bioanalysis. Rapid Commun Mass Spectrom 24:3221–3229
11. Meng M, Rohde L, Čápka V et al (2010) Fast chiral chromatographic method development and validation for the quantitation of eszopiclone in human plasma using LC/MS/MS. J Pharma Biomed Analy 53(4):973–982
12. Bennett PK, Meng M, Rohde L Software assisted chiral chromatographic method development for the quantitation of four chiral drugs in human plasma using LC/MS/MS, Presented at the 2009 ASMS conference, Philadelphia, PA
13. King R, Bonfiglio R, Fernandez-Metzler C et al (2000) Mechanistic investigation of ionization suppression in electrospray ionization. J Am Soc Mass Spectrom 11(11):942–950
14. Ahnoff M, Wurzer A et al. (2003) Characterization of serum albumin and lysoPCs as major contributors to plasma sample matrix effects on electrospray ionization efficiency, Presented at the 2009 ASMS conference, Montreal, Canada
15. Bennett PK, Van Horne KC (2003) Identification of the major endogenous and persistent compounds in plasma, serum, and tissue that cause matrix effects with electrospray LC/MS techniques, Presented at the 2003 AAPS annual meeting and exposition, Salt Lake City, Utah
16. Xia YQ, Jemal M (2009) Phospholipids in liquid chromatography/mass spectrometry bioanalysis: comparison of three tandem mass spectrometric techniques for monitoring plasma phospholipids, the effect of mobile phase composition on phospholipids elution and the association of phospholipids with matrix effects. Rapid Commun Mass Spectrom 23:2125–2138
17. Meng M, Bennett PK Source for imprecision resulting from ionization suppression from strongly retained phospholipids and dioctyl phthalate, Presented at 2004 ASMS conference, Nashville, TN
18. Little JL, Wempe MF, Buchanan CM (2006) Liquid chromatography-mass spectrometry/mass spectrometry method development for drug metabolism studies: examining lipid matrix ionization effects in plasma. J Chromatogr B 833:219–230
19. Du L, White RL (2008) Reducing glycerophosphocholine lipid matrix interference effects in biological fluid assays by using high-turbulence liquid chromatography. Rapid Commun Mass Spectrom 22:3362–3370
20. Ismaiel OA, Halquist MS, Elmamly MY et al (2008) Monitoring phospholipids for assessment of ion enhancement and ion suppression in ESI and APCI LC/MS/MS for chorpheniramine in human plasma and the importance of multiple source matrix effect evaluation. J Chromatogr B 875:333–343

21. Carter SJ, Čápka V, Viccarone S An orthogonal approach to increasing assay ruggedness at low limits of quantition in LC/MS/MS assays, Presented at the 2007 ASMS conference, Indianapolis, IN, June 2007
22. Voelker T, Tan L, Acheampong A et al Two dimensional PPE/SPE extraction coupled with three dimensional HPLC/UPLC/MS/MS for a 500 fg/mL LLOQ assay, Presented at the 2011 ASMS conference, Denver, June 2011
23. Tan A, Hussain S, et al (2009) Internal standard response variations during incurred sample analysis by LC-MS/MS: case by case trouble shooting. J Chromatogr B 877(2):3201–3209
24. Rocci ML, Devanarayan V, Haughey DB et al (2007) Confirmatory reanalysis of incurred bioanalytical samples. AAPS J 9(3):E336–E343
25. Zhou S, Song Q et al (2005) Critical review of development, validation, and transfer for high throughput bioanalytical LC-MS/MS methods. Current Pharma Analysis 1:3–14
26. Liu GW, Snapp HM, Ji QC, Arnold ME (2009) Strategy of accelerated method development for high-throughput bioanalytical assays using ultra high-performance liquid chromatography coupled with mass spectrometry. Anal Chem 81(22):9225–9232
27. Jemal M, Zheng O, Xia YQ (2010) Systematic LC-MS/MS bioanalytical method development that incorporated plasma phospholipids risk avoidance, usage of incurred sample and well thought-out chromatography. Biomed Chromatogr 24(1):2–19

Impact of Sampling Paper/Cards on Bioanalytical Quantitation via Dried Blood Spots by Liquid Chromatography-Mass Spectrometry

Jun Zhang, Ramona Rodila, Huaiqin Wu, and Tawakol A. El-Shourbagy

Abstract Dried blood spots (DBS) on paper as a sampling technique in newborn screening has been widely adopted by bioanalytical chemists for preclinical and clinical sample collection. DBS is based on the assumption that single size punch of DBS absorbs the same volume of blood regardless of physiological differences derived from test subject genders, disease states, nutrition, or hydration affecting blood viscosity, although it is well known that such differences have a fundamental impact on the quantitation accuracy of DBS as a sampling technique. There are multiple types of sampling media in either plain filter paper or chemically treated paper/cards available for DBS applications. Pretreated paper/cards contain chemicals such as denaturants, surfactants, and/or chelating agents to deactivate pathogens, enzymes and prevent the growth of biological organisms. In this chapter, the chemical constituents of various paper/cards are explored. The impact of paper/card type on analytical interference, evenness of DBS spots, and radial distribution of analyte are evaluated. The paper/card impact on the matrix effect was studied. It was found that the impregnated chemicals on the pretreated DBS paper/cards could be more than 50 % of the total paper/card weight. These water-soluble chemicals make the analyte distribution unpredictable across the dried blood spots and can interfere with LC-MS. Six compounds across a large Log D range were used to evaluate different types of paper/cards for understanding the impact of the sampling paper/cards on DBS quantitation accuracy. The results indicate that the impact is significant and an evaluation of sampling paper/card impact is necessary for most compounds.

J. Zhang, Ph.D. (✉) • R. Rodila • H. Wu • T.A. El-Shourbagy
Global Pharmaceutical Research and Development, Abbott Laboratories, Dept R46W, Building AP13A-2, 100 Abbott Park Road, Abbott Park, IL 60064-6126, USA
e-mail: jun.zhang@abbott.com

Q. Alan Xu and Timothy L. Madden (eds.), *LC-MS in Drug Bioanalysis*,
DOI 10.1007/978-1-4614-3828-1_3,

1 Introduction

Dried blood spots (DBS) were introduced by Guthrie and Susie [1] in 1963 as a sampling technique for screening metabolic diseases of neonates in a large population. The applications of DBS have been reviewed recently by two research groups [2, 3]. DBS offers many advantages over conventional sampling techniques for blood, plasma, or serum collection [4]. These advantages can be summarized as less invasive sampling, simpler storage and handling, less infection risks through using pretreated sampling cards, and smaller sample volume [5]. In addition to newborn screening, DBS has also been applied in therapeutic drug monitoring and pharmacokinetics studies [6–37]. All these applications were based on the sole assumption in which the same size punches from dried blood spots on sampling paper/cards absorb the same volume of blood regardless of any differences among test subject genders, matrix lots, disease states, or blood viscosity. Sampling paper/cards are the ultimate carrier of the target analyte in blood. There have been a few types of sampling paper/cards available specifically designed for this purpose. Hundred percent cotton filter paper, marketed under the names of plain filter paper, Guthrie paper, Shleicher & Scheull 903, S & S903, Whatman 903, or No. 545 filter paper over different times or by different vendors, has been used successfully over the last few decades for newborn screening. However, inconsistent extraction was randomly observed between low and high concentration spots via quality controls, but also between fresh and aged blood spots [16, 25]. Paper pretreatment with chemicals [25] and via impregnation [36] provided a solution for unbalanced extraction recoveries. The paper pretreatment via impregnation of other chemicals is commercialized as multiple products such as Whatman FTA series for lysing cells, denaturing proteins, and preventing the growth of microorganisms. FTA series (i.e., FTA, FTA Elute, DMPK cards) was designed with proprietary formulations so that the paper/cards could be used in the bioanalysis of drug metabolism and pharmacokinetic samples for better accuracy over plain filter paper. Although other factors such as hematocrit [37], blood volume, and blood distribution have direct impacts on the accuracy and precision of bioanalytical methods on DBS, in this article we present only the potential impact of the sampling paper/cards as we have learned from implementing DBS-LC-MS in our laboratory.

2 Experiments

2.1 Chemicals

Six proprietary compounds were from Abbott Laboratories (North Chicago, IL, USA). Two stable isotope labeled compounds and additional proprietary compound from Abbott Laboratories were used as the internal standards. HPLC grade methanol and acetonitrile were from Sigma-Aldrich (St. Louis, MO, USA). A.C.S. grade ammonium formate and formic acid were also from Sigma-Aldrich. Water was purified through a Milli-Q water purifier from Millipore (Billerica, MA, USA). Human blood was purchased from Biological Specialty Corporation (Colmar, PA, USA).

2.2 Sampling Paper/Cards

Nine kinds of sampling paper/cards were used. FTA Elute (Micro), FTA DMPK-B, FTA DMPK-A, Indicating FTA (pink), FTA Plant, Blood Stain, Protein Saver 903 (previously Guthrie Card or Schleicher & Scheull Card in different times), and 31 ET were from Whatman (now GE Healthcare, Piscataway, NJ, USA) and VWR 237 filter paper from VWR International (West Chester, PA, USA).

2.3 Instruments

Harris Uni-Core 1.00, 3.00, 6.00 mm cutters and cutting mats were from Whatman. A Sil-10 HTc autosampler and two LC-10AD pumps were from Shimadzu (Kyoto, Japan). An API-4000 mass spectrometer with turbo ionspray ionization probe was from AB Sciex (Foster City, CA, USA). The software for data acquisition and processing was Analyst ver. 1.5 from Applied Biosystems as well. An automated liquid handler was MicroLab AT 2 Plus from Hamilton (Reno, NV, USA). A MixMate vortexer was from Eppendorf (Hauppauge, NY, USA). A CR412 centrifuge was from Jouan (now Thermo Fischer). Column selection valve was from Valco Instruments (Houston, TX, USA). An Atlantis dc18 column (100 × 2.1 mm, 5 μm) was purchased from Waters (Milford, MA, USA).

2.4 Sample Preparation

Dried blood spot on different paper/cards were punched and dispensed into the wells of a 96-well plate, 20 μL of internal standard in approximately 50 % acetonitrile was added. Three hundred microliters of 70 % acetonitrile (v/v) was added and the plate was vortexed for 5 min at 800 rpm. Fifty microliters of 6 M ammonium formate was added and the plate was vortexed at 850 rpm for 10 min. The plate was then centrifuged at 3,500 rpm and 10°C for 5 min. The upper layer was transferred to a new plate and the samples were evaporated by a nitrogen stream at room temperature. The dried samples were reconstituted with 30 % acetonitrile and 5 μL of reconstituted sample were injected.

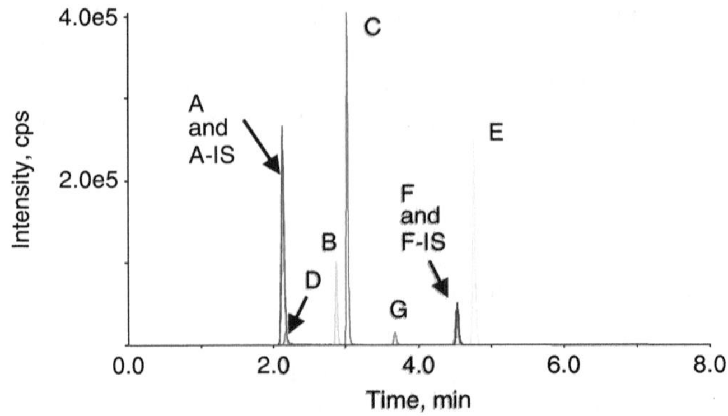

Fig. 1 Chromatogram of six compounds (Compounds A–F) with two labeled internal standards (IS) and one nonlabeled IS. A-IS is the labeled IS for Compounds A and D and F-IS is the labeled IS for Compound F. Compound G is the IS for Compounds C and E

2.5 *Chromatography*

Two mobile phases were used. Mobile phase A was 0.1 % formic acid and 5 % acetonitrile in water and mobile phase B was 0.1 % formic acid and 95 % acetonitrile in water. Chromatographic separation was achieved for all compounds using an 8.5 min gradient method. Initially, 7.5 % mobile phase B was used for system equilibration. After each injection, mobile phase B was run at 7.5 % for 1 min then it was linearly increased to 90 % over the course of 3 min, it was maintained at 90 % for 2.5 min and returned to 7.5 % in 0.1 min. The postgradient column equilibration was run for 1.9 min prior to the start of pretreatment for the next injection. Figure 1 shows a typical chromatogram of six compounds (Compounds A through F) and three internal standards eluted with the described chromatography conditions. Compound G was used as the internal standard for Compounds B, C, and E.

2.6 *Mass Spectrometry*

LC-MS/MS detection was performed using an AB Sciex API 4000™ triple quadrupole mass spectrometer equipped with a TURBO V™ ionspray ionization source operated in positive ion mode. The software used for instrument control was Analyst version 1.5. The spray voltage was set at 5,500 V and the source temperature was 550 °C. Other ion source parameter settings were 10 for curtain gas, 65 for GS1, 65 for GS2. The collision gas setting was 4 and the entrance potential was 10 V. Other compound specific parameters such as the declustering potential (DP), the collision energy (CE), and the cell exit potential (CXP) varied depending on the analyte. The DP ranged from 50 to 110 V, the CE ranged from 23 to 80 eV, and the CXP ranged from 5 to 15 V. Acquisition was performed in scheduled MRM mode with a 60 s window around the retention time of each analyte and a target scan time of 0.2 s.

Table 1 Measured amount of water-soluble chemicals on sampling paper/cards

Card type[a]	Chemicals on card (%)	Accuracy (%)
FTA Elute	53.2	101.2
FTA DMPK-B	50.9	102.8
FTA Plant	11.5	100.4
FTA DMPK-A	11.8	98.4
Indicating FTA (pink)	11.1	101.5
Blood Stain	2.5	98.5
Protein Saver 903	1.6	102.5
31ET	1.7	98.2
VWR 237	1.1	102.9

[a]All paper/cards were from Whatman except VWR 237 paper from VWR International

3 Results and Discussion

3.1 *Water-Soluble Chemicals on the Cards*

In order to estimate the amount of water-soluble components in the sampling paper/cards, two 6 mm disks from each type of paper were cut into preweighed test tubes and weighed using an analytical balance, the disks were wetted with 500 μL of distilled water, shaken well, and then weighed again. The water was carefully transferred off the washed paper disks into preweighed test tubes with a micropipette and weighed, then put under a stream of nitrogen gas for drying and reweighed when completely dried. The paper disks were also dried under nitrogen and reweighed. Table 1 shows the relative amounts of chemicals on the paper in percentage. The weight loss of the paper disks before and after water washing was confirmed by the weight of the water-soluble chemicals released into water which confirmed the quantitative transfer of the liquids between the vials. The ratio of weight of the chemicals released into water to the weight loss of the disk is presented as accuracy (%) in Table 1. FTA elute card appears to carry the most water-soluble chemicals (≥50 % of total card weight) among all sampling cards. FTA DMPK-A, FTA plant, and Indicating FTA cards appear to be the same or similar paper, but Indicating FTA paper carries a pink dye additive. Blood stain, 31ET, and VWR 237 paper/cards carry the least water-soluble chemicals. Actually, 31ET and VWR 237 are marketed as plain cotton fiber paper.

3.2 *Identities of Chemicals Impregnated on the Paper*

Based on the amount of water-soluble chemicals on the paper/cards, three types of cards (FTA Elute, Indicating FTA, and VWR 237) were used to explore the chemical identifications. The possible chemicals on the cards were checked with mass

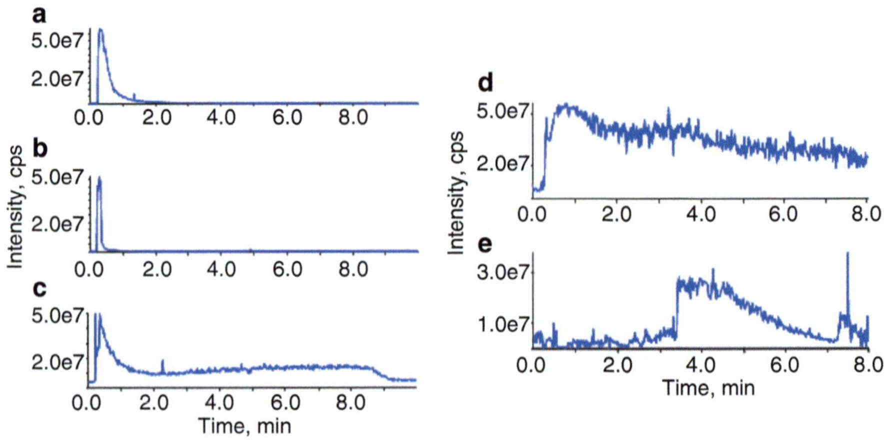

Fig. 2 (**a**) XIC of +Q1: 292.8–293.3 Da as EDTA from Indicating FTA Paper; (**b**) XIC of +Q1: 121.5–122.0 Da as Tris from Indicating FTA Paper; (**c**) XIC of +Q1: 59.8–60.3 Da as guanidine thiocyanate from FTA Elute Paper; (**d**) XIC of –Q1: 57.75–58.25 Da as guanidine thiocyanate on FTA Elute; (**e**) XIC of –Q1: 264.431–264.931 Da as SDS on Indicating FTA

spectrometry Q1 scanning in both positive and negative ionization modes. Two disks (6 mm i.d. each) free from biological matrix for FTA Elute, Indicating FTA and VWR 237 paper/cards were extracted. Two microliters of the reconstituted solution was injected into the liquid chromatograph system. More than ten chemicals were found to be present among pretreated paper/cards, but only a few typical chemicals such as *tris*-hydroxymethyl aminomethane and sodium dodecyl sulfate from Indicating FTA (DMPK-A), guanidine thiocyanate from FTA DMPK-B could be clearly identified by matching the *m/z* of their molecular ions. Figure 2a–e show the extracted ion chromatograms for the representative *m/z* of the various chemical components observed in the noted paper/cards. Figure 2a–c were from positive ionization scanning and Fig. 2d–e were from negative ionization scanning. VWR 237 paper appears to be significantly cleaner than FTA Elute and Indicating FTA paper/cards. According to the patent [38] covering the sampling cards, the coating materials on sampling cards may include impregnating agents such as polystyrene, a weak base, a chelating agent, an anionic detergent and optional uric acid/urate salt, protein denaturing agent, and a free radical trap, etc.

Most of the chemicals found on the FTA Elute and Indicating FTA paper/cards are believed not to be friendly with LC-MS because these chemicals affect the accurate quantitation of target analyte in the manners of ionization suppression, enhancement or matrix effect when they are co-eluted with the target analytes or even distortion of the chromatography by prolonged retention or accumulation. Therefore, the impacts of these water-soluble chemicals on the quantitation of the target analyte must be carefully evaluated in the development of any DBS-LC-MS method, a general approach is to minimize or eliminate the co-elution of these compounds by separating these chemicals from the target analyte using liquid chromatography. Depending on the compounds used to impregnate the cards that may be easily achievable or may become a challenge as in the case of anionic surfactants with common reversed phase columns. The anionic surfactants tend to be highly retained and difficult to clear from the chromatographic columns.

3.3 Paper/Card and Compound Dependence of Analyte Distribution in Blood Spots

In order to evaluate the impacts of different paper/cards on the analytical results, calibration standards and quality controls (QCs) were prepared in fresh human blood containing six compounds (Compound A through F) with different physical and chemical properties. Log$D$ was considered as a critical measure of the compound properties [39]. The log$D$ (at pH 7.4) is shown in Table 2. The purpose of using various compounds was to explore the possible correlations among the compounds, paper/card types, matrix effect, and spotting process related factors. Accurately controlled volumes (5 or 15 μL) of both calibration standards and QCs were spotted onto the sampling paper/cards with calibrated positive displacement pipettes and the spotted cards were air-dried at room temperature, a few sets of aliquots of the standards and QCs in blood were also refrigerated for future use. When the evaluation experiment was conducted a set of blood calibration standards and QCs (5 μL each), and/or a set of DBS standards and QCs were extracted through a salting-out assisted liquid–liquid extraction as described in Sect. 2. For the evaluation of analyte distribution in the dried blood spots, QCs were quantified against blood standards. Although concentrations of DBS QCs measured against blood standards is only semiquantitative due to the sample type difference, this approach was believed to be sufficient to provide the relative difference in concentration of an analyte across one spot. Certain volumes of blood quality controls of six compounds (A–F) as listed in Table 2 were spotted on three types of sampling paper/cards (FTA DMPK-B, FTA Elute, and VWR 237), only the results from two paper/cards (FTA Elute and VWR 237) and two compounds (Compounds A and F) are presented, to show the impact of the paper/card types on the compound quantitation.

3.3.1 Spotting Volume Impact on the Compound Quantitation on Dried Blood Spots on Different Paper/Cards

The impact of spotting volume has been evaluated. DBS cards were prepared by spotting increasing volumes of fresh blood QCs of six compounds from 5 to 50 μL and air-dried. Three millimeter (i.d.) punches were taken from the centers (visually located), extracted and quantified against calibration standards in fresh human blood. Three replicates were analyzed and the mean found concentration ± standard deviations for Compounds A and F are shown in Fig. 3a–d. The VWR 237 paper shows lower change in center spot concentration between the lowest and the highest spotting volumes than FTA Elute paper/cards. For example, the center concentration of Compound A for 50 μL spotting volume was approximately 120 % of that of the same

Table 2 Log *D* (7.4) of the compounds

Analyte	A	B	C	D	E	F
Log *D* 7.4	−0.32	−0.32	0.83	2.47	6.79	6.98

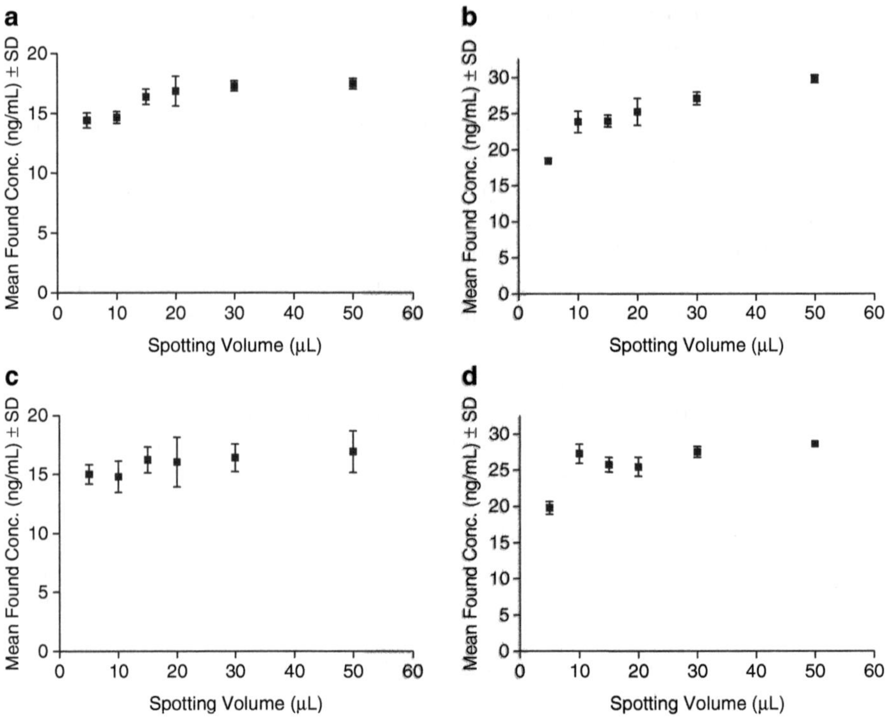

Fig. 3 Mean concentration (*center punches*) vs. spotting volume for (**a**) Compound A on VWR 237 paper; (**b**) Compound A on FTA Elute; (**c**) Compound F on VWR 237 paper; (**d**) Compound F on FTA Elute cards

compound for 5 μL spotting volume on the VWR 237 paper. On the other hand, the center concentration of Compound A for 50 μL spotting volume was approximately 170 % of that of the same compound for 5 μL spotting volume for the FTA Elute card. For Compound F with high log *D*, similar trends were observed, and VWR 237 paper show more consistent mean concentration with larger standard deviation, while the FTA Elute cards show more change in center spot concentration across different volumes with smaller standard deviation among replicates. Based on the observed trends, a conclusion can be drawn that a certain measure of control needs to be utilized for the spotted volume. Depending on the acceptance criteria of the individual assay, more or less variation of the spotted volume may be acceptable. The most accurate concentration will still be obtained when an accurate volume is spotted.

3.3.2 Paper/Card Dependent Distribution of Analytes on Dried Blood Spots

There are various factors that can influence the distribution of analytes in a dried blood spot. Water-soluble chemicals uniformly coated on DBS cards would redistribute when the blood was spotted. The redistribution of chemicals may depend on their properties, viscosity of blood, the volume spotted, and the technique used for spotting. Another factor is the viscosity of the blood. Viscosity is normally dependent on the blood composition (hematocrit, protein, lipid levels), and it can affect the physical spread of the blood spot in that the same volume of a less viscous blood will form a larger diameter spot than that of a more viscous blood sample. Viscosity, combined with the chemical redistribution on the sample cards, will increase the complexity of the analyte distribution.

In order to study the radial distribution of the six analytes on dried blood spots, smaller DBS punches (1 mm. i.d.) were extracted and quantified against blood standards. Six DBS 1 mm punches were taken from the centers of six separate spots and combined into one well of a 96-well plate. The next DBS punch was taken adjacent to the center one and each subsequent punch was taken adjacent to the previous one radially moving outward towards the edge as shown in Fig. 4. Additional punch was taken from just outside of the visual edge of the blood spot as a control sample to verify that analytes are not moving outside of the visual spot. All results from the control samples confirmed the absence of the tested compounds outside of the visual border of spots. Three types of paper/card and six compounds were used to evaluate the paper impact on the analyte distribution, but only FTA Elute and VWR 237 cards are presented here.

Figure 5 shows the radial distribution of analytes on FTA Elute cards and VWR 237 paper using two spotting volumes (15 and 50 μL). The general impression from the obtained results is that FTA Elute cards show a more significant change in concentration across the spot especially when crossing the line between the darker and lighter (halo) regions of the spot. The VWR 237 paper shows minimal change in distribution of the analytes across the spots. One observation relates to the distribution of the analytes within the dark center of the spots on the FTA Elute cards where the higher $\log D$ compound F shows a gradual change upward and then downward in concentration from center to the edge of the dark circle as shown in Fig. 5b (for both 15 and 50 μL spotting volumes), while the low $\log D$ compound A shows no obvious upward or downward change as shown in Fig. 5a. This was also seen for the rest of the compounds studied when looking at their $\log D$ values. It is possible that this phenomenon may be due to the redistribution of the water-soluble chemical components of the paper, as previously described, as well as to the specific properties of the analytes.

3.3.3 Paper/Card Dependent Matrix Effect

Accurate quantitation of analytes requires consistent measurements independent from matrix lot variation. Matrix lot to lot variation becomes critical for successfully implementing DBS. In addition to the analyte distribution across dried blood spots, matrix effect was also evaluated. In addition to the blood pool used for standards and

Fig. 4 High resolution punches with 1 mm (i.d.) punches

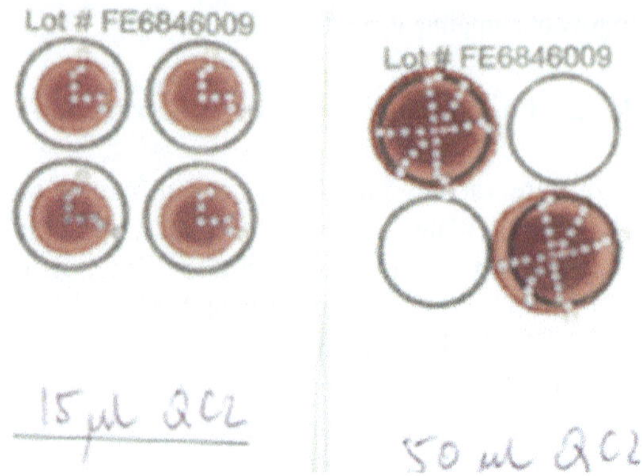

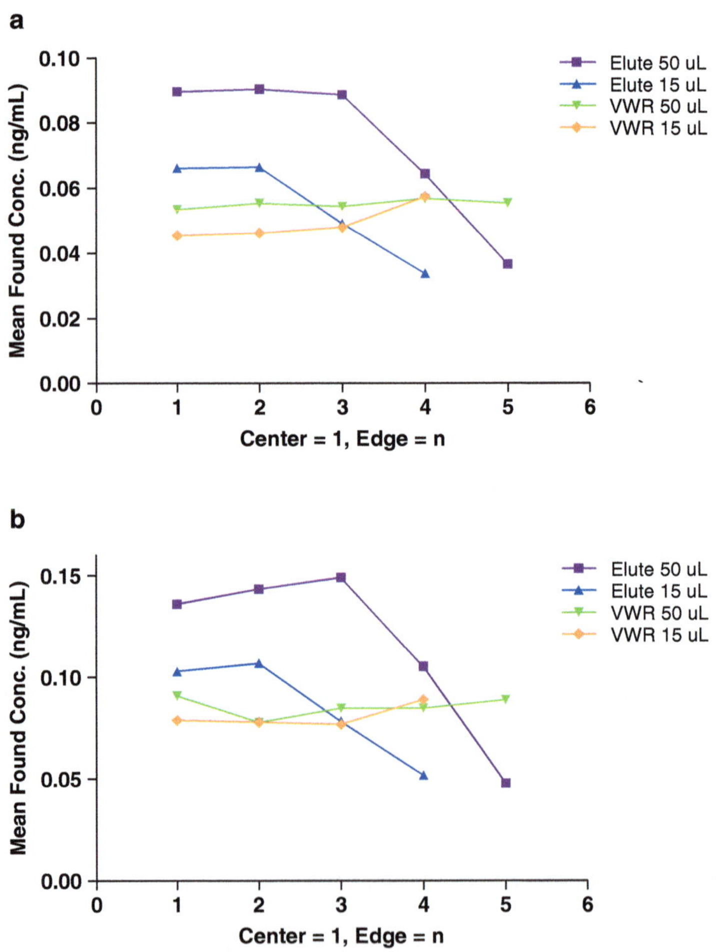

Fig. 5 Mean concentration vs. high resolution punch distance from the spot center for (**a**) Compound A on FTA Elute Cards and VWR 237 paper; (**b**) Compound F on FTA Elute Cards and VWR 237 paper

QCs preparation, two additional lots of blood were used to prepare two matrix effect (ME) samples at low and high concentrations. All ME samples, QCs and calibration standards in blood carrying six compounds were spotted on three types of paper/cards at the same time. The spotting volume was either 15 or 5 μL. The spot centers (3 mm i.d.) were punched for the samples prepared with 15 μL of blood but full spots (6 mm i.d.) were punched for the samples prepared with 5 μL of blood. The punches were extracted and analyzed with the procedure described earlier. QCs and ME samples were quantified against calibration standards that were prepared, spotted, and analyzed in the same fashion. Mean bias (%) was calculated for all evaluation sample lots (QCs or ME samples) and all concentration levels of each of the five compounds. Standard deviation error bars are used to reflect the variation of the individual bias among all concentration levels. Compounds A, B, C, E, and F on FTA Elute Cards and VWR 237 paper are presented in Fig. 6a–d showing the paper/card impact on the matrix effect. Compound D is not presented here because the analyte was eluted at the zone when severe compound-specific matrix ionization suppression was observed (chromatogram not shown). The acceptance criteria used was ±15 % for bias in matrix effect evaluation. Figure 6a, b indicate that FTA Elute cards have significant matrix effect compared to VWR 237 paper spot center punches for all analytes tested. However, when using accurate low volume spotting and a full punch, the overall matrix effects generated on FTA Elute cards and VWR 237 paper are similar. In other words, plain paper with either partial or full punch of the spot generates less matrix effect than treated paper/cards.

4 Conclusion

The paper/cards used for dried blood spot sampling can affect the accurate quantitative determination of analytes for pharmacokinetic evaluations. Various factors were evaluated using six analytes over a wide log *D* range. These were related to impregnated chemicals on the cards, variable volume, viscosity, and compound dependent distribution of the analytes. The widely used treated paper/cards generally showed a more complex and significant impact than the plain filter paper/cards. Chemicals on the treated paper/cards in conjunction with the specific properties of the analytes may cause variations in the analyte distribution that could provide a challenge, so evaluations of the sampling paper/card should be carefully conducted before a DBS method can be applied for accurate measurement of analytes in DBS-LC-MS or if pharmacokinetics is the end point of a study. The level of acceptable variation needs to be evaluated and decided before implementing such a method for quantitative work. Spotting on uniform plain filter paper was shown to reduce the variability with regard to matrix effects, thus resulting in a more accurate analyte determination. However, the most accurate matrix and paper/card independent values were derived from accurate volume spotting with full cut of the spots. In addition to the factors presented, hematocrit effect or any other components that lead to varying blood viscosity, the paper/card type, media evenness, and paper lot-to-lot variation should be considered as critical impact factors.

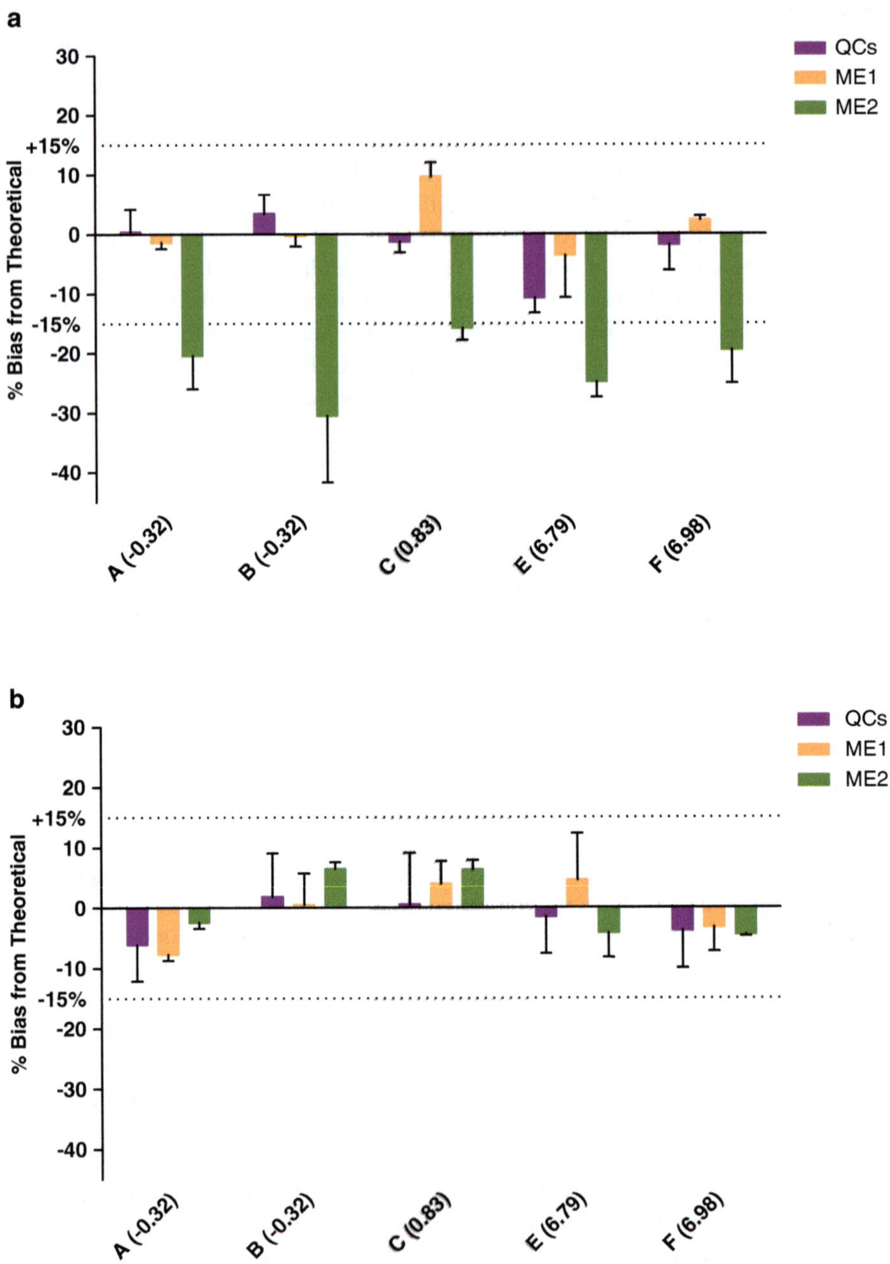

Fig. 6 (**a**) Mean bias (%) of ME samples from 15 μL spotting on FTA Elute cards and partial punches. (**b**) Mean bias (%) of ME samples from 15 μL spotting on VWR 237 paper and partial punches. (**c**) Mean bias (%) of ME samples from 15 μL spotting on FTA Elute cards and full punches. (**d**) Mean bias (%) of ME samples from 15 μL spotting on VWR 237 paper and full punches

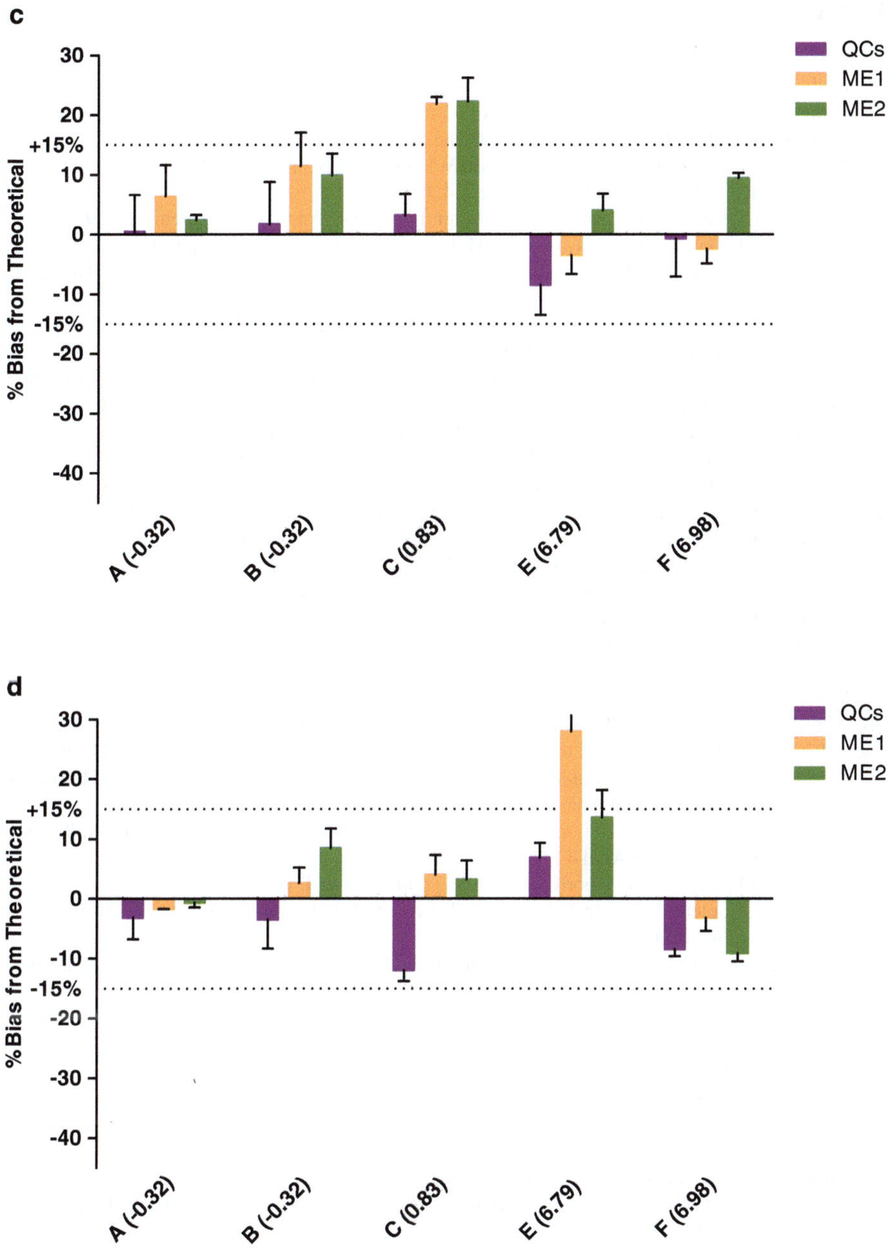

Fig. 6 (continued)

References

1. Guthrie R, Suzi A (1963) Pediatrics 32:338
2. Parker SP, Cubit WD (1999) J Clin Pathol 52:633
3. Gamlen TR, James HC, Batstone GF (1982) Scand J Clin Lab Invest 42:643
4. Mei JV, Alexander JR, Adam BW, Hannon WH (2001) J Nutr 131:1631S
5. Li W, Tse LF (2010) Biomed Chromatogr 24:49
6. Coombes EJ, Gamlen TR, Batstone GF, Leigh PN (1984) Ann Clin Biochem 21:519
7. Coombes EJ, Gamlen TR, Batstone GF, Holgate ST (1984) Clin Chim Acta 136:187
8. Lindström B, Ericsson Ő, Alva´n G, Rombo L, Ekman L, Rais M, Sjöqvist F (1985) Ther Drug Monit 7:207
9. Hibberd SG, Alveyn C, Coombes EJ, Holgate ST (1986) Br J Clin Pharmacol 22:337
10. Fujimoto T, Tawa R, Hirose S (1988) Chem Pharm Bull 36:1571
11. Bergqvist Y, Al Kabbani J, Krysén B, Berggren Palme I, Rombo L (1993) J Chromatogr 615:297
12. Ericsson Ő, Fridén M, Hellgren U, Gustafsson LL (1993) Ther Drug Monit 15:334
13. Bergqvist Y, Doverskog M, Al Kabbani J (1994) J Chromatogr B 652:73
14. Rønn AM, Lemnge MM, Angelo HR, Bygbjerg IC (1995) Ther Drug Monit 17:79
15. Bergqvist Y, Funding L, Kaneko A, Krysén B, Leek T (1998) J Chromatogr B 719:141
16. Craft NE, Haitema T, Brindle LK, Yamini S, Humphrey JH, West KP Jr (2000) J Nutr 130:882
17. Lindegårdh N, Funding L, Bergqvist Y (2001) J Chromatogr B 758:137
18. Green MD, Mount DL, Nettey H (2002) J Chromatogr B 767:159
19. Oliveira EJ, Watson DG, Morton NS (2002) J Pharm Biomed Anal 29:803
20. Malm M, Lindegårdh N, Bergqvist Y (2004) J Chromatogr B 809:43
21. Minzi OMS, Massele AY, Gustafsson LL, Ericsson Đ (2005) J Chromatogr B 814:179
22. Koal T, Burhenne H, Romline R, Svoboda M, Resch K, Kaever V (2005) Rapid Commun Mass Spectrom 19:2995
23. AbuRuz S, Millership J, McElnay J (2006) J Chromatogr B 832:202
24. Allanson AL, Cotton MM, Tettey JNA, Boyter AC (2007) J Pharm Biomed Anal 44:963
25. Blessborn D, Romsing S, Annerberg A, Sundquist D, Bjorkman A, Lindegardh N, Bergqvist Y (2007) J Pharm Biomed Anal 45:282
26. Hoogtanders K, van der Heijden J, Christiaans M, Edelbroek P, van Hooff JP, Stolk LML (2007) J Pharm Biomed Anal 44:658
27. Kolawole JA, Taylor RB, Moody RR (1995) J Chromatogr B 674:149
28. Lejeune D, Souletie I, Houze S, Le bricon T, Le bras T, Gourmet B, Houze P (2007) J Pharm Biomed Anal 43:1106
29. Ntale M, Mahindi M, Ogwal-Okeng JW, Gustafsson LL, Beck O (2007) J Chromatogr B 859:137
30. Barfield M, Spooner N, Lad R, Parry S, Fowles S (2008) J Chromatogr B 870:32
31. Sosnoff CS, Bernert JT (2008) Clin Chim Acta 388:228
32. ter Heine R, Rosing H, van Gorp ECM, Mulder JW, van der Steeg WA, Beijnen JH, Huitema ADR (2008) J Chromatogr B 867:205
33. Liang X, Li Y, Barfied M, Ji QC (2009) J Chromatogr B 877:799
34. Higashia T, Nishioa T, Uchida S, Shimada K, Fukushib M, Maedac M (2008) J Pharm Biomed Anal 4(8):177
35. Spooner N, Lad R, Barfield M (2009) Anal Chem 81:1557
36. van der Heijden J, de Beer Y, Hoogtanders K, Christiaans M, de Jong GJ, Neef C, Stolk L (2009) J Pharm Biomed Anal 50:664
37. Denniff P, Spooner N (2010) Bioanalysis 2:1385
38. Burgoyne LA. US Patent 5,496,562 (1996); 5,756,126 (1998) and 5,807,127 (1998).
39. Gao L, Cheng X, Zhang J, Burns DJ (2007) Rapid Commun Mass Spectrom 21:3497

Highly Sensitive Pharmaceutical and Clinical Analysis Using Selective Solid-Phase Extraction Coupled to Microflow Liquid Chromatography and Isotope-Dilution Mass Spectrometry

Haoying Yu, Xiaotao Duan, and Jun Qu

Abstract The capacity of ultrasensitive quantification is highly desirable in many clinical and pharmaceutical applications, such as the PK study of low-dose administration and quantification of low-abundance markers in circulating systems. Nonetheless, this task is highly challenging even for a LC-MS/MS approach, due to the extremely low levels of analytes and the highly complex matrices. Here we describe a robust method for the ultrasensitive quantification of therapeutic agents or clinical markers in highly complex biological systems. This is achieved by the combination of a selective solid-phase extraction (SPE) with a highly sensitive capillary LC (μLC)-MS/MS analysis. Comparing to a conventional LC-MS/MS, a μLC-MS/MS provides much higher sensitivity due to the lower peak dilution. SPE washing and elution conditions were optimized so that target drugs are selectively extracted from the biological matrix. By eliminating most undesirable matrix components, this selective SPE procedure enabled a high sample loading volume on the μLC column without compromising chromatographic performance and operational robustness, and helped to achieve ultralow detection limits. Sufficient μLC separation was employed in order to further improve analytical sensitivity and to decrease matrix effects. To show the application of this strategy, three paradigms, respectively, the quantifications of (1) corticosteroids after an intravitreous injection, (2) an anticancer agent after a low-dose treatment, and (3) several vitamin D (VitD) metabolites in plasma, are introduced in this chapter. Typically, low pg/mL LOQ were achieved, and the linearity, accuracy, and precision of these developed methods were excellent.

Collectively, this strategy enables the ultrasensitive, accurate, and robust LC-MS-based quantification of extremely low levels of drug/markers in biological samples, which were previously approachable only by ELISA or RIA methods.

H. Yu • X. Duan • J. Qu, Ph.D. (✉)
Department of Pharmaceutical Sciences, University at Buffalo, State University of New York, Cooke Hall 544, Amherst, NY 14260, USA
e-mail: junqu@buffalo.edu

Q. Alan Xu and Timothy L. Madden (eds.), *LC-MS in Drug Bioanalysis*,
DOI 10.1007/978-1-4614-3828-1_4,

1 Introduction

In many occasions of clinical, pharmaceutical, and biological analysis, a high analytical sensitivity is desirable. For example, certain low-dose regimens, and inhalation or intraocular injection (e.g., corticosteroids) may result in ultralow drug concentration in body fluids (e.g., low-pg/mL range in plasma). Furthermore, some highly active endogenous markers may present at very low levels in biological matrices. For instance, the circulating levels of an important vitamin D (VitD) metabolite, 1R, 25-dihydroxy-VitD$_3$, are extremely low, which constitute a daunting challenge for quantification. Given these importance of analysis of such compounds in clinical and pharmaceutical settings, a highly sensitive analytical method is necessary. LC-MS/MS is considered one of the most powerful methods for the clinical and pharmaceutical analysis. However, ultrasensitive quantification of drug/markers presenting at low-pg/mL levels in plasma remains challenging even for LC-MS-based methods, which usually carry typical sensitivities at high-pg/mL to low-ng/mL levels.

As electro-spray ionization (ESI) MS is a concentration-dependent detector, it seems to be plausible to employ low-flow-LC-MS for highly sensitive analysis. The following rationale suggests that the high sensitivity of analysis can be achieved by applying low-flow-LC-MS. During chromatographic separation, the dilution (D) of an injected sample ($D = C_{end}/C_{inj}$, where C_{end} is the concentration after chromatography and C_{inj} is the concentration injected) is given by:

$$D = \frac{\in \pi r^2 (1+k)(2\pi LH)^{1/2}}{V_{inj}}$$

where $\in$ is the column porosity, r is the column radius, k is the retention factor, L is the column length, H is the plate height, and V_{inj} is the injection volume [1]. If conditions are otherwise equal, D is in direct proportion to the square of column radius. Thus, compared to conventional HPLC, micro-LC increases the signal-to-noise ratio (S/N) drastically when ESI-MS/MS is employed as the detector, because micro-LC results in a much smaller dilution of peak concentration, and ESI-MS/MS is a concentration-sensitive detector [2, 3]. Nevertheless, the micro-LC-MS falls short in that the column loading capacity is proportionally small, which tends to counteract the gain in sensitivity achieved through using low flow rate. When analyzing highly complex biological samples, as in clinical and pharmaceutical analysis, micro-LC-MS analysis may encounter two primary problems when a relatively large amount of samples are loaded onto the column: (1) mass overloading, where chromatographic separation is severely compromised because a large amount of compounds are loaded onto a small-diameter column and result in nonlinear adsorption, and (2) volume overloading, where the volume of the sample injected is so large that the eluted peaks are markedly broadened. In order to circumvent these problems, users of micro-LC-MS often employ very small injection volumes (V_{inj}). As a result, while micro-LC-MS/MS greatly enhances the *absolute* sensitivity (i.e., sensitivity expressed in terms of the amount of the target on column) compared to conventional LC-MS/MS, the *relative* sensitivity (i.e., sensitivity expressed in terms of the concentration of the target in sample), which is more important for pharmaceutical/biological analysis, typically does not improve drastically unless a high V_{inj} is feasible. In order to achieve high concentration sensitivity for pharmaceutical and biomedical analysis, our lab has developed a suite of approaches using a high V_{inj} while overcoming the problems of mass/volume overloading [4–6].

Selective sample preparation strategies, such as a selective SPE procedure, was applied to simplify the complex biological matrices and to alleviate the mass overloading problem. SPE washing and elution conditions were optimized to reduce the amount of matrix components, as well as to decrease matrix effects and endogenous interferences from biological samples. Using such a SPE strategy, a high sample loading volume on the μLC column is feasible without causing mass overloading [4–6]. In one preliminary experiment, we observed when plasma samples were extracted using generic SPE methods, a relatively high V_{inj} of the extract easily exceeded micro-LC column capacity and deteriorated chromatography performance. By comparison, after a selective extraction, it was feasible to achieve high concentration sensitivity by employing a high V_{inj}, without mass overloading [4].

An on-column focusing strategy was employed to overcome volume overloading [4–6]. By using low-organic mobile phase for sample loading, the target compounds are focused on the front end of column prior to separation. For a typical μLC system that is running at low flow rates, a large V_{inj} leads to a long sample loading time and may result in undesirable consequences such as severe band-broadening and lag of gradient changes. Those problems can be avoided by applying on-column focusing strategy. In the above works, the combination of selective SPE and on-column focusing strategy enabled ultrasensitive quantification of clinically or pharmaceutically important compounds such as corticosteroids, VitD metabolites, and anticancer agents.

The detailed procedure of this method is described in the following three paradigms.

2 Paradigm 1: Ultrasensitive Quantification of Corticosteroids in Plasma After Intravitreous Injection

Low-dose regimens frequently result in very low corticosteroid concentrations in body fluids (e.g., low pg/mL range in plasma). PK analysis that is often necessary to manage these therapies is hindered by the inadequate sensitivity by most established analytical methods. In addition, certain drug delivery strategies, such as inhalation or intraocular injection, result in systemic levels too low to be detected by current methods. Furthermore, administration of some corticosteroid prodrugs, such as these in the forms of acetates/propionate, often results in sustained, low concentrations in plasma. Therefore, a highly sensitive and selective analytical approach is necessary for cases in which sustained low concentration of corticosteroids may be present systemically or in tissues.

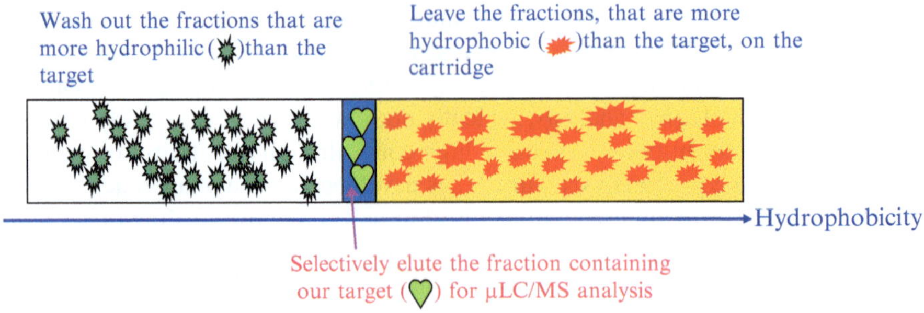

Fig. 1 Selective SPE rationale. The fraction that contains the target compound can be selectively collected, while simultaneously reducing the quantity of matrix components from complex sample, using optimized wash/elution strategy

Traditional analytical approaches such as GC-MS, HPLC fluorescence, and conventional LC-MS, are not sufficiently sensitive to quantify the low-pg/mL levels of corticosteroids in bodily fluids [4]. The purpose of this work was to develop an ultrasensitive and selective approach to quantify corticosteroids present at extremely low, but biologically meaningful concentration in body fluids. A selective solid-phase extraction (SPE) and reversed-phase capillary LC coupled to tandem mass spectrometry (μLC-MS/MS) strategy was employed to achieve a high concentration sensitivity.

2.1 *Experimental*

To concentrate target corticosteroids and to reduce or eliminate matrix components and interfering substances in plasma, a selective SPE procedure was developed. Figure 1 shows the rationale for selective SPE strategy. Oasis HLB cartridge was selected. Cartridges were conditioned with 1 mL of methanol followed by 1 mL of water. One-mL portions of pooled porcine plasma containing 200 pg/mL each of DEX, TACA, BUD, and DEX-AC were mixed 1:1 with 4 % phosphoric acid to displace the corticosteroids from plasma proteins and then respectively loaded onto 18 cartridges at a rate of approximately 1 mL/min, followed by washing with 1 mL 2 % phosphoric acid. The cartridges were divided into two groups: for group 1, individual cartridges were washed with 1 mL of 0.1 % formic acid in one of following concentrations of methanol: 20, 30, 35, 40, 45, 50, 55, 60, or 65 %. The absorbed corticosteroids were eluted from all nine cartridges with 1 mL methanol containing 0.1 % formic acid. For group 2, all nine cartridges were washed with 1 mL of methanol with 0.1 % formic acid, and the absorbed analytes were then elute with 1 mL of 0.1 % formic acid in one of the following concentrations of methanol: 60, 65, 70, 75, 80, 85, 90, 95, or 100 %. As an internal standard (I.S.), d7-TACA was then added to each eluate at a concentration of 100 pg/mL. The eluates were evaporated to dryness under a gentle nitrogen stream at 45 °C and reconstituted in 50 μL of 30 % acetonitrile. This optimization procedure was performed in duplicate, and the recoveries of the corticosteroids were determined by μLC-MS/MS using a V_{inj} of 0.4 μL.

For sample preparation, the I.S. (d7-TACA) was spiked into each 1 mL portion of porcine plasma sample to a concentration of 100 pg/mL; then mixed 1:1 (v:v) with 4 % phosphoric acid to release protein-bound drugs. SPE cartridges were preconditioned as described above, after sample loading, optimized SPE wash and elution methods were applied for sample extraction.

Sample analysis was performed by using an Applied Biosystems (Foster City, CA) API 3000 triple quadrupole mass spectrometer equipped with a TurboIonSpray source and an Agilent 1100 capillary HPLC system (Palo Alto, CA). The capillary HPLC system included a binary capillary pump with an active micro flow rate control system, an online degasser, and a microplate autosampler. The analytical column was a 300 μm I.D. × 150 mm Zorbax C18 Stablebond capillary column (pore size 100 Å and particle size 3.5 μm). The injection volume was 5 μL, and a needle ejection rate of 40 μL/min was used. The μLC flow rate was 6 μL/min. In order to minimize dead volume before the column, the autosampler was programmed to bypass the 8 μL sample loop 1.5 min after injection. The mobile phase consisted of (A) 2 mM ammonium acetate (adjusted to pH 3.2 with formic acid) in 10:90 acetonitrile–water, and (B) 2 mM ammonium acetate in 90:10 acetonitrile–water. The percentage of mobile phase B was held at 32 % for the first minute, increased to 80 % over 8 min, and then increased to100% over the following 1 min.

MRM conditions for the four corticosteroids and the I.S., including *m/z* of MRM pairs, collision energy, and orifice potential, were optimized by direct infusion of the compounds into the MS/MS at a concentration of 5 μg/mL. For μLC-MS/MS, the dwell time of each MRM transition was 200 ms, and the pause time for scan parameter changes was 8 ms. The flow rates of nebulizer gas (air) and curtain gas were 1 L/min and 0.6 L/min, respectively. The drying gas was turned off because of the low chromatographic flow rate. The pressure of target gas (N_2) for collisionally activated dissociation (CAD) was 4.8 mTorr. The ion spray voltage, orifice potential, and ring focus voltage were set at 5,000, 60, and 280 V, respectively.

2.2 *Results and Discussion*

Although μLC-MS/MS provides much lower peak dilution than conventional LC-MS/MS, the loading capacity of μLC columns is also correspondingly lower [1]. To some extent, this abrogates the advantage of higher sensitivity for μLC-MS/MS compared to conventional LC-MS/MS. In addition, a large number of matrix compounds may be retrieved if a generic SPE approach is used for highly complex samples such as those derived from plasma; this increases the likelihood that the sample may exceed the μLC column capacity unless a very low injection volume is used. In preliminary experiments, a generic SPE strategy was applied to analyze 1 mL of corticosteroid-spiked plasma. The chromatographic performance on μLC became unacceptable when more than 0.5 μL of an extracted plasma sample was injected on the 300 μm I.D. column. This degradation of μLC performance probably resulted from saturation of the capacity of the stationary phase of the column by the extracted matrix materials, thus modifying the chemical characteristics of the stationary phase and interfering with chromatographic separation.

To address this problem, it was necessary to develop a selective SPE procedure to extract the target corticosteroids more selectively from plasma and simultaneously remove unwanted plasma components to the extent possible. Reversed-phase SPE cartridges from several manufacturers were evaluated. The Oasis HLB cartridge appeared to exhibit the best performance for the extraction of the four corticosteroids. The plasma sample was mixed with 4 % phosphoric acid at a ratio of 1:1 to disrupt and/or inhibit drug–protein binding. Two wash steps were employed before elution: Wash-1 used 2 % phosphoric acid to remove residual plasma proteins, and Wash-2 used 0.1 % formic acid in a methanol–water mixture for the removal of compounds that were more polar than the target compounds. Elution employed a second, optimized methanol–water mixture (containing 0.1 % formic acid) in order to elute the target compounds efficiently and leave compounds less polar than the targets on the cartridges. Absolute analyte recoveries for the SPE extraction step were investigated using Wash-2 steps of 20–65 % methanol and dilution steps of 60–100 % methanol, and results are shown in Fig. 2 [4]. The final solvent system selected for the wash cycle was 0.1 % formic acid in 50 % methanol, and the elution solvent composition was 0.1 % formic acid in 85 % methanol in order to complement with multiple corticosteroids.

Compared to conventional plasma sample preparation methods such as protein precipitation and generic SPE, the selective SPE strategy described here has several advantages. Protein precipitation using organic solvents dilutes the sample, and the final solution contains a high percentage of organic solvent, which may cause peak broadening problems during chromatography. Generic SPE, which is not optimized for target analytes, nonspecifically extracts components from complex matrices and results in a highly complex extract that may easily exceed μLC column capacity if a relatively large injection volume was used. In contrast, the SPE strategy developed here selectively concentrated target compound while simultaneously simplifying the sample matrix greatly. As a result, this selective extraction approach enabled a relatively high injection volume of plasma samples, yet avoided over capacity and fouling of the μLC column. Representative chromatograms of corticosteroid extracted from plasma using generic SPE vs. the optimized selective SPE conditions are shown in Fig. 3 [4]. It was also observed that this approach improved the S/N of the target compounds in plasma samples, probably by both lowering chemical noise and decreasing the ion suppression effect.

The MS and MS/MS behaviors of the four corticosteroids under positive ion ESI was investigated using direct infusion of standard solutions into the mass spectrometer. In preliminary experiments, it was observed that all four corticosteroids formed strong $[M+Na]^{+}$and $[M+K]^{+}$ ions along with relatively weaker $[M+H]^{+}$ ions in methanol/acetonitrile/water. However, an acidic 2 mM ammonium acetate buffer

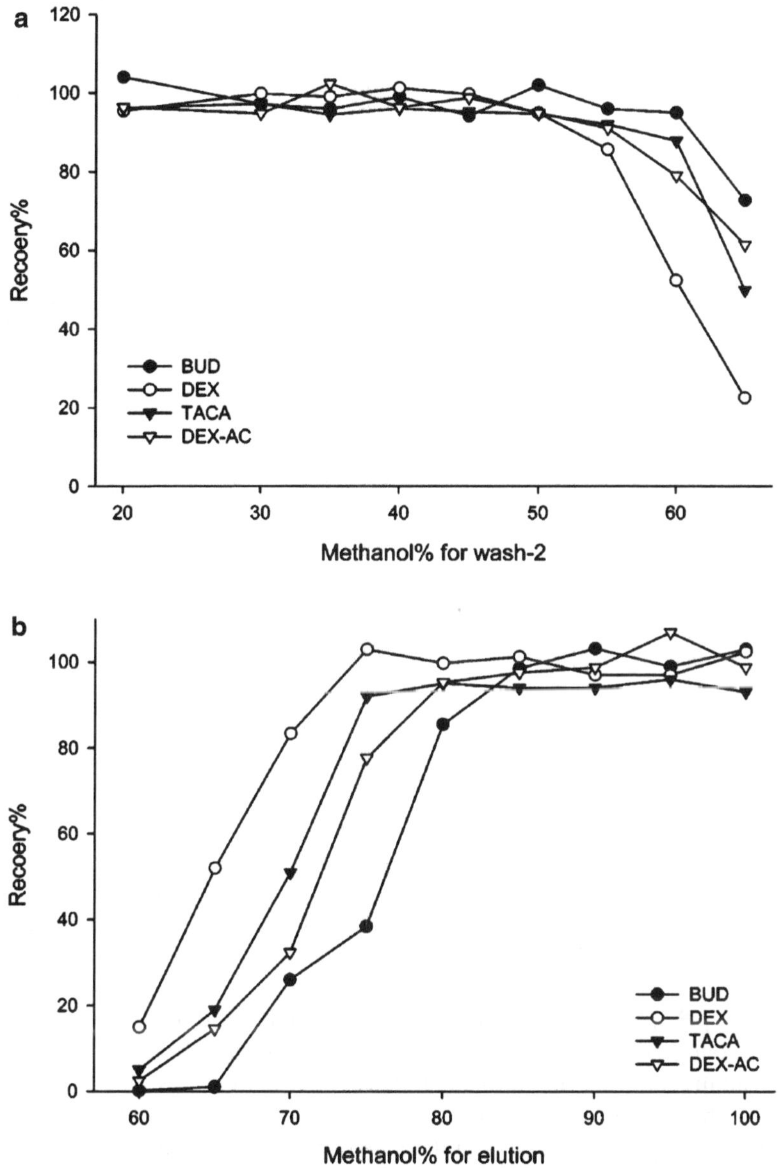

Fig. 2 Corticosteroid recovery by SPE extraction as a function of solvent composition. Solvents used in the Wash-2 step were composed of 0.1 % formic acid and the indicated solvent composition. Corticosteroid recovery vs. solvent composition is shown for the following conditions: (**a**) SPE cartridges were washed respectively with a range of 20–65 % methanol and eluted with 100 % methanol; (**b**) SPE cartridges were washed with 5 % methanol and eluted respectively with 60–100 % methanol (Reproduced with permission from American Chemical Society)

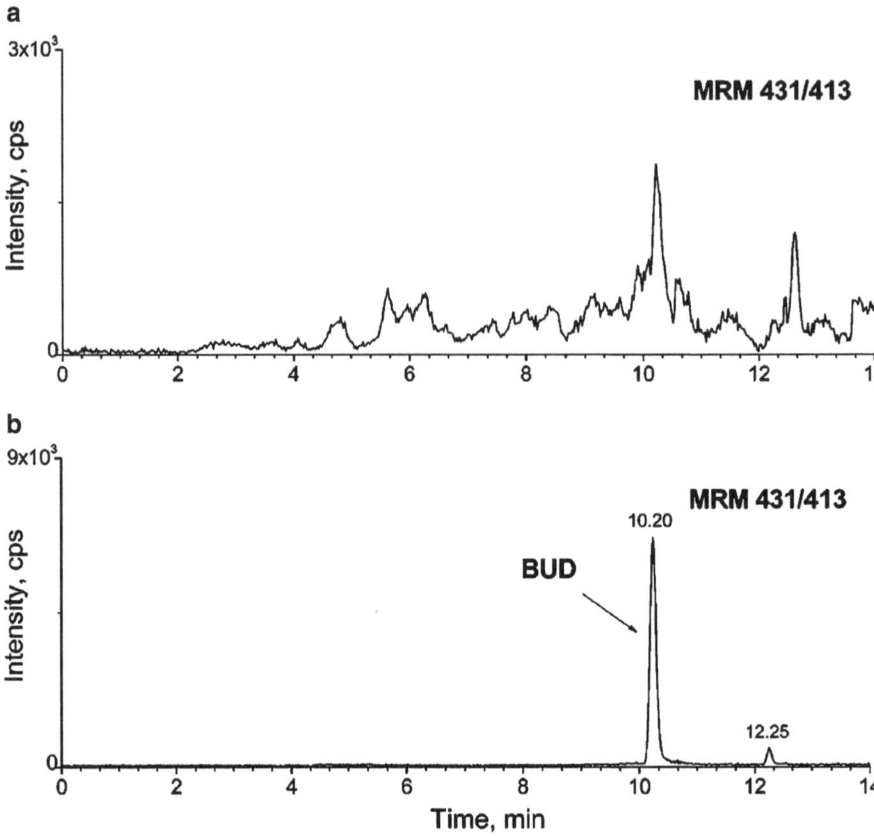

Fig. 3 Comparison of μLC/MS/MS chromatograms of budesonide (BUD) in plasma after (**a**) a generic SPE extraction and (**b**) a selective SPE extraction. The concentration of BUD was 20 pg/mL and the injection volume was 5 μL (Reproduced with permission from American Chemical Society)

adjusted to pH 3.2 using formic acid, noticeably suppressed both the $[M+Na]^+$ and $[M+K]^+$ ions, and the$[M+H]^+$ ions became predominant. Therefore, this acidified 2 mM ammonium acetate–formate mixture was selected as the mobile phase modifier. The collisionally activated dissociation (CAD) spectra of the four analytes were individually investigated, and the major fragments are shown in Table 1 [4]. The results are in agreement with previous publications, and the most abundant product ions were chosen for the MRM of each corticosteroid. The MS/MS conditions for each transition were optimized in order to achieve the maximum S/N, and the position of the spray needle was adjusted to achieve the optimal S/N for all compounds. The optimized collision energies for the MRM transitions are shown in Table 1 for each corticosteroid.

It was demonstrated previously that smaller inner-diameter (I.D.) columns with lower flow rates dramatically improve the sensitivity of LC/ESI-MS/MS. However, smaller I.D. columns also have proportionally lower loading capacities and may

Table 1 Major fragments, MRM transitions, and optimal collision energies of the four corticosteroids

Compound	(*m/z*)	(Precursor/product)	Energy (eV)
BUD	413, 147, 173, 225, 323	431/413	17
DEX	373, 355, 147, 237, 171, 337	393/373	16
TACA	415, 339, 213, 397, 147, 357	435/415	15
DEX-AC	415, 397, 337, 147, 319	435/415	15

The fragments are listed in order of relative abundance, from high to low (Reproduced with permission from American Chemical Society)

result in compromised ruggedness of analytical methods compared with larger I.D. columns. Sufficient μLC resolution was desirable for ultrasensitive quantification for the following reasons: first, plasma samples are highly complex and the concentration of target compounds is extremely low; interferences from the sample matrix may present a significant problem. Therefore, chromatographic separation of those interfering components from the target corticosteroids was necessary to ensure quantitative accuracy at low target concentrations. Second, the formation of both $[M+Na]^+$ and $[M+K]^+$ions was relatively high when minimal μLC separation was used, even when selective SPE cleanup was employed. However, with sufficient chromatographic separation, undesirable adduct ions were much reduced in intensity relative to $[M+H]^+$ in this study. This observation probably results from the fact that K^+ and Na^+ salts do not co-elute with the target compounds when sufficient chromatographic separation is obtained. Third, when using the minimal μLC separation, a severe ion suppression effect arising from the complex sample matrix was observed. This effect, which depresses the ionization of all target compounds, was demonstrated using an evaluation approach we established previously [7]. By achieving sufficient μLC separation, it was possible to minimize this effect and thus improve sensitivity.

In order to maximize the acceptable V_{inj} without causing peak broadening, a two-segment gradient was designed in this study: the first gradient step is the "trapping segment," and it employs a high percent of the aqueous mobile phase A. As a result, relatively hydrophobic compounds such as the corticosteroids are focused on the front end of the column during sample loading. The subsequent gradient steps constitute the "separation segment," in which a gradually increasing percentage of mobile phases B separate the compounds that were focused on the column during the trapping segment of the gradient. To determine the optimal gradient conditions in the separation segment, we investigated the chromatographic retention vs. S/N. Different partition ratios (k) of corticosteroids were achieved with five different gradient elution conditions, using an injection volume of 5 μL; Fig. 4 [4] shows the S/N as a function of μLC partition ratio (k). From this data, it is clear that the S/N of the target analytes improved considerably when their chromatographic retention increased, and for that beyond a certain extent of retention, the S/N approaches a maximum. Column-overloading was overcome to a great extent by applying this on-column focusing strategy.

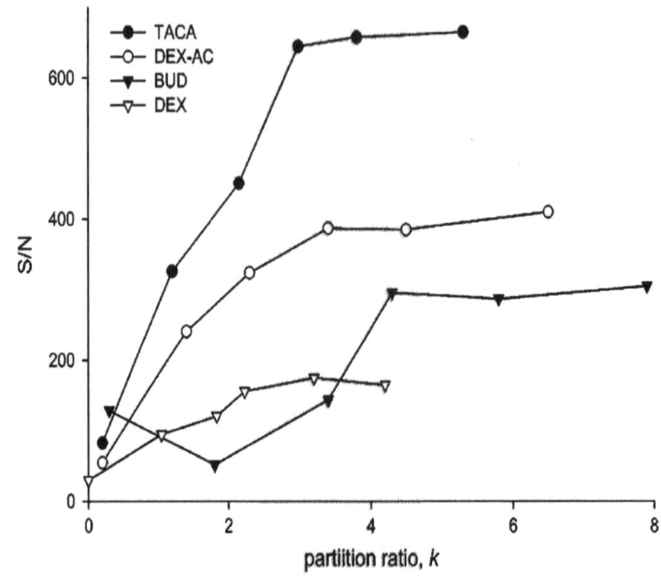

Fig. 4 Signal–noise ratio for various corticosteroids in plasma samples under conditions providing a range of chromatographic retentions (Reproduced with permission from American Chemical Society)

As a result of these studies, the optimal μLC separation conditions were selected based on three considerations: first, the peaks of target compounds must be resolved from endogenous interfering peaks; second, the S/N for all target compounds be maximized, the interfering peaks are separated from those of target compounds and do not contribute to the signals of targets, the run time must be as short as possible to ensure a reasonable assay throughput and prevent serious peak broadening. The final conditions are summarized in the experimental section.

Because of the smaller column volumes for capillary columns, the V_{inj} is held to a much smaller volume than for conventional columns to avoid column overcapacity and V_{inj} for the 0.3 mm I.D. column used here was usually 0.1–0.2 μL. Given that it would be difficult to concentrate the SPE eluate further without compromising quantitative performance, the ability to inject higher sample volumes would be advantageous. The hypothesis was that the complexity of plasma samples would be reduced considerably by the selective SPE approach employed, and therefore a considerably higher V_{inj} could be tolerated without compromising chromatographic separation if the proposed two-segment gradient was used.

To test this hypothesis, the μLC performance was evaluated as a function of V_{inj} for analyte-spiked plasma samples. Typical chromatograms are shown in Fig. 5 [4]. The retention times of target compounds increased only slightly as V_{inj} was increased,

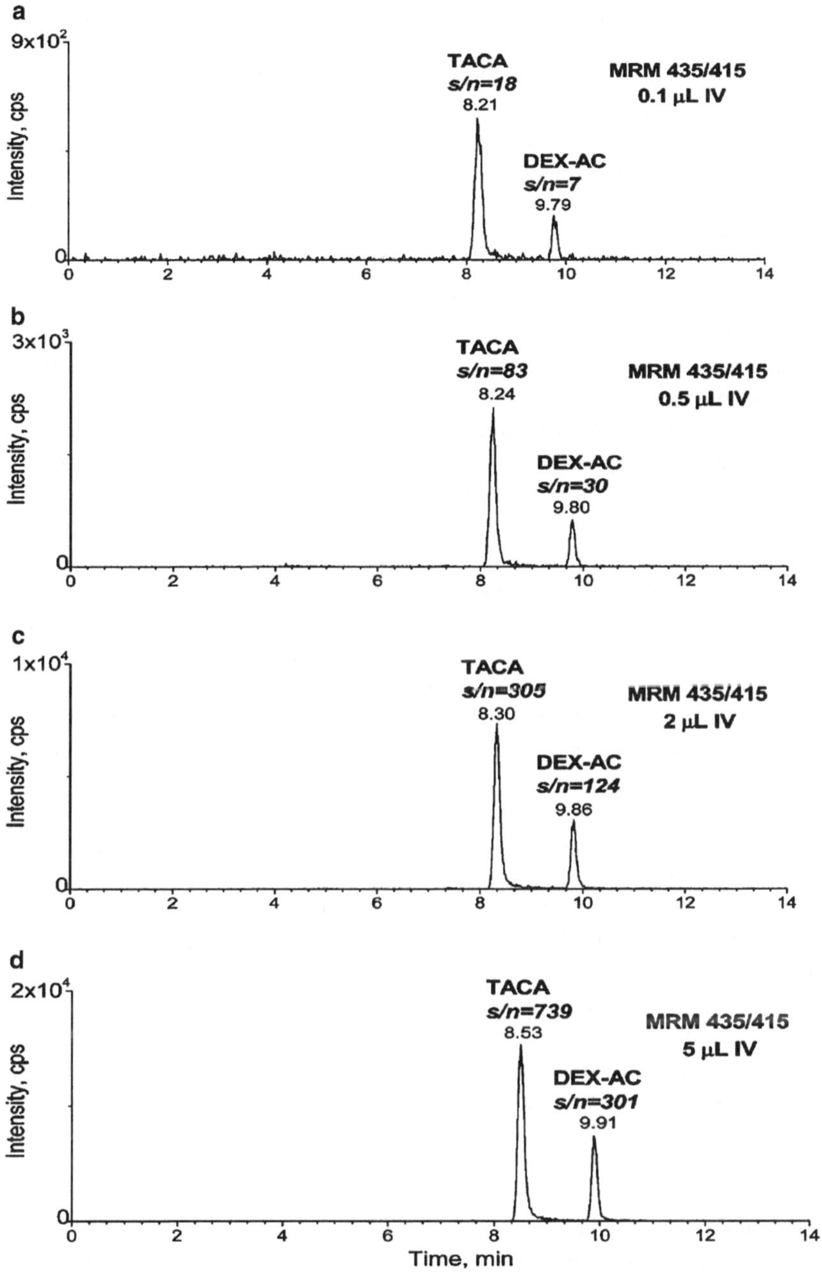

Fig. 5 Typical chromatograms showing injection volume (V_{inj}) optimization for plasma samples spiked with 50 pg/mL of TACA and DEX-AC and extracted by selective SPE. The V_{inj} was (**a**) 0.1 μL; (**b**) 0.5 μL; (**c**) 2 μL; (**d**) 5 μL. A 15 cm by 300 μm I.D. capillary column was used for separation; the manufacturer-recommended V_{inj} was 0.1–0.2 μL. Selective SPE and two-segment gradient separation enabled a large V_{inj} on a 300 μm I.D. capillary column without deterioration of chromatographic performance and with greatly increased sensitivity (Reproduced with permission from American Chemical Society)

but peak broadening or peak shape deterioration was not observed over the range of V_{inj} from 0.1 to 5.0 μL. Furthermore, the S/N for all targets increased in proportion with V_{inj} which indicates that the selective SPE procedure reduced the complexity of the plasma samples significantly, to a degree that would permit a high V_{inj} on the μLC column without causing column overcapacity. Furthermore, the utilization of separate "trapping" and "separation" gradient segments (described above) counteracted the tendency of increased V_{inj} to increase peak widths and thus rendered peak widths for the corticosteroids independent of loading times.

Figure 6 [4] shows typical chromatograms for 5 pg/mL of the corticosteroids in plasma under optimized μLC-MS/MS conditions. After approximately 200 injections of 5 μL (V_{inj}) plasma samples on a μLC column, the separation of the four corticosteroids remained highly reproducible, with neither loss of resolution nor increase in backpressure observed. This indicates that the injection of a relatively large volume of plasma-derived samples did not foul or damage the column, provided they were subjected to the selective SPE extraction procedure before injection.

Using the combination of selective SPE and μLC-MS/MS, detection limits in plasma, defined as S/N = 3, ranged from 0.2 to 1 pg/mL for each of the four corticosteroids (Table 3). These values are 100- to 500-fold lower than for most previously published LC-MS/MS methods and provides an advantage over conventional LC/MS/MS methods, in that low-pg/mL levels of corticosteroids can be detected in plasma. This level of sensitivity is necessary to quantify drug concentration ranges expected for the therapeutic strategies that this method was designed to support [13, 15–18]. Calibration curves for each of the corticosteroids showed good linearity over the concentration range of 5–5,000 pg/mL (r^2) 0.985–0.997, in Table 2 [4]. Method validation was carried out with spiked porcine plasma samples obtained from different sources. Accuracies for corticosteroid quantification were 88–107 % for plasma QC samples spiked with four different drug concentrations (Table 3). Intraday variations were in the range of 3.5–9.0 %, and interday variations were approximately 2.3–11.1 % (Table 3). Stability of the extracted drug was also good, with minimal loss of sample observed when the final extract was stored at −20 °C for up to 4 weeks. Validation of the lower limit of quantification (LOQ) was performed with plasma samples spiked with 5 pg/mL of each corticosteroid. Accuracy and precision of the LOQ for the four corticosteroids were satisfactory according to prevailing criteria. Matrix effects that could seriously compromise quantitative accuracy were not observed under the optimized SPE and μLC-MS/MS conditions. This is likely a result of selective SPE extraction combined with sufficient μLC separation.

To explore the utility of this approach, plasma samples obtained after injection of TACA into the posterior segment of the eye in pigs were analyzed [8]. The objective of this therapeutic strategy was to achieve a local, sustained, pharmacologically active concentration in the eye, with minimal systemic concentrations. In this application, the low water solubility of TACA contributes to its prolonged duration of action at the injection site; a previous human study showed that intraocular TACA maintained concentrations > 100 ng/mL in the eye for at least 700 h after a single 4-mg intravitreal TACA treatment [9]. However, TACA concentrations in plasma

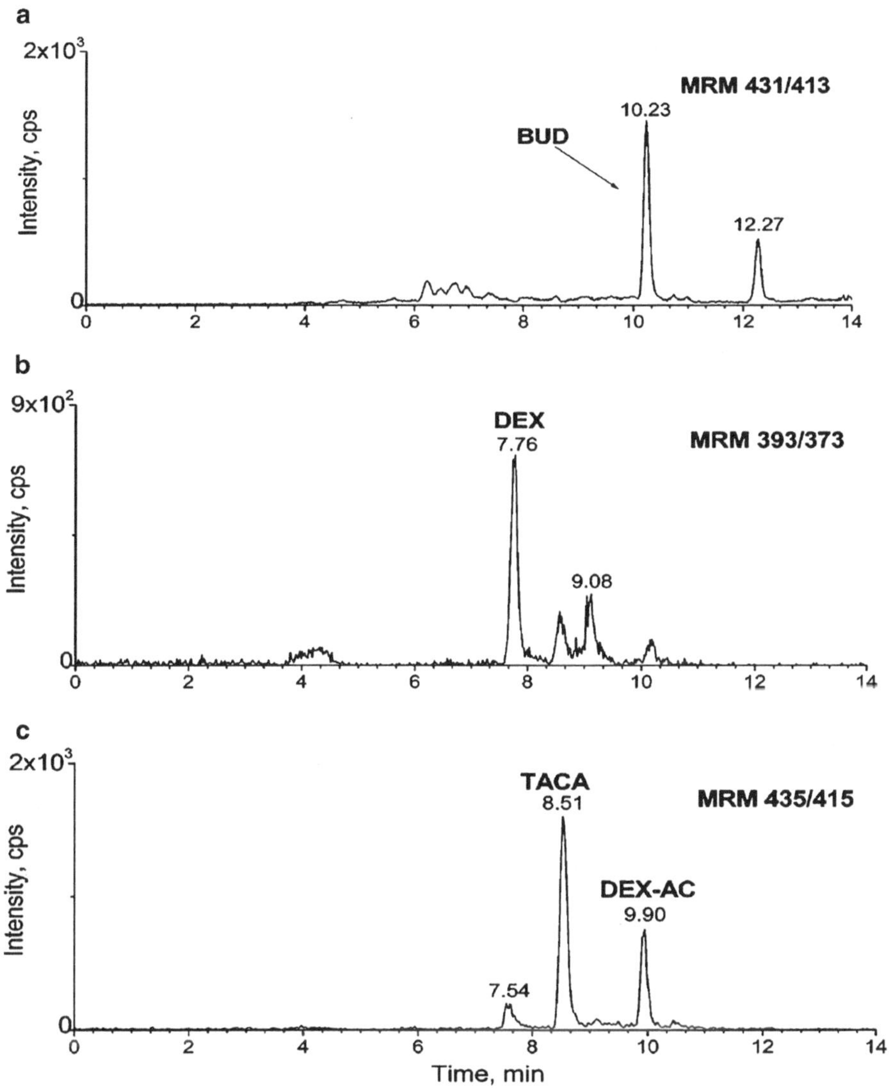

Fig. 6 LC/MS/MS chromatograms of 5 pg/mL (LOQ) of corticosteroids spiked in plasma: (**a**) budesonide; (**b**) dexamethasone; (**c**) triamcinoloneacetonide, and dexamethasone acetate (Reproduced with permission from American Chemical Society)

could not be evaluated, because of inadequate assay methods. A more recent human clinical study sought to measure plasma TACA concentrations after intraocular injections of doses of 20–25 mg. TACA plasma concentrations fell below the limits of detection in 17 out of 19 subjects (90 %) using an LC-MS/MS method that had an LOQ of 0.5 ng/mL. The plasma samples analyzed here were obtained from a group of 93 pigs that were administered TACA to the posterior segment of the eye via

Table 2 Linearity, detection limits, accuracies, and variability for four corticosteroids in plasma using an ultrasensitive μLC/MS/MS method

Compound	Linearity[a] (r^2)	Detect limit in plasma (pg/mL)	Accuracy[b] (SD)%	Intra-/interday variation[c] (%)
BUD	0.985	0.5	107(9), 99(5), 104(4), 102(5)	8.4/6.2
DEX	0.995	1	88(8), 93(10), 89(7), 96(4)	5.9/11.1
TACA	0.997	0.2	94(4), 97(2), 99(3), 102 (2)	3.5/2.3
DEX-AC	0.994	0.6	95(6), 93(5), 95(7), 93(6)	9.0/7.7

(Reproduced with permission from American Chemical Society)
[a]Calibration standards were prepared in porcine plasma in the concentration range of 5–200 pg/mL
[b]Accuracies were determined in triplicate for concentrations of 10,50,400, and 2,500 pg/mL in plasma
[c]Aliquots of plasma samples stored at −20 °C were analyzed six consecutive times in 1 day (intraday, $n=6$), and twice on three different days (interday, $n=6$)

Table 3 Sensitivity, quantification range, and linearity for the four metabolites

VitD metabolites	Sensitivity (LOD, pg/mL)[a]	Quantification range (ng/mL)	Linearity ($n=6$) (r^2)
25-Hydroxy vitamin D2	0.5	0.100–50.0	≥0.9924
25-Hydroxy vitamin D3	1.0	0.100–50.0	≥0.9967
1α, 25-Dihydroxy vitamin D3	0.5	0.005–2.50	≥0.9913
24(*R*), 25-Dihydroxy vitamin D3	1.0	0.050–25.0	≥0.9905

(Reproduced with permission from American Chemical Society)
[a]Defined as S/N = 3.

cannulation of the suprachoroidal space. Under these conditions, TACA concentrations in the posterior segment were sustained at therapeutic levels for more than 90 days [8].42 Plasma samples were collected serially for 90 days after suprachoroidal injection and analyzed for TACA concentration using the method developed here. The time course of plasma drug concentrations for the treatment group treated with 1.5 mg of TACA is shown in Fig. 7 [4]. The results showed that TACA plasma concentrations were sustained at high pg/mL levels for roughly the first 7 days and then decreased to the range of 1–10 pg/mL at around 30 days; this extremely low plasma concentration was maintained through the last time point, 90 days after treatment.

3 Paradigm 2: Ultrasensitive Quantification of Anticancer Drug in Low-Dose Treatment

3.1 Introduction

We applied the similar strategy for ultrasensitive quantification of a clinically important anticancer drug, paclitaxel. Although conventional high-dose therapeutic regiments of paclitaxel has been widely used for solid tumors treatment, severe side effects and acquired drug resistance are becoming major issue in clinical. Numerous

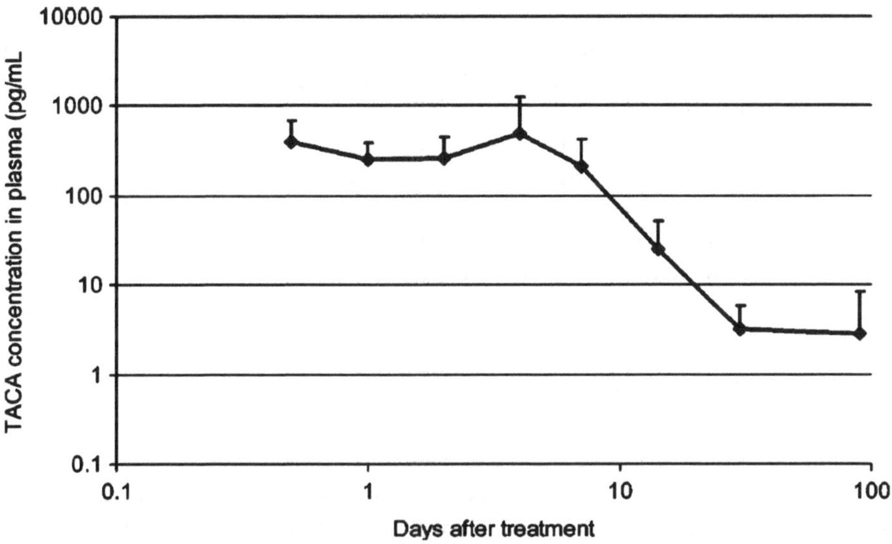

Fig. 7 Plasma TACA concentrations in pigs (*n*) (5 per time point) following suprachoroidal delivery of a 1.5-mg dose of TACA, measured using the ultrasensitive approach developed in this study (Reproduced with permission from American Chemical Society)

in vitro or preclinical experiments suggest that paclitaxel and other drugs can exert antivascular or antiangiogenic effects and some of these effects are observed at ultralow concentrations. In vivo, low-dose paclitaxel regimens employing more frequent administration have shown activity in cancer patients without significant side effects such as myelosuppression or acquired drug resistance. Interestingly, recent studies have shown that paclitaxel can exert antiangiogenic effects in the range of 0.1–100 pM, which is approximately 5,000-fold lower than concentrations required for antimitotic effects. In order to study the antiangiogenic mechanisms or pharmacokinetics (PK) of low-dose paclitaxel, highly sensitive methods are necessary to enable quantification of low pg/mL concentrations of drugs in complex biological matrices. The lower limits of quantification (LOQ) of paclitaxel using HPLC-UV are reported as 5–20 ng/mL [10], and for conventional LC-MS/MS are in the range of 0.1–0.25 ng/mL biological matrices. Recent preliminary work suggests the feasibility of quantification of paclitaxel in cell lysates at concentrations as low as 20 pg/mL. However, no published method provides sufficient sensitivity and operational robustness to quantify paclitaxel routinely in the low pg/mL range.

3.2 *Experimental*

In this study, A121a human ovarian cancer cells were incubated with low dose paclitaxel in four levels: 0.2, 0.8, 2, and 5 ng/mL respectively at 37 °C. Cells were harvested at five time points of 10, 30, 60, 180, and 360 min. The cells were lysated after three freeze–thaw cycles, and 3 % phosphoric acid was added in cell lysates in order to further release protein bound paclitaxel into cell lysates. Optimized SPE strategy was applied to cell lysates to clean up cell lysates matrices without losing any target compound.

In order to extract paclitaxel and docetaxel (the I.S.) selectively from cell lysates, while simultaneously reducing the quantity of matrix components that were also derived from the samples, the SPE wash and elution conditions were optimized in a manner similar as these described in the paradigm 1. Detailed experimental conditions can be found in a previous publication [5].

3.3 Results and Discussion

Optimization of the SPE procedure was carried out to meet three objectives. First, the overall conditions selected would provide the highest and most consistent absolute recovery of the target analyte. Second, the conditions for washing the SPE cartridges containing bound analyte would remove unwanted matrix components to the greatest extent possible without noticeably eluting the target analyte or I.S. Third, elution conditions would subsequently recover the analyte efficiently while minimizing the elution of less polar matrix components. To enable calculation of the absolute recovery of paclitaxel through the SPE procedure, cell lysates were spiked with known concentrations of paclitaxel, and the I.S. was added to the final eluate. Because of the relatively high protein binding of paclitaxel and the structurally similar I.S., phosphoric acid was added to the cell lysates to disrupt drug–protein binding before loading onto SPE cartridges [11, 12]. After the SPE cartridges were loaded with analyte-spiked cell lysates, they were washed in two steps. First, 3 % phosphoric acid in 35 % methanol ("Weak-wash" step) was used to remove residual cellular proteins and thus avoid precipitation and column blocking that could occur with subsequent washes that would employ higher-organic solvent conditions. Second, a higher concentration of methanol, identified through detailed optimization, was used to elute relatively polar matrix components without loss of the analyte ("Strong-wash" step). Finally, the target analyte was eluted quantitatively using a solvent composition optimized to recover it efficiently while leaving more hydrophobic matrix components on the cartridge. The results of absolute recovery of analyte through the SPE procedure are shown in Fig. 8 [5]. In preliminary investigations [13], paclitaxel recovery from the SPE procedure was very low even with 100% methanol. Here, after evaluating a number of mobile phase modifiers in the present study, paclitaxel was eluted completely in 95 % methanol with 0.1 % formic acid modifier (Fig. 8b); the polarity of paclitaxel increased by formic acid might be the reason. Based on the optimization results, 75 % methanol was chosen for Wash-2, and 0.1 % formic acid in 97.5% methanol was chosen for the elution.

The development of the chromatographic separation conditions was approached in terms of three interrelated variables: the column, the chromatographic separation

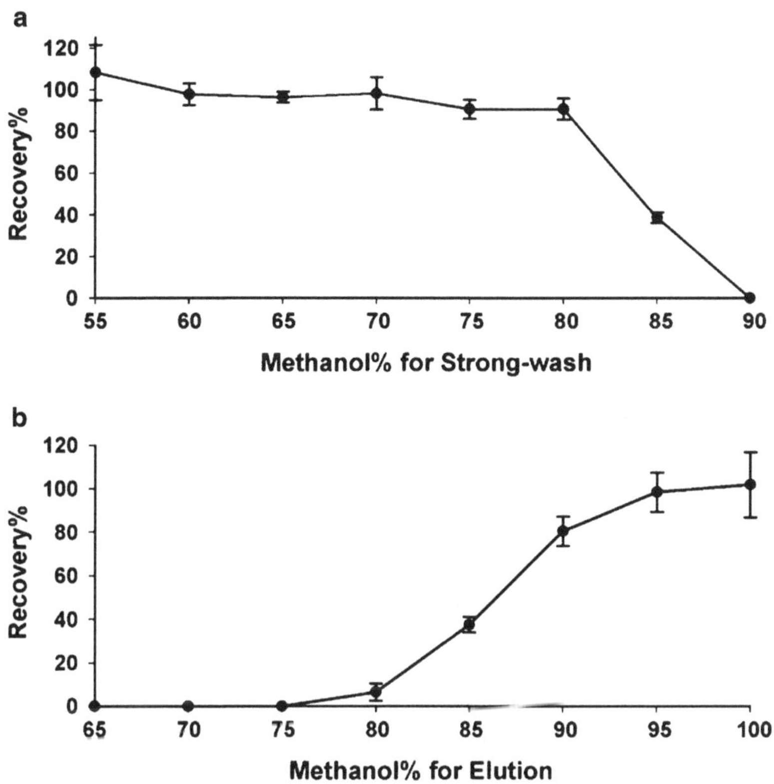

Fig. 8 Optimization of solvent composition for maximizing absolute recovery of paclitaxel after SPE extraction. Solvents used in the elution step were composed of 0.1 % formic acid in the indicated solvent composition. (**a**) Optimization of the strong-wash solvent: triplicate SPE cartridges were washed with a range of 55–95 % methanol; drug was eluted from all cartridges with 100%methanol and (**b**) optimization of the elution solvent: triplicate SPE cartridges were washed with 5 % methanol and eluted with methanol concentrations ranging from 60 to 100 % (Reproduced with permission from Elsevier)

strategy, and the injection volume. The use of smaller I.D. columns and lower flow rates can improve the sensitivity of ESI-MS/MS considerably [3]. However, loading capacities are proportionally lower too [14]. After balancing these considerations, a 0.5 mm I.D. μLC column was selected. Sufficient μLC resolution was desirable for two reasons. First, when minimal separation was employed, a severe ion suppression effect that originated from the sample matrix was observed. This suppression effect severely compromised sensitivity, but could be eliminated by increasing μLC separation. Second, it was found that the formation of both $[M+Na]^+$ and $[M+K]^+$ ions was relatively high when minimal μLC separation was used, even with the addition of acidified 2 mM ammonium acetate–formate as a mobile phase modifier. However, adequate chromatographic separation reduced these undesirable adduct ions and enhanced the formation of $[M+H]^+$. The most likely explanation of this observation is that K^+ and Na^+ salts do not co-elute with the target compounds when sufficient chromatographic separation is employed.

The selective SPE procedure reduced matrix components in the cell lysate samples and therefore permitted considerable concentration of the SPE eluate. Nonetheless, the matrix components that co-elute with the analyte during SPE limit the degree to which the eluate can be concentrated for μLC analysis. Large injection volumes can contribute significantly to the sensitivity of quantification, but may cause peak broadening as a result of both long sample injection times and the large volume of sample solvent. Furthermore, the smaller column volume selected here requires that V_{inj} be held to considerably smaller volumes in order to avoid column overcapacity and fouling [1]. To address these issues, it is necessary to develop an approach that would enable a high V_{inj} of the selectively extracted samples without causing peak broadening and column fouling. A two-segment gradient was designed, consisting of a focusing segment and a separation segment. The following theoretical rationale suggests the use of a low-organic mobile phase to load and focus a large V_{inj} on the column. The hypothesis was that the complexity of cell lysates would be reduced considerably by the selective SPE approach, and therefore a considerably higher V_{inj} could be tolerated without compromising chromatographic separation, provided the sample focusing strategy was also employed. To test this hypothesis, the μLC performance was evaluated as a function of V_{inj} using cell lysates spiked with the analyte. Typical chromatograms are shown in Fig. 9 [5]. The retention times of the target compounds increased only slightly as V_{inj} was increased, and peak broadening or peak shape deterioration was not observed over the range of V_{inj} from 0.2 to 8 μL. Furthermore, both the intensity and S/N for paclitaxel increased roughly in proportion with V_{inj}. This confirmed that the selective SPE procedure, in conjunction with the sample focusing strategy, permits a high V_{inj} on the μLC column without causing column overcapacity. A V_{inj} of 8 μL was selected for quantification of paclitaxel in cell lysates, which is 20- to 40-fold higher than the manufacturer's recommendation. Column durability was not compromised; after approximately 300 injections of 8 μL of cell lysate samples on the μLC column, the separation of the target analytes remained highly reproducible, with neither loss of resolution nor increase in backpressure. The subsequent gradient steps constitute the "separation" segment of the chromatographic strategy, in which a gradually increasing percentage of mobile phases B elutes the compounds that were focused on the column during the initial gradient step. To determine the optimal separation conditions, we investigated chromatographic retention vs. S/N for cell lysate samples spiked with 250 pg/mL of paclitaxel and I.S. The S/N of the target analyte improved considerably when their chromatographic retention increased, but beyond a certain point, the S/N approached a maximum. This behavior has been observed in the first example too, when a similar strategy was used for quantification of corticosteroids in plasma.

The validated quantification range is between 5 and 6,250 pg/mL with good linearity ($r^2 = 0.992$). An ultralow detection limit of 0.5 pg/mL (defined as S/N = 3) was achieved for paclitaxel in cell lysates. This detection limit is 200–500 folds lower than conventional LC-MS/MS methods published previously. The LOQ was

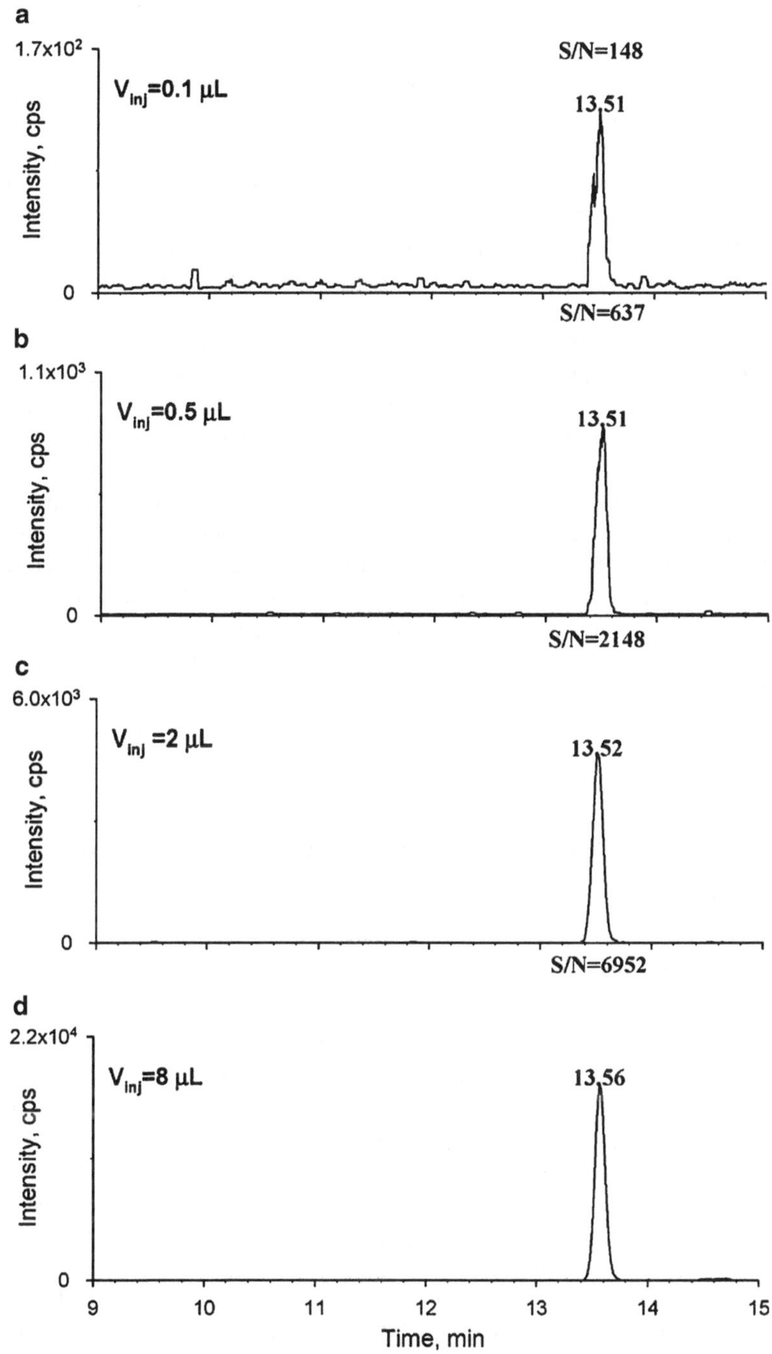

Fig. 9 Optimization of sample injection volume. Various injection volumes (V_{inj}) of spiked cell lysate samples were investigated to optimize V_{inj}. Lysates were spiked with 250 pg/mL of paclitaxel and extracted by selective SPE. The V_{inj} was (**a**) 0.1 μL; (**b**) 0.5 μL; (**c**) 2.0 μL; (**d**) 8.0 μL. A 15 cm × 0.5 mm I.D. capillary column was used for separation; the manufacturer-recommended V_{inj} for the column was 0.1–0.2 μL (Reproduced with permission from Elsevier)

validated at 5 pg/mL; representative chromatograms of cell lysates spiked with this concentration of drug are shown in Fig. 10c and d.

In this study, a LOQ of 20 pg/mL in cell lysates would be needed in order to quantify paclitaxel uptake by cells exposed to approx. 1 ng/mL drug, but the detection limits of published LC-MS/MS based methods were in the range of 0.1–0.25 ng/mL [10, 15–21]. Furthermore, concentrations of paclitaxel lower than 1 ng/mL are pharmacologically active, and therefore, a LOQ lower than 20 pg/mL is desirable.

A121a cells were exposed to paclitaxel at concentrations of 0.2, 0.8, 2, and 5 ng/mL. Paclitaxel uptake by cells was able to be quantified at each time point under four treatment conditions. The lowest measured intracellular accumulation was 6.5 pg/106 cells, observed 10 min after the addition of 0.2 ng/mL paclitaxel to cells. The detected concentration in this sample corresponded to approximately 9.6 pg/mL drug in the cell lysate, which is nearly twice the LOQ. Foremost concentrations of paclitaxel, it appeared that the intracellular drug concentrations increased rapidly with exposure time and reached a plateau within 1–3 h. However, for cells exposed to the lowest concentration (0.2 ng/mL), the maximum intracellular drug concentration apparently was not achieved within 6 h, which was longest time interval investigated. In future studies, these data will be expanded to include additional concentrations and exposure times, and analyzed according to cellular pharmacokinetic models such as those published previously [12].

4 Paradigm 3: Ultrasensitive Quantification of Vitamin D Metabolites in Human Plasma

4.1 Introduction

In this section, we show a paradigm for ultrasensitive quantification of endogenous compounds in circulation system by the combination of selective SPE with μLC-MS. VitD metabolites are clinically important and play a critical role in many important biological processes, including maintenance of calcium homeostasis, immunomodulation, and cell differentiation [22]. VitD itself is not biologically active and requires further metabolism that generates the active metabolites [23]. Studying the physiological actions of VitD metabolites would contribute greatly to the mechanism research, diagnosis, staging, and therapy of numerous diseases, such as multiple sclerosis, diabetes, cancer, osteoporosis, microbial infections, and cardiovascular diseases [22, 24]. For an example, the 25-hydroxyVitD (25(OH)VitD) metabolite is clinically used as the marker for VitD deficiency. One of its metabolite, the 1R, 25-dihydroxy-VitD (1,25(OH)2VitD), is considered the primary biologically active form of VitD and responsible for stimulating intestinal calcium absorption, modulating immune response, and maintaining calcium homeostasis [25]. Another dihydroxyl metabolite, the 24(*R*), 25-dihydroxy-VitD (24,25(OH)2VitD), has been

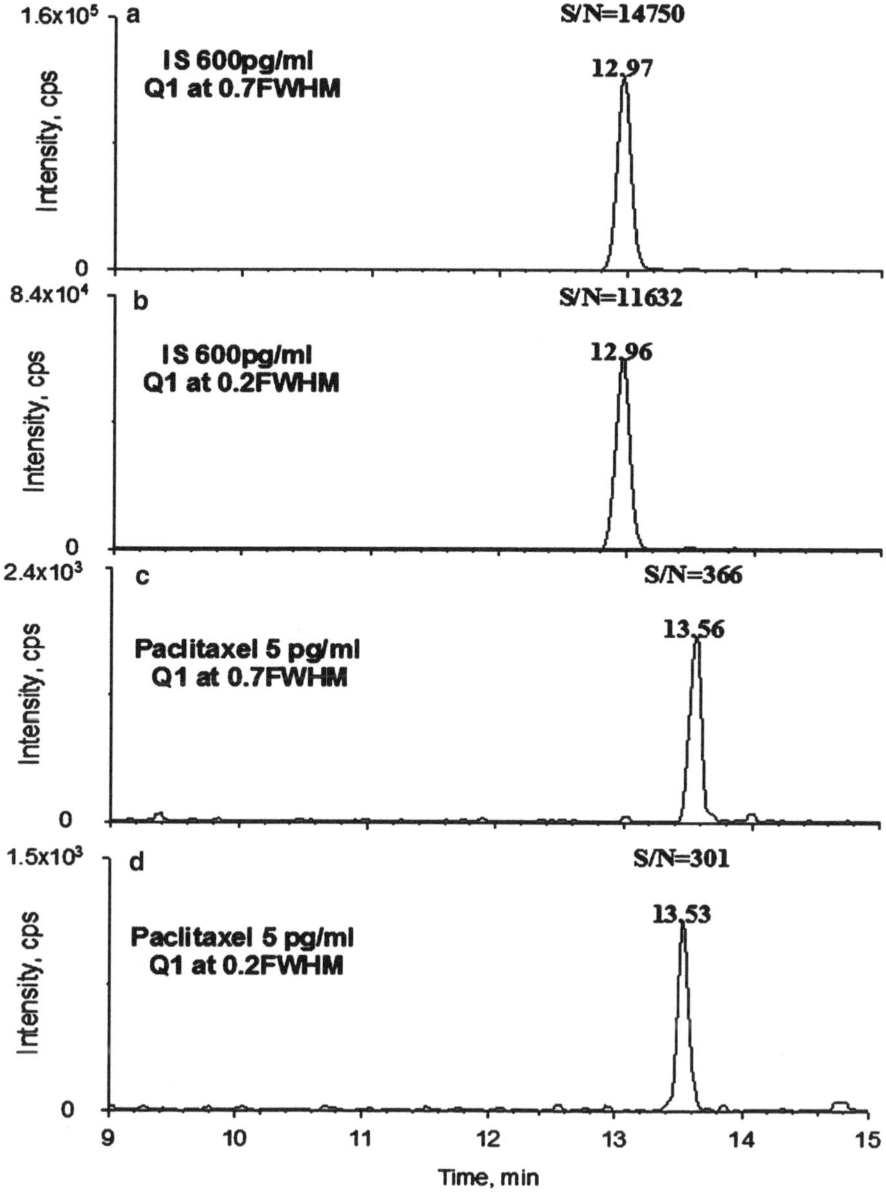

Fig. 10 Typical LC–MS/MS chromatograms of paclitaxel and docetaxel. Cell lysates were spiked with paclitaxel to a final concentration equaling the LOQ (5 pg/mL) and with the I.S. docetaxel at a final concentration of 300 pg/mL. Both SRM and HRSRM were evaluated. (**a**) SRM of docetaxel; (**b**) HR-SRM of docetaxel; (**c**) SRM of paclitaxel; (**d**) HR-SRM of paclitaxel (Reproduced with permission from Elsevier)

demonstrated as a critical component for healing processes in tissues and bones [26]. Therefore, the ability to quantify these circulating VitD metabolites would be highly valuable for the clinical studies of diseases that are associated with the deficiency and/or metabolic dysfunction of VitD.

A wide variety of methodologies have been developed for the quantification of VitD metabolites in serum/plasma samples. Among them, immunoassays including the radioimmunoassay (RIA) and enzyme immunoassay (EIA) are the most prevalent [24]. Nevertheless, the limitation of most of the RIA and EIA methods is that only one metabolite can be measured per assay therefore the selectivity, accuracy, and interbatch/lab reproducibility could be problematic [24, 27, 28]. Other methods include competitive protein binding assay, automated chemiluminescent protein-binding assay [29] and HPLC-UV [30]. More recently, liquid chromatography tandem mass spectrometry (LC-MS/MS) has emerged as a promising alternative. Although LC-MS/MS-based methods could readily measure 25(OH)VitD, which presents at relatively high levels in serum, quantification of the dihydroxyl metabolites can be highly challenging. For example, the 1,25(OH)2VitD3, which is highly active and of primary interest for the research of many diseases [14, 24], present at extremely low levels (low pg/mL) in human serum that are significantly lower than the detection limits of conventional LC-MS/MS methods. The relatively poor ionization efficiency of this metabolite, further compounds the problem. Although a recent study employed a derivatization procedure with conventional LC-MS/MS achieved a lower limit of quantification (LLOQ) of 25 pg/mL using 0.5 mL of human serum, the LLOQ achieved is not sufficiently sensitive for the measurement of 1,25(OH)2VitD3 in human serum, especially for those patients (e.g., multiple sclerosis) whose serum levels of VitD metabolites are expected to be lower than healthy subjects.

Multiple sclerosis is a neurodegenerative, chronic inflammatory disease, and has become the most prevalent neurological disorder affecting young adults in the USA [31]. Much evidence suggests that low levels of the VitD metabolites are potential risk factors for developing multiple sclerosis, probably owing to their immune modulating effects [22]. For instance, some epidemiological studies showed that low serum concentrations of 25(OH)VitD in adolescence are associated with an increased risk of developing multiple sclerosis and VitD supplementation and high serum concentrations of 25(OH)VitD are protective [32]. Nevertheless, the underlying mechanisms for the association of VitD metabolites and the development of multiple sclerosis remain unclear. In order to study this association, an ultrasensitive, selective and robust analytical method for quantification of the key circulating VitD metabolites in clinical samples is essential. In this study, four clinically related VitD metabolites were selected. They are 25(OH)VitD2 and 25(OH)VitD3, which are the indicators of VitD repletion; 1,25(OH)2VitD3 and 24,25(OH)2VitD3, which are biologically active metabolites. In this study, an ultrasensitive and selective-SPE/μLC-MS/MS analytical method was developed in order to quantify ultralow concentration of VitD metabolites in complex biological matrices and evaluated the method's applicability in a large-scale clinical analysis.

4.2 *Experimental*

The detailed experiment conditions are shown in a previous publication [6].

4.3 *Results and Discussion*

In order to achieve highest sensitivity of VitD metabolites detection, the efforts for method development were focused on three components of the analytical flow path (1) the selective SPE extraction procedure, (2) the ionization and SRM method for MS/MS, and (3) the μLC loading and separation method.

For the purpose of achieving a selective SPE extraction that eliminates matrix components to the greatest extent possible without loss of analytes, an extensive optimization of the wash/elute conditions was conducted, the procedure is similar as described in previous examples. In order to achieve high efficiencies of extraction and reproducibility of quantification, a protein precipitation with ACN/methanol prior to SPE extraction is found important to assist releasing VitD metabolites from serum protein and derivatization by PTAD plays a critical role to improve the stability of VitD metabolites.

In this study, a 0.5-mm-I.D. capillary column was chosen; however, the manufacturer-recommended injection volume (I.V.) for this μLC column is 0.2–0.4 μL. Nonetheless, the hypothesis is that the serum matrix has been reduced substantially by the selective SPE, and therefore, a markedly higher I.V. of the extracted sample can be tolerated without compromising chromatographic performance, provided an efficient sample focusing strategy was employed. To test this hypothesis, the evaluation of the μLC-MS/MS performance as a function of I.V. using serum samples fortified with the four analytes was performed. Typical chromatograms are shown in Fig. 11 [6]. The retention times of the analyte increased only slightly as the I.V. increased, but no appreciable change in peak width and peak shape was observed over the range of I.V. from 0.2 to 9.5 μL, while both the intensity and S/N increased roughly in proportion with I.V. This indicated the selective SPE procedure, in conjunction with the sample focusing approach, permits an I.V. that is more than 20-fold higher than manufacturer's recommendation, without causing column overcapacity. Therefore, an I.V. of 9.5 μL was selected for quantification of serum samples.

Sufficient chromatographic separation is critical in this work, for three reasons: first, the peak of target compounds must be separated from isomeric peaks, interfering signals and endogenous compounds that cause ion suppression. Second, given the close similarity in the structure and *m/z* among the analytes and I.S., it is necessary to resolve these compounds chromatographically to avoid various "cross talk" effects. Third, a sufficient μLC separation was found further helped to suppress the sodium-adducted precursors, a phenomenon also observed previously. Other considerations for developing the μLC method included the following: (1) upon complete elution of all analytes, it is necessary to perform an extensive cleaning of the column and spray tip, to minimize the memory effect and (2) a relatively high throughput is desirable for clinical analysis. The gradient conditions were optimized carefully to achieve the above objectives, and the optimal μLC conditions are described in Sect. 4.2. Under the optimized conditions, all chromatographic peaks of the target compounds were well defined and symmetric, without appreciable interference. Typical chromatograms of a patient serum sample are shown in Fig. 11 [6].

A robust analytical approach for clinical applications is highly desirable, because the analysis of large number of samples is often required, and that the sample matrices and concentrations of analytes among patients are often markedly different. One of the primary concerns of this study was whether the selective-SPE-μLC-MS/MS-based method is sufficiently robust for a large-scale clinical investigation. Therefore, we conducted an extensive assessment of the robustness of the quantitative method. The method robustness was evaluated when analyzing a large number of clinical samples (>500 injections in total). Each of the four "system control" (SC) samples, which were randomly selected from sera of multiple sclerosis patients, were analyzed once for every 48 injections, for totally ten times ($n=10$); the reproducibility of calculated concentrations, retention times, S/N, and peak shapes of all metabolites was assessed. The calculated concentrations (RSD% 2.4–12 %) and retention times (RSD% 0.5–2 %) of all analytes were highly reproducible, and no decline in sensitivity (RSD% 2–12 %) and symmetry of peaks (as indicated by the tailing factors, RSD% 6–13 %) was observed. Additionally, no noticeable deterioration in chromatographic performance, increase in back pressure, or fouling of the μLC was observed after more than 500 injections of serum samples. The high degree of robustness is attributable to the use of a selective SPE and a sufficient and reproducible μLC separation and that an effective cleaning procedure was employed for each run. In addition, the LLOQ adopted for each analyte is at least ten times higher than the LOD; the sufficiently high S/N at LLOQ contributes to the high robustness, accuracy, and reproducibility of the method, in that it renders the method well tolerable to fluctuations in analytical sensitivity, which could often be caused by such occasions as the higher chemical noises in certain samples and decrease in instrumental sensitivity during operation.

The method developed here holds several advantages that are important for clinical analysis of VitD metabolites. First, ultrahigh sensitivities, with LOD ranging from 0.5 to 1 pg/mL, were achieved for the four analytes (Table 3 [6]). Such a high level of sensitivity, which is comparable or superior to that of immunoassays, enabled a robust quantification of all analytes in human serum samples. Second, due to the use of selective SPE extraction and adequate μLC separation, the method is highly specific, as examined using structurally similar compounds. The excellent specificity helped to achieve a high quantitative accuracy. Third, the method is very robust and reproducible, as revealed by comprehensive evaluations during a large-scale clinical analysis. Additionally, the method is superior to immunoassay methods, in that it offers significantly better selectivity, accuracy, and the capacity of simultaneous quantification of multiple VitD metabolites. As a proof-of-concept, this method was applied to the analysis of serum samples from 281 multiple sclerosis patients and 22 healthy subjects. Because of the ultrahigh sensitivity achieved, the 1,25(OH)2VitD3 were detected in all samples, and for 100%of healthy subjects

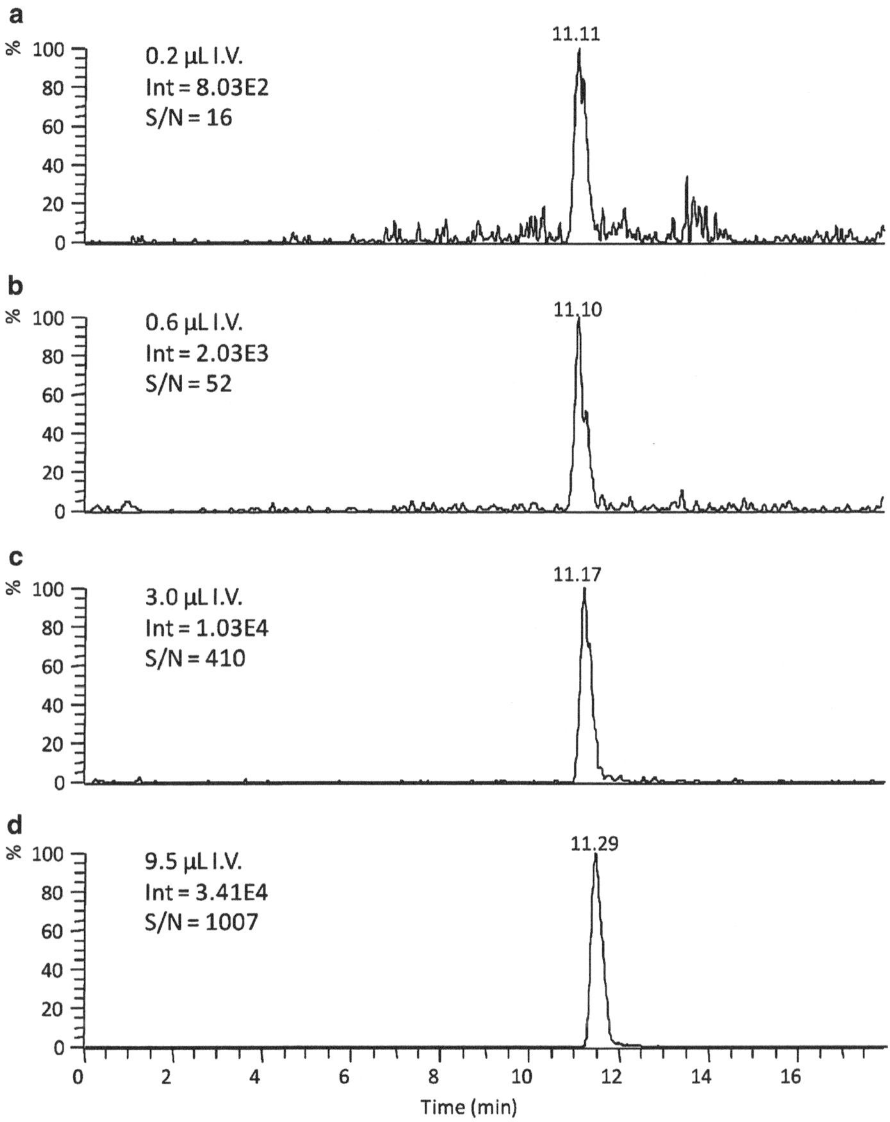

Fig. 11 Typical chromatograms showing injection volume (I.V.) optimization for pooled serum spiked with 100 pg/mL of 1,25(OH)2VitD3 and extracted by selective SPE. The I.V. was (**a**) 0.2 μL; (**b**) 0.6 μL; (**c**) 3 μL; and (**d**) 9.5 μL. A 15 cm × 0.5 mm I.D. capillary column (manufacturer suggestion I.V. was 0.2–0.4 μL) was used for separation. Using the selective SPE and a sample focusing approach before the separation, a large I.V. was enabled without compromising chromatographic performance and with substantially increased sensitivity (Reproduced with permission from American Chemical Society)

and 96 % of the patients, the serum levels were above a LOQ of 5 pg/mL. Apparently, conventional LC-MS/MS is not sufficiently sensitive for this task. The quantitative results revealed that the serum levels of both 1,25(OH)2VitD3 and 24,25(OH)2VitD3 are significantly lower in multiple sclerosis patients than in healthy subjects. However, with a cognizance that this study may be limited in the sample sizes, this finding may need to be examined in further clinical investigation. More mechanism-related studies are desired to elucidate how this altered VitD metabolism relates to multiple sclerosis development. The method developed here permitted, for the first time, the consistent and robust quantification of low-abundance dihydroxyl VitD metabolites such as the 1,25(OH)2VitD3 in clinical samples by LC-MS/MS-based methods. This work also demonstrated the applicability of the selective-SPE-μLC-MS/MS strategy in large-scale clinical analysis. Therefore, the method is well suited for the analysis of VitD metabolites in large-scale clinical studies, especially for these involving pathological conditions where the levels of the dihydroxyl VitD metabolites could be low.

5 Conclusion

In the above paradigms, selective SPE strategies were optimized and applied for preparations of highly complex biological samples. Because of this selective extraction procedure, the acceptable V_{inj} of biological samples on the μLC column was increased significantly. An additional benefit of this selective SPE strategy was an improved S/N resulting from decreased ion suppression. A two-segment gradient which included on-column focusing in first segment was used to maintain an optimal chromatographic separation. As a result, a high injection volume is feasible for μLC-MS analysis and sensitivity was increased, without compromising chromatographic performance or operational robustness. In injection volume optimization, the retention times of target compounds increased only slightly as V_{inj} was increased, but peak broadening or peak shape deterioration was not observed over a broad range of V_{inj} up to 9.5 μL. Furthermore, the S/N for all targets increased in proportion with V_{inj}. This indicates that the selective SPE procedure reduced the complexity of the plasma samples significantly, to a degree that would permit a high V_{inj} on the μLC column without causing column overcapacity. Furthermore, the utilization of separate "trapping" and "separation" gradient segments (described above) counteracted the tendency of increased V_{inj} to increase peak widths and thus rendered peak widths for the target compounds independent of loading times.

Compared with current LC-MS/MS methods, this approach increased assay sensitivity greatly (LOQ at low-pg/mL in plasma). The developed methods are also superior to some highly sensitive RIA methods, in that it offers significantly better selectivity and quantitative accuracy.

Collectively, the method has potentially widespread applications given the pervasive clinical investigation of diverse and highly potent pharmaceutical agents and endogenous markers. The combination of selective SPE and μLC-MS/MS represents a new approach for ultrasensitive quantification of compounds in highly complex matrices. This approach could be adapted for the analysis of other types of compounds in biological, pharmaceutical, and environmental applications in which highly sensitive quantification is desirable.

References

1. Vissers JPC, Claessens HA, Cramers CA (1997) J Chromatogr A 779:1–28
2. Wilm M, Shevchenko A, Houthaeve T, Breit S, Schweigerer L, Fotsis T, Mann M (1996) Nature 379:466–469
3. Emmett MR, Caprioli RM (1994) J Am Soc Mass Spectrom 5:605–613
4. Qu J, Qu Y, Straubinger RM (2007) Anal Chem 79:3786–3793
5. Yu HY, Straubinger RM, Cao J, Wang H, Qu J (2008) J Chromatogr A 1210:160–167
6. Duan XT, Weinstock-Guttman B, Wang H, Bang EJ, Li J, Ramanathan M, Qu J (2010) Anal Chem 82:2488–2497
7. Qu J, Wang YM, Luo GA (2001) J Chromatogr A 919:437–441
8. Olsen TW, Feng X, Warner K, Conston SR, Sierra DH, Folden DV, Smith ME, Cameron JD (2006) Am J Ophthalmol 142:777–787
9. Beer PM, Bakri SJ, Singh RJ, Liu WG, Peters GB, Miller M (2003) Ophthalmology 110:681–686
10. Alexander MS, Kiser MM, Culley T, Kern JR, Dolan JW, McChesney JD, Zygmunt J, Bannister SJ (2003) J Chromatogr B 785:253–261
11. Kumar GN, Walle UK, Bhalla KN, Walle T (1993) Res Commun Chem Pathol 80:337–344
12. Kuh HJ, Jang SH, Wientjes MG, Au JLS (2000) J Pharmacol Exp Ther 293:761–770
13. Gaspar JR, Qu J, Straubinger NL, Straubinger RM (2008) Analyst 133:1742–1748
14. Saito Y, Jinno K, Greibrokk T (2004) J Sep Sci 27:1379–1390
15. Stokvis E, Ouwehand M, Nan LGAH, Kemper EM, van Tellingen O, Rosing H, Beijnen JH (2004) J Mass Spectrom 39:1506–1512
16. Parise RA, Ramanathan RK, Zamboni WC, Egorin MJ (2003) J Chromatogr B 783:231–236
17. Hou WY, Watters JW, McLeod HL (2004) J Chromatogr B 804:263–267
18. Grozav AG, Hutson TE, Zhou X, Bukowski RM, Ganapathi R, Xu Y (2004) J Pharm Biomed 36:125–131
19. Guo W, Johnson JL, Khan S, Ahmad A, Ahmad I (2005) Anal Biochem 336:213–220
20. Wang LZ, Goh BC, Grigg ME, Lee SC, Khoo YM, Lee HS (2003) Rapid Commun Mass Spectrom 17:1548–1552
21. Gustafson DL, Long ME, Zirrolli JA, Duncan MW, Holden SN, Pierson AS, Eckhardt SG (2003) Cancer Chemother Pharm 52:159–166
22. Smolders J, Damoiseaux J, Menheere P, Hupperts R (2008) J Neuroimmunol 194:7–17
23. Houghton LA, Vieth R (2006) Am J Clin Nutr 84:694–697
24. Zerwekh JE (2008) Am J Clin Nutr 87:1087s–1091s
25. Norman AW, Okamura WH, Farachcarson MC, Allewaert K, Branisteanu D, Nemere I, Muralidharan KR, Bouillon R (1993) J Biol Chem 268:13811–13819
26. Seo EG, Norman AW (1997) J Bone Miner Res 12:598–606
27. Carter GD (2009) Clin Chem 55:1300–1302
28. Roth HJ, Schmidt-Gayk H, Weber H, Niederau C (2008) Ann Clin Biochem 45:153–159
29. Haddad JG (1971) J Clin Endocr Metab 33:992–995
30. Jones G (1978) Clin Chem 24:287–298
31. Khatri BO (2000) Ther Apher Off J Int Soc Apher Jpn Soc Apher 4:263–270
32. Munger KL, Levin LI, Hollis BW, Howard NS, Ascherio A (2006) JAMA 296:2832–2838

Pitfalls of LC-MS/MS in the Clinical Laboratory

Christoph Seger and Michael Vogeser

Abstract The technical maturation of liquid chromatography tandem mass spectrometry (LC-MS/MS) hyphenations brought this technology into most of the major clinical laboratories worldwide. It found its sound place amongst major basic routine technologies of laboratory medicine as enzyme based assays or immunoassays. LC-MS/MS extended the technological armamentarium of clinical laboratories significantly, both in analytical and economical terms. Especially in therapeutic drug monitoring, endocrinology, and toxicology, it became an indispensable routine tool.

Although well designed LC MS/MS assays generally outperform immunoassays due to their accuracy, sensitivity, precision, and inherent multiplexing capability, they are not free from analytical problems. Besides limitations in selectivity—isobaric analytes cannot be distinguished—sudden and unpredictable ion yield attenuations, often known as "ion suppression effect," have to be considered *the* Achilles heel of quantitative bio-analytical mass spectrometry. Ion yield attenuation is compromising both the accuracy of an assay and its precision. It can easily lead to gross errors in analyte quantification.

Co-medications or constituents found in pathologically altered patient specimen are major causes for both ion yield fluctuations. Special measures have to be taken to reduce such effect and cause has to be taken to evaluate these accuracy limiting interferences prior to bringing an LC-MS/MS assay into the highly regulated clinical routine environment.

Lacking assay accuracy may also stem from the fact, that most LC-MS/MS methods used in clinical laboratories are still locally designed laboratory-developed tests operating on very heterogeneous instrument configurations. Consequently,

C. Seger, Ph.D. (✉)
Institute of Medical and Chemical Laboratory Diagnostics (ZIMCL), University Hospital Innsbruck, Innsbruck, Austria
e-mail: Christoph.Seger@uki.at

M. Vogeser, M.D. (✉)
Institute of Clinical Chemistry, Hospital of the University of Munich,Munich, Germany
e-mail: Michael.Vogeser@med.uni-muenchen.de

Q. Alan Xu and Timothy L. Madden (eds.), *LC-MS in Drug Bioanalysis*,
DOI 10.1007/978-1-4614-3828-1_5,

assay heterogeneity and lacking traceability to reference procedures or materials leads to an increased imprecision in proficiency testing as well as to inaccurate result reporting if basic rules of assay validation and "post marketing" surveillance are violated.

The position of LC-MS/MS and its advantages / disadvantages compared to immunoassays will be discussed. Technical limitations and analytical problems of LC-MS/MS instrumentation will be critically evaluated in the light of technical development.

1 Introduction

Laboratory investigations play an essential role in medicine. Laboratory results are taken into consideration in about two thirds of all medical decisions in medical systems of industrialized countries today. The vast majority of clinical chemistry analyses are based on few analytical principles including photometry, ligand binding assays and potentiometry. For these standard methods complete automation has been achieved and multi-channel, random access analyzers realize several hundred analyses per instrument and hour on a very high level of user-friendliness. Consequently, clinical chemistry is very cost efficient today; typically clinical chemistry analyses contribute less than 5 % of all costs of tertiary care hospitals.

The development of photometric and ligand-binding based analytical methods in clinical chemistry between the 1960s and the 1990s has dramatically changed many fields in medicine and substantially enhanced the diagnostic potentials in patients' care globally. However, it must be recognized that these standard techniques still are compromised by important shortcomings: For many analytes cross-reaction with similar, naturally occurring or xenobiotic compounds or the impact of unspecified matrix compounds on signal generation limits the reliability of results, e.g. in the quantification of steroid hormones. Since the specificity of such tests often differs between different assays, method specific reference ranges must be interpreted in clinical decision making today. Although by principle extremely versatile, the development of new ligand binding tests in an industrial dimension is very expensive, resulting in the lack of routine methods for many desirable analytes, in particular for drug testing in therapeutic drug monitoring. Many small molecules can neither be addressed by photometric tests nor by ligand binding methods due to their particular molecular structure (e.g. methylmalonic acid). Current standard methods may be implemented on a multi-channel analyzer but are by principle single-target analyses which are not primarily reading out analyte profiles.

Mass spectrometric methods have a huge potential to overcome these limitations. While GC-MS has found a fundamental role in toxicology and occupational medicine it did not find a place in the routine clinical chemistry laboratory due to its very demanding handling not suitable for a 24 h/7 day service laboratory setup. The introduction of LC-MS has improved the practicability and robustness of highly specific and highly multiplexed mass spectrometric analyses very substantially.

Hyphenating liquid chromatography (with its inherent limited separation power) with simple low resolution molecular mass-selective detection devices as single stage quadrupole or ion trap MS instruments (LC-MS) was soon quickly recognized to offer only limited selectivity when used for quantitative target analyses of complex biological samples. High resolution instruments using time-of-flight (ToF) detection units did prove to be highly selective, but showed limited linear ranges; a prerequisite for most quantitative assays. The introduction of tandem mass spectrometry (MS/MS), which involves the coupling of two quadrupole mass filters with an interposed collision cell, initially seemed to overcome these limitations because the fragmentation pattern of target analytes became incorporated into analyte detection.

Summarizing, the essential strengths of LC-MS/MS technology for laboratory [1] medicine can be listed as:

1. *Specificity*. The potentially very high analytical specificity of tandem mass spectrometry as HPLC detector results from using the molecular mass of the analyte and its specific disintegration behaviour as detection principle.
2. *Wide range of applicability with good practicability*. In contrast to GC-MS as the "classical" mass spectrometry technique, the application of LC-MS/MS is not limited to volatile molecules (usually with molecular weights below 500 Da). Furthermore, aside from highly polar analytes (i.e. amino acids), sample preparation is usually simple and does not include derivatization techniques. Mass spectrometry detected LC assays are generally optimized to shorter runtimes. Hence, compared to GC-MS, far higher sample throughput can be realized.
3. *Flexibility*. New assays can typically be developed in-house with a high degree of flexibility and within a short time.
4. *Information rich detection*. A large number of quantitative or qualitative results can be obtained from a single analytical HPLC-MS/MS run, since due to the fast ion selection electronics, multiparametric, quasi parallel analyses can be performed with a mass spectrometer.

At present no MS manufacturer can offer complete solutions that meet the high analytical demands of a routine laboratory performing LC-MS/MS analytes. None of the mass spectrometers, let alone the complex LC-MS/MS instrument combination, is certified according to the in vitro diagnostic directive (IVDD) 98/79/EC of the European Community ("CE certified") or FDA approved. Generally there is no bidirectional connection of the MS control and evaluation software to the laboratory's electronic data processing. The use of self-generated scripts for the transmission of work lists to the MS or for retransmission of the measured values continues to be common practice. With few exceptions (IVD-CE certified assays by some providers) the manufacture of consumables (mobile solvents, precipitants) is still in the hands of the local laboratory.

Accordingly it is costly to establish and validate an "in house" LC-MS/MS installation that should track the prevailing recommendations and guidelines of international forums, such as CLSI (www.clsi.org), FDA (www.fda.gov) EMEA (www.ema.europa.eu), or ICH (www.ich.org). When selecting equipment (design qualification, DQ) and furnishing the workplace (installation qualification, IQ) it is necessary to take into account LC-MS/MS-specific requirements, which may not always be easy to implement in the routine laboratory [1]. Only after equipment has been successfully installed (possibly documented by an operation qualification; OQ) can a laboratory start to prepare an assay (performance qualification; PQ). Once the limitations of the assay are laid down, a subsequent validation and a risk analysis usually conclude the months-long process of a LC-MS/MS platform establishment. It must not be forgotten that a frequently observed considerable method bias towards immunological or other methods might necessitate a parallel measurement phase of several weeks or even months, to accustom the clients to the new measured value [2].

2 Application Fields

In many industrialized countries FIA-MS/MS is now used in the routine neonatal screening for inherited diseases of metabolism. Typically, these analyses are performed by a few specialized laboratories per country and a specified set of metabolites is traced in a semi-quantitative manner [3, 4]. Besides these applications, at least one HPLC-MS/MS system is available in most university hospital laboratories in industrialized countries (e.g. in clinical chemistry institutes, pharmacological and forensic institutes, institutes of occupational health, mass spectrometric core facilities). Here, they are often used for both research and routine analytical purposes, serving application fields like therapeutic drug monitoring (TDM; in particular quantification of immunosuppressants and new generation anticonvulsive and antipsychotic drugs), pharmacology, endocrinology, and toxicology. For more detailed information, the readership is referred to the literature [5–14]. Beyond university hospital laboratory centres only few tertiary care hospitals in Europe are equipped with a HPLC-MS/MS instrument at present. In addition, the majority of larger laboratory trusts are nowadays equipped with HPLC-MS/MS instruments, mostly in centralized core facilities. These instruments are predominantly used for TDM and endocrinology. HPLC-MS/MS methods replace more and more of the numerous conventional HPLC-UV and HPLC-FLD methods in such laboratories since they offer superior specificity, shorter runtimes, and less laborious method development and sample preparation.

Tandem mass spectrometer based protein quantification made tremendous technological advancements in the past years, transforming LC-MS/MS based proteomics from a mere qualitative screening biomarker discovery tool to a quantitative targeted biomarker verification oriented platform ("quantitative clinical proteomics") [15] useable in clinical practice [16, 17]. In this context—besides the specificity of MS/MS detection—the possibility to monitor a bundle of biomarkers simultaneously, distinguishes LC-MS/MS platforms remarkably from immunoassay based approaches [18–20]. Valid and inter-laboratory transferable concepts of target protein enrichment followed by tryptic breakdown to peptides amendable to LC separation are now well established [21, 22]. Protein quantification is usually carried out by assessing the MS/MS response of a set of carefully chosen reporter peptides by monitoring several MS/MS transitions. Applications range from single compound assay devoted to diverse analytes like parathyroid hormone (PTH) [23], thyroglobulin [24], insulin [25], haemoglobin A2 [26], Zn-α2 glycoprotein [27], ghrelin [28], to multiplex approaches devoted to cardiac markers [29, 30]. Since the methodological cornerstones of quantitative clinical proteomics—i.e. analyte enrichment, chromatographic separation, and MS/MS based quantification are following the same physical principles as in small molecule LC-MS/MS, causes of assay inaccuracy and imprecision are akin [31]. However, due to the different chemical nature of drugs and metabolites on the one and peptides and proteins on the other hand, additional specific analyte related challenges are encountered for each of the platforms.

3 Methodological Limitations

As soon as LC-MS/MS became more and more utilized in routine clinical laboratories, potential limitations in the analytical performance of this powerful technology became evident [32]. The selectivity of MS/MS detection was one aspect of this technology that was particularly overestimated during the first years of its application to clinical chemistry [33–35]. The recent debate about the inaccuracy of 25-OH-vitamin D results obtained by LC-MS/MS [36–38] was widely noticed and has highlighted the need for rigorous quality assurance in clinical mass spectrometry. Indeed, quality assurance is a particular challenge in clinical LC-MS/MS applications because the end users themselves implement and validate the methods. Only a few commercial LC-MS/MS assay kits are presently available—additionally, the instrument configurations used to run these assays are extremely heterogeneous.

3.1 Ionization Efficacy Modulation

Physico-chemical processes involved in ion generation and transfer under atmospheric conditions (atmospheric pressure ionization, API) are complex and modulated by a plethora of factors [39]. Generally, API operating under conventional chromatographic flow rates is a rather inefficient process. Even if sophisticated pneumatically assisted ion-sources are employed, only a minute fraction of target analytes become ionized and actually enter the high vacuum area of the mass analyzer [40]. If not operated at its optimum, a significant ion yield fluctuations on a timescale from seconds to minutes can be observed [41]. Hence, compared to UV- or FLD-detection the stability of LC-MS/MS signals has to be considered rather poor. This high degree of variation makes internal standardization mandatory for quantitative LC-MS/MS analyses.

The term "matrix effects" globally refers to the impact of constituents from the evaporated liquid (i.e. originating from the solvents and the sample) on the processes of de-clustering and ionization of analytes within the ion source region [42, 43]. If the ionization yield observed for an analyte in a complex matrix (e.g. supernatant of a serum specimen after protein precipitation) is lower than observed for the target analyte dissolved in a seemingly pure solvent, the term “ion suppression” is used; while “ion enhancement” describes the increase in ionization yield due to matrix constituents when compared to a “pure,” “neutral” or “inert” matrix. It should be noted, however, that “HPLC-grade” solvents such as methanol, acetonitrile or water also interact with analytes during ionization—on the one hand due to their inherent specific chemical properties or on the other hand by containing impurities [44, 45]. Thus, there exist no “matrix free” analyses in LC-MS/MS at all.

Factors that cause ion suppression or enhancement can include salts and hydrophilic small molecules which occupy or provide ions (i.e. hydrogen, sodium or ammonium ions) or compounds affecting the droplet formation as surface active compounds. Also late eluting sample constituents as phospholipids may cause ion suppression, in particular becoming more serious if gradient elution is applied. Matrix effects depend on the mode of ionization used as well as on the selected ion polarity. Generally, positive ion electrospray ionization (positive ESI) exhibits more pronounced effects, while more selective API techniques as negative ESI (which is applicable only for a minority of target analytes), atmospheric pressure chemical ionization (APCI), and atmospheric pressure photo ionization (APPI) methods are affected.

Ion suppression effects are investigated by most analysts during method development and validation by applying post-column infusion of dissolved target analyte into the eluate of the HPLC column via a T-piece (Fig. 1). This infusion generates a sustained background signal in the selected reaction monitoring (SRM—also known as multiple reaction monitoring—MRM) trace of the target analyte. Typically, in this set-up the injection of analyte free biological matrix (e.g. deproteinized plasma) by HPLC leads to a drop in the baseline SRM-signal generated by an infused analyte for seconds or up to several minutes. If the target analyte elutes during this period of baseline depression, a specific analyte is said to be subject to ion suppression. It shall not be overlooked, that this type of experiment is strictly qualitative and does not allow evaluating the impact of the observed ionization attenuation onto the final readout of the assay—the quantitative analysis result. Hence quantitative approaches as spiking experiments or the internal standard balanced matrix factor evaluation procedure should be added to any validation protocol.

Regarding the HPLC settings, it should be generally desired to develop chromatographic conditions which avoid elution of the target analyte during the period of ion suppression within a chromatographic run [46]. However, this approach is typically associated with increased run-times leading to diminished sample throughput and increased instrument costs per analyzed sample. The sample preparation protocol applied in a method in part determines the duration of ion suppression during a chromatographic run: typically laborious methods as solid phase extraction (SPE) or liquid–liquid extraction (LLE) achieve short periods of ion suppression associated with the sample solution solvent front, while simple protein precipitation protocols (only removing one dominant macromolecule class from the specimen) is

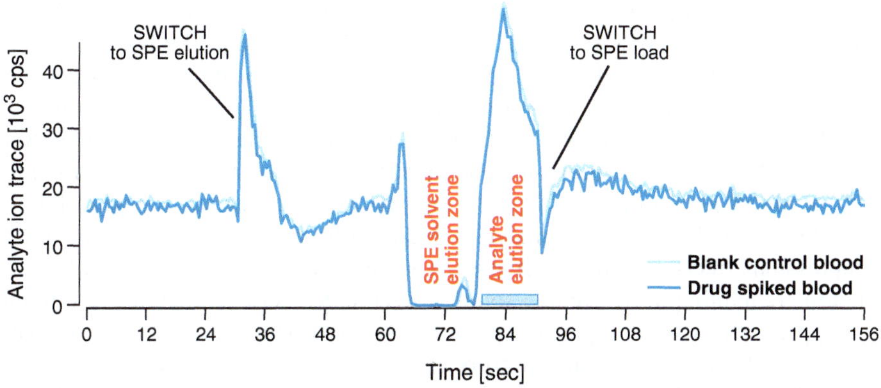

Fig. 1 Using a T-piece between HPLC and MS, a syringe pump can be utilized to infuse a constant flow of analytes at therapeutic concentrations whilst drug spiked whole blood samples are delivered via SPE-HPLC. Ion traces of the analytes (here everolimus at 6 ng/ml) are recorded. In this particular case, no ion yield attenuation due to the spiked drug can be observed, although strong effects can be seen in the solvent front elution zone shortly prior to the analyte elution time window

often associated with longer periods of ion suppression to be associated with remaining low-molecular weight matrix constituents.

Ion suppression is so far mainly considered in the context of sensitivity and the lower limit of quantification of an assay. But it has to be emphasized that short term variations in ion yields—particularly due to matrix components—can compromise the accuracy of analyses: Whenever the variation of ion yield has a differential impact on target analyte and internal standard, accuracy is compromised. This means that the reliability of LC-MS/MS analyses critically depends on (1) how similar the impact of ion suppression or ion enhancement on target analyte and internal standard compound is and on (2) how similar the matrices of calibrator samples and actual patients' samples are with respect to the modulation of ionization efficacy. This problem can be of relevance for an entire measuring series—if systematic differences in the ionization modulation properties of calibration materials and actual patients' samples are present—or it may non-systematically affect individual patients' samples as well.

In general, accuracy of LC-MS/MS analyses will be good if the physicochemical properties of target analyte and internal standard compound are very similar. Stable isotope labelled compounds are ideal internal standards since they have almost identical overall physico-chemical properties compared with their unlabeled counterpart—the analyte. In MS these two species can easily be distinguished by their molecular weight. In labelled compounds typically ^{1}H (hydrogen) is exchanged by ^{2}H (deuterium) or carbon ^{12}C is exchanged by ^{13}C in several positions of the molecule (ideally more than three atoms exchanged). It has to be stressed that the physico-chemical behaviour of labelled and unlabelled compound is *not completely* identical; the term isotope effects refers to these minor differences.

Nevertheless, isotope-dilution mass spectrometry (IDMS), and in particular IDMS-GC-MS are looked upon as “matrix independent” reference methods of extremely high accuracy, because all matrix-related effects are assumed to affect target analyte and labelled internal standard the similar manner [47]. However, it must be noted that substantial isotope effects in the ionization efficacy may be observed with LC-MS/MS, e.g. as previously described for the quantification of carvedilol [48] or piperaquine in plasma [49]. In the first case the internal standard did not completely co-elute with the analyte, which led to significant differences in ion suppression effects in some specimens. In the later case residual amounts of triethylamine remaining in the sample solution after solid-phase extraction were found to suppress the signals piperaquine and its six-fold deuterated internal standard differently, leading to significant errors in the analyte concentrations.

Furthermore, if high ionization temperatures are employed (particularly in APCI), hydrogen–deuterium exchange of deuterated internal standard compounds may occur during the ionization process [50]. For that reason, but also due to its typical location in the molecular “backbone” of an analyte (minimizing the influence on the electronegativity distribution of the scaffold), ^{13}C atoms are considered the more reliable label for isotope dilution internal standardization compared to deuterium.

Unfortunately, for the majority of small molecule LC-MS/MS analyses, stable isotope labelled internal standards are not available so far. In such cases, compounds with a very similar molecular structure typically serve as internal standard (“homologues” or “analogues”). Since the ionization properties are substantially determined by functional groups of a molecule, ionization behaviour may differ significantly—even between compounds with very similar over-all molecular structure. Differential clustering, e.g. with sodium, ammonium or formate ions often present in mobile phases may as well impact the parity of ionization yield between analyte and internal standard. Hence the availability of an appropriate homologue is crucial and critical for the development of reliable LC-MS/MS methods in TDM [51].

A fundamental requirement for LC-MS/MS calibration materials is that matrix effects exerted by these materials are most similar to the matrix effects exerted by actual patients’ sample materials. Lyophilisation, virus inactivation and other procedures applied during the industrial production of calibration and control materials, may notably impact the ionization behaviour of extracts from such samples and can result in differential matrix effects in calibrators and actual patients’ samples. If the internal standard peak areas found for calibration samples systematically differ from those found in patients samples, inappropriateness of the calibration materials should be suspected. However, we have previously observed that calibration materials from different commercial sources lead to inaccurate tacrolimus results in an instrument specific manner, without showing deviations in the internal standard peak area. This effect was most likely related to ionization enhancement affecting the target analyte but not the homologue internal standard (ascomycin) ionization *and* being restricted to calibrator samples. This resulted in systematically low tacrolimus results of clinical samples in one instrument for one specific calibrator lot [52].

Substantial matrix effects can be specific for individual patients' samples, e.g. due to co-administered drugs or other xenobiotics. It is in general tried to detect such samples by assessing the peak areas of the internal standard compound over a whole series: Outliers with respect to the peak area of the internal standard may indicate unusually pronounced matrix effects in an individual sample. In such cases it is speculative if the evident matrix effects affect analyte and internal standard to an identical degree, and it is up to the valuation of the analyst if quantitative results of such samples may be reported or not. Re-analysis after dilution can be useful in case of evident matrix effects—given sufficiently high analyte concentration in the respective sample.

Such decisions are always more or less arbitrary since in many routinely used LC-MS/MS instruments the "normal" variation of IS peak-areas between subsequent samples in a series can be substantial. Moreover, it is always uncertain if matrix effects in an individual sample might impact the target analyte but not at all the internal standard compound. This consideration underscores the importance of the appropriateness of internal standard compounds for reliable LC-MS/MS results. We believe, that matrix related modulation of ionization will remain an important issue in clinical LC-MS/MS and has to be approached with a multiple strategy spanning the whole life time of an assay:

- Matrix effects have to be detected by systematic experiments during method development.
- Matrix effects have to be minimized by adequate sample preparation and chromatography.
- Matrix effects have to be compensated by use of most appropriate internal standard materials.

3.2 *In-source Transformation*

API techniques as electrospray ionization (ESI) and atmospheric pressure chemical ionization (APCI) are generally looked upon as "soft." Nevertheless, if weak bonds are present, molecules can be disintegrated during the process of ionization before the ion enters the mass spectrometer. In bioanalysis, such "in source transformation" is particularly prominent if conjugated metabolites, i.e. if glucuronides or sulphates of target analytes are present in a sample [53]. Consequently, analytical inaccuracy can be easily introduced whenever a target analyte co-elutes together with its conjugate metabolites. (e.g. mycophenolic acid (MPA) with MPA-glucuronide) [34]. In such a case, analyte molecules may be generated from the conjugate metabolites within the ion source by fragmentation. By principle, this effect cannot be compensated by the subsequent SRM based mass spectrometric analysis of the once generated ions. Hence, in general the risk of in-source transformation related inaccuracy is particularly given whenever the selectivity of the chosen chromatographic process is too low to separate a target analyte from its more hydrophilic metabolites. Analyte specific tuning of ionization conditions favouring the ionization of non-conjugated analyte may reduce the impact of in source transformation, but cannot rule out its occurrence in individual samples. Hence, sufficient chromatographic resolution is the most reliable approach for this aim, albeit requiring additional analysis time.

In-source transformation effects can only be perceived during method development if extracts of biological samples are analyzed under extended chromatographic runs. Interference by in-source transformation will not become evident if only spiked quality control materials are used which do not contain relevant metabolites. If one or more additional peak in the SRM trace of the target analyte is observed in comparison to analyte reference solutions, chromatography may only be "accelerated" to such degree that these peaks still remain baseline separated from the analyte.

3.3 Isobaric Mass Transitions and Isomers

In *single-stage* LC-MS, potential inaccuracy due to isobaric constituents present in complex biological samples is well recognized, since isobaric compounds share one nominal molecular mass and are in general indistinguishable by mere molecular mass based detection. Isobars are either structural isomers of the target analyte sharing the same elemental formula or are structurally not related compounds sharing the same nominal molecular mass with or without sharing the same molecular formula.

Using LC-MS/MS, the probability that an analyte (or its internal standard compound) shares both precursor and product ion masses with other unrelated compounds present in the sample is far smaller than the probability of isobar effects in single stage LC-MS with mere molecular weight selective detection. Nevertheless, due to the extreme complexity of the metabolome and proteome, isobaric mass transitions are a ubiquitous potential source of analytical inaccuracy in clinical LC-MS/MS applications [54]. It must be noted, that shared mass transition of analyte and potential disturber may not necessarily be one of the most favoured mass transition of the latter: If the disturber is present in much higher concentrations compared to the target analyte, substantial problems may result even if only a less populated fragment ion of the disturber is shared by the target analyte.

With at least two different SRM-transitions available for the analysis of a compound, one transition can be used for quantification ("quantifier") and the other one for result confirmation ("qualifier"). If the branching ratio of quantifier to qualifier is found outside pre-defined ranges of acceptance, it is typically tried to apply a re-analysis with a more extensive chromatographic separation in order to overcome co-elution of interfering compound and analyte. This quantifier–qualifier-principle is used extensively in GC-MS and is now often applied for quantitative LC-MS/MS application too—especially in legally strictly regulated environments as forensic toxicology or pesticide and mycotoxin analysis in feed and foodstuff [55]. In contrast, in small molecule clinical LC-MS/MS applications this principle is only rarely applied at present [56, 57]. Besides the fact that for many analytes collision induced dissociation only generates one single dominant product ion of acceptable intensity, this is mainly for practical reasons: Branching ratios do reflect the fate of a molecule ion whilst travelling through the mass spectrometer. Although the kinetics of a single ion-molecule reaction is an universal property, the influence of instrument parameters (e.g. type and geometry of the collision cell, collision gas nature and distribution) is hardly predictable and can lead to distinctively different fragment ion yield distributions (=quantitative fragmentation patters) [58, 59]. Hence, SRM yields and ratios of SRM yields (branching ratios) also differ from instrument to instrument and even can show distinct matrix effects [60].

Consequently, branching ratios have to be assessed with matrix matches solutions of the target compound regularly and once accepted branching ratios have to be re-assessed on a regular basis. An alternative approach to branching ratios would be the confirmation of a quantitative result by the second SRM transition recorded (or ideally two separate quantifications for different transitions of the target analyte and the internal standard, respectively). Again, this second quantitative result has to be within set thresholds to define the result valid. Obviously, the accepted degree of deviation is arbitrary, and in case of discrepancies it is not evident which result might be correct. For these reasons also the "double quantification" strategy for result validation is used in only few laboratories, too.

If analyte and interfering compounds share the same nominal mass but differ in their elemental formula, their exact mass is different due to the pattern of naturally occurring isotopes of elements. Hence, MS instruments with high mass resolution can distinguish compounds with identical nominal mass but different elemental formula. Relying on sophisticated software algorithms, such platforms can be used to predict the elemental formula of unknown compounds based on a single accurate mass measurement—although some limitations have to be taken into account [61] and fragment ions have to be recorded to allow unequivocal analyte identification [62]. However, mass spectrometers offering such exact mass determination (e.g. TOF, FT-ICR and orbitrap instruments) are still rarely applied in quantitative clinical routine assays.

In the case of analytes with identical elemental formulas ("true isobars") as positional or geometrical isomers (e.g. 11-hydroxycortisol and 21-hydroxycortisol or testosterone and epi-testosterone)—a discrimination of analyte and interfering compound is not possible with any mass analyzer—even if enabling highest mass resolution. Potentially, the disintegration pattern of isomers may be different, allowing analytical discrimination but in most cases chromatographic baseline separation of analyte and isomer prior to their MS/MS detection is required for unequivocal quantitative measurement.

We previously demonstrated potential interference in the measurement of cyclosporine A due to an ion source decay of a cyclosporine A metabolite. This breakdown resulted in a fragment isobaric to cyclosporine D the widely used internal standard in cyclosporine A measurement [35]. This case, which has significant clinical relevance, demonstrates again, that isobaric metabolite related interferences in TDM can only be studied and disclosed in post-dose patients' samples.

In the early years of LC-MS/MS application in clinical laboratories, chromatographic separation was looked upon as rather unnecessary with tandem mass spectrometers being understood as extremely selective measuring devices. Thus, many LC-MS/MS methods with minimal degree of chromatographic resolution and analyte retention times close to the void time of the chromatographic systems ("dilute and shoot" approaches) have been described. However, from the issues discussed so far, the requirements of proper sample preparation and sufficient chromatographic separation prior to MS/MS detection have become evident.

In this chapter we have so far reviewed issues which can introduce inaccuracy into LC-MS/MS analyses and which are specifically related to this technology. It must be noted, however, that LC-MS/MS analyses in clinical pharmacology at present extensively involve manual handling steps which are prone to gross errors—independent of the analytical technology used. Such gross handling errors include: permutation of samples during sample preparation; permutation in the positioning of sample vials in autosampler racks (positive sample identification is not realized yet); incorrect transfer of data from report printouts to a LIMS (in most laboratories no LIMS connection of the chromatography software is implemented). In general it must be emphasized that the risk of gross errors is probably higher for LC-MS/MS compared to analyses run on standard clinical chemistry analyzers with positive sample identification from sample introduction to LIMS-transfer of the data. Consequently, progressive automation is essential to improve the reliability of LC-MS/MS analyses in clinical pharmacology.

Systematic error can also be introduced in the process of standardization; this is in particular true for research analyses in pharmacological research: Calibration materials and QC materials are typically prepared by spiking drug-free serum or plasma samples with solutions of the target analyte to desired concentration levels. The purity of drug compound specimens used for this process might be incorrect (e.g. due to sample breakdown). Furthermore some compounds require a high content of organic solvent in the stock solution for complete dissolvation; during the spiking of serum samples for the production of calibrator materials, the matrix of the spiking solution can induce local protein precipitation with loss of the analyte by co-precipitation. Such problem can be addressed by external quality assessment based on reference preparations or proficiency testing schemes which, however, are rarely available for innovative and research analytes.

3.4 Therapeutic Drug Monitoring: Immunosuppressants

A low therapeutic quotient, narrow therapeutic bandwidths, drastic side effects when leaving these, massive drug interactions, high intra- and inter-individual variability in metabolization, and the resulting loss of dosage-exposure relationship require tight control of the monitoring of whole blood-drug levels in the therapy with cyclosporine A, tacrolimus, sirolimus, everolimus. Particularly during the first few weeks following transplantation finding the right dosage is often difficult. Measurements must be fast, precise and true. Only the smallest possible laboratory errors—negligible imprecision and small inaccuracy—permit dosage adjustments at a narrow therapeutic target range. When systematic (e.g. calibration bias) and intra-individual (biologic) variability are added to this measurement dispersion entirely common in the practice, analysis platforms of this kind are no longer able to meet clinical requirements [63, 64]. The fact that modern therapeutic schemes (e.g. combination therapies, dosage reduction in cases of partial immunologic tolerance) [65, 66] mean an additional demand to measure lower drug levels more correctly than in the past, represents a challenge to be met by all commercial providers and manufacturers of "in-house assays" alike.

During the last years two distinctly different LC-MS/MS instrument configurations have emerged as suitable for immunosuppressant TDM. In each case sample preparation by precipitation of cellular components and serum proteins precedes chromatographic analysis. Similar to some immunoassays, a mixture of an aqueous zinc sulphate solution with methanol is used in many cases. The subsequent LC separation of the analytes also serves to purify the sample separating the analytes from hydrophilic non-precipitated matrix components (e.g. zinc sulphate) as well as from lipophilic matrix components (e.g. phospholipids). Sufficiently good assay results can be achieved with only one single, simple chromatographic step (LC-MS/MS) as long as important details are taken into account.

As stated further above, under all circumstances trueness, accuracy and sensitivity of the assay should be demonstrated on a sufficient number of patient samples. In our view, however, the use of an additional, also easily realizable chromatographic dimension (online-SPE-LC-MS/MS) [67, 68] without a doubt represents the analytical "state of the art" in immunosuppressant TDM. Nowadays tandem MS instruments are used almost exclusively for the detection and quantification of analytes. The detection of analytes is generally performed in the "selected reaction monitoring" (SRM, synonym MRM) mode. Depending on which instrumentation is used, an analysis can be completed within two to four minutes.

In spite of all the efforts with assay establishment described in the previous paragraphs and even taking into account the limitation of the method LC-MS/MS is an entirely valuable alternative to immunoassays. The statistics of the UK-NEQAS proficiency testing (PT) scheme (http://www.bioanalytics.co.uk), the largest and most significant PT scheme in immunosuppressant TDM, show that there are at least one hundred operational and active LC-MS/MS installations, corresponding to 20–50 % of the respective total [69]. The extraordinarily high number of challenges (per year 12 for cyclosporine, tacrolimus and sirolimus, 6 for everolimus, 4 for MPA) and the regular inclusion of patient samples (pool samples) make the UK-NEQAS results a valuable database to allow a critical evaluation of the capability of the analysis platforms. Since in the PT evaluation the current immunoassays as well as the LC-MS/MS platforms are gathered in separate sub-groups, such comparisons can be easily performed. The trueness analysis (comparison to the weighted analyte amount in standard addition samples or comparison to substance-specific measuring LC-MS/MS group) as well as the analysis of dispersion within one group (assessment of assay inaccuracy) permit valuable conclusions on the capability of measurement systems under routine conditions.

For instance, if comparing sirolimus assays it is striking that the CV of the LC-MS group is clearly above that of the sirolimus-CMIA group. An additional assay imprecision of approximately 7 % can be observed. It is independent from the analyte concentration; hence it does not represent a loss of assay sensitivity at a lowered analyte concentration, but an additional contribution to the measurement uncertainty, which only manifests itself in the total picture of the inter-laboratory test. The causes for this are unclear. However, one may speculate that, especially in the area of sirolimus TDM, where for several of the past years many of the laboratories have used chromatographic methods for lack of alternatives [70], outdated technologies (e.g. LC-MS instead of LC-MS/MS) and heterogeneous (in-house) calibration systems (single point vs. multipoint calibrations, different internal standards, etc.) are being used. Here, the medium-term goal of LC-MS/MS platform development must be an improvement of the inter-laboratory PT CV to <10 % in the therapeutic range.

4 Outlook/Conclusion

In summary LC-MS/MS has doubtlessly a high inherent potential for selectivity and accuracy. However, application of this technology is not automatically or necessarily translated into accurate results. Its pitfalls have to be recognized and must be addressed systematically. In particular interferences from in-source transformation of metabolites, differential matrix effects of analyte and internal standard and isobaric transitions can lead to inaccurate results of LC-MS/MS analyses. Further technological developments will probably help to make LC-MS/MS assays more robust towards such interferences, but clinical chemists have to remain watchful for inaccuracies also with powerful and fascinating technologies

References

1. Vogeser M, Seger C (2008) A decade of HPLC-MS/MS in the routine clinical laboratory-goals for further developments. Clin Biochem 41:649–662
2. Seger C, Tentschert K, Stöggl W, Griesmacher A, Ramsay SL (2009) A rapid HPLC-MS/MS method for the simultaneous quantification of cyclosporine A, tacrolimus, sirolimus and everolimus in human blood samples. Nat Protoc 4:526–534
3. Harms E, Olgemöller B (2011) Neonatal screening for metabolic and endocrine disorders. Dtsch Arztebl Int 108:11–22
4. Lehotay DC, Hall P, Lepage J, Eichhorst JC, Etter ML, Greenberg CR (2011) LC–MS/MS progress in newborn screening. Clin Biochem 44:21–31
5. Yang Z, Peng Y, Wang S (2005) Immunosuppressants: pharmacokinetics, methods of monitoring and role of high performance liquid chromatography/mass spectrometry. Clin Appl Immunol Rev 5:405–430
6. Seger C, Griesmacher A (2007) Some important aspects of implementing tandem mass spectrometry in a routine clinical laboratory environment. Biochem Med 17:29–51

7. Mauerer HH (2007) Current role of liquid chromatography-mass spectrometry in clinical and forensic toxicology. Anal Bioanal Chem 388:1315–1325
8. Peters FT (2011) Recent advances of liquid chromatography–(tandem) mass spectrometry in clinical and forensic toxicology. Clin Biochem 44:54–65
9. Kamel A, Prakash C (2006) High performance liquid chromatography/atmospheric pressure ionization/tandem mass spectrometry (HPLC/API/MS/MS) in drug metabolism and toxicology. Curr Drug Metab 7:837–852
10. Stokvis E, Rosing H, Beijnen JH (2005) Liquid chromatography-mass spectrometry for the quantitative bioanalysis of anticancer drugs. Mass Spectrom Rev 24:887–917
11. Baumann P, Hiemke C, Ulrich S, Eckermann G, Gaertner I, Gerlach M, Kuss HJ, Laux G, Müller-Oerlinghausen B, Rao ML, Riederer P, Zernig G (2004) The AGNP-TDM expert group consensus guidelines: therapeutic drug monitoring in psychiatry. Pharmacopsychiatry 37:243–265
12. Vogeser M, Parhofer KG (2007) Liquid chromatography tandem-mass spectrometry (LC-MS/MS) – technique and applications in endocrinology. Exp Clin Endocrinol Diabetes 115:555–570
13. Kushnir MM, Rockwood AL, Bergquist J (2010) Liquid chromatography-tandem mass spectrometry applications in endocrinology. Mass Spectrom Rev 29:480–502
14. Kushnir MM, Rockwood AL, Roberts WL, Yue B, Bergquist J, Meikle W (2011) Liquid chromatography tandem mass spectrometry for analysis of steroids in clinical laboratories. Clin Biochem 44:77–88
15. Hoofnagle AN (2010) Quantitative clinical proteomics by liquid chromatography-tandem mass spectrometry: assessing the platform. Clin Chem 56:161–164
16. Hortin GL (2005) Can mass spectrometric protein profiling meet desired standards of clinical laboratory practice? Clin Chem 51:3–5
17. Carr SA, Anderson L (2008) Protein quantitation through targeted mass spectrometry: the way out of biomarker purgatory? Clin Chem 54:1749–1752
18. Anderson NL (2010) The clinical plasma proteome: a survey of clinical assays for proteins in plasma and serum. Clin Chem 56:177–185
19. Hoofnagle AN, Wener MH (2009) The fundamental flaws of immunoassays and potential solutions using tandem mass spectrometry. J Immunol Methods 347:3–11
20. Hortin GL, Carr SA, Anderson NL (2010) Introduction: advances in protein analysis for the clinical laboratory. Clin Chem 56:159–161
21. Addona TA, Abbatiello SE, Schilling B, Skates SJ, Mani DR, Bunk DM et al (2009) Multi-site assessment of the precision and reproducibility of multiple reaction monitoring-based measurements of proteins in plasma. Nat Biotechnol 27:633–641
22. Anderson L, Hunter CL (2006) Quantitative mass spectrometric multiple reaction monitoring assays for major plasma proteins. Mol Cell Proteomics 5:573–588
23. Kumar V, Barnidge DR, Cheng LS, Twentyman JM, Cradic KW, Grebe SK, Singh RJ (2010) Quantification of serum 1–84 parathyroid hormone in patients with hyperparathyroidism by immunocapture in situ digestion liquid chromatography–tandem mass spectrometry. Clin Chem 56:306–313
24. Hoofnagle AN, Becker JO, Wener MH, Heinecke JW (2008) Quantification of thyroglobulin, a low-abundance serum protein, by immunoaffinity peptide enrichment and tandem mass spectrometry. Clin Chem 54:1796–1804
25. Rodriguez-Cabaleiro D, van Uytfanghe K, Stove V, Fiers T, Thienpont LM (2007) Pilot study for the standardization of insulin immunoassays with isotope dilution liquid chromatography/tandem mass spectrometry. Clin Chem 53:1462–1469
26. Daniel YA, Turner C, Haynes RM, Hunt BJ, Dalton RN (2007) Quantification of hemoglobin A2 by tandem mass spectrometry. Clin Chem 53:1448–1454
27. Bondar OP, Barnidge DR, Klee EW, Davis BJ, Klee GG (2007) LC-MS/MS quantification of Zn-alpha2 glycoprotein: a potential serum biomarker for prostate cancer. Clin Chem 53:673–678
28. Rauh M, Gröschl M, Rascher W (2007) Simultaneous quantification of ghrelin and desacyl-ghrelin by liquid chromatography-tandem mass spectrometry in plasma, serum, and cell supernatants. Clin Chem 53:902–910

29. Keshishian H, Addona T, Burgess M, Mani DR, Shi X, Kuhn E et al (2009) Quantification of cardiovascular biomarkers in patient plasma by targeted mass spectrometry and stable isotope dilution. Mol Cell Proteomics 8:2339–2349
30. Kuhn E, Addona T, Keshishian H, Burgess M, Mani DR, Lee RT et al (2009) Developing multiplexed assays for troponin I and interleukin-33 in plasma by peptide immunoaffinity enrichment and targeted mass spectrometry. Clin Chem 55:1108–1117
31. Keshishian H, Addona T, Burgess M, Kuhn E, Carr SA (2007) Quantitative, multiplexed assays for low abundance proteins in plasma by targeted mass spectrometry and stable isotope dilution. Mol Cell Proteomics 6:2212–2229
32. Vogeser M, Seger C (2010) Pitfalls associated with the use of liquid chromatography–tandem mass spectrometry in the clinical laboratory. Clin Chem 56:1234–1244
33. Allen KR (2006) Interference by venlafaxine ingestion in the detection of tramadol by liquid chromatography linked to tandem mass spectrometry for the screening of illicit drugs in human urine. Clin Toxicol 44:147–153
34. Vogeser M, Zachoval R, Spöhrer U, Jacob K (2001) Potential lack of specificity using electrospray tandem-mass spectrometry for the analysis of mycophenolic acid in serum. Ther Drug Monit 23:722–724
35. Vogeser M, Spöhrer U (2005) Pitfall in the high-throughput quantification of whole blood cyclosporin A using liquid chromatography-tandem mass spectrometry. Clin Chem Lab Med 43:400–402
36. Singh RJ (2008) Are clinical laboratories prepared for accurate testing of 25-hydroxy vitamin D? Clin Chem 54:221–223
37. Carter GD, Jones JC (2009) Use of a common standard improves the performance of liquid chromatography-tandem mass spectrometry methods for serum 25-hydroxyvitamin-D. Ann Clin Biochem 46:79–81
38. Carter GD (2009) 25-Hydroxyvitamin D assays: the quest for accuracy. Clin Chem 55:1300–1302
39. Kebarle P, Verkerk UH (2009) Electrospray: from ions in solution to ions in the gas phase, what we know now. Mass Spectrom Rev 28:898–917
40. Covey TR, Thomson BA, Schneider BB (2009) Atmospheric pressure ion sources. Mass Spectrom Rev 28:870–897
41. Lagerwerf FM, van Dongen WD, Steenvoorden RJJM, Honing M, Jonkman JHG (2000) Exploring the boundaries of bioanalytical quantitative LC–MS–MS. Trends Anal Chem 19:418–427
42. Annesley TM (2003) Ion suppression in mass spectrometry. Clin Chem 49:1041–1044
43. Taylor PJ (2005) Matrix effects: the Achilles heel of quantitative high-performance liquid chromatography-electrospray-tandem mass spectrometry. Clin Biochem 38:328–334
44. Keller BO, Sui J, Young AB, Whittal RM (2008) Interferences and contaminants encountered in modern mass spectrometry. Anal Chim Acta 627:71–81
45. Guo X, Bruins AP, Covey TR (2006) Characterization of typical chemical background interferences in atmospheric pressure ionization liquid chromatography-mass spectrometry. Rapid Commun Mass Spectrom 20:3145–3150
46. Gustavsson E, Andersson M, Stephanson N, Beck O (2007) Validation of direct injection electrospray LC-MS/MS for confirmation of opiates in urine drug testing. J Mass Spectrom 42:881–889
47. Thienpont LM, Van Uytfanghe K, Blincko S, Ramsay CS, Xie H, Doss RC et al (2008) State-of-the-art of serum testosterone measurement by isotope dilution-liquid chromatography-tandem mass spectrometry. Clin Chem 54:1290–1297
48. Wang S, Cyronak M, Yang E (2007) Does a stable isotopically labeled internal standard always correct analyte response? A matrix effect study on a LC/MS/MS method for the determination of carvedilol enantiomers in human plasma. J Pharm Biomed Anal 43:701–707
49. Lindegardh N, Annerberg A, White NJ, Day NP (2008) Development and validation of a liquid chromatographic-tandem mass spectrometric method for determination of piperaquine in

plasma stable isotope labeled internal standard does not always compensate for matrix effects. J Chromatogr B Analyt Technol Biomed Life Sci 862:227–236
50. Stokvis E, Rosing H, Beijnen JH (2005) Stable isotopically labeled internal standards in quantitative bioanalysis using liquid chromatography/mass spectrometry: necessity or not? Rapid Commun Mass Spectrom 19:401–407
51. Sauvage FL, Stanke-Labesque F, Gagnieu MC, Jourdil JF, Babany G, Marquet P (2009) Feasibility of ribavirin therapeutic drug monitoring in hepatitis C. Ther Drug Monit 31:374–381
52. Vogeser M (2008) Instrument-specific matrix effects of calibration materials in the LC-MS/MS analysis of tacrolimus. Clin Chem 54:1406–1408
53. Yan Z, Caldwell GC, Jones WJ, Masucci JA (2006) Cone voltage induced in-source dissociation of glucuronides in electrospray and implications in biological analyses. Rapid Commun Mass Spectrom 17:1433–1442
54. Sherman J, McKay MJ, Ashman K, Molloy MP (2009) How specific is my SRM? The issue of precursor and product ion redundancy. Proteomics 9:1120–1123
55. Careri M, Mangia A (2006) Validation and qualification: the fitness for purpose of mass spectrometry-based analytical methods and analytical systems. Anal Bioanal Chem 386:38–45
56. Rauh M, Gröschl M, Rascher W, Dörr HG (2006) Automated, fast and sensitive quantification of 17 alpha-hydroxy-progesterone, androstenedione and testosterone by tandem mass spectrometry with on-line extraction. Steroids 71:450–458
57. Carvalho VM, Nakamura OH, Vieira JG (2008) Simultaneous quantitation of seven endogenous C-21 adrenal steroids by liquid chromatography tandem mass spectrometry in human serum. J Chromatogr B Analyt Technol Biomed Life Sci 872:154–161
58. Oberacher H, Pavlic M, Libiseller K, Schubert B, Sulyok M, Schuhmacher R, Csaszar E, Köfeler HC (2009) On the inter-instrument and inter-laboratory transferability of a tandem mass spectral reference library: 1. Results of an Austrian multicenter study. J Mass Spectrom 44:485–493
59. Hopley C, Bristow T, Lubben A, Simpson A, Bull E, Klagkou K, Herniman J, Langley J (2008) Towards a universal product ion mass spectral library - reproducibility of product ion spectra across eleven different mass spectrometers. Rapid Commun Mass Spectrom 22:1779–1786
60. Kaufmann A, Butcher P, Maden K, Widmer M, Giles K, Uria D (2009) Are liquid chromatography/electrospray tandem quadrupole fragmentation ratios unequivocal confirmation criteria? Rapid Commun Mass Spectrom 23:985–998
61. Kind T, Fiehn O (2006) Metabolomic database annotations via query of elemental compositions: mass accuracy is insufficient even at less than 1 ppm. BMC Bioinform 7:234
62. Mylonas R, Mauron Y, Masselot A, Binz PA, Budin N, Fathi M et al (2009) X-Rank: a robust algorithm for small molecule identification using tandem mass spectrometry. Anal Chem 81:7604–7610
63. Steimer W (1999) Performance and specificity of monoclonal immunoassays for cyclosporine monitoring: how specific is specific? Clin Chem 45:371–381
64. Schütz E, Svinarov D, Shipkova M, Niedmann PD, Armstrong VW, Wieland E et al (1998) Cyclosporin whole blood immunoassays (AxSYM, CEDIA, and Emit): a critical overview of performance characteristics and comparison with HPLC. Clin Chem 44:2158–2164
65. Ekberg H, Tedesco-Silva H, Demirbas A, Vítko S, Nashan B, Gürkan A et al (2007) ELITE-Symphony Study. Reduced exposure to calcineurin inhibitors in renal transplantation. N Engl J Med 357:2562–2575
66. Srinivas TR, Meier-Kriesche HU (2008) Minimizing immunosuppression, an alternative approach to reducing side effects: objectives and interim result. Clin J Am Soc Nephrol 3:S101–S116
67. Ceglarek U, Lembcke J, Fiedler GM, Werner M, Witzigmann H, Hauss JP, Thiery J (2004) Rapid simultaneous quantification of immunosuppressants in transplant patients by turbulent flow chromatography combined with tandem mass spectrometry. Clin Chim Acta 346:181–190

68. Seger C, Tentschert K, Stöggl W, Griesmacher A, Ramsay SL (2009) A rapid HPLC-MS/MS method for the simultaneous quantification of cyclosporine A, tacrolimus, sirolimus and everolimus in human blood samples. Nat Protoc 4:526–534
69. Seger C, Vogeser M (2010) Immunosuppressant drug monitoring – a routine undertaking? J Lab Med 34:117–124
70. Holt D, Moreton M, Laamanen K, Johnston A (2005) A microparticle enzyme immunoassay to measure sirolimus. Transplant Proc 37:182–184

Therapeutic Drug Monitoring to Support Clinical Pharmacogenomics

Alan H.B. Wu and Kara L. Lynch

Abstract The implementation of pharmacogenomics can improve the efficacy of therapeutic drugs while reducing the incidence of side effects and drug toxicity. Therapeutic drug monitoring is well accepted and widely practiced for many drugs and is also relevant for drugs for which pharmacogenomic testing is needed. Tamoxefin is metabolized by CYP 2D6 to endoxifen, clopidogrel by CYP 2C19 to thiol-containing active metabolite, and opioid drugs by 2D6 to morphine and other metabolites. For these drugs, genetic testing can be used to predict efficacy for breast cancer outcomes, freedom from cardiovascular events, and adequate pain control, respectively. Therapeutic drug monitoring (TDM) can be used to determine drug compliance, especially for the opioids which have street value and can be diverted as a drug of abuse. Drug levels can be used to titrate drug dosage for individuals who are shown to be sub-therapeutic. TDM can also improve efficacy for tamoxifen for patients taking drug inhibitors, and be useful for determining the mechanism of clopidogrel resistance (i.e., pharmacokinetics vs. pharmacodynamics). Since there are no specific immunoassays for these drugs and metabolites for serum measurements, liquid chromatography/mass spectrometric methods will be necessary to implement TDM.

1 Introduction

"Personalized medicine" is a new medical approach that attempts to personalize medical treatment to the specific needs of the patient. Regarding drug therapeutics, it is a move from "one size fits all" to the "right drug" at the "right dosage" to the

A.H.B. Wu, Ph.D. (✉) • K.L. Lynch, Ph.D.
Department of Laboratory Medicine, University of California-San Francisco, San Francisco, CA, USA

San Francisco General Hospital, San Francisco, CA, USA
e-mail: wualan@labmed2.ucsf.edu

Q. Alan Xu and Timothy L. Madden (eds.), *LC-MS in Drug Bioanalysis*,
DOI 10.1007/978-1-4614-3828-1_6,

Table 1 Relevant pharmacogenomic drugs and possible phenotypic assessments

Drug	Genotype target	Phenotype target
Warfarin	CYP 2C9 and VKORC1[a]	Prothrombin time/INR[b]
Azothioprine	TPMT[c]	Red cell TPMT
Tamoxifen	CYP 2D6	Endoxifen
Clopidogrel	CYP 2C19	R-130964 metabolite
		Platelet aggregometry
Codeine	CYP 2D6	Morphine

[a]Vitamin K reductase epoxide complex
[b]International normalized ratio
[c]Thiopurine methyltransferase

"right person" at the "right time." Pharmacogenomics is a science that predicts therapeutic efficacy and toxicity avoidance by tailoring drug therapy according to an individual's genetic makeup. Polymorphisms in the genes that encode the enzymes necessary in the inter- and intracompartmental transportation, metabolism, and excretion of therapeutic drugs can have a major effect on the performance of a drug in clinical practice. Currently, genotyping for the family of cytochrome (CYP) microsomal enzyme systems is important in identifying individuals as slow, intermediate and ultrarapid metabolizers relative to the wild type. There are specific cytochrome isoenzymes that are responsible for the majority of the drugs that are detoxified by the liver. Homozygous subjects who are slow metabolizers produce enzymes that are defective and have low enzyme activity. Homozygous intermediate metabolizers produce enzymes that have reduced function. Alternately, an intermediate metabolizer could result from a heterozygous individual who has one wild-type copy and one null gene. Ultrarapid metabolizers have more "gene duplications," i.e., more than the usual two copies of the CYP enzyme.

An individual's germ line "genotype" is created at birth and remains largely unchanged during life. The major advantage of genotyping is that a determination can be made in the absence of the drug itself. However, while the coded protein is defined by the genotype, an individual's "phenotype" is dependent on genetic and nongenetic factors. The level of RNA expression for an individual can be variable in the amount of protein produced, and greatly affect the individual's phenotype. It has been known for many years that the expression of the CYP enzymes from the liver can be enhanced or inhibited by the presence of other drugs. Thus, for clinical pharmacogenomics, measurement of the concentration of the drug or its metabolites may be a better reflection of its enzyme activity. A disadvantage of drug measurements is that for an accurate assessment of concentrations, the individual must be regularly taking the drug at the prescribed dosage, and be at pharmacologic steady state. During the drug induction phase, this might expose the individual to unwarranted and unexpected side effects for some medications. For drugs with long half-lives, there may be a significant delay in the assessment of efficacy by therapeutic drug monitoring (TDM) levels. Clearly, the effective therapeutic management of many therapeutic drugs today is a combination of genotypes with phenotypes. Table 1 lists some of the drugs that warrant pharmacogenomic testing, and the relevant phenotypic targets.

2 LC-MS/MS to Support Clinical Pharmacogenomics

2.1 *Tamoxifen*

2.1.1 Pharmacology and Pharmacogenetics

Tamoxifen is a widely available hormonal adjuvant therapy for women with breast cancer, and is most effective on tumors that are positive for estrogen and progesterone receptors. While alternative drugs are available, such as the aromatase inhibitors, tamoxifen remains the drug of choice for premenopausal breast cancer patients. Tamoxifen exists as a prodrug that must be converted to endoxifen, the metabolite that has about 100-fold higher activity in blocking estrogen receptors than the parent drug [1]. The conversion of tamoxifen to endoxifen occurs through two metabolic pathways. The major pathway is catalyzed principally by CYP 3A4/5 and first to *N*-desmethyltamoxifen which has minimal biologic activity, and then to endoxifen by CYP 2D6. The minor pathway is conversion by CYP 2D6 to 4-hydroxytamoxifen, which has significant biologic activity but is present in low blood concentrations, and then to endoxifen by CYP 3A4/5 [2]. The rate-limiting step in either pathway is the enzyme CYP 2D6. Individuals who have CYP 2D6 genotypes that encode a null enzyme (e.g., *2D6 *3* through **8*) have reduced concentrations of endoxifen relative to wild-type individuals. Those who have a reduce enzyme activity genotypes (e.g., *2D6 *9, *10, *17,* and **29*) also have lower endoxifen concentrations. Retrospective clinical trials involving tamoxifen have shown that patients who are poor (2D6 **4/*4*) or intermediate (2D6 **10/*10*) have significantly shorter time to breast cancer recurrence and worse relapse-free survival [3, 4]. Other researchers have shown that there is a trend towards reduced endoxifen concentrations, with the highest concentrations seen in ultra (2D6 gene duplication) and extensive metabolizers (**1/*1*), successively lower levels with one or more copies of the reduced activity genes (**10*), and lowest concentrations with one or more copies of the null genes (**10*). The hypothesis is that worsening clinical outcomes are associated with decreasing endoxifen concentrations. Madlensky et al. recently showed that when patients are divided into quintiles, there is a threshold effect, i.e., only in the lowest endoxifen quintile of values was associated with poor outcomes [5]. As such, there are no consequences of having an excess endoxifen concentration, such as expected for individuals who have more than two copies of the 2D6 wild-type gene and are ultrarapid drug metabolizers.

2.1.2 Nongenetic Factors

While genotyping is an important first step in predicting individuals who will not benefit from tamoxifen treatment, there are nongenetic factors that also result in lowering of tamoxifen concentrations. The most widely studied are the class of drugs that are serotonin selective reuptake inhibitors (SSRI) such as paroxetine and fluoxetine [6]. These drugs are used to treat depression, hot flashes, and vaginal dryness and are potent inhibitors of CYP2D6 enzyme activity. Borges et al. showed that endoxifen concentrations among 2D6 wild-type patients who were 2D6 genotypes and were on SSRIs had endoxifen concentrations that were as low as poor metabolizers [7]. In a population-based study of 2,430 breast cancer patients treated with tamoxifen, those on a single SSRI had an increased risk of cancer deaths [8]. These and other investigators have strongly suggested the discontinuance of these drugs. Patients may not readily disclose their antidepressant drug use to their oncologists due to the stigma associated with depression. While the information for SSRIs is now well documented, there are many other potential 2D6 enzyme inhibitors that are taken by breast cancer patients. Herbal medications in particular are widely used and may have adverse effects on tamoxifen metabolism. While Wu et al. showed no difference with self-reported soy food intake, there are many other herbals that have not yet been tested [9]. In the absence of outcome evidence from clinical trials, there may be a role in routine therapeutic drug monitoring for the active metabolite. LC-MS measurement for the occult use of SSRI may also be warranted.

2.1.3 Analytical Assays and TDM Testing

LC/tandem MS assays for tamoxifen metabolites have been described by numerous investigators and used for research purposes [10–12]. Figure 1 shows a typical chromatogram for the analyses of these metabolites from a serum sample. Following chronic administration of a standard 20 mg tamoxifen per day dosage on presumably wild-type subjects, typical serum concentrations at steady state were 150 (±50) ng/mL for tamoxifen, 180 (±70) ng/mL for *N*-desmethytam, 2.5 (±1.2) ng/mL for 4-hydroxytam, and 5.0 (±2.5) ng/mL for endoxifen [10].

Assays for tamoxifen and its metabolites are currently not used for routine therapeutic drug monitoring. Because the half-life of tamoxifen is long, TDM for tamoxifen has the disadvantage over genotyping in that steady state concentrations are not reached for 2–3 months. A significant delay in prescribing the most effective adjuvant therapeutic regimen (i.e., tamoxifen vs. aromatase inhibitors) may have an impact on disease-free survival. Therefore, a combination of genotyping and phenotyping might be the best approach. Figure 2 shows a proposed algorithm combining the attributes of both strategies. This scheme has not been clinically validated nor is it endorsed by any clinical practice groups. An alternative use of TDM is to evaluate the effectiveness of increasing the tamoxifen dose from the standard 20 to 40 mg/day for individuals who are intermediate metabolizers. The National Cancer Institute is evaluating this protocol in a prospective trial [13].

2.2 *Clopidogrel*

2.2.1 Pharmacology and Pharmacogenomics

Clopidogrel is part of the thienopyridine class of antiplatelet drugs that are widely used to treat patients with cardiovascular disease. The American College of Cardiology has recommended antiplatelet medications for patients after angioplasty

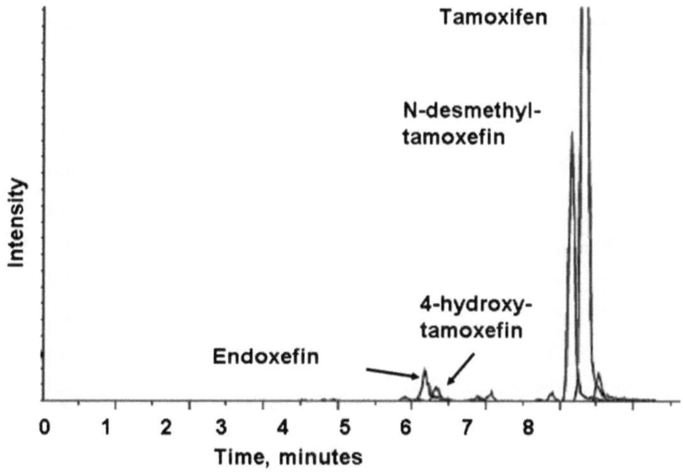

Fig. 1 Representative chromatograms of the LC/MS/MS analysis of tamoxifen and its principal metabolites from serum of a patient with breast cancer. EM, IM, PM, extensive, intermediate, and poor metabolizers, respectively, for 2D6. TDM, therapeutic drug monitoring. Tamoxifen 205 ng/mL, endoxifen 32.8 ng/mL, *N*-desmethyl tamoxefin 320 ng/mL, 4-ohydroxytamoxefin 5.6 ng/mL

Fig. 2 Proposed testing algorithm for tamoxifen therapy in breast cancer

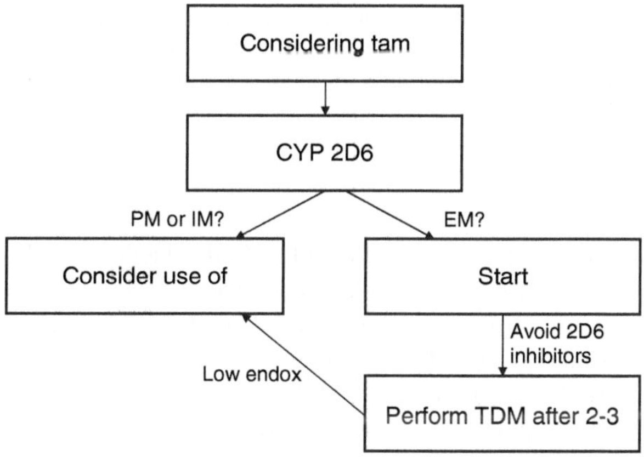

and stent placement [14]. The drug binds to the $P2Y_{12}$ platelet receptor thereby blocking the actions of agonists such as adenosine diphosphate (ADP). Like tamoxifen, clopidogrel is a prodrug that must be converted to an active metabolite for full pharmacologic action. Clopidogrel is first metabolized to 2-oxoclopidogrel by CYP 1A2, 2C19, and 2B6 and then to the R-130964 metabolite through CYP 2A, 2C9, 2C19, and 2B6 [15]. Of these enzymes, only the loss of function variants in CYP 2C19 and CYP 2C9 affect the pharmacokinetics to any appreciable extent. The effect for 2C19 polymorphism is greater than for 2C9 alone as individuals with a heterozygous **1/*2* genotype had no effect on the drug's kinetics [15] (Fig. 3).

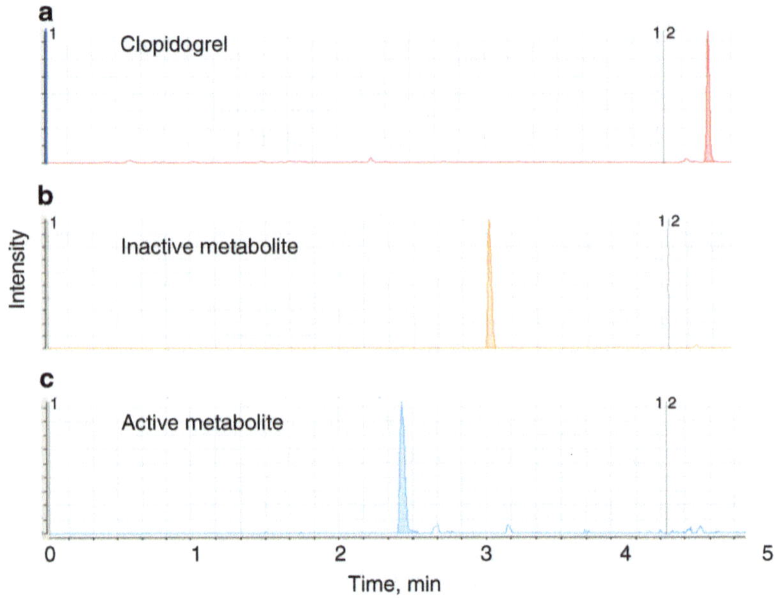

Fig. 3 Representative ion chromatograms for the LC/TOF MS analysis of (**a**) clopidogrel, (**b**) inactive metabolite, and (**c**) active metabolite. The latter is derivatized with 2-bromo-3′-methoxyacetophenone. Separate injections were necessary for the active vs. inactive metabolite due to the need to derivatize the biologically active clopidogrel form

There have been several clinical studies on the effect of CYP 2C19 carriers of at least one loss of function alleles. In the TRITON-TIMI39 study, carriers had a relative increase of 53 % in the incidence of death from cardiovascular disease, myocardial infarction, or stroke compared to noncarriers and a threefold higher risk of stent thrombosis [16]. In the Collet study among young patients, the hazard ratios (HR) were 3.69 and 6.02 ($p<0.05$) for cardiac events and stent thrombosis, respectively [17]. Similar findings were also reported by Giusti et al. (HR=2.36 and 2.59 for mortality and stent thrombosis, respectively, $p<0.05$) [18]. Based on these and other reports, the US Food and Drug Administration (FDA) issued a Black Box warning in March 2010 on the use of clopidogrel, addressing the need for pharmacogenomic testing [19]. While it was appropriate for the FDA to issue this warning, there is insufficient evidence to date to recommend the actions taken (e.g., higher clopidogrel dosing or use of an alternative medication) for individuals who are determined to be at risk [20]. Such guidance will likely follow with completion of ongoing randomized trials addressing these issues. GRAVITAS was a randomized clinical trial of 75 mg vs. 150 mg clopidogrel for individuals who are resistant to clopidogrel by platelet aggregometry, with 1 and 6-month outcomes recorded [21].

Recently, the CYP 2C19 **17* variant was identified as having increased transcriptional activity, resulting in ultrarapid metabolism. For clopidogrel, the consequence is an increased concentration of the active metabolite. Sibbing et al. showed that individuals with the **17* polymorphism were associated with an increased risk of bleeding [22]. Although not specifically mentioned in the FDA Black Box warning, testing for this variant is also indicated. Clinicians should be cautioned by clinical laboratorians who only test for the CYP **2* and **3* polymorphism, as the wild type (**1*) will be erroneously inferred in the absence of direct testing for the **17* variant.

2.2.2 Nongenetic Factors and Pharmacodynamics

Individuals with genetic variances to hepatic enzymes such as CYP 2C19 represent a pharmacokinetic mechanism towards drug resistance, i.e., insufficient concentrations of the active metabolite. These are a subset of individuals who have clopidogrel resistant as measured by functional testing. There are several methods to measure platelet function as an assessment of the pharmacodynamics of clopidogrel, including as light transmittance aggregometry, impedance measurement of whole blood aggregometry, vasodilator-stimulated phosphoprotein analysis (VASP) as measured by flow cytometric analysis, and various commercial assays and platforms such as the PFA100 (Siemens Healthcare, Deerfield, IL), VerifyNow (Accumetrics, Carlsbad, CA), and Impact-R (Cresier, Switzerland) [23]. Individuals can be resistant to clopidogrel even if they are wild type for CYP 2C19. The mechanisms include polymorphism in other metabolic enzymes such as CYP 2C9, drug–drug interactions, variable drug absorption or clearance, P2Y12 receptor variability such as an increase in the number of receptors or upregulation of alternate platelet activation pathways [24]. A combination of testing for the pharmacokinetic (CYP 2C19) and pharmacodynamic (platelet function testing) provides the most insight for a particular patient.

2.2.3 Analytical Assays and TDM Testing

The absence of a specific therapeutic algorithm based on pharmacogenomic and platelet function testing that has been endorsed by international cardiology societies has slowed the adoption of testing into routine clinical practice. An important parameter that would add to the understanding of clopidogrel resistance in a particular patient would be therapeutic drug monitoring for the active metabolite. While LC tandem MS assays have been described for the carboxylic acid metabolite [25, 26], this product is inactive and will not likely be clinically useful. More recently, an LC-MS assay has been developed for the active metabolite [27]. Accurate analysis requires production of the thiol group with alkylating agents such as *N*-ethylmaleimide added to the blood sample within a few minutes of collection.

Measurement of the concentration of the active metabolite can be used to assess clopidogrel efficacy for patients who have a pharmacokinetic mechanism for platelet resistance. Mega et al. showed that patients who are intermediate or slow metabolizers for CYP 2C19 have lower concentrations of the active metabolite than wild-type patients [16]. By monitoring the active metabolite concentration, a physician may be able to titrate the clopidogrel dosage for intermediate and poor metabolizers to match the levels seen in noncarriers. Assuming that these individuals do not also have a pharmacodynamic reason for their platelet resistance, increasing the drug dosage should reduce the rate of adverse cardiac events to that of wild types. A loading dose titration study was previously conducted by Bonello et al. using VASP testing as the therapeutic monitoring indicator [28]. Major adverse event rates were significantly lower in the VASP-guided group than controls with no increase in the incidence of major or minor bleeding. However, as much as 2,400 mg of clopidogrel was needed given, some eightfold higher than the recommended bolus dose of 300 mg. While this study was successful, many physicians will likely be hesitant to use this high dosage. In patients who are resistant to clopidogrel by platelet function tests and are wild type for CYP 2C19, a consideration can be made to switch to an alternate antiplatelet medication such as prasugrel, which is not affected by CYP 2C19 [15]. Figure 4 illustrates a proposed algorithm based on genotyping, functional testing, and phenotyping through therapeutic drug monitoring measurements. This scheme has not been clinically validated, nor is it endorsed by any clinical practice groups.

2.3 *Opioids for Pain Management*

2.3.1 Pharmacology and Pharmacogenomics

Opioid analgesics are used to treat moderate to severe pain. They function by decreasing pain perception and increasing pain tolerance. Opioids are highly prescribed; however, there is a large degree of variability in individual responses to opioids. The majority of opioids used in pain management are metabolized by CYP450 enzymes. *CYP2D6* is the primary enzyme responsible for the formation of the active metabolites of codeine, hydrocodone, dihydrocodeine, oxycodone, and tramadol. *CYP2D6* is the only noninducible CYP450 enzyme, thus genetic variation is the main source for interindividual differences in enzyme activity. *CYP2D6* is highly polymorphic. Over 90 distinct allelic variants have been identified [29]. These include single nucleotide polymorphisms, haplotype, and copy number variants. These variants result in a large degree of metabolic and phenotypic diversity within populations. *CYP2D6* variants can be categorized into ultrarapid metabolizers (UM), extensive (EM), intermediate (IM), and poor metabolizers (PM). An individual's highest functioning *CYP2D6* allele predicts his/her phenotypic activity. EMs are considered phenotypically normal and have at least one functional *CYP2D6* allele. UMs have multiple gene copy variants and may experience toxicity for opioids (i.e., codeine) due to increased levels of their active metabolites (i.e., morphine). IMs and PMs have a decreased ability to metabolize *CYP2D6* substrates compared to EMs, and may be at risk for adverse effects from higher plasma levels

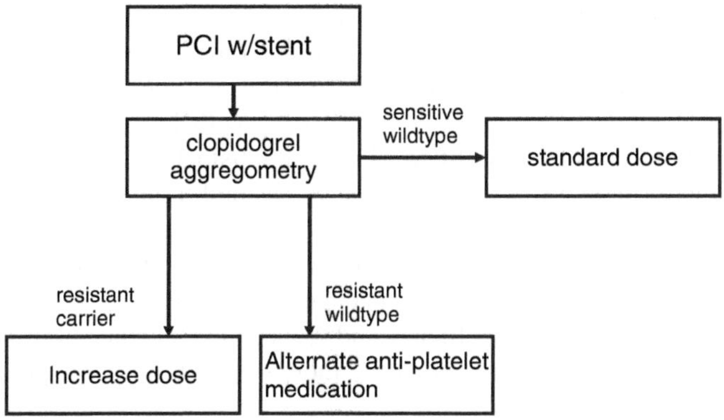

Fig. 4 Proposed testing algorithm for clopidogrel therapy for acute coronary syndromes

of the parent drug. However, IMs and PMs may experience a lack of efficacy from some opioids (i.e., codeine) because of their inability to form the active metabolite (i.e., morphine). The majority of individuals are *CYP2D6* EMs; however, 7–10 % of the Caucasian population and 1–4 % of other ethnic populations have nonfunctional alleles [29, 30].

CYP2D6 is the most well studied gene with respect to the pharmacogenetics of codeine metabolism. Approximately 50–70 % of codeine is glucuronidated to codeine-6-glucuronide by UGT2B7 and 10–15 % is *N*-demethylated to norcodeine by CYP3A4 [31]. Compared to codeine both metabolites have a similar affinity for the μ-opioid receptor. A smaller percentage of codeine (0–15 %) is *O*-demethylated to morphine which has a 200-fold increased affinity for the μ-opioid receptor compared to codeine. PMs may not experience adequate pain relief since they are unable to convert codeine to morphine, while UMs may experience morphine intoxication as a result of rapid conversion of codeine to morphine.

Multiple pharmacogenetic studies have shown that there is significant variability in both the pharmacokinetics and pharmacodynamics of codeine and that its analgesic effects are mostly dependent on metabolism to morphine [32–35]. However, many of these studies were small and had a limited sample-size. Large-scale studies are still needed to demonstrate impaired analgesic outcome in *CYP2D6* PMs. In 2002, Williams and colleagues investigated the postoperative analgesic efficacy in a pediatric population ($n=46$) by determining genotype, phenotype and morphine production from codeine [32]. They found that there was a significant relationship between phenotype and plasma morphine concentration after administration of codeine, however, no relationship was found between phenotype and analgesia. This could be a result of experimental cofounders such as coadministration with diclofenac. Another study found that *CYP2D6* UMs ($n=12$) had approximately 50 % higher plasma concentrations of morphine and its glucuronides compared with EMs ($n=11$) after administration of a single dose of 30 mg codeine [33]. Only half of the *CYP2D6* EMs felt sedation from the codeine compared to 91 % of the

CYP2D6 UMs. In 2009, Lotsch and colleagues conducted a study in 57 healthy Caucasians to determine if morphine formation from codeine could be predicted prior to codeine administration by using *CYP2D6* genotype- and phenotype-based prediction systems [34]. Most subjects (87.5 %) with low morphine production from codeine were correctly identified. However, satisfactory prediction (87.5 %) of high morphine formation was only achieved when combining genotype with phenotyping. There have been multiple case reports of life-threatening adverse events or fatalities in *CYP2D6* UMs [36–40]. The data from these pharmacogenetic studies suggest that implementation of pharmacogenetic testing for *CYP2D6* prior to codeine therapy could improve efficacy and reduce the incidence of drug toxicity if done in combination with therapeutic drug monitoring of morphine production.

2.3.2 Other Genetic and Nongenetic Factors

CYP2D6 is not the only polymorphic gene involved in the codeine and morphine metabolic pathway. Approximately 70 % of morphine is glucuronidated to morphine-3- and morphine-6-glucuronide, primarily by UGT2B7. The evidence concerning the influence of UGT2B7 polymorphism on codeine and morphine metabolism is conflicting and inconclusive. The most well studied SNP in UGT2B7 (UGT2B7*2) did not affect morphine metabolism in vitro [41, 42]. Other studies have identified additional variants that have an impact on morphine metabolism; however, these studies have not been reproduced in separate cohorts. It is possible that recently identified variants that alter mRNA splicing of UGT2B7 could significantly impact the codeine and morphine pathway [43].

Codeine is metabolized to norcodeine by CYP3A4. CYP3A4 is responsible for the metabolism of approximately 50–60 % of pharmaceuticals used today and is also important for the metabolism of steroid hormones. There are several known genetic variants in CYP3A4, but none of them have been shown to cause a phenotypic change in drug metabolism. However, coadministration of drugs that are CYP3A4 substrates, inducers or inhibitors can affect flux through the codeine pathway. In one case report of a life-threatening opioid intoxication, the authors attributed the observed toxicity to not only CYP2D6 ultrarapid metabolism but also inhibition of CYP3A4 activity by other medications [14]. Further studies in a larger study sample are required to determine the effects of coadministration of drugs that act on CYP3A4 for the metabolism of codeine.

The efflux transporter P-glycoprotein (P-gp), encoded by the ATP-binding cassette BI (ABCB1)/multiple drug resistance 1 (MDR1) gene, is responsible for the transport of many opioids, including morphine-3-glucuronide and morphine-6-glucuronide across the blood–brain barrier. Several studies suggest that the ABCB1:3435C>T variant may influence morphine efflux from the blood–brain barrier and result in variable analgesic response [44]. Also, the same studies have shown that a polymorphism (OPRM1 118A>G) in the μ-opioid receptor, encoded by the opioid receptor μ 1 (OPRM1) gene, is associated with opioid analgesia

[45–47]. In one study, daily opioid doses significantly decreased in a gene dose-dependent manner with the ABCB1 3435C>T variant [45]. The same study found that a tendency toward increased pain and the OPRM1 118A>G variant were associated in a dose dependent manner [45]. Another study found that pain relief variability was significantly associated with both polymorphisms [46].

2.3.3 Analytical Assays and Need for TDM Measurements for Drug Compliance

Serum or plasma assays for prescription opioids are not routinely used for therapeutic drug monitoring in the clinical laboratory. The current testing strategy includes urine screening with immunoassays followed by targeted confirmations with gas chromatography mass spectrometry (GC-MS) or liquid chromatography tandem mass spectrometry (LC-MS/MS). These assays are used to monitor drug use and compliance of prescribed opioids because of the long detection window of the metabolites in urine. However, serum and plasma assays for opioids have many advantages over urine because the opioid level in these matrices is reflective of the patient's clinical state at the time of blood collection. The levels could be used for titrating dose, determining steady-state concentrations or determining the drug and metabolite levels in cases of suspected opioid toxicity. However, the role of TDM in titration and monitoring opioids is not well defined. There are no current practice guidelines that have been established for clinical use. Many opioids meet the general criteria for TDM including a narrow therapeutic index, a poor relationship between drug dose and blood concentration, significant inter-individual variation and a serious consequence for overdosing in some individuals. In a study conducted by Lotsch and colleagues they concluded that in order for codeine therapy to be safe, prediction of morphine formation must be obtained by combining *CYP2D6* genotyping with phenotyping [34]. Phenotyping can account for the other genetic and nongenetic factors listed in Sect. 2.3.2.

Despite the fact that opioid TDM is not routine in the clinical laboratory, there are several published methods for the detection of opioids in serum or plasma using LC-MS/MS [48–52]. LC-MS/MS is capable of detecting polar and thermally labile compounds and thus has advantages over GC-MS for the analysis of opioids. Parent drugs and glucuronide metabolites can be quantitated in the same method. With LC-MS/MS, sample preparation is decreased because there is no need for hydrolysis. These LC-MS/MS methods are commonly used in forensic settings and for research purposes. They include anywhere from 6 to 24 opioids and their metabolites and have lower limits of quantitation down to approximately 0.5 ng/mL for many of the analytes. One method demonstrated an upper limit of quantitation at 2,500 ng/mL [48]. These methods are sufficient for determining opioid and metabolite levels and could be used for TDM in the clinical setting once further studies are conducted to establish appropriate guidelines.

Table 2 Advantages and disadvantages of LC-MS vs. immunoassay (where assays are available) for therapeutic drug monitoring

Attribute	LC-MS	Immunoassay
Analytical sensitivity	Highest	Sufficient for some tests
Analytical specificity	Very high	Questionable, antibody dependent
Operator expertise	High	Low
CLIA test category	Highly complex	Moderately complex
Instrument costs	High	High
Reagent costs	Low	High
Commercial availability	Lab developed test	Unavailable for most drugs of interest
Turnaround time	30–60 min	10–20 min
Random access	Limited	Widely practiced
Multiplex capability	Available	Not possible without multiple assays

3 Analytical Testing Platforms for Therapeutic Drug Monitoring

Pharmacogenomic testing for routine clinical practice is still in its infancy. Relabeling mandates and black box warnings issued by the Food and Drug Administration has accelerated the clinical interest in pharmacogenomic testing for select drugs. Critics to routine implementation have argued that genetic testing only accounts for some of the variability in drug response or prediction of adverse events. Therefore, other predictive phenotypic tools are needed, such as functional testing and therapeutic drug monitoring. The debate regarding the clinical significance of genotyping vs. phenotyping will continue for years to come. Each approach has its advantages and disadvantages. It is likely that optimal patient management will require a combination of both approaches. However, doing both testing may not be a cost-effective strategy unless it can be shown that the improvements in clinical outcomes justify additional costs. If therapeutic monitoring is important for the drugs discussed in this chapter, LC-MS appears to be the analytical method of choice. LC-MS enables detection of parent and metabolite concentration in the same analytical assay. Specific immunoassays for these and other drugs of pharmacogenomics interest are not commercially available. The analytical sensitivity of LC-MS is superior to immunoassays. Table 2 summarizes some of the advantages and disadvantages differences between these two analytical approaches for a variety of attributes.

References

1. Jin Y, Desta Z, Stearns V, Ward B, Ho H, Lee KH, Skaar T, Storniolo AM, Li L, Araba A, Blanchard R, Nguyen A, Ullmer L, Hayden J, Lemler S, Weinshilboum RM, Rae JM, Hayes DF, Flockhart DA (2005) CYP2D6 genotype, antidepressant use, and tamoxifen metabolism during adjuvant breast cancer treatment. J Natl Cancer Inst 97:30–39

2. Goetz MP, Rae JM, Suman VJ, Safgren SL, Ames MM, Visscher DW, Reynolds C, Couch FJ, Lingle WL, Flockhart DA, Desta Z, Perez EA, Ingle JN (2005) Pharmacogenetics of tamoxifen biotransformation is associated with clinical outcomes of efficacy and hot flashes. J Clin Oncol 23:9312–9318
3. Goetz MP, Knox SK, Suman VJ, Rae JM, Safgren SL, Ames MM, Visscher DW, Reynods C, Couch FJ, Lingle WL, Weinshilboum RM, Barr Fritcher EG, Barr Fritcher AM, Desta Z, Nguyen A, Flockhart DA, Perez EA, Ingle JN (2006) The impact of cytochrome P450 2D6 metabolism in women receiving adjuvant tamoxifen. Breast Cancer Res Treat 101:113–121
4. Lim HS, Lee HJ, Lee KS, Lee ES, Jang IJ, Ro J (2007) Clinical implications of CYP2D6 genotypes predictive of tamoxifen pharmacokinetics in metastatic breast cancer. J Clin Oncol 25:3837–3845
5. Madlensky L, Natarajan L, Tchu S, Pu M, Mortimer J, Flatt SW, Parker BA, Wu AHB, Pierce JP (2010) Tamoxifen metabolite concentrations, CYP2D6 genotype and breast cancer outcomes. J Clin Oncol 89(5):718–725
6. Sterns V, Johnson MD, Rae JM, Morocho A, Novielli A, Bhargava P, Hayes DF, Desta Z, Flockhart DA (2003) Active tamoxifen metabolite plasma concentrations after coadministration of tamoxifen and the selective serotonin reuptake inhibitor paroxetine. J Natl Cancer Inst 95:1758–1764
7. Borges S, Desta Z, Li L, Skaar TC, Ward BA, Nguyen A, Jin Y, Storniolo AM, Nikoloff DM, Wu L, Hillman G, Hayes DF, Stearns V, Flockhart DA (2006) Quantitative effect of CYP2D6 genotype and inhibitors on tamoxifen metabolism: implication for optimization of breast cancer treatment. Clin Pharmacol Ther 80:61–74
8. Kelly CM, Juurlink DN, Gomes T, Duong-Hua M, Pritchard KI, Austin PC, Faszat LF (2010) Selective serotonin reuptake inhibitors and breast cancer mortality in women receiving tamoxifen: a population based cohort study. BMJ 340:c693
9. Wu AH, Pike MC, Williams LD, Spicer D, Tseng CC, Churchwell MI, Doerge DR (2007) Tamoxifen, soy, and lifestyle factors in Asian American women with breast cancer. J Clin Oncol 25:3024–3030
10. Lee KH, Ward BA, Desta Z, Flockhart DA (2003) Jones DR (2003) Quantification of tamoxifen and three metaoblites in plasma by high-performance liquid chromatography with fluorescence detection: application to a clinical trial. J Chromatogr B 791:245–253
11. Furlanut M, Franceschi L, Pasqual E, Bacchetti S, Poz D, Giorda G, Cagol PP (2007) Tamoxifen and its main metabolites serum and tissue concentration in breast cancer women. Ther Drug Monit 29:349–352
12. Gjerde J, Kisanga ER, Hauglid M, Holm PI, Mellgren G, Lien EA (2005) Identification and quantification of tamoxifen and four metabolites in serum by liquid chromatography-tandem mass spectrometry. J Chromatogr A 1082:6–14
13. National Cancer Institute. Studying blood samples from women with breast cancer or ductal carcinoma in situ who are receiving tamoxifen. NCT00764322. http://www.clinicaltrials.gov/ct2/show/NCT00764322?term=tamoxifen+pharmacogenomics+and+university+of+north+carolina&rank=1
14. ACC/AHA/SCAI 2005 (2006) Guideline Update for Percutaneous Coronary Intervention—Summary Article. A Report of the American College of Cardiology/American Heart Association Task Force on Practice Guidelines (ACC/AHA/SCAI Writing Committee to Update the 2001 Guidelines for Percutaneous Coronary Intervention). Circulation 113:156–175
15. Brandt JT, Close SL, Iturria SJ, Payne CD, Farid NA, Ernest CS, Lachno DR, Salazar D, Winters KJ (2007) Common polymorphisms of CYP2C19 and CYP2C9 affect the pharmacokinetic and pharmacodynamic response to clopidogrel but not prasugrel. J Thromb Haemost 5:2429–2436
16. Mega JL, Close SL, Wiviott SD, Shen L, Hockett RD, Brandt JT, Walker JR, Antman EM, Macias W, Braunwald E, Sabatine MS (2009) Cytochrome P-450 polymorphisms and response to clopidogrel. N Engl J Med 360:354–362
17. Collet JP, Hulot JS, Pena A, Villard E, Esteve JB, Silvain J, Payot L, Brugier D, Cayla G, Beygui F, Bensimon G, Funck-Bretano C, Montalescot G (2009) Cytochrome P450 2C19

polymorphism in young patients treated with clopidogrel after myocardial infarction: a cohort study. Lancet 373:309–317

18. Giusti B, Gori AM, Marcucci R, Saracini C, Il S, Paniccia R, Buonamici P, Antoniucci D, Abbate R, Gensi GF (2009) Relation of cytochrome P450 2C19 loss-of-function polymorphism to occurrence of drug-eluting coronary stent thrombosis. Am J Cardiol 103:806–811
19. FDA Drug Safety Communication: reduced effectiveness of Plavix (clopidogrel) in patients who are poor metabolizers of the drug. http://www.fda.gov/Drugs/DrugSafety/Postmarket DrugSafetyInformationforPatiensandProviders/ucm203888.htm
20. Holmes DR, Dehmer GJ, Kaul S, Leifer D (2010) ACCF/AHA clopidogrel clinical alert: approaches to the FDA "Boxed Warning". J Am Coll Cardiol 56:321–341
21. Price MJ, Berger PB, Angiolillo DJ, Teirstein PS, Tanquay JF, Kandzari DE, Cannon CP, Topol EJ (2009) Evaluation of individualized clopidogrel therapy after drug-eluting stent implantation in patients with high residual platelet reactivity: design and rationale of the GRAVITAS trial. Am Heart J 157:818–824
22. Sibbing D, Koch W, Gebhard D, Schuster T, Braun S, Stegherr J, Morath T, Schomig A, von Beckerath N, Kastrati A (2010) Cytochrome 2C19*17 allelic variant, platelet aggregation, bleeding events, and stent thrombosis in clopidogrel-treated patients with coronary stent placement. Circulation 121:512–518
23. Bouman HJ, Parlak E, Van Werkum JWV, Breet NJ, Ten Cate H, Hackeng CM, Ten Berg JM, Taubert D (2009) Which platelet function test is suitable to monitor clopidogrel responsiveness? A pharmacokinetic analysis on the active metabolite of clopidogrel. J Thromb Haemost 8:482–488
24. Wiviott SD, Antman EM (2004) Clopidogrel resistance. A new chapter in a fast-moving story. Circulation 109:3064–3067
25. Ksycinska H, Rudzki P, Bukowska-Kiliszek M (2006) Determination of clopidogrel metabolite (SR26334) in human plasma by LC-MS. J Pharm Biomed Anal 41:533–539
26. Shi BS, Yoo SD (2007) Determination of clopidogrel in human plasma by liquid chromatography/tandem mass spectrometry: application to a clinical pharmacokinetic study. Biomed Chromatogr 21:883–889
27. Takahashi M, Pang H, Kawabata K, Farid NA, Kurihara A (2008) Quantitative determination of clopidogrel active metabolite in human plasma by LC-MS/MS. J Pharm Biomed Anal 48:1219–1224
28. Bonello L, Camoin-Jau L, Arques S, Boyer C, Panagides D, Wittenberg O, Simeoni MC, Barragan P, Dignat-George F, Paganelli F (2008) Adjusted clopidogrel loading dosese according to vasodilator-stimulated phosphoprotein phosphorylation index decrease rate of major adverse cardiovascular events in patients with clopidogrel resistance. J Am Coll Cardiol 51:1404–1411
29. Daly AK, Brockmoller J, Broly F, Eichelbaum M, Evans WE, Gonzalez FJ, Huang JD, Idle JR, Ingelman-Sundberg M, Ishizaki T, Jacqz-Aigrain E, Meyer UA, Nebert DW, Steen VM, Wolf CR, Zanger UM (1996) Nomenclature for human CYP2D6 alleles. Pharmacogenetics 6(3):193–201
30. Ingelman-Sundberg M, Sim SC, Gomez A, Rodriguez-Antona C (2007) Influence of cytochrome P450 polymorphisms on drug therapies: pharmacogenetic, pharmacoepigenetic and clinical aspects. Pharmacol Ther 116(3):496–526
31. Coffman BL, Rios GR, King CD, Tephly TR (1997) Human UGT2B7 catalyzes morphine glucuronidation. Drug Metab Dispos 25:1–4
32. Williams DG, Patel A, Howard RF (2001) Pharmacogenetics of codeine metabolism in an urban population of children and its implications for analgesic reliability. Br J Anaesth 86(3):413–421
33. Kirchheiner J, Schmidt H, Tzvetkov M, Keulen J (2007) LotschJ, Roots I, Brockmoller J. Pharmacokinetics of codeine and its metabolite morphine in ultra-rapid metabolizers due to CYP2D6 duplication. Pharmacogenomics J 7(4):257–265
34. Lotsch J, Rohrbacher M, Schmidt H, Doehring A, Brockmoller J, Geisslinger G (2009) Can extremely low or high morphine formation from codeine be predicted prior to therapy initiation? Pain 144:119–124

35. He YJ, Brockmoller J, Schmidt H, Roots I, Kirchheiner J (2008) CYP2D6 ultrarapid metabolism and morphine/codeine ratios in blood: was it codeine or heroin? J Anal Toxicol 32:178–182
36. Gasche Y, Daali Y, Fathi M, Chiappe A, Cottini S, Dayer P, Desmeules J (2004) Codeine intoxication associated with ultrarapid CYP2D6 metabolism. N Engl J Med 351(27):1356–1358
37. Koren G, Cairns J, Chitayat D, Gaedigk A, Leeder SJ (2006) Pharmacogenetics of morphine poisoning in a breastfed neonate of a codeine-prescribed mother. Lancet 368(9536):704
38. Ciszkowski C, Madadi P, Phillips MS, Lauwers AE, Koren G (2009) Codeine, ultrarapid-metabolism genotype, and postoperative death. N Engl J Med 361(8):827–828
39. Voronov P, Przybylo HJ, Jagannathan N (2007) Apnea in a child after oral codeine: a genetic variant – an ultra-rapid metabolizer. Paediatr Anaesth 17(7):684–687
40. Madadi P, Koren G, Cairns J, Chitayat D, Gaedigk A, Leeder JS, Teitelbaum R, Karaskov T, Aleksa K (2007) Safety of codeine during breastfeeding: fatal morphine poisoning in the breastfed neonate of a mother prescribed codeine. Can Fam Physician 53(1):33–35
41. Court MH, Krishnaswamy S, Hao Q, Duan SX, Patten CJ, Von Moltke LL, Greenblatt DJ (2003) Evaluation of 3′-azido-3′-deoxythymidine, morphine, and codeine as probe substrates for UDP-glucuronosyltransferase 2B7 (UGT2B7) in human liver microsomes: specificity and influence of the UGT2B7*2 polymorphism. Drug Metab Dispos 31:1125–1133
42. Bhasker CR, McKinnon W, Stone A, Lo AC, Kubota T, Ishizaki T, Miners JO (2000) Genetic polymorphism of UDP-glucuronosyltransferase 2B7 (UGT2B7) at amino acid 268; ethnic diversity of alleles and potential clinical significance. Pharmacogenetics 10:679–685
43. Innocenti F, Lui W, Frackenthal D, Ramirez J, Chen P, Ye X, Wu X, Zhang W, Mirkov S, Das S, Cook E Jr, Ratain MJ (2008) Single nucleotide polymorphism discovery and functional assessment of variation in the UDP-glucuronosyltransferase 2B7 gene. Pharmacogenet Genomics 18:683–697
44. Lotsch J, Skarke C, Liefhold J, Geisslinger G (2004) Genetic predictors of the clinical response to opioid analgesics. Clin Pharmacokinet 43(14):983–1013
45. Lötsch J, von Hentig N, Freynhagen R, Griessinger N, Zimmermann M, Doehring A, Rohrbacher M, Sittl R, Geisslinger G (2009) Cross-sectional analysis of the influence of currently known pharmacogenetic modulators on opioid therapy in outpatient pain centers. Pharmacogenet Genomics 19(6):429–436
46. Campa D, Gioia A, Tomei A, Poli P, Barale R (2008) Association of ABCB1/MDR1 and OPRM1 gene polymorphisms with morphine pain relief. Clin Pharmacol Ther 83(4):559–566
47. Coulbault L, Beaussier M, Verstuyft C, Weickmans H, Dubert L, Trégouet D, Descot C, Parc Y, Lienhart A, Jaillon P, Becquemont L (2006) Environmental and genetic factors associated with morphine response in the postoperative period. Clin Pharmacol Ther 79(4):316–324
48. Coles R, Kushnir MM, Nelson GJ, McMillin GA, Urry FM (2007) Simultaneous determination of codeine, morphine, hydrocodone, hydromorphone, oxycodone, and 6-acetylmorphine in urine, serum, plasma, whole blood, and meconium by LC-MS-MS. J Anal Toxicol 31(1):1–14
49. Bogusz MJ, Maier RD, Erkens M, Driessen S (1997) Determination of morphine and its 3- and 6-glucuronides, codeine, codeine-glucuronide and 6-monoacetylmorphine in body fluids by liquid chromatography atmospheric pressure chemical ionization mass spectrometry. J Chromatogr B Biomed Sci Appl 703(1–2):115–127
50. Schänzle G, Li S, Mikus G, Hofmann U (1999) Rapid, highly sensitive method for the determination of morphine and its metabolites in body fluids by liquid chromatography-mass spectrometry. J Chromatogr B Biomed Sci Appl 721(1):55–65
51. Al-Asmari AI, Anderson RA (2007) Method for quantification of opioids and their metabolites in autopsy blood by liquid chromatography-tandem mass spectrometry. J Anal Toxicol 31(7):394–408
52. Dahn T, Gunn J, Kriger S, Terrell AR (2010) Quantitation of morphine, codeine, hydrocodone, hydromorphone, oxycodone, oxymorphone, and 6-monoacetylmorphine (6-MAM) in urine, blood, serum, or plasma using liquid chromatography with tandem mass spectrometry detection. Methods Mol Biol 603:411–422

Liquid Chromatography-Mass Spectrometry for the Determination of Antidepressants and Some of their Major Metabolites in Human Biological Matrices

Ana de Castro, Angelines Cruz, and Manuel López-Rivadulla

Abstract Due to the high prescription of antidepressants and their frequent involvement in clinical and forensic intoxications, reliable analytical techniques for identification and quantification of these therapeutic drugs should be available in clinical and toxicological laboratories. Improvements in liquid chromatography-mass spectrometry LC-MS(MS) technology over the last two decades have favored this technique to be one of the most commonly employed for this purpose, as it combines the high selectivity and sensitivity of the mass spectrometer with the great versatility of the liquid chromatographic separation. In this chapter, LC MS(MS) applications for antidepressants determination are reviewed, detailing typical sample preparation techniques used for these therapeutic drugs, as well as common chromatographic and mass spectrometric characteristics. In addition, an LC-MS/MS method for the most common antidepressants used in clinical practice is described as an example of a whole method development and validation.

1 Introduction

Antidepressants were first introduced into the market in the 1950s with the serendipitous discovery of the antidepressant effect of two drugs initially evaluated for other medical uses: Iproniazide, a monoamine oxidase inhibitor (MAOI), and Imipramine, a tricyclic antidepressant (TCA). Since then, a whole new generation of chemically and pharmacologically unrelated compounds have been introduced, which appear to be safer and better tolerated due to a more specific mechanism of action. These include selective serotonin reuptake inhibitors (SSRIs), serotonin and

A. de Castro • A. Cruz • M. López-Rivadulla (✉)
Forensic Toxicology Service, Institute of Forensic Sciences, University of Santiago de Compostela, 15782, Santiago de Compostela, Spain
e-mail: manuel.lopez-rivadulla@usc.es

Q. Alan Xu and Timothy L. Madden (eds.), *LC-MS in Drug Bioanalysis*,
DOI 10.1007/978-1-4614-3828-1_7, © Springer Science+Business Media, LLC 2012

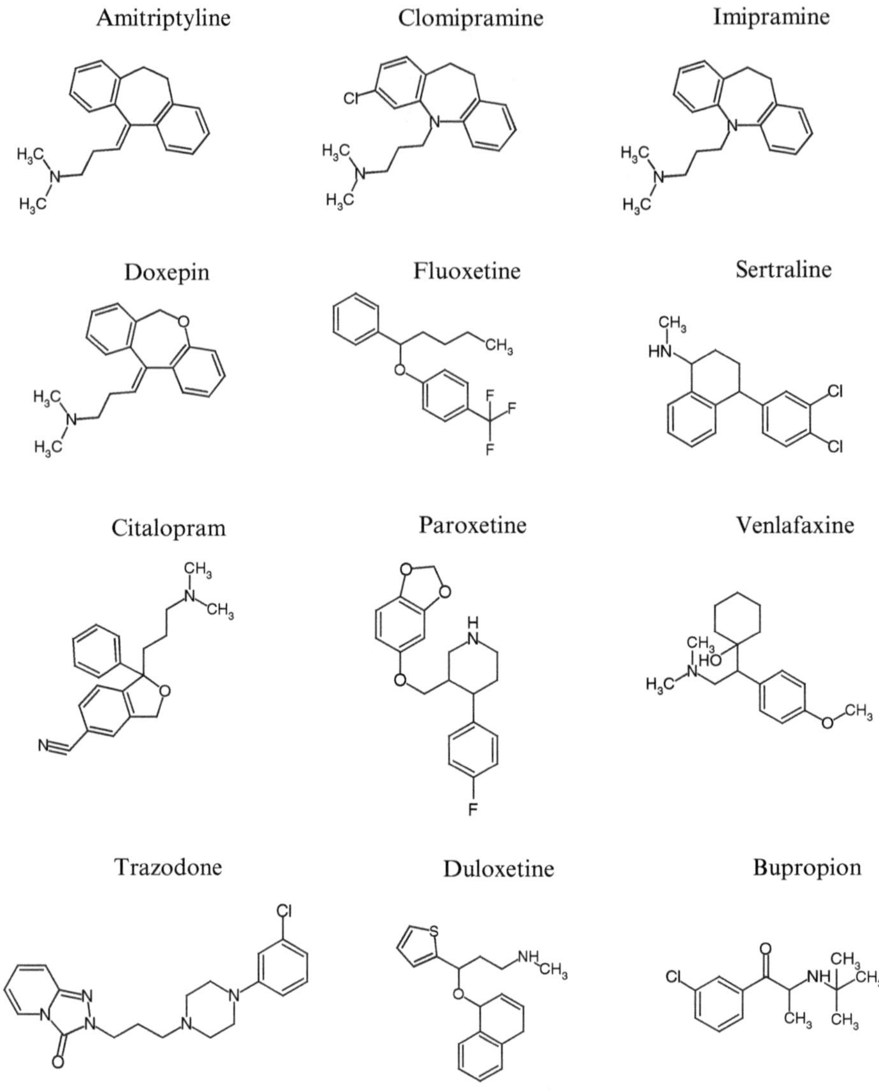

Fig. 1 Chemical structure of some antidepressants commonly used in clinical practice

noradrenalin reuptake inhibitors (SNaRIs), noradrenergic and specific serotoninergic antidepressants (NaSSAs), and noradrenalin reuptake inhibitors (NaRIs).

Figure 1 shows the chemical structures of some of the most common antidepressants used in clinical practice.

The main indication for antidepressants is depressive disorders, which have received increased attention owing to the growing recognition of their high prevalence. Other antidepressant uses include treatment of anxiety disorders, attention deficit hyperactivity disorders, nocturnal enuresis or psychosomatic disorders developed in several illnesses such as chronic pain, fibromyalgia, irritable bowel syndrome, or chronic fatigue syndrome.

Due to their high prescription, these therapeutic drugs are easy to acquire by depressed patients, who are prone to suicide attempts. This explains why antidepressants are among the most frequent therapeutic drug classes involved in forensic and clinical intoxications, mainly associated to voluntary intoxications [1–4]. Antidepressants with sedative side effects, like TCAs, could also be potentially used in drug facilitated sexual assault (DFSA) crimes [5]. Therefore, analytical techniques for the reliable identification and quantification of antidepressants should be available in clinical and forensic toxicology laboratories in order to perform a competent toxicological report. In addition, some antidepressants impair cognitive and psychomotor functions, and may increase the risk of driving accidents when under their influence [6–8]. For this reason, the Guidelines for Research in Drugged Driving elaborated by international experts in order to harmonize research in this field recommended the inclusion of sedative antidepressants in the panel of drugs to be analyzed in specimens collected from the roadside [9]. In addition to toxicological applications, analytical methodologies are also required to monitor plasmatic concentrations of some antidepressants. Therapeutic drug monitoring (TDM) of TCA is widely accepted [10–13] because of their narrow therapeutic window, and the development of severe cardiotoxicity and CNS toxicity close to the upper therapeutic concentrations; moreover, the enzymes involved in their metabolization show genetic polymorphism, being in part responsible for the interindividual variability in the plasmatic concentrations achieved at a given dose. Although new generations of antidepressants are less toxic and have wider therapeutic ranges due to a more selective mechanism of action, TDM of their plasmatic levels could be justified in special situations (assess compliance in nonresponder patients, hepatic or renal impairment, polymedicated patients, poor metabolizers, etc.) [11–13]. Antidepressant determination also is required to perform pharmacokinetic, bioavailability, and bioequivalence studies.

Different immunoassays are commercialized for the analysis of some antidepressants [14–17]. Although useful for fast identification of these analytes, these techniques have several limitations: TCA are the main targeted analytes, it is not possible to differentiate between structurally related antidepressants, and several substances can cross-react with antidepressant assays [18–20]. Therefore, positive results must be confirmed with more specific techniques, usually using chromatography based procedures coupled to different detectors. Mass spectrometry is one of the most common detectors employed in clinical and toxicological analysis because of its high selectivity and sensitivity. Gas chromatography coupled to mass spectrometry (GC-MS) is a robust and well-established technique, and it is considered the golden standard for general unknown screening. Several GC-MS analytical methodologies have been described for identification and quantification of antidepressants in different biological matrices [21–25]; however, in most cases, antidepressants have to be derivatized because of their relatively high polarity. In the 1990s, the development of atmospheric pressure ionization interfaces (API) allowed the successful hyphenation of liquid chromatography to mass spectrometry

(LC-MS). LC-MS combines the power of MS detection with the versatility of LC, which allows chromatographic separation of thermally unstable analytes, as well as compounds with a wide range of polarities without performing derivatization processes. For this reason, LC-MS and LC-MS/MS applications for the analysis of antidepressants have significantly grown over the last decade.

2 Sample Preparation for Antidepressants LC-MS(MS) Analysis

In the early stages of LC-MS, it was thought that the high selectivity provided by this technique could effectively eliminate intereferences caused by endogenous matrix compounds, and thus, the need for sample cleanup [26]. However, nowadays, it is well known that one limitation associated to LC-MS analysis is the suppression or enhancement on the analyte response when coeluting undetected compounds compete with the analyte in the ionization process [27, 28]. This phenomenon, generically known as matrix effect, may affect several validation parameters, such as the limit of detection, linearity, precision, and accuracy. Therefore, matrix effect may diminish method sensitivity and the reliability of quantitative results [29]. One of the main strategies to overcome matrix effect is to minimize the presence of these coeluting interferences through a more effective sample cleanup; thus, this step should be always taken into consideration when developing an LC-MS method.

Most of the published analytical methods for the determination of antidepressants were developed in plasma, serum, whole blood, and urine, which are the most useful matrices for clinical and toxicological analysis of these therapeutic compounds. However, albeit few, some LC-MS methodologies have also been described for the analysis of several antidepressants in hair [30–33], oral fluid [34, 35], breast milk [36], or typical forensic matrices such as gastric content, bile, vitreous humor, brain, liver, lung, and/or muscle [37–40].

Some biological matrices, such as hair and internal organs, require a special pretreatment prior to extraction. Hair samples must be washed to avoid external contamination, including an initial organic solvent, followed by aqueous washes [41]. After cutting or powdering, the hair matrix is usually disintegrated to extract the analytes from its inner structure. Ultrasonication in methanol (MeOH) [31, 32] or incubation in HCl [33] has been employed to extract antidepressants from the hair structure. Internal organs should be initially homogenized, and this process was usually performed in water or basic buffers by means of a blender or an ultrasonicator [37–40]. In addition, urine hydrolysis is sometimes performed to break glucuronide conjugates or some metabolites. β-Glucuronidase was employed for urine hydrolysis of mirtazapine [42] and bupropion metabolites conjugates [43, 44], and HCl was used for hydrolysis of 4-hydroxy-3-metoxy paroxetine metabolite [45], prior to sample extraction.

Several sample preparation or extraction procedures have been described for the analysis of antidepressants, from the most simple protein precipitation to online solid phase extraction (SPE). The optimum extraction technique depends on the analytical requirements, biological matrix, chromatographic separation, or the need for high throughput.

Several authors described a minimum sample pretreatment by plasmatic protein precipitation using acetonitrile (ACN), MeOH, acid solutions, or a mixture of them, for the analysis of one antidepressant and its metabolite(s) [46–52]. Single analyte procedures are useful for TDM or to perform pharmacokinetic, bioequivalence and pharmacogenomic studies, where the targeted antidepressant is known; however, this is an unrealistic situation in clinical and forensic toxicology, and for those applications, multianalyte procedures are preferable. Only Kirchherr et al. [53] used protein precipitation with ACN/MeOH as the only sample treatment procedure for the determination of 48 psychopharmaceuticals in human serum, including the main antidepressant drugs. No matrix effect was reported for any of the analytes, except for olanzapine, for which a 185% signal enhancement was observed.

However, protein precipitation usually produces severe matrix effect [54–56]. Therefore, in most LC-MS published methods, more extensive sample extraction procedures were used, being liquid–liquid (LLE) and SPE the most frequent techniques.

2.1 Liquid–Liquid Extraction (LLE)

In LLE, analytes are isolated from the biological matrix by means of an organic solvent immiscible with aqueous solution. Due to the basic character of antidepressants, samples are initially alkalinized with NaOH, NH_4OH or a basic buffer to achieve a pH >8.5. The polarity range of the analytes included in the analytical method, as well as the biological matrix, determine the organic solvent of choice. Hexane, dichloromethane, butyl chloride, butyl acetate, isoamylic acid, or a mixture of them, have been used for the extraction of several antidepressants from the biological samples [39, 57–61]. Apart from those previously mentioned, other solvents employed for the extraction of only one antidepressant and/or metabolite(s) were diethylether, ethyl acetate or methyl *tert*-butylether [62–64]. After centrifugation, the organic layer is evaporated to dryness and reconstituted in a small volume for LC-MS analysis. In some cases, the organic layer was reextracted before evaporation by addition of aqueous acid solution to obtain cleaner extracts. De Santana et al. [42] and Halvorsen et al. [65] determined one or several antidepressants, respectively, in whole blood and/or plasma, using liquid-phase microextraction (LPME). LPME is a minituarized LLE procedure, where analytes are extracted from the biological matrix through an organic solvent impregnated in the pores of a hollow fiber and into a micro-liquid phase (acceptor solution) inside the fiber, which is subsequently injected into the LC-MS system. The main advantage of this technique is that it enables simultaneous analyte preconcentration and sample cleanup, using small sample and solvent volumes.

2.2 Solid Phase Extraction (SPE)

Different SPE procedures have been described using sorbents based on reversed-phase mechanism (C8, C18, hydrophilic–lipophilic balance), where the sample is loaded in basic conditions to guarantee analyte retention. After washing the cartridge, elution is performed with different organic solvents, sometimes in acid conditions [33, 66–71]. Due to the basic properties of antidepressants, mixed-mode sorbents simultaneously acting by reversed-phase and cation-exchange mechanisms, allow for a more selective extraction by removing nonbasic endogenous material [32, 34, 35, 43, 72, 73]. To allow analyte retention by cation-exchange mechanism, the sample should be initially conditioned with acidic solutions; after the washing step(s), analytes are eluted using alkalinized organic solvents. Common elution solvents are MeOH, ACN, dichloromethane, 2-propanol, or mixtures of them. The extraction process can be facilitated using semiautomated SPE robots [32, 66]. Moreover, online SPE-LC-MS instruments have also been employed for antidepressant analysis [72]. In these instruments, SPE extraction of the sample is performed in parallel to the chromatographic separation of the previous one, allowing complete automation of the whole analytical method, and the consequent sample high throughput.

2.3 Online Extraction

Online extraction techniques by means of supports coupled to the chromatographic system have also been applied to the determination of antidepressants [74–78]. These systems allow for direct sample injection, since large molecules from the biological matrix can readily pass through the column, while the analytes of interest can be retained under aqueous conditions, and subsequently eluted using high organic solvents composition. There are different supports, such as the so-called restricted access media (RAM), large particle size (LPZ) and monolithic supports, which can be used in the single column configuration or, more frequently, in the column-switching approach.

2.4 Solid-Phase Microextraction (SPME)

Another alternative technique, solid-phase microextraction (SPME), was used for the determination of fluoxetine [79] and several TCAs [80]. SPME is a miniaturized and solvent-free technique, where analytes are extracted from the sample by adsorption on a thin polymer coating fixed to the solid surface of a fiber, located inside an injection needle or a capillary. Its main disadvantage is that special strategies are needed to couple SPME to the LC-MS analysis.

3 Chromatographic Separation of Antidepressants by LC-MS(/MS)

LC is a very versatile technique that allows separation of nonvolatile, thermolabile and high polar analytes, without the need to perform derivatization processes. LC coupled to traditional detectors such as UV requires separation of interferences from the analyte of interest, and between them when several analytes are included in the analytical method; this is usually traduced in long chromatographic run times. The hyphenation of the LC system to the MS significantly reduced the chromatographic run, as there is no need for complete chromatographic resolution of the analytes included in the method. However, as previously mentioned, although endogenous interferences are not observed when analyte specific masses (*m/z*) or transitions are monitored, they can cause ion suppression or enhancement if they coelute with the analyte(s) of interest. Therefore, efficient chromatographic separation can minimize matrix effects and increase precision and accuracy of the assay.

In LC, several parameters such as the mobile phase, stationary phase, and column temperature can be optimized for a specific analytical application.

3.1 Mobile Phase

Mobile phase options are quite restricted, as only volatile buffers are suitable for LC-MS. In addition, ion-pairing agents traditionally used in LC to improve peak shape and retention time such as trifluoroacetic acid (TFA), have shown to produce ion suppression, and they are not recommended for LC-MS analysis [27]. Therefore, although few applications for specific antidepressants employ TFA or its ammonium salt due to sensitivity enhancement [48, 81–85], aqueous phases in most LC-MS analytical methods are composed by formic or acetic acid in water, or its ammonium buffers. Although acid mobile phases are by far the most common, basic aqueous mobile phases (pH ranging from 8 to 10) have been used for specific applications in order to increase antidepressant retention time or to couple the online SPE elution to chromatographic analysis [57, 72, 86, 87]. Organic phase composition was typically ACN and/or MeOH.

3.2 Stationary Phase

Antidepressant separation was usually performed by reversed-phase chromatography with typical C8 or C18 alkyl chain columns, although phenyl [30, 59] or cyano [48, 64, 84] stationary phases were also employed. As an exception, hydrophilic interaction liquid chromatography (HILIC), a variation of normal phase chromatography, was employed in two analytical methods for duloxetine [38] and paroxetine [85] determination, respectively. HILIC columns allow adequate retention of polar analytes poorly retained by reversed-phase chromatography, while still retaining less polar analytes; moreover, the high organic composition of the mobile phase increases electrospray efficiency, thus providing higher sensitivity. Chiral stationary phases were employed for enantioselective analysis of antidepressants marketed as a racemic mixture such as fluoxetine, citalopram, venlafaxine, mirtazapine, or bupropion [42, 43, 82, 88, 89].

Analytical methods for the determination of one antidepressant and/or its metabolite(s) were usually performed in isocratic mode, with total run times from seconds to a few minutes. However, as previously mentioned, multianalyte procedures are preferable, particularly if the method is intended for clinical or forensic analysis. Gradient separation was usually applied when the most common antidepressants were included in the methodology; however, total chromatographic run times varied widely, from 5 to 40 min [57, 76], depending on column length, extraction technique (offline vs. online techniques), biological matrix or the specific application of the method.

4 Mass Spectrometry Characteristics for Antidepressants LC-MS(MS) Analysis

4.1 Atmospheric Pressure Ionization Interfaces

Several LC-MS interfaces have been developed since 1974, when Arpino et al. [90] described the first attempt to couple the LC system to the mass spectrometer. Some of them were commercialized [91], but it was the development of atmospheric pressure ionization interfaces (API) which actually lead to the great expansion of LC-MS applications.

Within API interfaces, electrospray ionization (ESI) is the preferable ionization method for polar analytes, and it was used in the great majority of LC-MS methods for antidepressants analysis. However, ESI interfaces are more susceptible to matrix effects than atmospheric pressure chemical ionization (APCI) [54, 92, 93] due to differences in the ionization process. ESI is based on liquid phase reactions, where ion suppression is more likely than in gas phase due to high concentrations of non-volatile materials present in the spray with the analyte [92]. Because of the lower susceptibility for matrix effects, although few, there have been authors who selected APCI as ionization mode in spite of the lower sensitivity for these analytes [59, 62, 74, 75]. Although ESI and APCI are by far the most popular ionization methods, other alternative interfaces were used for antidepressant analysis. Shinozuka et al. [69] used sonic spray ionization (SSI), a variant of ESI, where ionization is produced by high sonic gas velocity during pneumatic nebulization, instead of the electric field and capillary high temperatures applied in ESI. Atmospheric pressure photoionization (APPI) is a novel interface originally developed to widen the group of analytes to be determined by LC-MS towards less polar compounds that are not efficiently ionized by ESI or APCI. Theron et al. [61] compared ESI and APPI ionization for the analysis of venlafaxine and its metabolite *O*-desmethylvenlafaxine in plasma and water solutions. They concluded that the linearity of response in water, which is lost at high concentrations with ESI due to droplet surface saturation or limited amount of excess charge on the droplet, could be increased with APPI ionization. Moreover, APPI was less susceptible to ion suppression than ESI, with which signal response and calibration range was dramatically reduced in plasma compared to water, while similar results were obtained with APPI in both experiments. Regardless of the ionization method, positive ionization mode was employed in all cases, except for sulfate conjugates, which are more efficiently ionized in negative mode [45, 73].

4.2 Mass Analyzer

With regard to the mass analyzer, quadrupoles (Q) were usually employed due to its suitability for quantitative analysis and its relatively low prices, affordable for most laboratories. Ion trap mass spectrometers (IT) were employed in some but few LC-MS methodologies [65, 69, 74, 75]. Although these analyzers are less robust for quantitative analysis than Q, MS^n spectrums can be obtained by successive selection and fragmentation of specific product ions. Franceski et al. [94] used Q-IT tandem mass spectrometer, combining the advantages of both analyzers, for the determination of fluoxetine and its metabolite norfluoxetine. A time-of-flight (TOF) mass analyzer was used for fragmentation pathway elucidation of some TCAs [58] and sertraline [62], structural elucidation of bupropion metabolites in urine [45], and fluoxetine quantitation [95]. TOF analyzers allow accurate mass measurement with an assignation of four decimal digits, which dramatically reduces the possible elemental formula of the detected analyte; however, TOF is more expensive and has a narrower linear dynamic range than that achieved with Q and, therefore, the later is preferable for quantitative analysis of target analytes.

5 LC-MS Applications for Antidepressant Quantitative Analysis

Quantitative LC-MS analysis is usually performed by selection of the pseudomolecular ion of the analyte of interest in the selected ion monitoring (SIM) mode. In single quadrupole instruments, selection of fragment ions (*m/z*) is also possible by promoting collision induced dissociation (CID) reactions in the entrance to the mass spectrometer. Tandem LC-MS/MS instruments enhance selectivity and signal-to-noise ratio compared to single quadrupole instruments by operating on multiple reaction monitoring (MRM) mode. In MRM mode, specific precursor-to-product ion transitions can be monitored by fragmentation of the analyte in the collision cell located between the two analyzers, and subsequently monitoring a selected fragment. According to the European Commission Decision 2002/657/EC [96], quantitative LC-MS(MS) methods for confirmation purposes should include a minimum number of fragments (*m/z*) or precursor-to-product ion transitions, which depends on the resolution of the mass analyzer. For low resolution analyzers, at least the parent ion (*m/z*) and two fragments (*m/z*) in SIM mode, or two precursor-to-product ions transitions in MRM mode, should be monitored.

Quantitative analytical methods should be validated in order to guarantee the quality, accuracy, and precision of the results. This is especially important in forensic and clinical toxicology, where the correct interpretation of the toxicological findings depends on the reliability of the analytical method. Minimum parameters that should be validated in quantitative procedures include selectivity, calibration model, lower limit of quantification (LLOQ), precision (repeatability and intermediate precision), accuracy (bias), and stability. If the method is applied to specimens where the analyte(s) concentration(s) is above the validated range, dilution integrity evaluation should also be included in the validation process. In addition to these parameters, in LC-MS methods it is essential to assess suppression or enhancement of the analyte signal due to matrix effects. Additional parameters that might be evaluated are recovery, limit of detection (LOD), reproducibility, and robustness. A review on validation experiments and acceptance criteria for LC-MS analysis has been published by Peters et al. [29, 97].

Another requirement for qualitative or quantitative analysis is the use of internal standards (IS) to compensate for sample preparation or chromatographic variability. This is of particular importance in LC-MS analysis, as an adequate IS can also compensate for the negative influence of matrix effects on method precision and accuracy. Stable-isotope-labeled ISs are the most appropriate for this purpose. If a specific deuterated analogue is not commercially available, it could be substituted for deuterated substances with similar physicochemical properties to the analyte of interest. However, the use of other marketed pharmaceuticals for this purpose should be avoided, as it cannot be excluded that the patient to be monitored has taken that drug.

LC-MS(MS) quantitative methods for antidepressant determination should fulfill special requirements regarding linearity range, LLOQ or the need for high throughput, depending on the specific application for which they were developed. Conditions and studied validation parameters for selected LC-MS(MS) methodologies for antidepressant determination are shown in Table 1.

Several LC-MS and LC-MS/MS methods were developed in plasma for only one antidepressant and, sometimes, its major metabolite(s) to perform pharmacokinetic, bioavailability, or bioequivalence studies. Analytical methods developed for these purposes require very low LLOQ values and, usually, narrow linear ranges covering the low range of the therapeutic concentrations are validated. In this context, several methodologies were described for the determination of fluoxetine [94, 95, 98–100], paroxetine [44, 71, 85, 101, 102], venlafaxine [48, 61, 64, 86, 103, 104], sertraline [62, 68, 83], citalopram [46, 89] and escitalopram [105], mianserine [106, 107], mirtazapine [42], trazodone [84], nefazodone [51, 81], duloxetine [47, 50, 73], and bupropion [43]. Deuterated analogues of the analyte of interest or of other drugs were employed by few authors as IS [43, 61, 73, 81, 85, 99]; however, in most of these methods, another antidepressant or other therapeutic drug was used for this purpose.

Table 1 Selected LC-MS(MS) methods for antidepressants analysis in serum, plasma, and whole blood

Author, year [Ref]	Compounds	IS	Sample	Extraction procedure	Stationary phase	Mobile phase	IM/analyzer (detection mode)	Method validation
Zhang et al. 2000 [58]	Doxepine, imipramine, desipramine, amitriptyline, trimipramine	Imipramine-d3	Plasma (1 mL)	LLE: hexane	SB-C18 Mac Mod rapid resolution (15×2.1 mm, 3 μm)	3 mM ammonium acetate pH 3.3/ACN	ESI+/TOF (full scan)	– Selectivity – LLOQ = 1–2 ng/mL – Linearity: 1/2 to 50/100 ng/mL; $r^2 > 0.99$ Intra-/inter-day imprecision: %CV < 18% – Intra-/inter-day inaccuracy: MRE ± 15% – Relative recovery >69%
Kollroser et al., 2002 [74]	Amitriptyline, melitracem, dibenzepin, doxepin, dusolepin, opipramol, nortriptyline	lofepramine	Plasma (1 mL)	Online preparative column: Oasis HLB (50×1 mm, 30 μm)	Symmetry C18 (150×3 mm, 5 μm) Symmetry C18 guard column (20×3.9 mm, 5 μm)	Formic acid 0.1%/ACN	APCI+/IT (daughter full scan, *m/z* 100–450)	– LOD = 5 ng/mL – LLOQ = 10–50 ng/mL – Linearity: 10/50 to 800/1500 ng/mL; $r^2 > 0.99$ – Intra-/inter-day imprecision: CV <15% – Intra-/inter-day inaccuracy: MRE ± 15% – Absolute recovery >90% – Application to real specimens
Kollroser et al., 2003 [75]	Citalopram, fluvoxamine, paroxetine		Plasma (1 mL)	Online preparative column: Oasis HLB (50×1 mm, 30 μm)	Symmetry C18 (150×3 mm, 5 μm) Symmetry C18 guard column (20×3.9 mm, 5 μm)	Formic acid 0.1%/ACN	APCI+/IT (daughter full scan, *m/z* 100–400)	– Selectivity – LOD = 5 ng/mL – LLOQ = 10–20 ng/mL – Linearity: 10/20 to 600/800 ng/mL; $r^2 > 0.99$ – Intra-/inter-day imprecision: CV <15% – Intra-/inter-day inaccuracy: MRE ± 15% – Absolute recovery >86% – Matrix effect: %CV <10% ($n = 6$)

(continued)

Table 1 (continued)

Author, year [Ref]	Compounds	IS	Sample	Extraction procedure	Stationary phase	Mobile phase	IM/analyzer (detection mode)	Method validation
Gutteck et al., 2003 [108]	Amitriptyline, citalopram, clomipramine, desipramine, dibenzepin, doxepin, escitalopram, fluvoxamine, imipramine, opipramol, reboxetine, trimipramine (and some metabolites) + five neuroleptic drugs	Imipramine-d3, clomipramine-d3, doxepin-d3, chlorohaloperidol	Serum (1 mL)	LLE: *n*-hexane/DCM (4:1) or DCM, and different reconstitution volumen depending on analytes group	Silice Uptisphere C18 (125 × 2 mm, 5 μm) Guard column (8 × 2 mm)	50 mM ammonium formate pH 4/ ACN	ESI+/Q (SIM)	– Selectivity – LLOQ = 1.2–54 nmol/L – Linearity: 2.3/540 to 46/5,400 nmol/L; $r^2 > 0.99$ Intra-/inter-day imprecision: %CV < 15% – Intra-/inter-day inaccuracy: MRE ± 15% – Proccess efficiency: 60–103% – Application to real specimens
Halvorsen et al., 2003 [65]	Amitriptyline, clomipramine,doxepin, trimipramine, citalopram, fluoxetine, fluvoxamine, paroxetine	Paroxetine	Plasma and whole blood (0.05 mL)	LPME: 0.2M formic acid(aceptor)/ dodecyl acetate (organic solvent)	Inertsil 3 ODS-3 RP (50 × 3mm, 3 μm)	5 mM ammonium formate pH 2.7/ACN	ESI+/IT (full scan, *m/z* 260-335)	– LLOQ = 1-106 ng/mL – Intra-/inter-day imprecision: %CV = 11-36% – Intra-/inter-day inaccuracy: MRE ± 15% – Relative recovery: 9-75% – Matrix effect (post-column infusion experiment)
Juan et al., 2005 [66]	Fluoxetine, citalopram, venlafaxine, paroxetine	Fluvoxamine	Plasma (0.5 mL)	SPE: Oasis HLB (Auto Science AP-0IP vacuum pump)	Macherey-Nagel C18 (250 × 4.6 mm, 5 μm)	30 mM ammonium acetate (0.06% formic acid)/ ACN	ESI+/Q (SIM)	– Selectivity – LOD = 0.1-0.5 ng/mL – Linearity: 5-1000 ng/mL; $r^2 > 0.99$ – Intra-/inter-day imprecision: %CV < 15% – Intra-/inter-day inaccuracy: MRE ± 15% – Absolute recovery >73% – Carry over – Stability (20°C/24h and -20°C/3 months) – Application to real specimens

Shinozuka et al., 2006 [69]	Imipramine, desipramine, amitriptyline, nortriptyline, trimipramine, clomipramine, lofepramine, amoxapine, dosulepin, maprotiline, mianserin, setiptiline, trazodone, fluvoxamine, paroxetine, milnacipram, sulpiride, tandspirone, methylphenidate, melitracen	Diazepam	Plasma (1 mL)	SPE: Oasis HLB	Inertsil C8 (150×2 mm, 5 μm)	10 mM ammonium acetate pH5/ MeOH/ACN	SSI+/IT (full scan, *m/z* 60-500)	– LOD = 30–630 ng/mL – LLOQ = 100–1,000 ng/mL – Linearity: 100–1,000 ng/mL; $r^2 > 0.99$ – Intra-day imprecision: %CV < 15% – Intra-day inaccuracy: MRE ± 15% – Relative recovery >69%
Kirchherr et al., 2006 [53]	Amitriptyline, imipramine, trimipramine, clomipramine, citalopram, fluoxetine, fluvoxamine, paroxetine, sertraline, venlafaxine, reboxetine, viloxazine, doxepin, maprotiline, mianserine, mirtazapine, moclobemide, trazodone, opipramol (and some metabolites) + antipsychotics	Clonidine, dihydromethylrisperidone, metabenztiazurone	Serum (0.1 mL)	Protein precipitation (ACN/MeOH) Collected supernatat volume and dilution factor depending on therapeutic range of analytes group	Chromolith Speed ROD C18 monolithic column (50×4.6 mm, 5 μm)	5 mM ammonium acetate pH 3.9/MeOH	ESI+/QqQ (MRM)	– LLOQ = 0.03–28.3 ng/mL – Linearity: 1 to 20/50/100 ng/mL; 5/10 to 200/500/1,000 ng/mL, 100–10,000 ng/mL (depending on therapeutic range); $r^2 > 0.99$ – Intra-/inter-day imprecision: %CV < 15% – Intra-/inter-day inaccuracy: MRE ± 15% – Absolute recovery >90%
Sauvage et al., 2006 [76]	Amitriptyline, imipramine, desipramine, clomipramine, trimipramine, amoxapine, doxepin, maprotiline, mianserin, dothiepin, fluoxetine, paroxetine, sertraline, citalopram (and some metabolites)	Mianserine-d3, amitriptyline-d6, clomipramine-d3, imipramine-d3	Serum (0.1 mL)	Protein precipitation (ACN) Online SPE: Cyclone (50×0.5 mm, 50 μm)	XTerra MS C18 (50×2.1 mm, 5 μm)	0.1% formic acid/ACN (0.1% formic acid)	ESI+/QqQ (MRM)	– Linearity: 10–500 ng/mL; $r^2 > 0.99$ – Intra-/inter-day imprecision: %CV < 15% – Intra-/inter-day inaccuracy: MRE ± 15% – Matrix effect (post-column infusion experiment) – Dilution integrity (1/2 and 1/10): %CV <15% – Carry over – External proficiency test – Application to real specimens

(continued)

Table 1 (continued)

Author, year [Ref]	Compounds	IS	Sample	Extraction procedure	Stationary phase	Mobile phase	IM/analyzer (detection mode)	Method validation
Alves et al., 2007 [80]	Amitriptyline, nortrityline, imipramine, desipramine	Clomipramine	Plasma (1 mL)	Online SPME (PDMS/DVB, 60 µm)	Zorbax XDB RP-18 (150 × 2.1 mm, 5 µm) RP8 guard column	ammonium acetate/ACN	ESI+/Q (SIM)	– Selectivity – LOD = 0.1 ng/mL – LLOQ = 50 ng/mL – Linearity: 50–500 ng/mL; $r^2 > 0.985$ – Intra-/inter-day imprecision: %CV < 15% – Absolute recovery = 1.2–7.7%
Titier et al., 2007 [60]	Amitriptyline, clomipramine, trimipramine, imipramine, doxepin, mianserin, maprotiline, dosulepine, amoxapine, toloxatone, moclobemide, viloxazine (and some metabolites) + two other pharmaceuticals	Methylrisperidone	Whole blood (1 mL)	LLE: hexane/isoamylic alcohol (99:1); organic layer back-extracted with HCl 0.02N	Xterra RP18 (100 × 2.1 mm)	4 mM ammonium formate pH 3.2/ACN	ESI+/QqQ (MRM)	– Selectivity – LLOQ = 2 ng/mL – Linearity: 5–100 ng/mL; $r^2 > 0.99$ – Intra-/inter-day imprecision: %CV < 22% – Inaccuracy: MRE ± 15% – Relative recovery = 13–82% – Matrix effect <16% – Dilution integrity (1/10 and 1/50): %CV <15%, MRE ± 15% – Application to real specimens

Castaing et al., 2007 [109]	Fluoxetine, paroxetine, sertraline, fluvoxamine, citalopram, milnacipram, venlafaxine, mirtazapine (and some metabolites)	Methylrisperidone	Whole blood (0.5 mL)	LLE: hexane/isoamylic alcohol (99:1); organic layer back-extracted with HCl 0.05N	Xterra RP18 (100 × 2.1 mm)	4 mM ammonium formate pH 3.2/ACN	ESI+/QqQ (MRM)	– Selectivity – LLOQ = 5–20 ng/mL – Linearity: 5/20 to 500/2,000 ng/mL; $r^2 > 0.99$ – Intra-/inter-day imprecision: %CV < 15% – Inter-day inaccuracy: MRE ± 15% – Relative recovery = 10–85% – Matrix effect <20% – Dilution integrity (1/10): MRE ± 15% – Application to real specimens
de Castro et al., 2007 [72]	Amitriptyline, nortriptyline, imipramine, desipramine, trazodone, fluoxetine, paroxetine, fluvoxamine, sertraline	Imipramine-d3, desipramine-d3, clomipramine-d3, fluoxetine-d6, paroxetine-d6	Plasma (0.05 mL)	Online SPE: Oasis MCX Prospekt cartridge	Gemini C18 (150 × 2 mm, 5 µm) Gemini C18 guard column (4 × 2 mm, 5 µm)	10 mM ammonium hydrogencarbonate pH 10/ACN	ESI+/QqQ (MRM)	– Selectivity – LOD = 0.5–1 ng/mL – LLOQ = 10 ng/mL – Linearity: 10–1,000 ng/mL; $r^2 > 0.99$ – Intra-/inter-day imprecision: %CV < 15% – Inaccuracy: MRE ± 15% – Relative recovery >99% – Matrix effect <20% – Autosampler (6 ± 2 °C/24 and 48 h) and freeze/thaw stability – Application to real specimens

(continued)

Table 1 (continued)

Author, year [Ref]	Compounds	IS	Sample	Extraction procedure	Stationary phase	Mobile phase	IM/analyzer (detection mode)	Method validation
Santos-Neto et al., 2008 [77]	Fluoxetine, imipramine, desipramine, amitriptyline, nortriptyline	Fluoxetine-d5	Plasma/Urine (0.2 mL)	LiChrospher ADS-C18 capillary RAM extraction precolumn (60 Å, 25 µm)	YMC ODS-AQ capillary column (120 Å, 5 µm)	20 mM ammonium formate pH 5.4/ ACN	ESI+/QqQ (MRM)	In plasma: – LLOQ = 1 ng/mL – Linearity: 1–250 ng/mL; $r^2 > 0.99$ – Intra-/inter-day imprecision: %CV < 20% – Inter-day inaccuracy: MRE ± 20% – Relative recovery >80% – Matrix effect (post-column infusion experiment) – Carry over: 0.3% signal from the previous injection
Breaud et al., 2010 [78]	Amitriptyline, nortrityline, imipramine, desipramine	Amitrityline-d3, nortriptyline-d3, imipramine-d3, desipramine-d3	Plasma (10 µL)	Turbulent-flow, Cyclone-P column (0.5 × 50 mm)	Hypersil Gold C18 (50 × 3 mm, 5 µm)	0.1% formic acid/ ACN (0.1% formic acid)	ESI+/QqQ (MRM)	– Selectivity – LOD = 2–3 ng/mL – LLOQ = 9–18 ng/mL – Linearity: 69/74 to 1131/1155 ng/mL; $r^2 > 0.99$ – Intra-/inter-day imprecision: %CV < 20% – Intra-day inaccuracy: MRE ± 15% – External proficiency test: ±2 SD of target concentration – Matrix effect <20% – Carry over – Application to real specimens

Remane et al., 2010 [59]	Imipramine, desipramine, amitriptyline, trimipramine, clomipramine, amoxapine, atomoxetine, bupropion, citalopram, cyclobenzapride, dosulepin, doxepin, maprotiline, mianserin, mirtazapine, moclobemide, trazodone, fluoxetine, fluvoxamine, paroxetine, opipramol, reboxetine, sertraline, tianeptin, venlafaxine, viloxazine (and some metabolites) + other pharmaceuticals	Citalopram-d6, trimipramine-d3, diazepam-d5, ketamine-d3, norclozapine-d8, nordiazepam-d5, zolpidem-d6	Plasma (0.5 mL)	LLE: butyl acetate/ethyl acetate (50:50)	TF Hypersil GOLD Phenyl (100 × 2.1 mm, 1.9 μm)	10 mM ammonium formate (0.1% formic acid)/ ACN	APCI+/QqQ (MRM)	– Relative recovery = 8–84% – Matrix effect = 1 – 1,082% – Proccess efficiency = 4 – 423%

LLE liquid–liquid extraction, *SPE* solid phase extraction, *LPME* liquid-phase microextraction, *ESI* electrospray ionization, *APCI* atmospheric pressure chemical ionization, *SSI* sonic spray ionization, *Q* quadrupole, *QqQ* triple quadrupole, TOF: time-of-flight, *IT* ion trap, *IS* internal standard, *CV* coefficient of variation, *MRE* mean relative error, *LOD* limit of detection, *LLOQ* lower limit of quantification

Although it is not recommended, the use of commercially available pharmaceuticals as IS could be only justified if the method is applied to a controlled administration study in which the presence of that IS in the biological matrix can be ruled out.

Analytical methods for TDM should cover therapeutic ranges, and LLOQ requirements are not as low as those for pharmacokinetic applications. LC-MS methodologies for the determination of one specific antidepressant could also be applied in these cases. However, as previously stated in this chapter, multianalyte procedures are preferable because they are simpler, faster, cheaper, and allow for the determination of several analytes using the same sample aliquot. Moreover, multianalyte procedures are not only useful, but also required in clinical or forensic applications, where the target antidepressant is initially unknown, and several of them could be involved in the intoxication.

LC-MS(MS) methods including several antidepressants will be described in more detail. Some of these methods include analytes belonging to the same antidepressant group. Alves et al. [80] developed an LC-MS method for the determination of some TCAs in plasma with a total run time of 18 min. Kollroser et al. described two LC-IT-MS procedures in plasma for the determination of seven TCAs [74] and three SSRIs [75], respectively, using an online preparative column. Total analysis time per sample was 12 and 6 min, respectively. The main disadvantage of these methods is that a therapeutic drug was used as IS, hindering its applicability to TDM or clinical and forensic toxicology analysis. Also Juan et al. [66] used fluvoxamine as IS for the determination of several SSRIs; however, the authors argue that this drug was not commercially available in their region, and that patients' history was studied in order to exclude fluvoxamine intake. Breaud et al. [78] developed a LC-MS/MS method for two TCAs and their main metabolites by turbulent flow LC-MS/MS, which allowed analyte determination from only 10 μL of plasma, with a total analysis time of 3.5 min from the second injection. Zhang et al. [58] used a LC-TOF-MS instrument for accurate mass measurement of TCAs in 18 s, using reference standards and plasma samples fortified with these analytes; however, validated linearity ranges were below the upper therapeutic concentrations and, therefore, evaluation of dilution integrity would be needed in order to analyze specimens from patients on antidepressant treatment or from intoxication cases. More details on validation experiments performed in these methods are shown in Table 1.

Other authors described LC-MS or LC-MS/MS methods in serum, plasma or whole blood for the determination of several analytes belonging to different antidepressant groups. LC-MS methods were developed by Halvorsen et al. [65] and Gutteck et al. [108] using an ion trap and a single quadrupole mass analyzer, respectively, with chromatographic elution of the analytes included in each method within 10 min. Analytes included in Gutteck's et al. method were divided in four groups depending on their therapeutic ranges, and slightly different extraction and chromatographic procedures were applied to each group. Shinozuka et al. [69] extended the number of antidepressants using a LC-IT-MS instrument, and chromatographic separation was performed in 30 min. Tandem LC-MS/MS instruments were also employed in several occasions [53, 59, 60,

72, 77, 109]. Kirchherr et al. [53] employed protein precipitation of 0.1 mL of serum for the determination of 48 psycotherapeutics, including the most common antidepressants, with a total chromatographic run of 8 min. Also Remane et al. [59] developed a multianalyte procedure for the determination of 136 analytes from different drug classes using a generic LLE procedure; however, only matrix effect, recovery, and process efficiency were evaluated for method validation. Santos-Neto et al. [77] developed a LC-MS/MS method for the analysis of some antidepressants in plasma and urine using a monolithic column for simultaneous online sample preparation and chromatographic separation, with a total analysis time of 8 min. Sauvage et al. [76] extended the number of antidepressants determined in 0.1 mL of serum using turbulent flow LC-MS/MS, and decreased the total analysis time to 6 min. De Castro et al. [72] developed an online SPE-LC-MSMS procedure using 0.05 mL of plasma, where the complete analytical process for the determination of nine antidepressants was performed in 20 min. Table 1 shows the parameters evaluated for method validation, as well as analytes employed as IS, in each of these methods.

Although most LC-MS(MS) methods for antidepressants were applied to plasma, serum, or whole blood specimens, some analytical methodologies were also described in other biological matrices. Petsalo et al. [45] identified 20 bupropion metabolites in urine by accurate mass measurement using a LC-TOF-MS instrument. De Santana [42] applied a LC-MS/MS method with SPME extraction to urine samples to assess cumulative urinary excretion of mirtazapine and two metabolites. LC-MS/MS methods were also developed to compare enantiomers disposition of reboxetine [110], and bupropion and its metabolite [43] in plasma and urine specimens. Bupropion and some metabolites were also determined in urine by LC-MS/MS in a fatal overdose case where this antidepressant was involved [70]. LC-MS/MS was also applied to the analysis of other typical postmortem specimens, including gastric content, bile, vitreous humor, cerebrospinal fluid, brain, liver, lungs, kidney, or muscle, usually for the determination of one antidepressant and its main metabolite [38–40]. Goeringer et al. [57] extended the number of analytes included in the method to seven antidepressants, some metabolites and one antipsychotic in different forensic specimens. Unfortunately, validation performed in most of these methods was poor and, sometimes, only described for some of the analyzed specimens.

Few authors described antidepressant analysis in alternative specimens, such as hair or oral fluid. LC-CID-MS and MS/MS mass spectra libraries for identification of several drugs were employed by Müller et al. [32] for the detection of maprotiline, citalopram, and their desmethyl metabolites in authentic hair specimens; extracted ions chromatograms were employed for subsequent antidepressant quantification. Also Klys et al. [33] applied LC-MS/MS to the analysis of blood, urine, and hair specimens in a fatal case due to clomipramine overdose in combination with alcohol. Blood clomipramine and norclomipramine concentrations explained the fatal outcome, and hair analysis confirmed that the deceased was on clomipramine treatment for, at least, 12 months prior to his death. With regard to oral fluid analysis, de Castro et al. [34] developed and validated a

LC-MS/MS procedure for the determination of the main marketed antidepressants in plasma and oral fluid to assess the possible correlation between the concentrations found in these two biological matrices. Also Coulter et al. [35] validated a LC-MS/MS methodology for several antidepressants and two other psychopharmaceuticals in oral fluid for its application to driving under the influence of drugs (DUID) cases.

6 LC-MS/MS Method for the Determination of Nine Antidepressants and Some of their Main Metabolites in Oral Fluid and Plasma

An LC-MS/MS method [34] will be described in detail in order to serve as an example of method development and validation. This method allows the determination of some of the most common antidepressants used in clinical practice and some of their main metabolites in oral fluid and plasma samples.

Plasma is the main biological sample used in clinical and toxicological analysis, as concentrations found in this matrix are correlated to the pharmacological effect, as well as to the side and toxic effects. However, oral fluid has also been employed in some specific applications because of the advantages associated to this alternative specimen: easy, painless, and noninvasive collection, which does not require qualified personnel, it represents the free analyte fraction, and it has a window of detection similar to that in plasma. Within the possible applications of oral fluid analysis, two are of special relevance:

1. Detection of subjects driving under the influence of sedative antidepressants effects.
2. TDM: As previously stated in this chapter, TDM is recommended for TCAs and, in special situations, for the new generations of antidepressants. However, several factors can affect diffusion of the analytes from plasma to oral fluid (pH, oral contamination, collection with or without stimulation). Therefore, the correlation between antidepressant plasma and oral fluid concentrations should be studied before using this alternative specimen for TDM purposes.

An example of method application to a study to assess the possible correlations between antidepressant concentrations in plasma and oral fluid will also be discussed in this section.

6.1 LC-MS/MS Method

A Waters Alliance 2795 Separation Module with a Waters Alliance series column/heater (Waters Corp., Milford, MA) was employed, using a Sunfire C18 (20 mm × 2.1 mm, 3.5 μm) Intelligent Speed™ column (Waters Corp., Milford, MA) for chromatographic separation at 26 °C. Mobile phase was ammonium formate buffer (pH 3.0; 2 mM) and ACN at a flow rate of 0.4 mL/min, applying the following gradient: 15% ACN for 0.5 min, increased to 50% over 3.5 min, and increased again to 70% at minute 5. With these conditions, all compounds eluted within 5 min, with a total run time of 8 min. Figure 2 shows the chromatograms of the 13 antidepressants in oral fluid at the LLOQ (2 ng/mL).

For the detection, a tandem mass spectrometer Quattro Micro™ API ESCI (Waters Corp., Milford, MA) with a triple quadrupole was employed. The instrument was operated in electrospray in the positive ionization mode (ESI+) with the following optimized parameters: capillary voltage, 0.5 kV; source block temperature, 130 °C; nebulization and desolvation gas (nitrogen) heated at 400 °C and delivered at 800 L/h, and as cone gas at 50 L/h; collision cell pressure, 3×10^{-6} bar (argon). Data was recorded in the multiple reaction monitoring (MRM) mode by selection of the two most intense precursor-to-product ion transitions for each analyte, except for the ISs, for which only one transition was monitored. The most intense transition for each analyte was used for quantitative purposes. Table 2 shows MRM transitions, cone voltages and collision energies used for the analysis of the antidepressants included in the LC-MS/MS method.

6.2 *Oral Fluid and Plasma Extraction Procedure*

Solid phase extraction (SPE) was performed with an ASPEC XL automated system (Gilson, Middletown, USA) and mixed mode OASIS MCX cartridges 60 mg 3 cm^3 (Waters Corp., Milford, USA). Before SPE, 1 mL of sodium acetate buffer pH 3.6 and 50 μL of IS mixture (nortriptyline-d3, clomipramine-d3, paroxetine-d6, norfluoxetine-d6, and fluoxetine-d6) at 0.2 mg/L in oral fluid and 0.4 mg/L in plasma were added to 0.2 mL of sample. The applied SPE procedure is summarized in Fig. 3.

6.3 *Method Validation*

LC-MS/MS method validation was performed as follows:

6.3.1 Selectivity

Selectivity of the method was evaluated by analysis of blank oral fluid and plasma specimens from ten healthy subjects. In addition, potential exogenous interferences were assessed by analysis of authentic plasma specimens containing other common therapeutic drugs like benzodiazepines and/or drugs of abuse. No endogenous or exogenous interferences were found in the monitored MRM channels in any of the analyzed specimens.

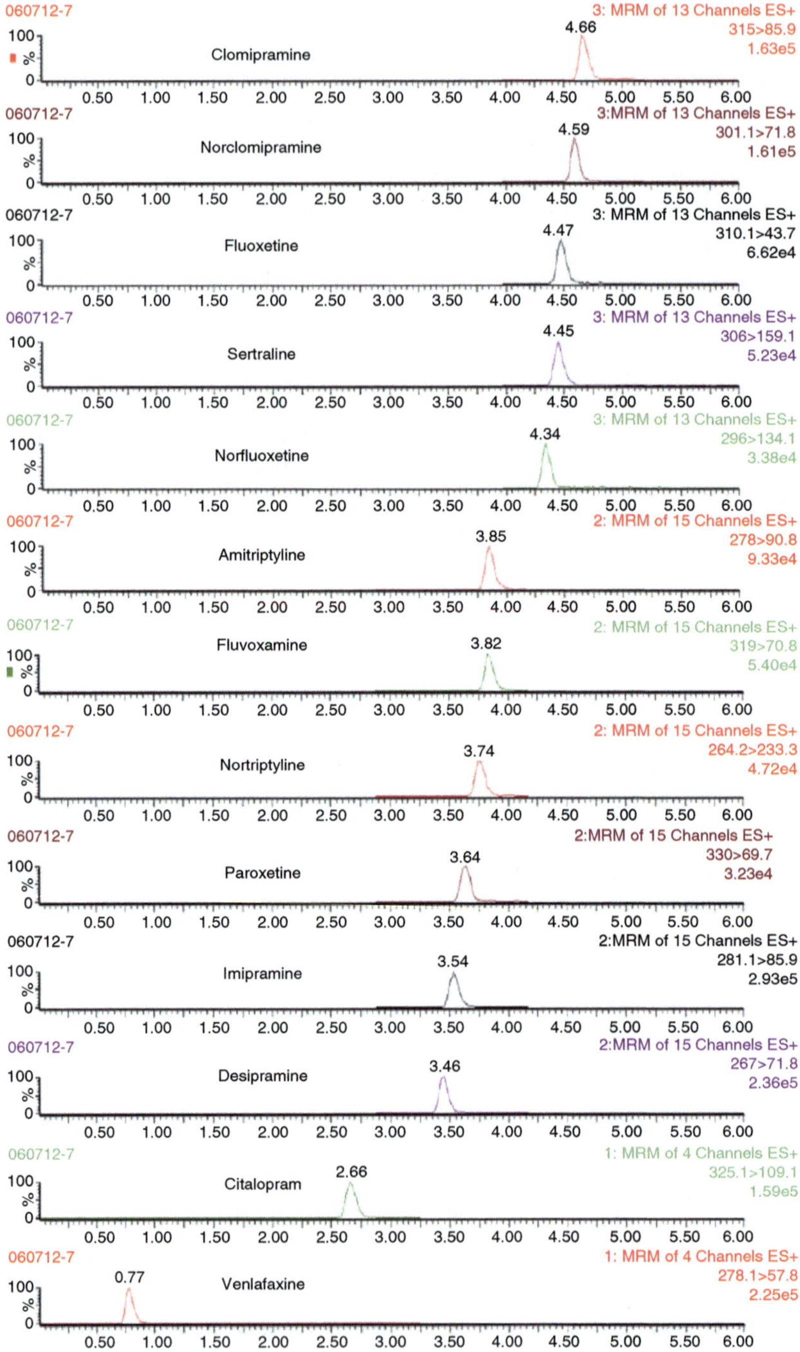

Fig. 2 Chromatograms of the quantitative transitions for the 13 antidepressants included in the LC-MS/MS method in oral fluid at the LLOQ (2 ng/mL) (*Source*: From de Castro A et al. (2008) J Pharm Biomed Anal. 48:183. With permission)

Table 2 MRM transitions, cone voltages (CV), and collision energies (CE) used in the analysis of the antidepressants and the deuterated internal standards included in the LC-MS/MS method

Compound	CV	Transition	CE
Venlafaxine	25	278.1>57.8	18
		278.1>260.3	12
Citalopram	35	325.1>109.1	26
		325.1>262.1	20
Desipramine	25	267.0>43.8	40
		267.0>71.8	15
Imipramine	25	281.1>57.6	40
		281.1>85.9	16
Paroxetine	35	330.0>69.7	30
		330.0>192.3	20
Nortriptyline	25	264.2>90.8	22
		264.2>233.3	16
Fluvoxamine	25	319.0>70.8	18
		319.0>86.8	18
Amitriptyline	30	278.1>90.8	24
		278.1>233.2	18
Norfluoxetine	15	296.0>29.8	10
		296.0>134.1	6
Sertraline	18	306.0>159.1	26
		306.0>275.1	12
Fluoxetine	22	310.1>43.7	12
		310.1>148.2	8
Norclomipramine	25	301.1>71.8	18
		301.1>270.2	16
Clomipramine	22	315.0>57.8	32
		315.0>85.9	20
Imipramine-3	25	284.1>89.0	15
Paroxetine-d6	35	336.1>76.0	30
Nortriptyline-d3	30	267.1>233.3	15
Norfluoxetine-d6	15	302.2>140.1	6
Fluoxetine-d6	25	316.1>43.7	12
Clomipramine-d3	30	318.1>88.9	18

Underlined transitions were used as quantifiers (*Source*: From de Castro A et al. (2008) J Pharm Biomed Anal. 48:183. With permission)

6.3.2 Calibration Model

Calibration curves ($n=6$, six different days) were generated at eight concentration levels in the range from 2 to 500 ng/mL and 2–1,000 ng/mL for oral fluid and plasma, respectively. Coefficient of determination (r^2) was >0.99 for all analytes using a $1/x$ weighted linear regression, except for fluvoxamine, for which a quadratic response was observed.

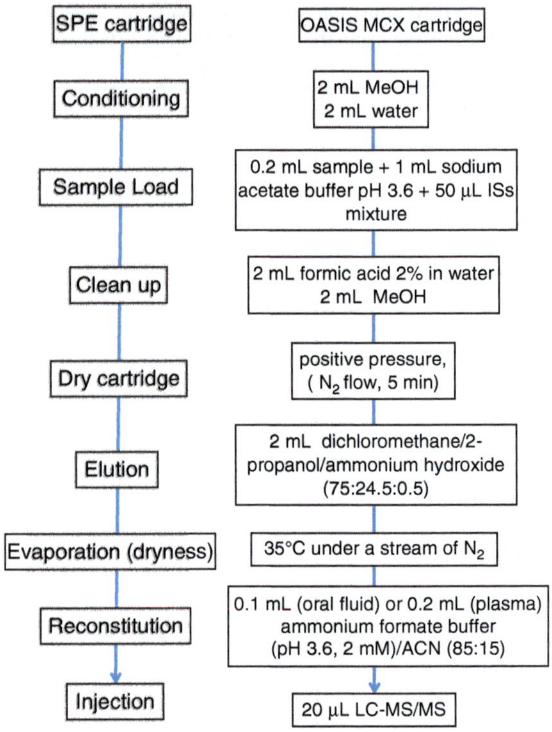

Fig. 3 Solid-phase extraction (SPE) procedure applied to plasma and oral fluid samples for LC-MS/MS determination of nine antidepressants and some metabolites

6.3.3 Limit of Detection (LOD) and Limit of Quantification (LLOQ)

LOD was defined as the lowest concentration for which the two monitored transitions could be detected, and the peak area of the quantifier transition was, at least, three times the background noise. LOD was 0.5 ng/mL for all analytes. LLOQ was defined as the lowest concentration that could be quantified with appropriate imprecision (coefficient of variation (%CV) <20%) and inaccuracy (mean relative error (MRE) ± 20%). LLOQ was 2 ng/mL in oral fluid, and 2–4 ng/mL in plasma for most of the analytes. Carryover after automated SPE of clomipramine at 1,000 ng/mL was half of the signal obtained at 2 ng/mL; therefore, LLOQ in plasma was increased to 10 ng/mL for this analyte.

6.3.4 Intra- and Interday Imprecision and Inaccuracy

Intraday imprecision and inaccuracy were evaluated at low, medium and high concentrations within the validated concentration ranges for each matrix by analysis of five replicates on the same day. *Interday imprecision and inaccuracy* were assessed by analysis of five replicates at the same concentrations on five different days. MRE was ±15% and %CV <17% in all cases.

6.3.5 Relative Ions Intensity

Intraday imprecision of *relative ions intensity* for each analyte was evaluated by analysis of five replicates at four concentration levels analyzed on the same day, and interday imprecision on this parameter was calculated at the same concentration levels ($n = 5$ each) analyzed on five different days. %CV for the relative ions intensities was <17% in all cases, except for norfluoxetine, for which %CV was <26%.

6.3.6 Extraction Recovery

Extraction recovery was assessed at low and high concentrations by comparing analyte-to-IS peak area ratio when standards were added before extraction ($n=5$) to that obtained when added after extraction ($n=5$). IS mixture was added after extraction in all cases. Recoveries ranged from 49 to 72 % for all analytes.

6.3.7 Matrix Effect

Matrix effect was initially evaluated by postcolumn infusion of a mixture containing the analytes and the ISs (1 μg/mL, 10 μL/min) in "T" with the effluent of the chromatographic system. Chromatograms after the injection of extracted blank plasma ($n=6$) and oral fluid samples ($n=6$) from different sources were compared with the chromatograms after the injection of mobile phase (no matrix effect). Quantification of matrix effect was performed comparing average peak areas of blank plasma ($n=6$) and oral fluid samples ($n=6$) fortified with the analytes (100 ng/mL) and ISs after extraction to those obtained when the same amount of analytes and ISs were added to a clean tube, evaporated and reconstituted in mobile phase. Matrix effect was <15% for all analytes, except for norfluoxetine (signal enhancement <45%) and paroxetine (signal enhancement <30%). Nevertheless, inclusion of the deuterated analogues for these two analytes compensate for the possible errors in imprecision, inaccuracy and recovery.

6.3.8 Stability Study

Antidepressant *stability* after three freeze–thaw cycles was evaluated in triplicate at low and high concentrations. Calculated concentrations in the samples subjected to these conditions (stability samples) were compared to those obtained in freshly prepared samples (control samples). Stability and control samples were quantified with a calibration curve prepared on the day of the analysis. All analytes were stable under these conditions, except sertraline, for which a slight signal decrease was observed at 250 ng/mL in oral fluid (MRE = −33.4%; %CV = 6.0%).

6.4 Oral Fluid-Plasma Preliminary Correlation Study

Initially, plasma and oral fluid specimens from patients ($n = 21$) on different antidepressant treatment were collected twice to assess if any of the studied analytes was likely to show a good correlation. The best results were obtained for venlafaxine (%CV for plasma/oral fluid concentrations ratio ($R_{OF/PL}$) <27%). Therefore, the study was extended for this antidepressant by analysis of oral fluid and plasma specimens from five patients on venlafaxine treatment collected on four occasions. Daily doses of venlafaxine retard formulations were 75 mg for two patients, and 150 mg for the remaining participants. Collection of oral fluid (direct spitting into polypropylene tubes) and plasma (heparinized tubes) specimens was performed, when possible, before the next dose to ensure the drug was in the elimination phase. The dose and the time of collection was the same on the four different occasions for each patient. For the analysis, oral fluid and plasma specimens were centrifuged at 14×10^3 rpm, and 0.2 mL of the supernatant were extracted. In addition, correlation between the concentrations in the plasmatic free fraction and in oral fluid was also evaluated. Plasmatic proteins were eliminated by filtering 0.5 mL of plasma samples using Microcon filter devices Ultracel YM-3 (Millipore Corp., Billerica, MA, USA).

Table 3 shows plasmatic and oral fluid concentrations, ratios between those concentrations and calculated coefficients of determination for the five participants on venlafaxine treatment. Results were not analyzed interindividually due to the low number of patients and differences in the daily dosage. As expected, oral fluid concentrations were higher than those found in plasma in all cases. This is explained by venlafaxine's weak basic properties, which favors oral fluid accumulation due to the slightly lower pH of this matrix compared to that in plasma. Intraindividually, plasmatic concentrations (C_{PL}) were similar in the four analyzed specimens, but higher variability was found in oral fluid (C_{OF}). Also, a great variability was observed in plasmatic free fraction concentrations ($C_{PL\text{-}FF}$), which could be favored by some retention of the analytes in the filter used to eliminate the plasmatic proteins. Correlation between C_{OF} vs. C_{PL} or $C_{PL\text{-}FF}$ on the four different days was analyzed by linear regression. Only r^2 values for one patient were >0.6; therefore, our data indicates that a correlation between venlafaxine concentrations in oral fluid and plasma or the plasmatic free fraction is unlikely. Nevertheless, a strong conclusion would require the extension of the study, including a higher number of patients on the same daily dose, collecting more samples from each patient, and standardizing the interval time between drug administration and sample collection. Although our data do not support oral fluid analysis for antidepressant TDM, this alternative matrix could be employed in special situations to assess patients' noncompliance or to detect impaired drivers under antidepressant influence.

Table 3 Results from venlafaxine study

Patient	Venlafaxine daily dose (mg)		Min.	Max.	Mean	%CV		Min.	Max.	Mean	%CV	r^2
1	150	C_{PL}	74.7	125.7	97.3	21.8	$R_{OF/PL}$	2.4	5.5	4.0	30.9	0.631
		$C_{PL\text{-}FF}$	39.0	134.1	84.9	53.0	$R_{OF/PL\text{-}FF}$	4.0	6.8	5.0	25.6	0.848
		C_{OF}	182.5	531.2	397.7	39.6						
2	150	C_{PL}	52.4	72.6	60.6	14.8	$R_{OF/PL}$	4.2	7.7	5.5	28.3	0.239
		$C_{PL\text{-}FF}$	37.2	46.2	42.1	11.2	$R_{OF/PL\text{-}FF}$	6.5	8.7	7.7	12.9	0.332
		C_{OF}	285.6	404.7	323.9	16.8						
3	150	C_{PL}	25.1	44.9	33.8	24.7	$R_{OF/PL}$	2.3	8.6	4.3	69.6	0.027
		$C_{PL\text{-}FF}$	6.8	14.0	11.2	27.7	$R_{OF/PL\text{-}FF}$	7.3	23.3	12.5	58.8	0.101
		C_{OF}	58.8	265.0	140.9	63.1						
4	75	C_{PL}	40.0	51.1	44.8	11.6	$R_{OF/PL}$	3.6	6.3	4.8	24.2	0.245
		$C_{PL\text{-}FF}$	13.8	20.9	17.0	17.3	$R_{OF/PL\text{-}FF}$	6.9	16.9	13.1	33.6	0.252
		C_{OF}	144.4	289.2	215.7	27.4						
5	75	C_{PL}	17.5	25.7	22.5	15.6	$R_{OF/PL}$	3.4	9.0	6.9	35.7	0.178
		$C_{PL\text{-}FF}$	8.3	16.6	12.4	37.7	$R_{OF/PL\text{-}FF}$	4.9	25.0	14.4	57.2	0.028
		C_{OF}	82.1	214.7	156.8	41.6						

Venlafaxine concentrations in plasma (C_{PL}), in the plasmatic free fraction ($C_{PL\text{-}FF}$) and in oral fluid (C_{OF}), plasmatic or plasmatic free fraction/oral fluid concentration ratios ($R_{OF/PL}$ or $R_{OF/PL\text{-}FF}$, respectively) and the calculated coefficients of determination (r^2) for C_{OF} vs. C_{PL} or $C_{PL\text{-}FF}$ ($n=4$) (*Source*: From de Castro A et al. (2008) J Pharm Biomed Anal. 48:183. With permission)

References

1. Drasch G, Dahlmann F, von Meyer L et al (2008) Frequency of different anti-depressants associated with suicides and drug deaths. Int J Legal Med 122:115–121
2. Jonsson A, Holmgren P, Ahlner J (2004) Fatal intoxications in a Swedish forensic autopsy material during 1992–2002. Forensic Sci Int 143:53–59
3. Ohberg A, Vuori E, Klaukka T et al (1998) Antidepressants and suicide mortality. J Affect Disord 50:225–233
4. Shields LB, Hunsaker DM, Hunsaker JC III et al (2006) Toxicologic findings in suicide: a 10-year retrospective review of Kentucky medical examiner cases. Am J Forensic Med Pathol 27:106–112
5. Jones G, Singer P (2001) Miscellaneous prescription and over-the-counter medications. In: LeBeau M, Mozayani A (eds) Drug-facilitated sexual assault. A forensic handbook, 1st edn. Academic Press, San Diego, CA
6. Verster JC, Mets MA (2009) Psychoactive medication and traffic safety. Int J Environ Res Public Health 6:1041–1054
7. Brunnauer A, Laux G, Geiger E et al (2006) Antidepressants and driving ability: results from a clinical study. J Clin Psychiatry 67:1776–1781
8. Ramaekers JG (2003) Antidepressants and driver impairment: empirical evidence from a standard on-the-road test. J Clin Psychiatry 64:20–29
9. Walsh JM, Verstraete AG, Huestis MA et al (2008) Guidelines for research on drugged driving. Addiction 103:1258–1268
10. Mann K, Hiemke C, Schmidt LG et al (2006) Appropriateness of therapeutic drug monitoring for antidepressants in routine psychiatric inpatient care. Ther Drug Monit 28:83–88
11. Mitchell PB (2004) Therapeutic drug monitoring of non-tricyclic antidepressant drugs. Clin Chem Lab Med 42:1212–1218
12. Lundmark J, Bengtsson F, Nordin C et al (2000) Therapeutic drug monitoring of selective serotonin reuptake inhibitors influences clinical dosing strategies and reduces drug costs in depressed elderly patients. Acta Psychiatr Scand 101:354–359
13. Burke MJ, Preskorn SH (1999) Therapeutic drug monitoring of antidepressants: cost implications and relevance to clinical practice. Clin Pharmacokinet 37:147–165
14. Hackett LP, Dusci LJ, Ilett KF (1998) A comparison of high-performance liquid chromatography and fluorescence polarization immunoassay for therapeutic drug monitoring of tricyclic antidepressants. Ther Drug Monit 20:30–34
15. Meenan GM, Barlotta S, Lehrer M (1990) Urinary tricyclic antidepressant screening: comparison of results obtained with Abbott FPIA reagents and Syva EIA reagents. J Anal Toxicol 14:273–276
16. Asselin WM, Leslie JM (1990) Use of the EMITtox serum tricyclic antidepressant assay for the analysis of urine samples. J Anal Toxicol 14:168–171
17. Banger M, Hermes B, Hartter S et al (1997) Monitoring serum concentrations of clomipramine and metabolites: fluorescence polarization immunoassay versus high performance liquid chromatography. Pharmacopsychiatry 30:128–132
18. Benitez J, Dahlqvist R, Gustafsson LL et al (1986) Clinical pharmacological evaluation of an assay kit for intoxications with tricyclic antidepressants. Ther Drug Monit 8:102–105
19. Yuan CM, Spandorfer PR, Miller SL et al (2003) Evaluation of tricyclic antidepressant false positivity in a pediatric case of cyproheptadine (periactin) overdose. Ther Drug Monit 25:299–304
20. Caravati EM, Juenke JM, Crouch BI et al (2005) Quetiapine cross-reactivity with plasma tricyclic antidepressant immunoassays. Ann Pharmacother 39:1446–1449
21. Maurer HH, Bickeboeller-Friedrich J (2000) Screening procedure for detection of antidepressants of the selective serotonin reuptake inhibitor type and their metabolites in urine as part of a modified systematic toxicological analysis procedure using gas chromatography-mass spectrometry. J Anal Toxicol 24:340–347

22. Lacassie E, Ragot S, Gaulier JM et al (1999) A specific dosage method for the analysis of 24 antidepressants using gas chromatography-mass spectrometry (GC/SM). Acta Clin Belg Suppl 1:20–24
23. Way BA, Stickle D, Mitchell ME et al (1998) Isotope dilution gas chromatographic-mass spectrometric measurement of tricyclic antidepressant drugs. Utility of the 4-carbethoxyhexafluorobutyryl derivatives of secondary amines. J Anal Toxicol 22:374–382
24. Salgado-Petinal C, Lamas JP, Garcia-Jares C et al (2005) Rapid screening of selective serotonin re-uptake inhibitors in urine samples using solid-phase microextraction gas chromatography-mass spectrometry. Anal Bioanal Chem 382:1351–1359
25. Eap CB, Bouchoux G, Amey M et al (1998) Simultaneous determination of human plasma levels of citalopram, paroxetine, sertraline, and their metabolites by gas chromatography-mass spectrometry. J Chromatogr Sci 36:365–371
26. Matuszewski BK, Constanzer ML, Chavez-Eng CM (2003) Strategies for the assessment of matrix effect in quantitative bioanalytical methods based on HPLC-MS/MS. Anal Chem 75:3019–3030
27. Annesley TM (2003) Ion suppression in mass spectrometry. Clin Chem 49:1041–1044
28. Taylor PJ (2005) Matrix effects: the Achilles heel of quantitative high-performance liquid chromatography-electrospray-tandem mass spectrometry. Clin Biochem 38:328–334
29. Peters FT, Drummer OH, Musshoff F (2007) Validation of new methods. Forensic Sci Int 165:216–224
30. Thieme D, Rolf B, Sachs H et al (2008) Correlation of inter-individual variations of amitriptyline metabolism examined in hairs with CYP2C19 and CYP2D6 polymorphisms. Int J Legal Med 122:149–155
31. Smyth WF, Leslie JC, McClean S et al (2006) The characterisation of selected antidepressant drugs using electrospray ionisation with ion trap mass spectrometry and with quadrupole time-of-flight mass spectrometry and their determination by high-performance liquid chromatography/electrospray ionisation tandem mass spectrometry. Rapid Commun Mass Spectrom 20:1637–1642
32. Muller C, Vogt S, Goerke R et al (2000) Identification of selected psychopharmaceuticals and their metabolites in hair by LC/ESI-CID/MS and LC/MS/MS. Forensic Sci Int 113:415–421
33. Klys M, Scislowski M, Rojek S et al (2005) A fatal clomipramine intoxication case of a chronic alcoholic patient: application of postmortem hair analysis method of clomipramine and ethyl glucuronide using LC/APCI/MS. Leg Med (Tokyo) 7:319–325
34. de Castro A, Concheiro M, Quintela O et al (2008) LC-MS/MS method for the determination of nine antidepressants and some of their main metabolites in oral fluid and plasma. Study of correlation between venlafaxine concentrations in both matrices. J Pharm Biomed Anal 48:183–193
35. Coulter C, Taruc M, Tuyay J et al (2010) Antidepressant drugs in oral fluid using liquid chromatography-tandem mass spectrometry. J Anal Toxicol 34:64–72
36. Lobo ED, Loghin C, Knadler MP et al (2008) Pharmacokinetics of duloxetine in breast milk and plasma of healthy postpartum women. Clin Pharmacokinet 47:103–109
37. Goeringer KE, McIntyre IM, Drummer OH (2001) Postmortem tissue concentrations of venlafaxine. Forensic Sci Int 121:70–75
38. Avella J, Lehrer M, Katz M et al (2004) Two cases involving clomipramine intoxication. J Anal Toxicol 28:504–508
39. Gronewold A, Dettling A, Haffner HT et al (2009) Doxepin and nordoxepin concentrations in body fluids and tissues in doxepin associated deaths. Forensic Sci Int 190:74–79
40. Wenzel S, Aderjan R, Mattern R et al (2006) Tissue distribution of mirtazapine and desmethylmirtazapine in a case of mirtazapine poisoning. Forensic Sci Int 156:229–236
41. Society of Hair Testing (2004) Consensus on hair analysis. http://www.soht.org/. Accessed 2 Dec 2010
42. de Santana FJ, Bonato PS (2008) Enantioselective analysis of mirtazapine and its two major metabolites in human plasma by liquid chromatography-mass spectrometry after three-phase liquid-phase microextraction. Anal Chim Acta 606:80–91

43. Coles R, Kharasch ED (2007) Stereoselective analysis of bupropion and hydroxybupropion in human plasma and urine by LC/MS/MS. J Chromatogr B Anal Technol Biomed Life Sci 857:67–75
44. Petsalo A, Turpeinen M, Tolonen A (2007) Identification of bupropion urinary metabolites by liquid chromatography/mass spectrometry. Rapid Commun Mass Spectrom 21:2547–2554
45. Segura M, Ortuno J, Farre M et al (2003) Quantitative determination of paroxetine and its 4-hydroxy-3-methoxy metabolite in plasma by high-performance liquid chromatography/electrospray ion trap mass spectrometry: application to pharmacokinetic studies. Rapid Commun Mass Spectrom 17:1455–1461
46. Jiang T, Rong Z, Peng L et al (2010) Simultaneous determination of citalopram and its metabolite in human plasma by LC-MS/MS applied to pharmacokinetic study. J Chromatogr B Anal Technol Biomed Life Sci 878:615–619
47. Ma N, Zhang BK, Li HD et al (2007) Determination of duloxetine in human plasma via LC/MS and subsequent application to a pharmacokinetic study in healthy Chinese volunteers. Clin Chim Acta 380:100–105
48. Patel BN, Sharma N, Sanyal M et al (2008) Liquid chromatography tandem mass spectrometry assay for the simultaneous determination of venlafaxine and O-desmethylvenlafaxine in human plasma and its application to a bioequivalence study. J Pharm Biomed Anal 47:603–611
49. Rudberg I, Hermann M, Refsum H et al (2008) Serum concentrations of sertraline and N-desmethyl sertraline in relation to CYP2C19 genotype in psychiatric patients. Eur J Clin Pharmacol 64:1181–1188
50. Senthamil Selvan P, Gowda KV, Mandal U et al (2007) Determination of duloxetine in human plasma by liquid chromatography with atmospheric pressure ionization-tandem mass spectrometry and its application to pharmacokinetic study. J Chromatogr B Anal Technol Biomed Life Sci 858:269–275
51. Yao M, Srinivas NR (2000) Simultaneous quantitation of d7-nefazodone, nefazodone, d7-hydroxynefazodone, hydroxynefazodone, m-chlorophenylpiperazine and triazole-dione in human plasma by liquid chromatographic-mass spectrometry. Biomed Chromatogr 14:106–112
52. Zhao RK, Cheng G, Tang J et al (2009) Pharmacokinetics of duloxetine hydrochloride enteric-coated tablets in healthy Chinese volunteers: a randomized, open-label, single- and multiple-dose study. Clin Ther 31:1022–1036
53. Kirchherr H, Kuhn-Velten WN (2006) Quantitative determination of forty-eight antidepressants and antipsychotics in human serum by HPLC tandem mass spectrometry: a multi-level, single-sample approach. J Chromatogr B Anal Technol Biomed Life Sci 843:100–113
54. Dams R, Huestis MA, Lambert WE et al (2003) Matrix effect in bio-analysis of illicit drugs with LC-MS/MS: influence of ionization type, sample preparation, and biofluid. J Am Soc Mass Spectrom 14:1290–1294
55. Bonfiglio R, King RC, Olah TV et al (1999) The effects of sample preparation methods on the variability of the electrospray ionization response for model drug compounds. Rapid Commun Mass Spectrom 13:1175–1185
56. Naidong W, Bu H, Chen YL et al (2002) Simultaneous development of six LC-MS-MS methods for the determination of multiple analytes in human plasma. J Pharm Biomed Anal 28:1115–1126
57. Goeringer KE, McIntyre M, Drummer OH (2003) LC-MS analysis of serotonergic drugs. J Anal Toxicol 27:30–35
58. Zhang H, Heinig K, Henion J (2000) Atmospheric pressure ionization time-of-flight mass spectrometry coupled with fast liquid chromatography for quantitation and accurate mass measurement of five pharmaceutical drugs in human plasma. J Mass Spectrom 35:423–431
59. Remane D, Meyer MR, Peters FT et al (2010) Fast and simple procedure for liquid–liquid extraction of 136 analytes from different drug classes for development of a liquid chromatographic-tandem mass spectrometric quantification method in human blood plasma. Anal Bioanal Chem 397:2303–2314

60. Titier K, Castaing N, Le-Deodic M et al (2007) Quantification of tricyclic antidepressants and monoamine oxidase inhibitors by high-performance liquid chromatography-tandem mass spectrometry in whole blood. J Anal Toxicol 31:200–207
61. Theron HB, van der Merwe MJ, Swart KJ et al (2007) Employing atmospheric pressure photoionization in liquid chromatography/tandem mass spectrometry to minimize ion suppression and matrix effects for the quantification of venlafaxine and O-desmethylvenlafaxine. Rapid Commun Mass Spectrom 21:1680–1686
62. Chen X, Duan X, Dai X et al (2006) Development and validation of a liquid chromatographic/tandem mass spectrometric method for the determination of sertraline in human plasma. Rapid Commun Mass Spectrom 20:2483–2489
63. Djordjevic S, Kovacevic I, Miljkovic B et al (2005) Liquid chromatographic-mass spectrometric method for the determination of fluoxetine and norfluoxetine in human plasma: application to clinical study. Farmaco 60:345–349
64. Liu W, Cai HL, Li HD (2007) High performance liquid chromatography-electrospray ionization mass spectrometry (HPLC-MS/ESI) method for simultaneous determination of venlafaxine and its three metabolites in human plasma. J Chromatogr B Anal Technol Biomed Life Sci 850:405–411
65. Halvorsen TG, Pedersen-Bjergaard S, Reubsaet JLE et al (2003) Liquid-phase microextraction combined with liquid chromatography-mass spectrometry. Extraction from small volumes of biological samples. J Sep Sci 26:1520–1526
66. Juan H, Zhiling Z, Huande L (2005) Simultaneous determination of fluoxetine, citalopram, paroxetine, venlafaxine in plasma by high performance liquid chromatography-electrospray ionization mass spectrometry (HPLC-MS/ESI). J Chromatogr B Anal Technol Biomed Life Sci 820:33–39
67. Bhatt J, Jangid A, Venkatesh G et al (2005) Liquid chromatography-tandem mass spectrometry (LC-MS-MS) method for simultaneous determination of venlafaxine and its active metabolite O-desmethyl venlafaxine in human plasma. J Chromatogr B Anal Technol Biomed Life Sci 829:75–81
68. Jain DS, Sanyal M, Subbaiah G et al (2005) Rapid and sensitive method for the determination of sertraline in human plasma using liquid chromatography-tandem mass spectrometry (LC-MS/MS). J Chromatogr B Anal Technol Biomed Life Sci 829:69–74
69. Shinozuka T, Terada M, Tanaka E (2006) Solid-phase extraction and analysis of 20 antidepressant drugs in human plasma by LC/MS with SSI method. Forensic Sci Int 162:108–112
70. Mercerolle M, Denooz R, Lachatre G et al (2008) A fatal case of bupropion (Zyban) overdose. J Anal Toxicol 32:192–196
71. Bhatt M, Shah S, Shivprakash (2010) Solid-phase extraction and analysis of paroxetine in human plasma by ultra performance liquid chromatography-electrospray ionization mass spectrometry. Biomed Chromatogr 24:209–215
72. De Castro A, Ramírez Fernandez MM, Laloup M, Samyn N, De Boeck G, Wood M, Maes V, López-Rivadulla M (2007) High-throughput on-line solid-phase extraction-liquid chromatography-tandem mass spectrometry method for the simultaneous analysis of 14 antidepressants and their metabolites in plasma. J Chromatogr A 1160:3–12
73. Satonin DK, McCulloch JD, Kuo F et al (2007) Development and validation of a liquid chromatography-tandem mass spectrometric method for the determination of the major metabolites of duloxetine in human plasma. J Chromatogr B Anal Technol Biomed Life Sci 852:582–589
74. Kollroser M, Schober C (2002) Simultaneous determination of seven tricyclic antidepressant drugs in human plasma by direct-injection HPLC-APCI-MS-MS with an ion trap detector. Ther Drug Monit 24:537–544
75. Kollroser M, Schober C (2003) An on-line solid phase extraction liquid chromatography tandem mass spectrometry method for the analysis of citalopram, fluvoxamine and paroxetine in human plasma. Chromatographia 57:133–138
76. Sauvage FL, Gaulier JM, Lachatre G et al (2006) A fully automated turbulent-flow liquid chromatography-tandem mass spectrometry technique for monitoring antidepressants in human serum. Ther Drug Monit 28:123–130

77. Santos-Neto AJ, Bergquist J, Lancas FM et al (2008) Simultaneous analysis of five antidepressant drugs using direct injection of biofluids in a capillary restricted-access media-liquid chromatography-tandem mass spectrometry system. J Chromatogr A 1189:514–522
78. Breaud AR, Harlan R, Di Bussolo JM et al (2010) A rapid and fully-automated method for the quantitation of tricyclic antidepressants in serum using turbulent-flow liquid chromatography-tandem mass spectrometry. Clin Chim Acta 411:825–832
79. Queiroz ME, Oliveira EB, Breton F et al (2007) Immunoaffinity in-tube solid phase microextraction coupled with liquid chromatography-mass spectrometry for analysis of fluoxetine in serum samples. J Chromatogr A 1174:72–77
80. Alves C, Santos-Neto AJ, Fernandes C et al (2007) Analysis of tricyclic antidepressant drugs in plasma by means of solid-phase microextraction-liquid chromatography-mass spectrometry. J Mass Spectrom 42:1342–1347
81. Mao Y, Huang MQ, Xia YQ et al (2007) High-throughput quantitation of nefazodone and its metabolites in human plasma by high flow direct-injection LC-MS/MS. J Pharm Biomed Anal 43:1808–1819
82. Liu W, Wang F, Li HD (2007) Simultaneous stereoselective analysis of venlafaxine and O-desmethylvenlafaxine enantiomers in human plasma by HPLC-ESI/MS using a vancomycin chiral column. J Chromatogr B Anal Technol Biomed Life Sci 850:183–189
83. Patel BN, Sharma N, Sanyal M et al (2009) Analysis of second-generation antidepressant drug, sertraline and its active metabolite, N-desmethyl sertraline in human plasma by a sensitive and selective liquid chromatography-tandem mass spectrometry method. J Chromatogr B Anal Technol Biomed Life Sci 877:221–229
84. Patel BN, Sharma N, Sanyal M et al (2008) High throughput and sensitive determination of trazodone and its primary metabolite, m-chlorophenylpiperazine, in human plasma by liquid chromatography-tandem mass spectrometry. J Chromatogr B Anal Technol Biomed Life Sci 871:44–54
85. Naidong W, Eerkes A (2004) Development and validation of a hydrophilic interaction liquid chromatography-tandem mass spectrometric method for the analysis of paroxetine in human plasma. Biomed Chromatogr 18:28–36
86. Wei Z, Bing-Ren X, Cai-Yun W (2007) Liquid chromatography-mass spectrometry method for the determination of venlafaxine in human plasma and application to a pharmacokinetic study. Biomed Chromatogr 21:266–272
87. Choong E, Rudaz S, Kottelat A et al (2009) Therapeutic drug monitoring of seven psychotropic drugs and four metabolites in human plasma by HPLC-MS. J Pharm Biomed Anal 50:1000–1008
88. Shen Z, Wang S, Bakhtiar R (2002) Enantiomeric separation and quantification of fluoxetine (Prozac) in human plasma by liquid chromatography/tandem mass spectrometry using liquid-liquid extraction in 96-well plate format. Rapid Commun Mass Spectrom 16:332–338
89. Rocha A, Marques MP, Coelho EB et al (2007) Enantioselective analysis of citalopram and demethylcitalopram in human and rat plasma by chiral LC-MS/MS: application to pharmacokinetics. Chirality 19:793–801
90. Arpino P, Baldwin MA, McLafferty FW (1974) Liquid chromatography-mass spectrometry. II. Continuous monitoring. Biomed Mass Spectrom 1:80–82
91. Niessen WMA (1999) History of LC-MS. In: Niessen WMA (ed) Liquid chromatography-mass spectrometry, 2nd edn. Marcel Dekker, Inc., New York
92. King R, Bonfiglio R, Fernandez-Metzler C et al (2000) Mechanistic investigation of ionization suppression in electrospray ionization. J Am Soc Mass Spectrom 11:942–950
93. Smeraglia J, Baldrey SF, Watson D (2001) Matrix effects and selectivity issues in LC-MS-MS. Chromatographia 55:S95–S99
94. Franceschi L, Faggiani A, Furlanut M (2009) A simple method to monitor serum concentrations of fluoxetine and its major metabolite for pharmacokinetic studies. J Pharm Biomed Anal 49:554–557
95. Saber AL (2009) On-line solid phase extraction coupled to capillary LC-ESI-MS for determination of fluoxetine in human blood plasma. Talanta 78:295–299

96. European Union Decision 2002/657/EC (17/8/2002) (2002) Off J Eur Commun 221:8–36
97. Peters FT (2006) Method validation using LC-MS. In: Polettini A (ed) Applications of LC-MS in toxicology, 1st edn. Pharmaceutical, London
98. Moraes MO, Lerner FE, Corso G et al (1999) Fluoxetine bioequivalence study: quantification of fluoxetine and norfluoxetine by liquid chromatography coupled to mass spectrometry. J Clin Pharmacol 39:1053–1061
99. Li C, Ji Z, Nan F et al (2002) Liquid chromatography/tandem mass spectrometry for the determination of fluoxetine and its main active metabolite norfluoxetine in human plasma with deuterated fluoxetine as internal standard. Rapid Commun Mass Spectrom 16:1844–1850
100. Sutherland FC, Badenhorst D, de Jager AD et al (2001) Sensitive liquid chromatographic-tandem mass spectrometric method for the determination of fluoxetine and its primary active metabolite norfluoxetine in human plasma. J Chromatogr A 914:45–51
101. Massaroti P, Cassiano NM, Duarte LF et al (2005) Validation of a selective method for determination of paroxetine in human plasma by LC-MS/MS. J Pharm Pharm Sci 8:340–347
102. Zhu Z, Neirinck L (2002) High-performance liquid chromatography-mass spectrometry method for the determination of paroxetine in human plasma. J Chromatogr B Anal Technol Biomed Life Sci 780:295–300
103. Qin F, Li N, Qin T et al (2010) Simultaneous quantification of venlafaxine and O-desmethylvenlafaxine in human plasma by ultra performance liquid chromatography-tandem mass spectrometry and its application in a pharmacokinetic study. J Chromatogr B Anal Technol Biomed Life Sci 878:689–694
104. Rajasekhar D, Kumar IJ, Venkateswarlu P (2009) Rapid high-performance liquid chromatography-tandem mass spectrometry method for simultaneous measurement of venlafaxine and O-desmethylvenlafaxine in human plasma and its application in comparative bioavailability study. Biomed Chromatogr 23:1300–1307
105. Singh SS, Shah H, Gupta S et al (2004) Liquid chromatography-electrospray ionisation mass spectrometry method for the determination of escitalopram in human plasma and its application in bioequivalence study. J Chromatogr B Anal Technol Biomed Life Sci 811:209–215
106. Xu P, Li HD, Chen BM et al (2008) Determination of mianserin in human plasma by high performance liquid chromatography-electrospray ionization mass spectrometry (HPLC-ESI/MS): application to a bioequivalence study in Chinese volunteers. J Pharm Biomed Anal 47:994–999
107. Chauhan B, Rani S, Guttikar S et al (2005) Analytical method development and validation of mianserin hydrochloride and its metabolite in human plasma by LC-MS. J Chromatogr B Anal Technol Biomed Life Sci 823:69–74
108. Gutteck U, Rentsch KM (2003) Therapeutic drug monitoring of 13 antidepressant and five neuroleptic drugs in serum with liquid chromatography-electrospray ionization mass spectrometry. Clin Chem Lab Med 41:1571–1579
109. Castaing N, Titier K, Receveur-Daurel M et al (2007) Quantification of eight new antidepressants and five of their active metabolites in whole blood by high-performance liquid chromatography-tandem mass spectrometry. J Anal Toxicol 31:334–341
110. Fleishaker JC, Mucci M, Pellizzoni C et al (1999) Absolute bioavailability of reboxetine enantiomers and effect of gender on pharmacokinetics. Biopharm Drug Dispos 20:53–57

The Analysis of Antipsychotic Drugs in Human Biosamples by LC-MS

Eva Saar, Dimitri Gerostamoulos, Olaf H. Drummer, and Jochen Beyer

Abstract Over the last 60 years extensive development has occurred in the pharmacological treatment of mental illnesses with currently over 40 different compounds being prescribed.

Antipsychotic drugs (APs) are today amongst the most commonly prescribed drug classes with "second generation" or "atypical" APs accounting for the majority of prescriptions worldwide. As noncompliance is a common problem amongst patients treated with APs, these drugs are frequently subject to therapeutic drug monitoring (TDM). LC-MS techniques have facilitated the detection of these drugs which are frequently associated with very low blood concentrations particularly after IM depot injections. Multiple methods for the detection and quantification of APs in human biological specimens have been published using LC-MS, particularly in the last decade, as this instrumentation has been available for routine laboratories. This chapter reviews published multianalyte methods and highlights the important criteria when developing an analytical method for the detection of APs, potential pitfalls, and the importance of appropriate method validation, in order to assure reproducibility of results.

E. Saar • D. Gerostamoulos • O.H. Drummer • J. Beyer (✉)
Department of Forensic Medicine, Victorian Institute of Forensic Medicine and Monash University, 57-83 Kavanagh St, Southbank 3006, VC, Australia
e-mail: jochenb@vifm.org

Q. Alan Xu and Timothy L. Madden (eds.), *LC-MS in Drug Bioanalysis*,

DOI 10.1007/978-1-4614-3828-1_8,

1 Introduction: Treatment of Psychotic Illnesses

1.1 History

Prior to the 1950s, electroconvulsive therapy and psychosurgery were considered suitable treatments for patients suffering from mental illnesses. Due to a lack of knowledge surrounding the pathophysiology of psychotic disorders such as schizophrenia, the only pharmacological "treatment" at this time consisted of sedation with barbiturates and drug-induced epileptic seizures [1]. The first notable development in drug-therapy for people suffering from bipolar disorder—a disorder characterized by alternating manic and depressive episodes, was the discovery of lithium as a suitable treating agent in the late-1940s [2]. While the mechanism of action of lithium still remains unclear, its effectiveness in the treatment of bipolar disorders has been accepted worldwide and it is still considered the leading prophylactic treatment for this condition, even 60 years after its discovery [1].

In the 1950s, a more evidence-based approach to antipsychotic drug (AP) therapy was undertaken when structural variations of antihistamines were produced by a French scientist (Paul Charpentier), in order to make use of the "unwanted" sedative side effect produced by these drugs. Initially used to lower body temperature in patients undergoing cardiac surgery, chlorpromazine (Fig. 1I) was the first drug with antipsychotic properties successfully used in clinical trials [3].

1.2 Typical APs

Since its official release in 1952, Chlorpromazine has been considered the prototype of so-called "typical" APs. Chlorpromazine has a characteristic phenothiazine structure, which formed the basis of other APs synthesized in subsequent years, such as fluphenazine, perphenazine, promazine, promethazine, thioridazine, trifluoperazine, and triflupromazine (Fig. 1II–VIII). These APs formed the largest subgroup of all drugs commonly referred to as "Typical APs." They are also known as "First Generation APs". Other subgroups within the typical APs are also characterized by their chemical structures, such as the butyrophenones (e.g droperidol, haloperidol, trifluperidol, melperone, and pipamperone (Fig. 1IX–XIII), the thioxanthenes (e.g., flupentixol, zuclopenthixol, chlorprothixene, and thiothixene (Fig. 1XIV–XVII)), diphenylbutylpiperidines (e.g., pimozide, fluspirilene, penfluridol (Fig. 1XVIII–XX)), indoles (i.e., molindone (Fig. 1XXI)), and others (e.g., loxapine (Fig. 1XXII)).

Despite having varying chemical structures, all typical APs have a significant affinity to dopamine (DA) receptors, mainly the D_2-type, while also showing minor antagonism at muscarinic, 5-HT, adrenergic (α), and histaminergic (H1) receptors. By blocking receptors in the prefrontal cortex and the limbic area of the brain, both areas which are linked with mood and emotional behavior, an improvement in positive symptoms is achieved. However, the same action of APs in other

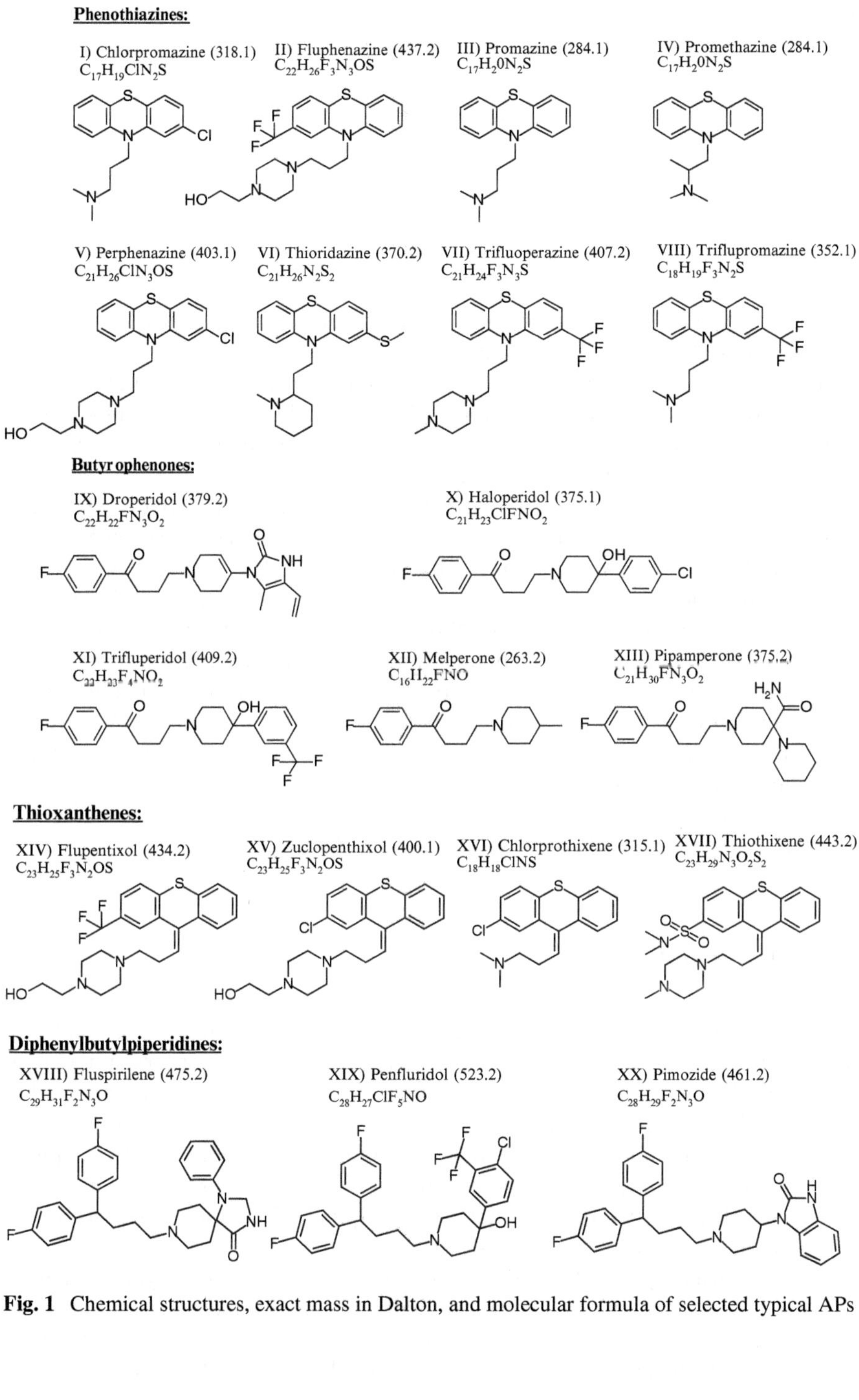

Fig. 1 Chemical structures, exact mass in Dalton, and molecular formula of selected typical APs

Indoles:

XXI) Molindone (276.2)
$C_{16}H_{24}N_2O_2$

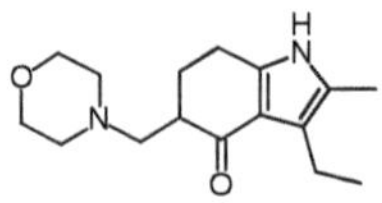

Others:

XXII) Loxapine (327.1)
$C_{18}H_{18}ClN_3O$

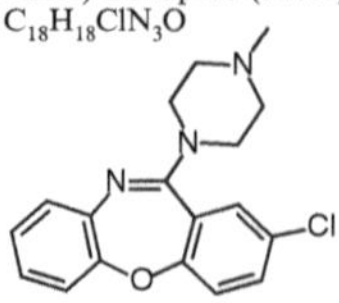

Fig. 1 (continued)

cerebral areas such as the striatum, which is associated with motor control, has been shown to lead to serious side effects.

Patients treated with typical APs are likely to suffer parkinsonian symptoms such as mobility difficulties (tremor, bradykinesia, postural instability). The blockage of DA receptors leads to a decrease in DA, ultimately causing symptoms consistent with patients suffering from Parkinson's disease.

The use of typical APs is associated with a number of side effects which can outweigh their positive outcomes at times. Neuroleptic Malignant Syndrome (NMS), although difficult to distinguish from other disorders [4], is characterized by elevated temperature, changed mental status, and severe muscle rigidity. While recent studies suggest that the prevalence of NMS has decreased from around 2.4 % [5] to 0.01–0.02 % [6], this is most likely due to more conservative prescription patterns of typical APs and a greater awareness of the illness [7]. The mortality rate associated with NMS was reported to be as high as 20 % at one stage [5, 8]. However, in the last two decades, mortality rates have fallen below 10 % due to early recognition and improved management [6].

Additional problems associated with typical APs include cardiotoxicity [9], seizures, and an increased risk of sudden cardiac death [10, 11].

1.3 Atypical APs

Due to the broad range of side effects associated with typical APs, combined with their inability to improve all symptoms of psychotic disorders, a new generation of APs was introduced in the 1970s. These APs are generally referred to as "second generation" or atypical APs. This group includes indoles (e.g., ziprasidone and sertindole (Fig. 2I–II)), benzamides (e.g., amisulpride, sulpiride (Fig. 2III–IV)) diazepines/oxazepines/thiazepines (e.g., clozapine, olanzapine, quetiapine (Fig. 2V–VII)), and others (e.g., aripiprazole, risperidone, buspirone, paliperidone, zotepine (Fig. 2VIII–XII)).

Clozapine, a tricyclic dibenzodiazepine derivative, was the first atypical AP to be approved by the FDA in 1989. It was originally thought that increased affinity to the $5HT_{2A}$ receptor, in combination with a lower or no affinity to the D_2 receptors, might define a compound as an "atypical" AP [12, 13]. However, studies have shown that

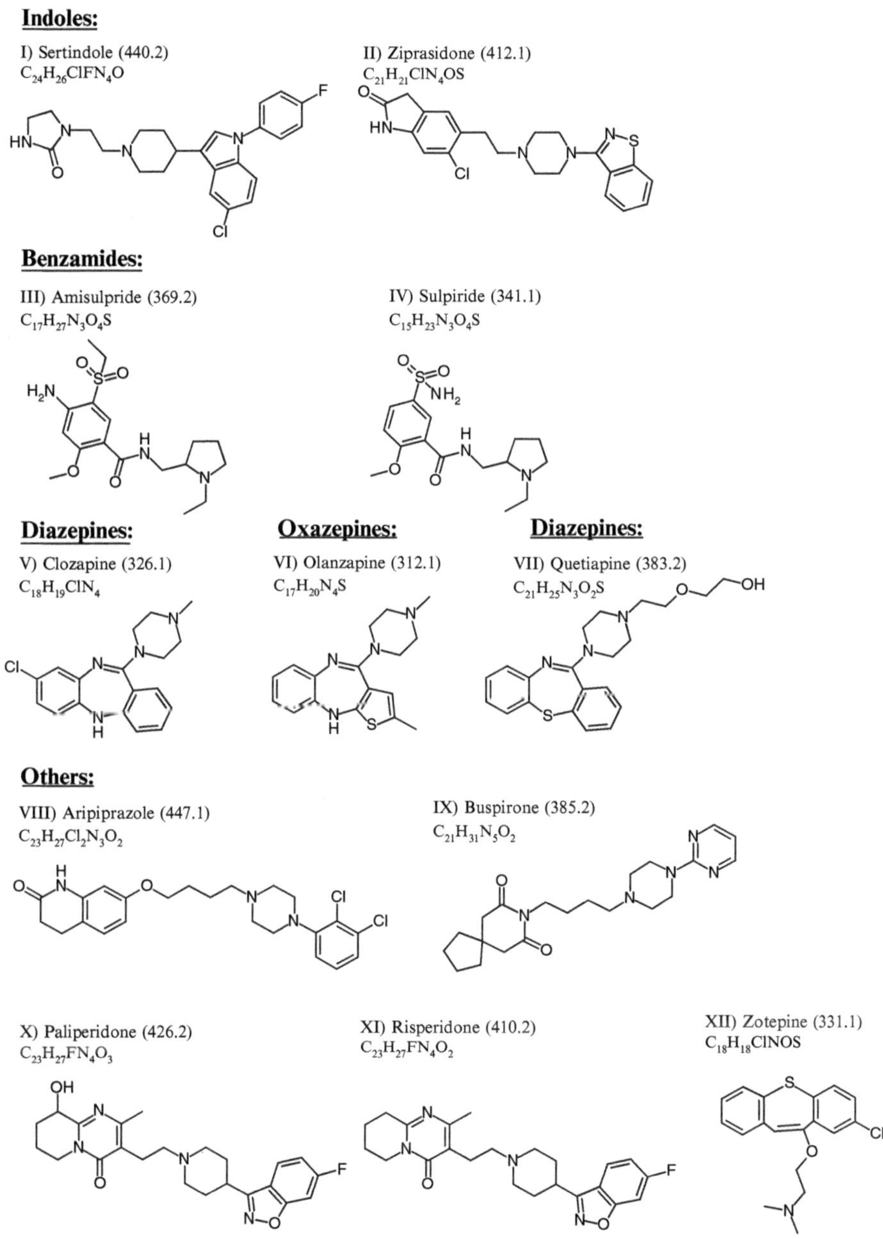

Fig. 2 Chemical structures, exact mass in Dalton, and molecular formula of selected atypical APs

selective 5HT antagonists do not have antipsychotic properties [14]. Some antagonism at the D_2 receptor appears to be mandatory for any antipsychotic effect. More recent studies have presented an alternative to this theory.

While typical APs bind with high affinity to the D_2 receptors, these newer atypical APs possess only moderate affinity to D_2 receptors and high dissociation constants. They initially occupy D_2 receptors and then rapidly dissociate to allow normal DA neurotransmission, which ultimately decreases extrapyramidal side effects [15, 16].

Although it was initially thought that atypical APs may be linked with fewer side effects than their typical counterparts, this is unlikely to be the case; they have shown to have a similar risk of sudden cardiac death [17] as the first generation APs and appear to cause even more problems in regard to metabolic syndrome (e.g., obesity, type 2 diabetes mellitus) [18].

As different APs possess different advantages and disadvantages (not just depending on the group they belong to but on their individual properties) it is usual practice to select antipsychotic medication depending on a patient's individual needs. Therefore, combination therapy is a common occurrence amongst patients treated with APs [19] highlighting the need for detection methods that consist of a large variety of these drugs.

1.4 Off-label Use

The practice of prescribing pharmaceuticals for an unapproved indication is called "off-label use." While this article mainly highlights the use of APs for psychotic disorders such as schizophrenia, it must be noted that there is a significant off-label use of these drugs worldwide [20]. This is largely caused by heterogeneous indications of different APs in different countries. Fleischhacker et al. [21] present an overview of the indications of relevant atypical APs and haloperidol in 10 European countries. Variations in the indication and labeling of APs should be considered in analytical screening procedures applied in clinical and forensic toxicology.

1.5 Administration of APs

The two main administration routes for APs are peroral (P.O.) and intramuscular (I.M.). P.O. administration results in a significant first-pass effect; therefore there is considerable loss of drug after the first liver passage. I.M preparations are usually synthesized by esterification of the drugs with fatty acids [22] and can be divided into short-acting I.M. antipsychotic medications (SAIM) and long acting I.M. antipsychotic medications (LAIM). SAIM are used for the treatment of agitation and aggressive behavior of patients experiencing an acute psychotic phase [23, 24]. LAIM are referred to as "depot" injections. I.M. formulations have several advantages such as a higher bioavailability due to the lack of first-pass metabolism and are preferred in the treatment of patients where compliance issues are likely.

1.6 Toxicological Considerations

The volume of distribution (V_D) quantifies the distribution of a drug between plasma and the rest of the body after oral or parental application. Drugs which only distribute in blood show a V_D of ~0.05 L/kg, whereas drugs distributed in extracellular water have a V_D of ~0.2 L/kg; the total body water volume is 0.55 L/kg. APs are lipid soluble weak bases which are easily taken up into body fat and organs, therefore generally show a large V_D. Plasma protein binding, also referred to as the "fraction bound" (Fb), is the percentage of a drug bound to plasma proteins after its admission. Only the unbound fraction of a drug will cause a pharmacological effect and can be detected using analytical methods. Despite their high V_D, most common APs show a high Fb. Both parameters significantly contribute to low or undetectable blood concentrations.

Therapeutic blood concentrations described in this chapter are based on studies that relate to drug concentrations measured at steady state of patients treated with recommended daily doses of a drug. However, the same dose of a drug can result in considerably different plasma concentrations in different individuals, depending on factors such as diet, lifestyle, comedication, and genetic makeup resulting in altered absorption, distribution, and elimination of drugs.

The terminal elimination half-life ($t_{1/2}$) of a drug is not dose-dependent and can range from a few days up to several weeks for APs ($t_{1/2}$ of up to 16 weeks with a very large variation has been described for Flupentixol-decanoate [25]). Importantly, the time required to reach steady state depends only on $t_{1/2}$ of a compound. Steady state is usually reached after five $t_{1/2}$. It is usually not advisable to measure plasma concentrations before this equilibrium is reached except for drugs with very long $t_{1/2}$ such as depot-formulations. This is done in order to ensure that patients with impaired metabolism or excretion are not at risk of reaching above-therapeutic drug concentrations at their initial dosage regimen.

Table 1 shows the V_D, $t_{1/2}$, Fb, and therapeutic blood concentrations of common APs.

2 Detection of APs Using LC-MS Techniques

2.1 Instrumentation

Until approximately a decade ago, gas chromatography mass spectrometry (GC-MS) was considered the *gold standard* of analysis and quantification of drugs and their metabolites in biological matrices [26]. In order to be suitable for GC-analysis, a compound needs to be volatile and thermally stable to such an extent that it can be transferred into a gas phase. Unfortunately, several APs, such as risperidone are unstable and are not amenable to direct GC analysis due to their chemical properties. In addition, standard GC-MS instruments do often not provide sufficient sensitivity to detect the low concentrations in which certain

Table 1 Pharmacokinetic parameters of common APs

Drug	Concentrations expected following therapeutic use (mg/L)	$t_{1/2}$ (h)[a]	V_D (L/kg)[b]	Fb[c]
9-OH-Risperidone	0.01–0.1	23	U/K	U/K
Amisulpride	0.05–0.4	11–27	13–16	0.17
Aripiprazole	0.05–0.35	60–90	4.9	0.99
Bromperidol	0.001–0.02	15–35	U/K	U/K
Buspirone	0.001–0.01	3–12	5–6	0.95
Chlorpromazine	0.03–0.3	7–119	10–35	0.98
Chlorprothixene	0.02–0.2	8–12	11–23	U/K
Clozapine	0.2–0.8	6–17	2–7	0.95
Droperidol	0.005–0.05	1.3–2.7	1.5–2.5	0.85–0.90
Flupentixol	0.001–0.015	19–39 (HCl) 5–113 days (decanoate)	14.1	U/K
Fluphenazine	0.002–0.02	13–58 (HCl) 3–4 days (enanthate) 5–12 days (decanoate)	220	0.99
Fluspirilene	U/K	21 days	U/K	U/K
Haloperidol	0.005–0.05	14–41 (lactate) 14–28 (decanoate)	18–30	0.90
Levomepromazine	0.015–0.06	15–30	30	U/K
Loxapine	0.01–0.1	3–4	U/K	0.91–0.99
Melperone	0.05–0.4	2–4	7–10	U/K
Mesoridazine	0.15–1	2–9	3–6	0.99
Molindone	~0.5	1.2–2.8	3–6	U/K
Olanzapine	0.01–0.1	21–54	10–20	0.93
Penfluridol	0.004–0.025			
Perazine	0.1–0.23	8–15	U/K	0.96
Pericyazine	0.005–0.06	U/K	U/K	U/K
Perphenazine	0.0006–0.0024	8–12	10–35	>0.90
Pimozide	0.015–0.02	28–214	11–62	0.99
Pipamperone	0.1–0.4	12–30	U/K	U/K
Prochlorperazine	0.01–0.5	14–27	13–32	>0.90
Promazine	0.01–0.4	7–17	27–42	0.84
Promethazine	0.05–0.4	10–20	9–19	0.93
Quetiapine	0.07–0.17	2.7–9.3	8–12	0.83
Risperidone	0.01–0.1	3–20	0.7–2.1	0.90
Sertindole	0.05–0.5	U/K	20–40	0.99
Sulpiride	0.05–0.4	4–11	2.7	0.40
Thioridazine	0.2–2	26–36	18	0.96
Thiothixene	U/K	12–36	U/K	0.91–0.99
Trifluoperazine	0.001–0.05	7–18	U/K	>0.90
Trifluperidol	U/K	U/K	U/K	U/K
Triflupromazine	0.03–0.1	U/K	U/K	U/K
Ziprasidone	0.05–0.12	2–8	1.5–2.3	0.99
Zotepine	0.005–0.3	12–30	50–168	0.97
Zuclopenthixol	0.005–0.1	12–26 (HCl) 18–24 (acetate) 19 days (decanoate)	15–20	0.98

[a]Terminal elimination half-life
[b]Volume of distribution
[c]Blood to plasma concentration ratio
U/K = unknown

APs are present in biological specimens (see Table 1 for therapeutic concentrations of common APs).

In recent years, liquid chromatography (LC) instruments combined with a variety of MS mass analyzers (quadrupole, Time-of-Flight (TOF), Ion Trap) have become more affordable and widespread in clinical and forensic toxicology. Since HPLC does not require volatilization or derivatization, this has led to the widespread use of LC-MS for the detection of APs in biological matrices. More and more laboratories are now able to provide such testing as part of a routine toxicological service including therapeutic drug monitoring (TDM).

2.2 *Biosamples*

Analysis of APs in human specimens is particularly important in both a clinical and forensic setting: In a clinical environment, AP analysis is necessary in order to monitor patient compliance and to maintain drug concentrations within the recommended therapeutic range of the respective drug. In a medicolegal setting involving postmortem analysis, the determination of APs is essential in assisting to establish whether these drugs played a role in a person's death. The absence of prescribed APs in a clinical case may also indicate noncompliance, ultimately resulting in an altered mental state and an increased risk of sudden death. In forensic cases, this noncompliance may be especially relevant in cases where behavioral disturbances lead to the death of a person by another, e.g., homicides.

Urine is the specimen of choice for screenings to show the presence of a particular compound qualitatively. Therefore, published methods for detection in urine [27–30], as well as the less common specimens, namely, saliva [31], and cerebrospinal fluid [32] are rare and not reviewed in this book chapter. Hair is the specimen of choice when a larger surveillance window is required. Segmental hair analysis in particular can give a good indication of the history of drug use in an individual. While hair analysis is frequently used as a tool in the analysis of drugs of abuse, only a limited number of methods using LC-MS(/MS) technology targeting APs in hair have been published to date [29, 33–37]. Most of these methods show no or very limited method validation and are not used in routine toxicological testing.

2.3 *Approaches to the Detection of APs in Human Biological Matrices*

Blood is most commonly used for the analysis of APs as drug concentrations in blood represent a more accurate correlation with pharmacological effects than in any other specimen. Plasma or serum is usually obtained from clinical samples, as blood cells can easily be separated from liquid components to create a "cleaner" sample and reduce matrix effects. Plasma and serum are the specimens of choice in TDM environments, and it is therefore not surprising that the majority of the published methods for the analysis of APs use plasma. In a postmortem setting, whole blood is routinely used for analysis, as plasma cannot be obtained in the majority of cases due to postmortem lysis and other forms of decomposition [38]. One needs to be aware that once a method has been validated for use in a particular matrix (e.g., plasma) it cannot be transferred to another matrix (e.g., whole blood) without sufficient cross-validation, as factors such as matrix effects are likely to vary.

Whilst there has been a large number of single analyte procedures published over the last decade, multianalyte procedures targeting APs are still limited. To the authors' knowledge, only three currently published methods include at least 15 APs [39–41]. Kratzsch et al. [40] include 15 APs and three metabolites in their analytical method using plasma. Method validation has been carried out; however, matrix effects were not investigated in this method published in 2003, as they were not considered an essential part of method validation at the time. Kirchherr et al. [39] cover 22 APs and over 20 antidepressants in their LC-MS/MS method suitable for detection and quantification of APs in serum. The only drawback of this comprehensive method is the use of therapeutic drugs (and metabolites) as internal standard (IS). This is not good practice and can cause significant problems including overestimation of drug concentrations, which is discussed later in this chapter. A comprehensive method for the detection and quantification of APs was published in 2010 by Saar et al. [41]. This fully validated multianalyte procedure covers 30 APs. It will therefore be used as an example method to highlight the important components of AP analysis methods and potential pitfalls.

TDM laboratories largely focus on targeted APs and therefore only include relevant APs of interest in their respective methods. However, in a patient-population where noncompliance is a problem and polypharmacy is a common occurrence, it is advisable to include more APs in the analytical methodology. In the case of a prescribed drug not being present or at a subtherapeutic concentration, it may be helpful to ensure no other APs (or indeed other relevant drugs) are present in the patient's sample.

Sample volumes ranging from 25 μL [31] at the lower end up to 1 mL of specimen are commonly used. Saar et al. [41] chose a volume of 100 μL as this provided easy handling, and was likely to be still obtained in cases where only limited sample volume was available but was also sufficient to reach the LLOQ for all drugs incorporated in the method.

An important step in AP analysis is appropriate sample preparation. Frequently used extraction techniques for the analysis of APs include liquid-liquid extraction (LLE) and solid-phase extraction (SPE). Whilst SPE procedures show the advantage of being more automated and requiring less time of the analyst, this can be outweighed by technical issues such as blockage, causing delays especially when dealing with more complex matrices like postmortem whole blood. Less common approaches include protein-precipitation [39, 42, 43] and direct injection [27, 44]. In order to manage large volumes of samples, several authors describe sample preparation in a 96-well format [45, 46]. Sample preparation for LC analysis is considered to be less time-consuming than for GC analysis as no derivatization of the compounds is required.

After extraction of the drugs of interest from the sample matrix, the selection of the most appropriate ionization conditions is required. Electrospray Ionization (ESI) is used in the vast majority of published methods whilst Atmospheric-Pressure Chemical Ionization (APCI) is less common [40, 45, 47, 48]. Since all APs carry at least one nitrogen-group, ionization generally takes place in positive mode. In 2007 following a study using morphine, Dams et al. [49] concluded that ESI is more prone to matrix effects than APCI, but both ionization techniques are affected by this potentially significant analytical issue. Accordingly, new techniques should not be accepted for use unless sufficient matrix effect studies have been conducted as part of the method validation, using one of the two internationally accepted approaches either by Bonfiglio et al. [50] or Matuszewski et al. [51].

The use of an appropriate IS is mandatory when developing new LC-MS(/MS) detection methods. The IS must not be in therapeutic use, as the possibility of a patient being comedicated with this drug can never be fully excluded. Ideally a deuterated IS from a different drug class should be incorporated, rather than taking the risk of overestimating the peak area of the IS by using a nondeuterated compound from the same class [52]. Examples for recommended IS are 9OH-risperidone-d_4 [53], haloperidol-d_4 [41, 54], and olanzapine-d_3 [32].

All published methods for the detection and quantification of APs in biological matrices use reversed phase (RP) stationary phases with C_8 or C_{18} chains. The predominantly hydrophobic APs are prone to bind to the hydrophobic column in a polar mobile phase. A decrease in polarity of the mobile phase by introduction and increase of a nonpolar solvent, results in desorption of the ionized compounds from the stationary phase over time and elution of the APs of interest. Isocratic and gradient elution are both common, using different combinations of an aqueous buffer and an organic solvent. Run times vary from 2 to 20 min. Shorter run times may seem beneficial in laboratories with a large sample throughput but can also create two major problems. First, Saar et al. [41] highlight the importance of sufficient run-times in order to distinguish compounds included in the method that show identical Q1 masses and share the first two transitions despite having different chemical structures such as the butyrophenone derivatives haloperidol (Fig. 1X) and pipamperone (Fig. 1XIII). Making a clear distinction between these two typical APs is only possible by having a sufficiently long run time or adding a third transition. Second, unknown compounds or matrix components can co-elute and cause ion-suppression or enhancement, resulting in an over- or underestimation of a concentration [55].

Multiple Reaction Monitoring (MRM) methods are most commonly used in analytical methods as they provide the opportunity for fast and simple detection and quantification. However, a large number of published methods do not fulfill the international requirement of at least two MRM transitions for reliable identification of an analyte [56–58] which can cause problems; particularly misidentification. Several methods that allow for high sample throughput due to short run times monitor only one transition per compound, as peak widths would be too small in order to include a sufficient number of data points per peak in the given cycle time.

All analytical methods must be validated in accordance with international guidelines [59–61] prior to application. As discussed in the first part of this book, minimum criteria that need to be met to in order to satisfy these guidelines are: selectivity, matrix effects, extraction efficiency, process efficiency, processed sample stability, linearity, accuracy, precision, and freeze–thaw stability.

Table 2 shows a selection of published TDM methods. All methods presented use plasma or serum, with volumes ranging from 0.06 mL [46] to 0.5 mL [45, 62]. Berna et al. include several olanzapine metabolites in their detection method, while most other methods target one or very few analytes. Appropriate IS (i.e., deuterated analogs of APs or nondeuterated drugs which are not in therapeutic use) are used in three of the five presented methods. However, one method uses prazosin [62], a sympatholytic antihypertensive drug, while the other uses sulpiride [63]. This can cause problems including underestimation of the target drug as previously discussed in this chapter. Liquid–liquid extraction, solid phase extraction, and protein-precipitation are used in the presented methods, some in combination with a 96-well plate format in order to accommodate for a high sample throughput. RP columns are common with one approach applying a chiral column [53] in order to separate the two enantiomers (+) 9-OH-risperidone and (−) 9-OH-risperidone. Both isocratic and gradient elution are applied for drug separation. All methods use MRM mode for compound identification; however, only one transition is monitored in all presented methods. International guidelines recommend to include a minimum of two transitions per analyte in order to lower the risk of misidentification of a drug. The "validation data" presented in Table 2 shows which validation experiments have been performed by the respective authors. Before adapting a method for use in a laboratory, one should check the method validation for completeness (according to the international guidelines presented earlier in this chapter) and perform additional validation experiments where applicable.

2.4 Stability of Antipsychotic drugs

It is desirable to also investigate long-term stability of samples under different storage conditions prior to release of a method. This is particularly important when samples are not analyzed promptly after admission to a laboratory, or reanalysis of specimens is required due to additional tests becoming necessary. There is limited data available on the stability of APs in stored blood samples. The only comprehensive study investigating the stability of 30 commonly prescribed APs in whole blood was published in 2011 [64]. Four different storage temperatures (20, 4, −20, −60 °C) and two concentration levels were included

Table 2 Overview of selected TDM methods

Author	Matrix	Sample [mL]	Drugs	IS	"Work Up"	Stationary phase	Mobile phase	Detection mode	Validation data	Ref.
Berna et al.	Plasma + serum	0.5	Olanzapine, *N*-desmethyl-olanzapine, 2-hydroxymethyl-olanzapine, olanzapine-*N*-oxide, olanzapine-10-*N*-glucuronide	LY017222 (structural analog)	96-well SPE	MetaChem monochrom HPLC (4.6 × 150 mm, 5 μm)	Gradient with 10 mM ammonium acetate and propane-1-ol in MeOH	APCI, positive mode, MRM, MS/MS	Linearity, LLOQ, accuracy, precision, extraction efficiency, matrix effects, selectivity, F/T and LT stability, processed sample stability	[45]
Cabovska et al.	Plasma	0.2	Risperidone, 9-OH-risperidone	R068808 (structural analog), (+)9-OH-risperidone; (-)9-OH-risperidone	LLE (MTBE)	Chiralcel OJ column (50 mm × 4.6)	10 mM ammonium acetate in 50:50 EtOH/PrOH (v/v) and hexane	ESI, positive mode, MRM, MS/MS	Matrix effects, precision, accuracy, selectivity, recovery, ion suppression, processed sample stability	[53]
Chew et al.	Plasma	0.5	Buspirone	Prazosin	SPE	Luna C18 (50 mm × 2.5 μm)	Gradient with TFA in dH_2O and TFA in ACN	ESI, positive mode, MRM, MS/MS	LLOQ, accuracy, precision, extraction recovery, matrix effects	[62]
Gschwend et al.	Plasma	0.25	Amisulpride	Sulpiride	LLE (diisopropylether: dichloromethane, 1:1)	Phenomenex Synergi Polar-RP analytical column (75 mm × 4.6 mm, 4 μm)	Isocratic with 5 mM ammonium formate/ACN	ESI, positive mode, MRM, MS/MS	Linearity, selectivity, recovery, precision, accuracy, F/T and LT stability, processed sample stability	[63]

(continued)

Table 2 (continued)

Author	Matrix	Sample [mL]	Drugs	IS	"Work Up"	Stationary phase	Mobile phase	Detection mode	Validation data	Ref.
Hasselstrom et al.	Serum	0.06	Clozapine, quetiapine, ziprasidone	Clozapine-d3, quetiapine-d8, ziprasidone-d8	Zinc sulfate, MeOH, 96-well plate	Zorbax SB-C8 (2.0 × 50 mm, 1.8 μm)	Gradient with formic acid in water and formic acid in MeOH	ESI, positive mode, MS/MS, MRM	Selectivity, recovery, ion suppression, LLOQ, precision, trueness, LT stability, processed sample stability	[46]

IS internal standard, *SPE* solid phase extraction, *LLE* liquid–liquid extraction, *MTBE* methyl-*t*-butyl-ether, *MeOH* methanol, *EtOH* ethanol, *PrOH* propanolol, *APCI* atomic pressure chemical ionization, *LLOQ* lower limit of quantification, *F/T* freeze–thaw, *LT* long-term in this study. Whilst low storage temperatures such as −20 and −60 °C appeared to be preferable for storage of samples containing APs, the low temperatures did not guarantee stability for all investigated drugs. The commonly used atypical AP olanzapine (Fig. 2VI) has been associated with severe stability issues not only in stored whole blood [43, 48], serum [65], and plasma [66], but also under the conditions of analysis [41]. Oxidation of the thiophene-ring has been suggested as a possible cause of instability but this theory has not been confirmed. A loss greater than 70 % of the initial olanzapine concentration was observed under all four storage temperatures in less than a week. Addition of ascorbic acid as an antioxidant has been suggested in order to prevent degradation, but is not always practical. Ziprasidone showed losses of ~85 % after storage for ten weeks at 20 °C, whereas storage at lower temperatures was not affected by degradation. One needs to be aware that inappropriate storage of samples containing ziprasidone—even for a short time—can result in significant concentration changes.

3 Conclusions

When analyzing APs in human biological specimens, various factors need to be considered. Due to the large number of APs currently on the market, multianalyte methods are preferred over single-analyte methods. Analytical methods need to be sensitive enough to cover the low therapeutic range in which APs can be present. Also, validation in accordance with international guidelines needs to be undertaken in order to obtain reproducible results. When adapting a previously developed method for use in a different laboratory, sufficient cross-validation must be carried out prior to use. Sample integrity needs to be preserved by appropriate storage and it is advisable to have knowledge about concentration changes of APs after different storage times. When interpreting results, it is not sufficient to compare obtained concentrations to target ranges of a drug, but cofactors such as sampling time, dosage history, half-life of a drug, comedication, and genetic variations amongst individuals, need to be taken into consideration.

References

1. Jones AW (2011) Perspectives in drug discovery 6. Antipsychotics. TIAFT Bull 41(2):16–19
2. Cade JF (1949) Lithium salts in the treatment of psychotic excitement. Med J Aust 2(10):349–352
3. Delay J, Deniker P, Harl JM (1952) Therapeutic method derived from hiberno-therapy in excitation and agitation states. Ann Med Psychol (Paris) 110(2(2)):267–273
4. Caroff SN, Mann SC (1993) Neuroleptic malignant syndrome. Med Clin North Am 77(1):185–202
5. Caroff SN (1980) The neuroleptic malignant syndrome. J Clin Psychiatry 41(3):79–83
6. Keck PE Jr, Caroff SN, McElroy SL (1995) Neuroleptic malignant syndrome and malignant hyperthermia: end of a controversy? J Neuropsychiatry Clin Neurosci 7(2):135–144

7. Stubner S, Rustenbeck E, Grohmann R, Wagner G, Engel R, Neundorfer G, Moller HJ, Hippius H, Ruther E (2004) Severe and uncommon involuntary movement disorders due to psychotropic drugs. Pharmacopsychiatry 37(Suppl 1):S54–S64. doi:10.1055/s-2004-815511
8. Carbone JR (2000) The neuroleptic malignant and serotonin syndromes. Emerg Med Clin North Am 18(2):317–325
9. Welch R, Chue P (2000) Antipsychotic agents and qt changes. J Psychiatry Neurosci 25(2):154–160
10. Ray WA, Meredith S, Thapa PB, Meador KG, Hall K, Murray KT (2001) Antipsychotics and the risk of sudden cardiac death. Arch Gen Psychiatry 58(12):1161–1167. doi:yoa20305 [pii]
11. Straus SM, Bleumink GS, Dieleman JP, van der Lei J, 't Jong GW, Kingma JH, Sturkenboom MC, Stricker BH (2004) Antipsychotics and the risk of sudden cardiac death. Arch Intern Med 164(12):1293–1297. doi:10.1001/archinte.164.12.1293, 164/12/1293 [pii]
12. Meltzer HY (1989) Clinical studies on the mechanism of action of clozapine: the dopamine-serotonin hypothesis of schizophrenia. Psychopharmacology (Berl) 99(Suppl):S18–S27
13. Kapur S, Zipursky RB, Remington G, Jones C, DaSilva J, Wilson AA, Houle S (1998) 5-ht2 and d2 receptor occupancy of olanzapine in schizophrenia: a pet investigation. Am J Psychiatry 155(7):921–928
14. Duinkerke SJ, Botter PA, Jansen AA, van Dongen PA, van Haaften AJ, Boom AJ, van Laarhoven JH, Busard HL (1993) Ritanserin, a selective 5-ht2/1c antagonist, and negative symptoms in schizophrenia. A placebo-controlled double-blind trial. Br J Psychiatry 163:451–455
15. Seeman P (2002) Atypical antipsychotics: mechanism of action. Can J Psychiatry 47(1):27–38
16. Seeman P, Tallerico T (1999) Rapid release of antipsychotic drugs from dopamine d2 receptors: An explanation for low receptor occupancy and early clinical relapse upon withdrawal of clozapine or quetiapine. Am J Psychiatry 156(6):876–884
17. Ray WA, Chung CP, Murray KT, Hall K, Stein CM (2009) Atypical antipsychotic drugs and the risk of sudden cardiac death. N Engl J Med 360(3):225–235. doi:360/3/225 [pii] 10.1056/NEJMoa0806994
18. Muench J, Hamer AM (2010) Adverse effects of antipsychotic medications. Am Fam Physician 81(5):617–622
19. Ananth J, Parameswaran S, Gunatilake S (2004) Antipsychotic polypharmacy. Curr Pharm Des 10(18):2231–2238
20. Barbui C, Ciuna A, Nose M, Patten SB, Stegagno M, Burti L, Amaddeo F, Tansella M (2004) Off-label and non-classical prescriptions of antipsychotic agents in ordinary in-patient practice. Acta Psychiatr Scand 109(4):275–278. doi:283 [pii]
21. Fleischhacker WW (2003) Second-generation antipsychotics: Discrepancies between licensed indications, evidence base and actual use. Psychopharmacology (Berl) 169(2):207–210. doi:10.1007/s00213-003-1496-9
22. Barnes TR, Curson DA (1994) Long-term depot antipsychotics. A risk-benefit assessment. Drug Saf 10(6):464–479
23. Milton GV, Jann MW (1995) Emergency treatment of psychotic symptoms. Pharmacokinetic considerations for antipsychotic drugs. Clin Pharmacokinet 28(6):494–504
24. Jones B, Taylor CC, Meehan K (2001) The efficacy of a rapid-acting intramuscular formulation of olanzapine for positive symptoms. J Clin Psychiatry 62(Suppl 2):22–24
25. Jorgensen A, Andersen J, Bjorndal N, Dencker SJ, Lundin L, Malm U (1982) Serum concentrations of cis(z)-flupentixol and prolactin in chronic schizophrenic patients treated with flupentixol and cis(z)-flupentixol decanoate. Psychopharmacology (Berl) 77(1):58–65
26. Maurer HH (1992) Systematic toxicological analysis of drugs and their metabolites by gas chromatography-mass spectrometry. J Chromatogr 580(1–2):3–41
27. Arinobu T, Hattori H, Iwai M, Ishii A, Kumazawa T, Suzuki O, Seno H (2002) Liquid chromatographic-mass spectrometric determination of haloperidol and its metabolites in human plasma and urine. J Chromatogr B Analyt Technol Biomed Life Sci 776(1):107–113. doi:S1570023202001757 [pii]

28. De Meulder M, Remmerie BM, de Vries R, Sips LL, Boom S, Hooijschuur EW, van de Merbel NC, Timmerman PM (2008) Validated lc-ms/ms methods for the determination of risperidone and the enantiomers of 9-hydroxyrisperidone in human plasma and urine. J Chromatogr B Analyt Technol Biomed Life Sci 870(1):8–16. doi:S1570-0232(08)00281-X [pii] 10.1016/j.jchromb.2008.04.041
29. Josefsson M, Kronstrand R, Andersson J, Roman M (2003) Evaluation of electrospray ionisation liquid chromatography-tandem mass spectrometry for rational determination of a number of neuroleptics and their major metabolites in human body fluids and tissues. J Chromatogr B Analyt Technol Biomed Life Sci 789(1):151–167. doi:S1570023203002071 [pii]
30. Kumazawa T, Seno H, Watanabe-Suzuki K, Hattori H, Ishii A, Sato K, Suzuki O (2000) Determination of phenothiazines in human body fluids by solid-phase microextraction and liquid chromatography/tandem mass spectrometry. J Mass Spectrom 35(9):1091–1099. doi:10.1002/1096-9888(200009)35:9<1091::AID-JMS31>3.0.CO;2-M [pii] 10.1002/1096-9888(200009)35:9<1091::AID-JMS31>3.0.CO;2-M
31. Flarakos J, Luo W, Aman M, Svinarov D, Gerber N, Vouros P (2004) Quantification of risperidone and 9-hydroxyrisperidone in plasma and saliva from adult and pediatric patients by liquid chromatography-mass spectrometry. J Chromatogr A 1026(1–2):175–183
32. Josefsson M, Roman M, Skogh E, Dahl ML (2010) Liquid chromatography/tandem mass spectrometry method for determination of olanzapine and n-desmethylolanzapine in human serum and cerebrospinal fluid. J Pharm Biomed Anal 53(3):576–582. doi:S0731-7085(10)00221-9 [pii] 10.1016/j.jpba.2010.03.040
33. McClean S, O'Kane EJ, Smyth WF (2000) Electrospray ionisation-mass spectrometric characterisation of selected anti-psychotic drugs and their detection and determination in human hair samples by liquid chromatography-tandem mass spectrometry. J Chromatogr B Biomed Sci Appl 740(2):141–157
34. Muller C, Vogt S, Goerke R, Kordon A, Weinmann W (2000) Identification of selected psychopharmaceuticals and their metabolites in hair by lc/esi-cid/ms and lc/ms/ms. Forensic Sci Int 113(1–3):415–421. doi:S0379073800002024 [pii]
35. Nielsen MK, Johansen SS, Dalsgaard PW, Linnet K (2010) Simultaneous screening and quantification of 52 common pharmaceuticals and drugs of abuse in hair using uplc-tof-ms. Forensic Sci Int 196(1–3):85–92. doi:S0379-0738(09)00528-3 [pii] 10.1016/j.forsciint.2009.12.027
36. Thieme D, Sachs H (2007) Examination of a long-term clozapine administration by high resolution segmental hair analysis. Forensic Sci Int 166(2–3):110–114. doi:S0379-0738(06)00267-2 [pii] 10.1016/j.forsciint.2006.04.015
37. Weinmann W, Muller C, Vogt S, Frei A (2002) Lc-ms-ms analysis of the neuroleptics clozapine, flupentixol, haloperidol, penfluridol, thioridazine, and zuclopenthixol in hair obtained from psychiatric patients. J Anal Toxicol 26(5):303–307
38. Skopp G (2009) Postmortem toxicology: artifacts. In: Wiley encyclopedia of forensic science. doi:10.1002/9780470061589.fsa417
39. Kirchherr H, Kuhn-Velten WN (2006) Quantitative determination of forty-eight antidepressants and antipsychotics in human serum by hplc tandem mass spectrometry: a multi-level, single-sample approach. J Chromatogr B Analyt Technol Biomed Life Sci 843(1):100–113. doi:S1570-0232(06)00445-4 [pii] 10.1016/j.jchromb.2006.05.031
40. Kratzsch C, Peters FT, Kraemer T, Weber AA, Maurer HH (2003) Screening, library-assisted identification and validated quantification of fifteen neuroleptics and three of their metabolites in plasma by liquid chromatography/mass spectrometry with atmospheric pressure chemical ionization. J Mass Spectrom 38(3):283–295. doi:10.1002/jms.440
41. Saar E, Gerostamoulos D, Drummer OH, Beyer J (2010) Identification and quantification of 30 antipsychotics in blood using lc-ms/ms. J Mass Spectrom 45(8):915–925. doi:10.1002/jms.1783
42. Bhatt J, Subbaiah G, Singh S (2006) Liquid chromatography/tandem mass spectrometry method for simultaneous determination of risperidone and its active metabolite 9-hydroxyrisperidone in human plasma. Rapid Commun Mass Spectrom 20(14):2109–2114. doi:10.1002/rcm.2537

43. Nielsen MK, Johansen SS (2009) Determination of olanzapine in whole blood using simple protein precipitation and liquid chromatography-tandem mass spectrometry. J Anal Toxicol 33(4):212–217
44. Kollroser M, Schober C (2002) Direct-injection high performance liquid chromatography ion trap mass spectrometry for the quantitative determination of olanzapine, clozapine and n-desmethylclozapine in human plasma. Rapid Commun Mass Spectrom 16(13):1266–1272. doi:10.1002/rcm.718
45. Berna M, Shugert R, Mullen J (1998) Determination of olanzapine in human plasma and serum by liquid chromatography/tandem mass spectrometry. J Mass Spectrom 33(10):1003–1008. doi:10.1002/(SICI)1096-9888(1998100)33:10<1003::AID-JMS716>3.0.CO;2-P [pii] 10.1002/(SICI)1096-9888(1998100)33:10<1003::AID-JMS716>3.0.CO;2-P
46. Hasselstrom J (2011) Quantification of antidepressants and antipsychotics in human serum by precipitation and ultra high pressure liquid chromatography-tandem mass spectrometry. J Chromatogr B Analyt Technol Biomed Life Sci 879(1):123–128. doi:S1570-0232(10)00727-0 [pii] 10.1016/j.jchromb.2010.11.024
47. Moody DE, Laycock JD, Huang W, Foltz RL (2004) A high-performance liquid chromatographic-atmospheric pressure chemical ionization-tandem mass spectrometric method for determination of risperidone and 9-hydroxyrisperidone in human plasma. J Anal Toxicol 28(6):494–497
48. Berna M, Ackermann B, Ruterbories K, Glass S (2002) Determination of olanzapine in human blood by liquid chromatography-tandem mass spectrometry. J Chromatogr B Analyt Technol Biomed Life Sci 767(1):163–168
49. Dams R, Huestis MA, Lambert WE, Murphy CM (2003) Matrix effect in bio-analysis of illicit drugs with lc-ms/ms: Influence of ionization type, sample preparation, and biofluid. J Am Soc Mass Spectrom 14(11):1290–1294. doi:S1044030503005749 [pii]
50. Bonfiglio R, King RC, Olah TV, Merkle K (1999) The effects of sample preparation methods on the variability of the electrospray ionization response for model drug compounds. Rapid Commun Mass Spectrom 13(12):1175–1185. doi:10.1002/(SICI)1097-0231(19990630)13:12<1175::AID-RCM639>3.0.CO;2-0 [pii] 10.1002/(SICI)1097-0231 (19990630)13:12<1175::AID-RCM639>3.0.CO;2-0
51. Matuszewski BK, Constanzer ML, Chavez-Eng CM (2003) Strategies for the assessment of matrix effect in quantitative bioanalytical methods based on hplc-ms/ms. Anal Chem 75(13):3019–3030
52. Maurer HH (2005) Advances in analytical toxicology: the current role of liquid chromatography-mass spectrometry in drug quantification in blood and oral fluid. Anal Bioanal Chem 381(1):110–118. doi:10.1007/s00216-004-2774-z
53. Cabovska B, Cox SL, Vinks AA (2007) Determination of risperidone and enantiomers of 9-hydroxyrisperidone in plasma by lc-ms/ms. J Chromatogr B Analyt Technol Biomed Life Sci 852(1–2):497–504. doi:S1570-0232(07)00105-5 [pii] 10.1016/j.jchromb.2007.02.007
54. Roman M, Kronstrand R, Lindstedt D, Josefsson M (2008) Quantitation of seven low-dosage antipsychotic drugs in human postmortem blood using lc-ms-ms. J Anal Toxicol 32(2):147–155
55. Remane D, Meyer MR, Wissenbach DK, Maurer HH (2010) Ion suppression and enhancement effects of co-eluting analytes in multi-analyte approaches: systematic investigation using ultra-high-performance liquid chromatography/mass spectrometry with atmospheric-pressure chemical ionization or electrospray ionization. Rapid Commun Mass Spectrom 24(21):3103–3108. doi:10.1002/rcm.4736
56. U.S. Department of Health and Human Services. Guidance for industry: mass spectrometry for confirmation of the identity of animal drug residues, final guidance. Available at: http://www.fda.gov/AnimalVeterinary/GuidanceComplianceEnforcement/GuidanceforIndustry/default.htm . Accessed 15 Jan 2012.
57. (WADA) TWA-DA (2010) Identification criteria for qualitative assays document td2003idcr
58. Union CotE (2002) Validation concepts for pesticide residues in food of animal origin 2002/657/ec
59. Peters FT, Maurer HH (2002) Bioanalytical method validation and its implications for forensic and clinical toxicology - a review. Accred Qual Assur 7(11):441–449

60. Peters FT, Drummer OH, Musshoff F (2007) Validation of new methods. Forensic Sci Int 165(2–3):216–224. doi:S0379-0738(06)00323-9 [pii] 10.1016/j.forsciint.2006.05.021
61. Shah VP, Midha KK, Findlay JW, Hill HM, Hulse JD, McGilveray IJ, McKay G, Miller KJ, Patnaik RN, Powell ML, Tonelli A, Viswanathan CT, Yacobi A (2000) Bioanalytical method validation–a revisit with a decade of progress. Pharm Res 17(12):1551–1557
62. Chew WM, Xu MJ, Cordova CA, Chow HH (2006) Quantification of a cytochrome p450 3a4 substrate, buspirone, in human plasma by liquid chromatography-tandem mass spectrometry. J Chromatogr B Analyt Technol Biomed Life Sci 844(2):235–239. doi:S1570-0232(06)00542-3 [pii] 10.1016/j.jchromb.2006.07.005
63. Gschwend MH, Arnold P, Ring J, Martin W (2006) Selective and sensitive determination of amisulpride in human plasma by liquid chromatography-tandem mass spectrometry with positive electrospray ionisation and multiple reaction monitoring. J Chromatogr B Analyt Technol Biomed Life Sci 831(1–2):132–139. doi:S1570-0232(05)00890-1 [pii] 10.1016/j.jchromb.2005.11.042
64. Saar E, Gerostamoulos D, Drummer OH, Beyer J (2011) Assessment of the stability of 30 antipsychotic drugs in stored blood specimens. Forensic Sci Int. doi:S0379-0738(11)00104-6 [pii] 10.1016/j.forsciint.2011.02.022
65. Olesen OV, Linnet K (1998) Determination of olanzapine in serum by high-performance liquid chromatography using ultraviolet detection considering the easy oxidability of the compound and the presence of other psychotropic drugs. J Chromatogr B Biomed Sci Appl 714(2):309–315
66. Zhou Z, Li X, Li K, Xie Z, Cheng Z, Peng W, Wang F, Zhu R, Li H (2004) Simultaneous determination of clozapine, olanzapine, risperidone and quetiapine in plasma by high-performance liquid chromatography-electrospray ionization mass spectrometry. J Chromatogr B Analyt Technol Biomed Life Sci 802(2):257–262. doi:10.1016/j.jchromb.2003.11.037 S1570023203010109 [pii]

Therapeutic Drug Monitoring of Targeted Anticancer Therapy. Tyrosine Kinase Inhibitors and Selective Estrogen Receptor Modulators: A Clinical Pharmacology Laboratory Perspective

Laurent Decosterd, Elyes Dahmane, Marine Neeman, Thierry Buclin, Chantal Csajka, Amina Haouala, and Nicolas Widmer

Abstract In the last decade, a new era of cancer therapy has emerged, and the treatment of several cancers has shifted from cytotoxic and nonspecific chemotherapy to chronic oral treatment with targeted molecular therapies. Most oral anticancer-targeted drugs approved at present are tyrosine kinase inhibitors (TKIs) and some of them are accompanied with diagnostic test aiming at preselecting patients who are more likely to respond to anticancer treatment, constituting vivid examples of the emerging field of personalized medicine. In that context, since most TKIs are also characterized by an important interindividual variability in their pharmacokinetics, renewed efforts for treatment optimization should be made for targeting adequate drug exposure in patients, increasing thereby the likelihood of optimal clinical response and tolerability of anticancer treatment. This can be done through the Therapeutic Drug Monitoring (TDM) approach, whereby the careful selection

L. Decosterd (✉)
Innovation and Development Unit, Service of Biomedicine, University Hospital Center and University of Lausanne, Laboratory BH 18-218, Center Hospitalier Universitaire Vaudois, 1011 Lausanne, CHUV Switzerland
e-mail: LaurentArthur.Decosterd@chuv.ch

E. Dahmane • C. Csajka
Division of Clinical Pharmacology, Service of Biomedicine, University Hospital Center and University of Lausanne, Lausanne & School of Pharmaceutical Sciences, University of Geneva and Lausanne, Geneva, Switzerland

M. Neeman • A. Haouala
Innovation and Development Unit, Division of Clinical Pharmacology, Service of Biomedicine, University Hospital Center and University of Lausanne, Lausanne & School of Pharmaceutical Sciences, University of Geneva and Lausanne, Geneva, Switzerland

T. Buclin • N. Widmer
Division of clinical Pharmacology, Service of Biomedicine, University Hospital Center and University of Lausanne, Lausanne, Switzerland

Q. Alan Xu and Timothy L. Madden (eds.), *LC-MS in Drug Bioanalysis*,
DOI 10.1007/978-1-4614-3828-1_9,

of TKI dosage is adapted to each patient according to individual plasma levels, contributing to minimize the risk of major adverse reactions and to increase the probability of efficient, long-lasting, therapeutic response. This chapter reviews the bioanalytical developments by chromatography and mass spectrometry in the field of targeted anticancer therapy, across the growing family of recent FDA-approved oral TKIs as well as for tamoxifen and its active metabolites, being in fact the most widely used targeted anticancer agent. The text also provides an introduction to existing pharmacokinetics–pharmacodynamics knowledge in the field of targeted anticancer therapy, and the rationale for a TDM program for TKIs.

1 Introduction

Cancer has become one of the most common diseases in developed countries. It is the leading cause of death among men and women aged younger than 85 years in the United States, above cardiovascular problems [1]. Fortunately, considerable medical advances have been achieved in the field of cancer chemotherapy in the last few decades, notably via the progressive optimization and improved management of toxicities of approved anticancer drugs, or by the continuous discovery of novel agents. Nowadays, treatment of many cancers relies on cytotoxic chemotherapy regimens, sometimes in combination with radiation therapy and surgery. Standard cytotoxic drugs used for cancer therapy have generally a narrow therapeutic index, are nonspecific, as they target ubiquitous cell division mechanisms. In that context, considerable research efforts have been pursued for finding specific treatments of cancer [2], with limited success until 2000s, with the notable exception of all-trans retinoic acid in Acute Promyelocytic Leukemia, and tamoxifen in breast cancer. These two drugs, designed for binding to retinoic acid receptors (RAR) and estrogen receptors (ER), respectively, can actually be considered the first clinically used targeted anticancer agents that have been associated with high rates of treatment success and control of the disease for prolonged period of time.

In the last decade, a new era of cancer therapy has emerged, and the treatment of several cancers has shifted from cytotoxic and nonspecific chemotherapy to chronic oral treatment with targeted molecular therapies. These treatments are characterized by unique mechanisms of action and are highly specific for single or multiple key cellular biological pathways responsible per se or implicated in the cancer process [3]. Targeted therapy via protein kinase inhibitors is directed against (onco) proteins allowing the modulation of various signaling pathways and is therefore characterized by more limited nonspecific toxicities. At present, 12 new oral targeted anticancer agents have been approved by FDA (Table 1), over 20 compounds are in Phase I and II trials, and many more at various stages of preclinical development [3]. Except for one agent (vemurafenib), all oral anticancer-targeted drugs approved by FDA are tyrosine kinase inhibitors (TKIs), and will be collectively designated thereafter as the generally accepted acronym TKIs.

Table 1 FDA-approved targeted tyrosine kinase inhibitors

DCI name (trade name)	Year of approval	Target	Indication, cancer
Imatinib (Gleevec®, Glivec®)	2001	BCR-ABL, c-KIT, PDGRF	Philadelphia-positive-chronic myelogenous leukemia (CML) and acute lymphoblastic leukemia (ALL)
			Myelodisplasic syndrome-myeloproliferative disorders (MDS/MPD)
			Aggressive systemic mastocytosis (ASM)
			Hypereosinophilic syndrome (HES), chronic eosinophilic leukemia (CEL)
			Dermato-fibrosarcoma protuberans (DFSP)
			CD17-positive gastrointestinal stromal tumors (GIST)
Gefitinib (Iressa®)	2003	EGFR	Non-small cell lung cancer (NSCLC)
Erlotinib (Tarceva®)	2004	EGFR	NSCLC
			Pancreatic cancer
Sorafenib (Nexavar®)	2005	VEGFR, PDGRF, RAF, Mek, Erk	Hepato-cellular carcinoma (HCC)
			Renal cell carcinoma (RCC)
Sunitinib (Sutent®)	2006	FLT3, PDGFR, VEGFR, KIT	RCC, GIST
			Pancreatic NET
Dasatinib (Sprycel®)	2006	Src, ABL	Philadelphia-positive CML, ALL
Lapatinib (Tyverb®)	2007	EGFR, HER2	HER2-positive breast cancer
Nilotinib (Tasigna®)	2007	BCR, ABL	Philadelphia-positive CML
Pazopanib (Votrient®)	2009	VEGRF 1,2,3	RCC
Vandetanib (Caprelsa®)	2011	VEGFR, EGFR	Thyroid cancer
Vemurafenib (Zelboraf®)	2011	B-RAF	Melanoma with B-RAF V600E mutation
Crizotinib (Xalkori®)	2011	ALK, hepatocyte growth factor receptor (HGFR; cMet)	Anaplastic lymphoma kinase (ALK)-positive NSCLC

These therapeutic agents are about to revolutionize cancer treatment, and some of them have allowed to transform deadly malignancies into chronically manageable conditions. Nevertheless, primary or secondary drug resistance, persistence of cancer stem cells, and drug adverse effects still limit their ability to stabilize or even cure malignant diseases in the long term. In addition, poor tolerance and therapeutic failure are not uncommon, and relapse is a nearly inevitable consequence of treatment interruption. The appropriate management of oncologic patients therefore requires careful monitoring of these novel treatments [4], for which most clinicians have at present a limited experience.

All these drugs, some of them accompanied with diagnostic tests aiming at preselecting patients who are more likely to respond to anticancer treatment [5], constitute vivid examples of the emerging field of personalized medicine [6, 7]. In that context, since most TKIs are also characterized by an important interindividual variability in their pharmacokinetics (PK), increasing efforts for treatment optimization should be made for targeting adequate drug exposure in patients, increasing thereby the likelihood of optimal clinical response and tolerability of anticancer treatment. This can be done through the Therapeutic Drug Monitoring (TDM) approach, whereby the careful selection of TKI dosage is adapted to each patient according to individual plasma levels, contributing to minimize the risk of major adverse reactions and to increase the probability of efficient, long-lasting, therapeutic response [8, 9].

Conversely, although clinically used for more than 30 years, it has been less than a decade ago that several publications have reported that the clinical efficacy of tamoxifen, the first and most widely used targeted therapy for estrogen-sensitive breast cancer, may depend on the formation of the active metabolites 4-hydroxy-tamoxifen and 4-hydroxy-*N*-desmethyl-tamoxifen (endoxifen) [10]. Large interindividual variability in endoxifen plasma concentrations has been observed and related both to genetic and environmental (i.e., drug-induced) factors altering CYP450s metabolizing enzymes activity [11]. Since endoxifen is considered to be responsible for an important part of the in vivo pharmacological activity of tamoxifen [10], there is a growing interest for monitoring endoxifen plasma concentrations in breast cancer patients. Whether this would constitute a valid approach to optimize individual tamoxifen dosage remains however to be formally demonstrated in randomized clinical trials (RCTs).

Implementation of a routine TDM program for both TKIs and tamoxifen/endoxifen necessitates the access to suitable instrumental technology, bioanalytical expertise, and definite knowledge in clinical pharmacokinetics for drug level interpretation leading possibly to dosage adjustment. The analytical results, integrated with the clinical observations, may influence the therapeutic intervention and in turn, clinical outcome. Reliability of analytical methods is therefore a critical issue, justifying the efforts and time devoted to their thorough validation and to extensive characterization of their performance (i.e., precision, accuracy, robustness, and turnaround time). Initially, high performance liquid chromatography techniques coupled to ultraviolet detection (HPLC-UV) have been developed for the measurement of imatinib [12] and other TKIs in biological fluids. At present, however, because of its unsurpassed selectivity and sensitivity, HPLC or Ultra performance liquid chromatography (UPLC) coupled to tandem triple quadrupole Mass Spectrometry (LC-MS/MS) has become the method of choice for drug plasma level measurements and is extensively applied for early and more recent TKIs. These powerful analytical technologies are becoming accessible to an increasing number of Academic Hospital Centers for TDM clinical service and research projects. These LC-MS/MS assays can bring invaluable information on patients' drug exposure and contribute, in conjunction with patient's pharmacogenetic tests as well as tumor genetic profiles determination, to the reinforcement and refinement of the personalized anticancer drug prescription.

This chapter reviews the bioanalytical developments by mass spectrometry in the field of targeted anticancer therapy, across the growing family of recent FDA-approved oral TKIs as well as tamoxifen and its active metabolites. The text also provides an introduction to existing pharmacokinetics–pharmacodynamics knowledge in the field of targeted anticancer therapy.

2 New Targets, New Drugs, and New Strategies for Improved Tolerability and Enhanced Clinical Response of the Anticancer Therapy

2.1 New Targets, News Drugs

The first prominent example of TKIs, imatinib, has revolutionized the treatment and prognosis of chronic myelogenous leukemia (CML) and gastrointestinal stromal tumors (GIST) [13, 14]. However, imatinib treatment is not devoid of toxicity, and resistance occurs. It is becoming increasingly recognized that the response is influenced not only by the genetic heterogeneity of drug target determining the tumor's sensitivity (BCR-ABL for CML, and c-KIT for GIST) but also by patient's genetic background and environmental factors that influence drug disposition and overall exposure in the body. Indeed, imatinib drug exposure was found to be a predictor of clinical response in CML [15, 16] and in GIST [17–20].

Following imatinib, other TKIs, including sunitinib, nilotinib, dasatinib, sorafenib, and lapatinib have been developed and are now used for treating various hematological malignancies, solid tumors including GIST [21], advanced renal cell carcinoma (RCC), hepatocellular carcinoma (HCC), and breast cancer [22], and have shown promising activity in other tumors as well [23]. In addition, there was a renewed interest for the EGRF inhibitors gefitinib and erlotinib for the treatment of non-small cell lung cancer (NSCLC), when it was discovered that a patient subgroup, with tumors harboring specific activating mutations of the EGFR genes, was likely to respond better [24, 25]. In 2011, three additional TKIs have been approved by the FDA: vandetanib for the treatment of thyroid cancer [26], vemurafenib against melanoma with B-RAF V600E mutation [27], and crizotinib for anaplastic lymphoma kinase (ALK)-positive NSCLC [28] (Table 1).

Despite their important specificity, toxicity and side effects similar to those of the standard cytotoxic chemotherapeutic approaches can also occur with signal transduction inhibitors.

Whereas toxicities encountered with TKIs treatment are generally less severe than those encountered with conventional cytotoxic approaches, they can, however, significantly impact the safety and quality of life in the long term, jeopardizing treatment adherence. The clinical responses for TKIs may also not always be optimal, calling for a renewed effort for exploring novel avenues and strategies to improve tolerability and therapeutic response.

2.2 New Strategy: Therapeutic Drug Monitoring, General Criteria

During the past decades, it has been established that the therapeutic use of selected drugs could be optimized by an individualization of their dosage, based on blood concentration measurement [29, 30]. As previously mentioned, such a feedback strategy, termed TDM, is now current practice for drugs such as antibiotics, antiepileptics, immunosuppressant drugs, antifungals [31], and, more recently, anti-HIV drugs [32, 33].

TDM is generally considered for drugs with large interindividual but limited intraindividual pharmacokinetic variability with both consistent concentration–efficacy and concentration–toxicity relationships. The sources of variability in drug response are multifactorial, and apart from genetics, other factors such as patient's pathophysiological conditions, environment, drug–drug interactions, food, drinking and smoking habits, medication errors, and poor compliance, may have an important impact on drug pharmacokinetics and/or pharmacodynamics, thereby affecting the therapeutic outcome [29, 30]. Information provided by TDM is particularly useful for drugs with a narrow therapeutic index, subjected to physiologic, genetic, and environmental influences and used for prolonged periods.

In oncology patients, maintaining circulating drug concentrations over a given threshold appears to be crucial to ensure optimal pharmacological action as exposure to suboptimal drug levels during chronic therapy substantially increases the risk of therapeutic failure, due to the progressive selection of cancer cell clones. On the other hand, excessive drug concentrations may be associated with intolerance and adverse drug reactions, leading in term to frequent therapeutic treatment interruption.

While careful monitoring is normally recommended for any type of treatment, their interest varies according to the clinical situations. Short-term treatments generally require little, if any, blood drug level monitoring. For long-term treatments, the interest of TDM is probably limited if all patients respond similarly to the standard regimen. Alternately, in the presence of a significant interindividual variability in response to treatment, the determination of circulating drug concentration in patients' blood may provide clinically useful information for patients' tailored treatment optimization [34].

Like any diagnostic test, the measurement of drug plasma level is, however, justified only when the information provided is of potential therapeutic benefit and has been demonstrated in clinical trials. The clinical value of plasma level monitoring depends on how precisely the treatment outcome can be defined. On the other hand, when a precise therapeutic end point is difficult to define, monitoring of drug levels may be of considerable therapeutic assistance for clinicians [35].

2.3 Therapeutic Drug Monitoring in Conventional Cytotoxic Chemotherapy

In oncology, drug dosage individualization for conventional cytotoxic anticancer therapy is performed according to mg/m^2 or mg/kg. However, even after dose adjustment, the pharmacokinetic variability observed for many cytotoxic chemotherapeutic agents remains important. The TDM approach is still limitedly used for conventional cytotoxic therapy [36], notably because of the lack of established therapeutic ranges for drug plasma concentrations. Pharmacokinetics-pharmacodynamics studies have shown that TDM of some cytotoxic drug improves the management of therapeutic response and hematological toxicity. For a few drugs (busulfan, cytosine-arabinoside, 5-fluorouracile, and methotrexate), minimal concentration (C_{min}) levels have been found to be predictive of clinical response [37]. Nowadays, in high-dose methotrexate chemotherapy regimen, methotrexate plasma levels are monitored in order to adjust the tetrahydrofolate (i.e., leucovorin) dose administered in prevention of methotrexate renal and systemic toxicity (i.e., methotrexate overdosing prevention by "leucovorin rescue") [30, 37].

2.4 Therapeutic Drug Monitoring Program for Targeted Anticancer Therapy

Fixed dosing is still standard practice for TKIs in the medical oncologist community. While most standard anticancer chemotherapy regimens are administered through short *i.v.* cycles, targeted drugs such as imatinib and the more recent TKIs are orally administered and must be taken in the long term, if not indefinitely. Moreover, they are metabolized mostly by cytochromes P450, in particular the isoenzymes CYP3A4/5, whose activity is known to present a large interindividual variability and influenced by co-medications, organs diseases, diet, and environmental factors, as well as genetic background. Some TKIs are also substrates of drug transporters, such as efflux pumps (e.g., P-glycoprotein; P-gp) or uptake pumps (e.g., human organic cation transporter 1; hOCT1) [18, 22, 38–41]. Finally, as they are extensively bound to circulating proteins in plasma (such as for instance, imatinib on α1-acid glycoprotein [42]), only a small fraction of drug as free drug [43] is likely to enter cells to exert its pharmacological action. The wide inter-individual pharmacokinetic variability of TKIs has been clearly demonstrated in several studies [42, 44–52] and there are some preliminary evidences of its consequences on treatment response (see also Sect. 2.5) [15, 16, 19, 20, 47, 48, 53–55]. Identified factors affecting drug disposition include genetic polymorphisms of drug metabolizing enzymes and efflux and influx transporters, age, gender, weight, diet, smoking habit, alcohol consumption, renal and liver function, concomitant diseases, and co-medications. A given dose can therefore yield very different circulating concentrations between patients, favoring the selection of resistant cellular clones in case of subtherapeutic drug exposure, or increasing the risk of adverse drug reactions at excessive plasma levels.

There are therefore several strong lines of arguments for monitoring plasma levels of current and probably other newer TKIs drugs to come [48]. The initiation of such a TDM program for TKIs must also comprise a comprehensive investigation on their concentration–effect relationships, which is still lacking for a majority of them. At present, although not yet formally validated, the TDM for TKIs should probably be considered in special clinical situations such as in case of less-than-expected initial clinical response, disease recurrence, adverse drug reactions, drug interaction problems, doubt on patient compliance, and in further defined clinical conditions (pediatrics, renal and hepatic failure, etc.). However, further extensive evaluation should be carried out in well-conducted clinical trials before systematic TDM can be integrated into cancer patient's standard of care [56]. The recent experience with the TDM of anti-HIV drugs that has been adopted in the current medical practice without a rigorous evaluation by RCTs of its impact on clinical response and toxicity should be avoided. Most TKIs have just been introduced in the clinical practice, and such window of opportunity should not be missed. The formal demonstration of the clinical usefulness of TDM in RCT for the first major TKI imatinib may therefore constitute the initial step opening the way of a generalized TDM program for all subsequent TKIs for the optimal management of anticancer-targeted therapy [16, 17, 19, 20].

2.5 Pharmacokinetic Variability, Concentration–Efficacy and Concentration–Toxicity Relationships for TKIs

This section presents an overview of the existing pharmacokinetics–pharmacodynamics knowledge in the field of targeted anticancer therapy for the TKIs approved or in late phase of clinical development. An excellent comprehensive review of the clinical pharmacokinetics of the first eight TKIs has been already published [57]. Addressing these aspects is relevant because besides significant variability in pharmacokinetics, relationships between concentrations and efficacy and/or toxicity are amongst the principal characteristics that must be met for considering a formal TDM program.

2.5.1 Imatinib

Pharmacokinetic Variability

Imatinib is characterized by an important interpatient pharmacokinetic variability, yielding trough plasma concentrations spreading over between 40 and 80 % under standard dosing regimens [15, 16, 19, 20, 42, 50, 58]. The intraindividual variability is lower and does not exceed 30 % [42]. The high interindividual variability in pharmacokinetics has been mostly related to differences in the distribution and metabolism of this drug. Imatinib distribution is mainly influenced by plasma protein concentrations, as approximately 95 % of the drug binds to albumin and α-1 acid glycoprotein [18, 59]. The levels of α-1 acid glycoprotein are known to be altered (i.e., increased) in case of infections, and in acute and chronic conditions (cancer, etc.). Active transport mechanisms are also responsible for imatinib tissue uptake (via the carrier human organic cation hOCT-1) and efflux from tissues and cells (P-glycoprotein [P-gp]) [18, 60, 61]. The expression and activity of these drug transporters is modulated both by genetic factors (polymorphisms affecting function and expression) and environmental influences. As previously mentioned, imatinib is metabolized by CYP3A4, whose activity is genetically determined, as well as likely to be inhibited or induced by various environmental factors (coadministered drugs [62, 63] or food). Low concentrations have been described with the CYP3A4-inducing agents rifampicin, antiepileptics drugs, and St. John's Wort [63]. Alternately, excessive plasma concentration of imatinib, associated with clinical toxicity, has been reported during the coadministration of voriconazole, a known CYP3A4 inhibitor [64]. Finally, poor adherence has been recognized as an important additional source of pharmacokinetic variability [8].

Concentration–Effect Relationship

Several studies have described a relationship between imatinib trough plasma concentrations and clinical response. Initially, Picard et al. [16] have shown that trough plasma concentrations of imatinib are significantly higher in patients with complete cytogenetic response (CCR, defined by the complete disappearance of the Philadelphia positive cells) and a major molecular response (MMR, defined by a 3 log decrease of *BCR-ABL* transcripts). A trough level above 1,002 ng/ml was then recommended for CML patients [16]. In the landmark IRIS study [65], Larson et al. [15] have retrospectively observed higher imatinib trough concentrations 1 month after treatment initiation in patients who showed a complete cytogenetic response and major molecular response, in comparison to patients without cytogenetic or molecular response (cut-off: 1,000 ng/ml). These findings were confirmed in other studies. Whereas target values may differ between studies, most authors acknowledged the potential clinical value of TDM of imatinib in CML patients [66–70].

In GIST, important findings are also emerging from studies examining the relationships between imatinib PK and response to treatment [9]. A pharmacokinetic analysis from a clinical trial of imatinib in patients with unresectable or metastatic GIST reveals a correlation between imatinib total exposure and clinical response. Trough levels over 1,100 ng/ml predicted a better overall benefit rate [17]. Widmer et al. [19, 20] showed that free trough level was correlated with a clinical benefit in GIST patients, with responders having higher free levels than non-responders. Target levels might further depend on tumor genetics [20].

The threshold of effective concentrations of imatinib for optimal clinical response remains to be clearly defined and validated in a prospective clinical trial. It remains also possible that different thresholds exist for different levels of response and cancer cell genetic profiles [58].

Widmer et al. [19, 20] demonstrated that both total (in GIST) and free drug exposure (in CML and GIST) correlated with the occurrence and number of side effects. Moreover, the study of Larson et al. [15] showed that during the first 3 months of imatinib treatment, the types and grades of emerging adverse events were similar among patients, except for fluid retention, nausea, musculoskeletal pain, rash, myalgia, and anemia, which were more frequently reported by patients with higher imatinib concentrations. Based on the overall 5-year data, only fluid retention, rash, myalgia, and anemia were more frequently reported by the patients with higher imatinib concentrations. These studies therefore suggest that some, but not all, adverse events may be related to elevated imatinib plasma concentrations [34].

TDM for imatinib probably represents therefore a clinically useful tool for providing valuable information for clinicians to investigate the absence of expected clinical response, the occurrence of toxicity, drug–drug interaction problems, and to assess patients' short-term adherence. At present, the level of proof for imatinib TDM varies between "recommended" and "potentially useful" [34]. Whether TDM is also beneficial for the other TKIs remains to be established, but can be anticipated considering their pharmacokinetics characteristics and metabolic pathways and the drug interaction potentials (see below).

2.5.2 Nilotinib

Pharmacokinetic Variability

The interpatient variability in exposure to nilotinib is 32–64 % for exact reasons remaining yet to be explained [44, 57]. In the phase I dose escalation study, a saturation of nilotinib serum levels was observed with doses ranging from 400 to 1,200 mg daily. With the administration of daily doses at the steady-state level, the peak concentration and the area under the concentration–time curve increased among patients receiving 50–400 mg of the drug and reached a plateau among patients receiving more than 400 mg. A possible explanation might be that nilotinib gastrointestinal absorption saturates at doses exceeding 400 mg [71].

Concentration–Effect Relationship

The relationships between nilotinib plasma concentration and clinical efficacy (or toxicity) have not been studied yet. Irrespective of nilotinib PK-PD per se, Saglio et al. [72] showed that nilotinib at a dose of either 300 or 400 mg twice daily was superior to imatinib in patients with newly diagnosed chronic-phase Philadelphia chromosome-positive CML. At 12 months, the rates of major molecular response for nilotinib were nearly twice that observed for imatinib. The rates of complete cytogenetic response by 12 months were also significantly higher for nilotinib than for imatinib [72]. No data have been published for nilotinib concentration–toxicity relationships nor plasma target values to be achieved for optimal clinical response.

The limited PK-PD information available at present for nilotinib does not exclude, however, that this drug may be a good candidate for TDM.

2.5.3 Dasatinib

Pharmacokinetic Variability

Dasatinib interpatient and inter-occasion variability is important and ranges from 32 to 118 %. A substantial proportion of the inter-occasion variability is supposedly related to the drug bioavailability. The origin of the interpatient variability has not been elucidated yet, but is presumably related to dasatinib CYP3A-mediated metabolism, characterized by high variability in activity and expression [46].

Concentration–Effect Relationship

Preclinical and clinical investigations have demonstrated that dasatinib is active against imatinib-resistant BCR-ABL variants CML and has further improved the treatment of CML [73]. Moreover, dasatinib, as compared with imatinib, induced significantly higher and faster rates of complete cytogenetic response and major molecular response. Since achieving complete cytogenetic response within 12 months has been associated with better long-term, progression-free survival, dasatinib may improve the long-term outcomes among patients with newly diagnosed chronic-phase CML [54]. However, no clear concentration–efficacy relationship has been proposed yet, which again does not exclude a role for TDM in the future.

In the study of Wang et al., dasatinib trough levels appear to correlate strongly with toxicity but not with efficacy. The lowest trough concentration was achieved with the lowest dose regimen (100 mg once daily) which has been shown to have the optimal therapeutic index among the regimens tested [74].

2.5.4 Sunitinib

Pharmacokinetic Variability

Sunitinib interpatient variability in pharmacokinetics is also significant, of approximately 40 %, which is unexplained yet [47].

Concentration–Effect Relationship

The results of a meta-analysis [47] indicate that increased exposure to sunitinib in patients with advanced solid tumors, including patients with GIST and metastatic RCC, is associated with improved clinical outcomes, as well as some increased risks of adverse effects. This analysis indicates that increased exposure to sunitinib is associated with longer time to tumor progression, longer overall survival, a higher probability of a response, and greater tumor-size decreases. A sunitinib 50-mg starting dose has been proposed to provide clinical benefit with acceptably low risk of adverse events [47]. Based on preclinical data [75] and a phase I study [76], a target plasma concentration of 50 ng/ml (parent drug *plus* metabolite SU12662) was defined for sunitinib, even though no formal TDM study has, to the best of our knowledge, been initiated yet for this latter drug.

Houk et al. [47] have shown that increased exposure to sunitinib is associated with increased risk of adverse effects generally mild to moderate in severity. Faivre et al. [76] have found dose-limiting toxicities at plasma concentrations of sunitinib *plus* SU12662 higher than 100 ng/ml.

2.5.5 Sorafenib

Pharmacokinetic Variability

Sorafenib pharmacokinetics shows a large interpatient variability [49, 57]. The large interpatient variability is supposed to be the result of slow dissolution of the drug in the gastrointestinal tract and of the existence of an entero-hepatic circulation [51].

Concentration–Effect Relationship

No information on sorafenib concentration–toxicity relationships is available at present. Again, the absence of any PK-PD data does not preclude any interest for a formal TDM for sorafenib.

A study has shown in patients with metastatic RCC and hepatocarcinoma, given the standard regimen of sorafenib (800 mg daily), that toxicity occurrence may be related to high plasma sorafenib exposure [77]. However, an upper plasma level was not determined.

2.5.6 Lapatinib

Pharmacokinetic Variability

Lapatinib variability is large (68 %) and not significantly reduced by the coadministration of food (52 %) [45, 57]. M. Ratain and E. Cohen [78] have suggested that a lower dose of lapatinib could be administered if taken with food, to take advantage of the increased absorption of lapatinib in the presence of high fat meals, or if taken with grapefruit juice, a known CYP3A inhibitor, which should result in an overall reduction in treatment cost. However, they strongly recommended that this approach should not be done without a formal pharmacokinetic assessment.

Concentration–Effect Relationship

Burris et al. [53] showed in patients with metastatic solid tumors treated with lapatinib at doses ranging from 500 to 1,600 mg once daily that clinical responses were generally associated with doses in the middle of the range examined. Clinical response was more often associated with doses of 900–1,200 mg daily. However, due to the limited response data, it was not possible to adequately characterize the relationship between clinical response and drug exposure, which would be a prerequisite before assessing the potential role of TDM in lapatinib dosage individualization.

Finally, relationships between lapatinib plasma concentration and clinical toxicity have not been yet formally studied.

2.5.7 Miscellaneous TKIs

Pharmacokinetic Variability

Very large variations have been demonstrated for gefitinib exposures and for the recent TKI axitinib (evaluated for metastatic melanoma, renal cell and thyroid cancer, and NSCLC), with variation in drug exposure ranging between 113 % and 39–94 % for gefitinib and axitinib, respectively [48]. In that context, an assay for phenotyping patients' CYP3A activity has been proposed for predicting gefitinib systemic exposure and helping at drug dosage selection [79]. Erlotinib interpatient pharmacokinetic variability is also important (60 %) and, as yet, unexplained [52].

Information on the clinical pharmacokinetics is also available of the recently approved TKIs. Vandatenib pharmacokinetics, studied in healthy volunteers and patients [80, 81] was found to be both influenced by patient's renal function and vulnerable to drug–drug interactions [82, 83]. An important pharmacokinetic variability is noticeable in the mean steady state PK profiles published for verumafenib (formerly PLX4032) [27].

Several new TKIs are in advanced stage of clinical development, including bosutinib [84] and bafetinib [85], the third-generation TKIs against imatinib-, nilotinib-, and dasatinib-resistant CML; the multi-targeted kinase inhibitor pazopanib, approved for advanced or metastatic RCC [86]; and neratinib with antitumor activity in HER2–positive breast cancer. Bosutinib, pazopanib and neratinib are all substrates of the CYP3A enzymatic system, and their plasma exposure is increased when coadministered with potent CYP3A inhibitors [86–88], potentially requiring dose adjustment for neratinib [87].

Finally, pharmacokinetics and metabolic studies are also available for TKIs at various stages of clinical development, including vatalanib [89–91], cediranib [92–94], and motesanib [95].

Concentration–Effect Relationship

Statistically significant associations were demonstrated between the 5 and 10 h post-dosing plasma concentrations of erlotinib and survival in patients receiving 150 mg erlotinib daily for advanced recurrent and/or metastatic squamous cell cancer of the head and neck [55]. Based on in vitro data, a target plasma concentration of erlotinib higher than 420 ng/ml has been proposed [96]. Moreover, vatalanib trough plasma concentration, AUC and maximal concentration were positively correlated with likelihood of response in metastatic liver lesions on imaging [48]. Finally, target efficacious plasma levels based on in vitro data have been proposed for crizotinib, a TKI recently approved for locally advanced or metastatic NSCLC positive for anaplastic lymphoma kinase (ALK) [97, 98].

Erlotinib area under the curve was positively correlated with the occurrence of skin toxicity in two independent studies, in patients with non-small cell lung cancer [99, 100]. Moreover, Mohamed et al. [101] showed that the occurrence of skin rash was associated with significantly improved survival for advanced NSCLC patients who failed prior chemotherapy, upon treatment with gefitinib. Hypertension is another example of group effect of VEGF inhibitors, which might be due to inhibition of vascular relaxation, decreased production, and rarefaction of nitric oxide [102]. Indeed, the rise in diastolic blood pressure during treatment with the VEGF1/2/3 inhibitor axitinib was a predictor of longer survival in patients with various malignancies [48].

In conclusion, most recent TKIs share with imatinib the same large interindividual pharmacokinetic variability with, at least for a few of them, some reports of concentration–efficacy and concentration–toxicity relationships, calling for further extensive evaluation of the TDM approach. The development of analytical methods allowing to confidently quantifying TKIs in biological fluids is a prerequisite prior to the implementation of any clinically useful TDM Service.

2.6 *Analytical Methods by LC Tandem MS for the Bioanalysis of TKIs*

This section reviews the analytical methods using liquid chromatography coupled to tandem mass spectrometry that has been developed for measuring the concentration of TKIs in various human biological samples.

2.6.1 Methods of Quantification for Single TKIs

Quantification of TKIs in patients' plasma samples is at present principally performed by liquid chromatography-mass spectrometry (LC-MS) after suitable plasma pretreatment, which implies most generally a protein precipitation with an organic solvent, or a liquid–liquid extraction (LLE) using a non-miscible phase (sometimes adjusted at a pH that takes advantage of the mostly basic nature of TKIs) or, alternately, using either an off-line or online solid phase extraction (SPE) step.

Most analytical methods published to date using liquid chromatography coupled to mass spectrometry (LC-MS) have focused on the assay in human biological fluids, generally plasma, of a single TKI, namely imatinib [103–110] (see Table 2). Most proposed methods use generic protein precipitation by acetonitrile (ACN) [104–106, 108, 109] of whole blood [107] or plasma [106, 108, 109], prior to imatinib quantification in supernatants. A semiautomated protein precipitation step within a 96-well plate format is also described [104]. A methodology using LLE with hexane-ethylacetate (30:70, v/v), followed by evaporation and reconstitution in acetonitrile/water/formic acid (30:70:0.1 %, v/v/v) [103] or in 4 mM ammonium formate buffer/methanol (1:1, v/v) [110], is an alternative method for sample preparation that was performed by two other groups. The described sample extraction procedures for the bioanalysis of the closely related TKI nilotinib involved plasma protein precipitation by acetonitrile [111], as well as LLE with methyl *tert*-butyl ether, evaporation and reconstitution in acetonitrile/0.2 % formic acid (1:9, v/v) [112].

The two methods published for the assay of sunitinib have used LLE of plasma with methyl *tert*-butyl ether solvent, evaporation and reconstitution in acetonitrile/water/formic acid (20:80:0.1, v/v/v) [113] and acetonitrile [114].

Two articles have described analytical methods for the determination of sorafenib plasma concentrations, involving a similar protein precipitation step with acetonitrile [115, 116]. There is at present only one published method for the quantification of lapatinib in human plasma after off-line SPE onto C18 cartridge, followed by evaporation and reconstitution in ACN/5.0 mM ammonium formate pH 3/formic acid (1,000:50:1, v/v/v) [117].

An assay has been described for the determination of vandetanib in human plasma and in cerebrospinal fluid. The assay consists in an LLE with *tert*-butyl methyl ether in the presence of ammonium hydroxide, followed by evaporation of the top organic layer and reconstitution in ACN/10 mM ammonium formate pH 5, prior to reversed-phase LC tandem MS [118].

Analytical methods have been published also for the more recent TKIs in early or late phase of clinical development, which comprise vatalanib [119] and axitinib [120]. Abbas et al. have developed an analytical method for measurement of the anti-CML bosutinib in plasma of healthy subjects [88]. Samples were extracted from plasma by LLE with carbonate buffer pH 10 and 1-chlorobutane (1:10, v/v) prior to evaporation and reconstitution in water/methanol solution (50:50, v/v).

The assays of these TKIs in plasma involve mostly reversed-phase liquid chromatography. The chromatographic principles and separation mechanisms are the same for High Performance Liquid Chromatography (HPLC) and Ultra Performance Liquid Chromatography (UPLC), while speed, sensitivity, and resolution are improved from UPLC [121]. The main advantage of UPLC is a significant reduction of analysis time, resulting in a decrease in solvent consumption, turnaround time

Table 2 Description and performance of the reported methods for quantification of TKIs

Drugs analyzed	Volume of biological sample	Analytical method	LOQ (ng/ml)	Accuracy at LOQ (% bias)	Precision at LOQ (CV%)	Reference
Imatinib	200 µl Human plasma	ESI-TSQ	10	–14.7	5.27	Titier et al. [110]
Imatinib	200 µl Human plasma	ESI-TSQ	10	1.68	5.75	Awidi et al. [103]
Imatinib	250 µl Human red blood cells plasma	ESI-TSQ	2.1 1.8	–4.3 to 0.8 0.2 to 2	3.78 to 5.02 5.74 to 2.39	Guetens et al. [106]
Imatinib	200 µl Whole human blood and leukemia cells	Turbo ion spray-TSQ	0.03	0.9 –1.9	8.8 13.2	Klawitter et al. [107]
Imatinib and CGP 74588	200 µl human plasma	ESI-TSQ	1 2	–	–	Boddy et al. [105]
Imatinib and CGP 74588	200 µl human plasma	ESI-single quadrupole	30	–	–	Parise et al. [108]
Imatinib and CGP 74588	100 µl human plasma	ESI-TSQ	1	–	–	Rochat et al. [109]
Imatinib and CGP 74588	200 µl human plasma	ESI-TSQ	4	–0.8 to 4	4.19 to 5.64	Bakhtiar et al. [104]
Nilotinib	200 µl Human plasma Serum	ESI-single quadrupole	5	–7.9 1.6	3.6 7.8	Parise et al. [111]
Nilotinib	100 µl Human serum	ESI-TSQ	2.5	–	–	Tanaka et al. [112]
Dasatinib	100 µl Human plasma	ESI-linear trap	–	–	–	Christopher et al. [126]
Bosutinib	100 µl Human plasma	Turbo ion spray-TSQ	1	–	–	Abbas et al. [88]
Sunitinib	200 µl Human plasma	ESI-TSQ	0.6	–	–	Minkin et al. [114]
Sunitinib	100 µl Human plasma	ESI-TSQ	0.2	9.5	11.7	De Bruijn et al. [113]
Sorafenib	50 µl Human plasma	ESI-TSQ	5	–4.33	7.45	Jain et al. [115]
Sorafenib	100 µl Human plasma	ESI-TSQ	7.3	2.3	3.3	Zhao et al. [116]
Lapatinib	100 µl Human plasma	Turbo ion spray-TSQ	15	–	11	Bai et al. [117]

Vatalanib	50 μl Human plasma	ESI-TSQ	10.2	6.25	8.38	Lankheet et al. [119]
Axitinib	100 μl Human plasma	ESI-TSQ	0.2	–	–	Sparidans et al. [120]
Neratinib	250 μl Human plasma	ESI-TSQ	3			Abbas et al. [87]
Vandetanib	100 μl Human plasma 100 μl Human CSF	Turbo ion spray-TSQ	1 0.25	8.5 –0.2	3.4 2.6	Bai et al. [118]
Imatinib, dasatinib, nilotinib	250 μl Human plasma	ESI-single quadrupole	78.1 62.5 62.5	–	–	De Francia et al. [133]
Imatinib Erlotinib Gefitinib	100 μl Human plasma	ESI-TSQ	4.4 4.5 4.8	–3.8 to 10.6	2.1 to 3.2	Chahbouni et al. [135]
Imatinib, dasatinib, nilotinib, sunitinib, sorafenib, and lapatinib	100 μl Human plasma	ESI-TSQ	1–10	–6.3 to 6.7	3.4 to 9.9	Haouala et al. [122]
Erlotinib, imatinib, lapatinib, nilotinib, sorafenib, and sunitinib	100 μl Human plasma	ESI-TSQ	2.2 for sunitinib 3.8–12.6 for other TKI			Gotze et al. [137]
Bosutinib, erlotinib, gefitinib, and vemurafenib		ESI-TSQ	1, 10, 10, and 80, resp.			Neeman [139]
Dasatinib and lapatinib	5×10^4 cells/mL lung cancer cell-lines	ESI-TSQ	50 100	–	8 13	Roche et al. [141]

(continued)

Table 2 (continued)

Drugs analyzed	Volume of biological sample	Analytical method	LOQ (ng/ml)	Accuracy at LOQ (% bias)	Precision at LOQ (CV%)	Reference
Gefitinib, erlotinib, sunitinib and sorafenib	20 μl Human whole blood, serum, plasma and cell culture medium	Turbo ion spray-TSQ	5	3.7 to 13.5	1.9 to 12	Honeywell et al. [136]
Imatinib	PBMC	ESI-TSQ	10	–	–	Widmer et al. [42]
Nilotinib, dasatinib, and bosutinib	2×10^6 K562 cells	ESI-TSQ	0.1–1	–	–	Hegedus et al. [142]
Imatinib, nilotinib, dasatinib, sunitinib, and sorafenib	9×10^5 K562 cells	ESI-TSQ	1–10	−12.6 to 10.1 %	9.1 to 11.3 %	Haouala et al. [143]
Imatinib, dasatinib, and nilotinb	PBMC	HPLC-MS	0.25 ng			D'Avolio et al. [144]

and costs [121]. However, UPLC has been applied only for quantification of the TKIs sunitinib [113] and axitinib [120].

A publication describes a column switching procedure for an imatinib assay, involving a C8 extraction column, followed after activation of the switching valve by the back-flushing of imatinib onto a C18 analytical column [107]. The run time was 10 min. The other published methods use HPLC with C18 columns with analytical times of 2 min [106], 3 min [103], 6 min [110], 10 min [107], 14 min [108], and 20 min [109]. C8 columns were used by two groups with analytical times varying between 2.5 min [104] and as much as 40 min [105].

Nilotinib was analyzed by HPLC onto a C18 column [111, 112], with a reported run time of 15 min [111].

Interestingly, the two methods published for quantification of sunitinib used either HPLC [114] or UPLC [113] but the analytical time periods were of similar duration (run time of 3 min and 4 min, respectively). Of note, we observed during the course of our own method development [122] the presence of two peaks with the same molecular mass/signal transition for sunitinib, which, to the best of our knowledge, has not been reported elsewhere [113, 114, 123]. The phenomenon was known, however, and is due to a *Z–E* isomerization reaction of sunitinib [124]. Previous studies by the sunitinib manufacturer have shown that *E* isomer can be generated from the *Z* isomer in a reversible manner in solution [124]. The rate of interconversion between the *Z–E* configurations in solution is dependent on a number of factors, most notably exposure to light. In our studies [122], we found that both isomers could be detected in the pharmaceutical preparation (tablet) at ratios of about 1:2, as well as in patients' plasma samples (variable ratios).

The determination of sorafenib plasma concentrations was performed by HPLC onto a C8 column (4 min run time) [115] and C18 column (6 min run time) [116]. Similarly, HPLC C18 columns have been used for the quantification in plasma of lapatinib [117], vatalanib [119] (3 and 8 min run time, respectively), and also for bosutinib [88], whereas UPLC was used for the TKI in development of axitinib (1.2 min run time) [120].

Detection of imatinib is performed by triple quadrupole mass spectrometer with an electrospray ionization (ESI) interface operated in positive ion mode [103, 104, 106, 107, 109, 110]. Except the methods published by Parise et al. for imatinib and its main metabolite [108], and for nilotinib [111], where a single quadrupole mass spectrometer was used, most TKIs are analyzed in plasma by atmospheric pressure ionization (electrospray or turbo ion spray) coupled to triple stage mass spectrometer. Expectedly, higher limit of quantifications for imatinib (30 ng/ml) [108], and nilotinib (5 ng/ml) [111], are obtained for the assays using single quadrupole MS (see Table 2).

In general, triple stage quadrupole mass spectrometer with ESI in positive ion mode is perceived as the most appropriate MS technique available at present for small—mostly basic—molecules and was used for the assay of nilotinib [112], sorafenib [115, 116], lapatinib [117], sunitinib [113, 114], bosutinib [88], vatalanib [119] axitinib [120], vandetanib [118] neratinib [87], and crizotinib [125]. The latter assay for crizotinib was developed for preclinical experiments and does not contain information on its analytical performance, and was subsequently adapted for pharmacokinetic studies in cancer patients [125].

So far, validated LC-MS/MS methods published for the assay of dasatinib in human plasma also comprise the analysis of metabolites [46, 126, 127].

Finally, to the best of our knowledge, there are, as yet, no analytical method validation reports for the assay of the latest TKIs pazopanib, bafetinib, cediranib, and motesanib in human biological samples.

2.6.2 Methods for Quantification of TKIs and Metabolites

Up to now, most investigations on the pharmacokinetic–pharmacodynamic aspects of TKIs therapies have focused almost exclusively on concentrations of the parent TKI drug in plasma, considering it as the best pharmacokinetic marker of anticancer drug exposure and, in case of higher levels, of toxicity. However, drug metabolites resulting from complex mutual genetic and environmental influences can also contribute to treatment outcome. The metabolite profile can be considered as a snapshot on the phenotypic pattern of the metabolizing activity in a patient at a given time. Unfortunately, integration of this aspect with pharmacokinetics has attracted little attention so far in the field of TKIs therapy. Distinct metabolite profiling patterns per se could play an important role in the toxicity, tolerability, and outcome of targeted anticancer therapy.

In that context, the LC–MS/MS technology makes it possible to determine in patients' plasma not only the parent drug but also metabolites. Such an approach has been applied for the quantification of imatinib and its main active *N*-desmethyl metabolite CGP 74588 in plasma [104, 108], for monitoring imatinib metabolites profile in patients' plasma [109], and for metabolism studies on dasatinib [46, 126, 127]. Assays enabling the quantification of sunitinib and its n-desethyl metabolite SU12662 [113] and, more recently, dasatinib and two active metabolites [127] have also been published. Overall, exposures of pharmacologically active metabolites in patients suggested that they are not expected to contribute significantly for the in vivo activity.

2.6.3 Methods for Multiplex Quantification of TKIs

Plasma Measurements

Mass spectrometry detection qualifies for the simultaneous measurement of arrays of structurally unrelated anticancer-targeted agents in a single analytical run. Multiplex analyses offer, therefore, the advantage of the establishment of calibration curves for several TKIs simultaneously, resulting in an overall reduction in analytical time, turnaround time, and costs [128, 129]. Analytical methods using a simplified extraction procedure followed by simultaneous quantification of multiple TKIs are more efficient for rapidly providing TDM results allowing real-time processing of blood samples from patients receiving different single-drug or combined regimens and for maximizing laboratory's resource utilization.

Thus, the development and validation of enhanced throughput methods with simple extraction procedure followed by LC-MS/MS are of high interest for the simultaneous analysis of every major anticancer-targeted agent [130, 131], which in the future may possibly be used also in combination therapy [132].

In that context, an assay limited to the antileukemic drugs imatinib, dasatinib, and nilotinib was proposed in 2009 implying plasma protein precipitation procedure followed by reversed-phase chromatography. TKIs detection was made by an ESI interface, coupled to positive SIM mode single quadrupole mass spectrometer [133]. However, single quadrupole MS analysis is probably not sensitive enough for the accurate quantification of very low plasma levels of dasatinib. In fact, the lower limit of quantification for dasatinib reported in this study is 62.5 ng/ml, which corresponds to peak plasma concentrations rather than trough dasatinib levels [134] (see Table 2). This suggests that such an assay, because of the insufficient sensitivity provided by a single quadrupole mass spectrometer, is of limited clinical usefulness for a formal therapeutic monitoring of dasatinib.

At about the same time, our laboratory has reported the development and validation of an LC tandem MS assay for as much as six TKIs simultaneously. The proposed LC-MS/MS method allows the simultaneous determination of clinically relevant ranges of concentrations for the six major TKIs currently in use imatinib, dasatinib, nilotinib, sunitinib, sorafenib, and lapatinib [122]. Plasma is purified by acetonitrile protein precipitation followed by reversed-phase chromatographic separation. Analyte quantification is performed by electrospray ionization–triple quadrupole mass spectrometry by selected reaction monitoring (SRM) detection using the positive mode. This was the first broad-range LC-MS/MS assay covering the major currently in-use TKIs.

Various methodologies have been proposed since then for multiple TKIs assays. The measurement of the first three marketed TKIs gefitinib, erlotinib, and imatinib was carried out by liquid–liquid extraction of human plasma, using hexane-ethyl acetate (30:70, v/v) as extracting solvent. The reconstituted extracts in the organic upper phase were subjected to reversed-phase HPLC and the TKIs were detected by electrospray triple quadrupole mass spectrometry, operated in the positive mode [135]. A multiplex analysis of TKIs used for the treatment of solid tumors, (gefitinib, erlotinib, sunitinib, and sorafenib) was also proposed using plasma protein precipitation with acetonitrile, supernatant injection into reversed-phase column, and TKIs detection/quantification by a triple quadrupole mass spectrometer equipped with a turbo-spray ionization operating in positive multi-reaction-monitoring-mode [136]. Just recently, Götze et al. have published a multiplex assay for the determination of erlotinib, imatinib, lapatinib, nilotinib, sorafenib, and sunitinib that was proposed for routine clinical application [137]. Finally, an assay allowing the determination of as much as nine TKIs simultaneously (imatinib, its metabolite, nilotinib, lapatinib, erlotinib, sorafenib, dasatinib, axitinib, gefitinib, and sunitinib) has been recently developed by Bouchet et al., using 96-well SPE plates and UPLC tandem MS [138].

Recently, we have adapted our previous methodology for the multiplex assay of TKIs [122] for the assay of additional current and newer TKIs possibly analyzed simultaneously. Briefly, a modification of the gradient program and the adjustment of the composition of the mixture (15 % of MeOH in Ammonium formate 20 mM pH 2) used for diluting the supernatants (obtained after plasma protein precipitation with acetonitrile) were carried out to account for the chromatographic behavior as well as different solubilities of early (bosutinib, gefinib) and late (verumafenib, formerly PLX4032) eluting drugs (Fig. 1) [139]. Such adaptation allowed the simultaneous measurement in plasma of gefitinib, erlotinib, the third generation anti-CML agent bosutinib, as well as vemurafenib. This latter drug attracted much interest lately, because of its impressive clinical effect against melanoma harboring the BRAF V600E mutation [27]. First applications of this methodology to patients samples confirmed the wide interindividual variability and—fairly unpredictable—pharmacokinetics of erlotinib and verumafenib, and gave preliminary insights on clinical consequences of its pharmacokinetic variability. For example, very high plasma concentrations of erlotinib were found in a female patient who developed a grade 2 rash, confirming the known relationships between erlotinib plasma levels and the incidence of cutaneous adverse drug reactions [140].

Cellular Measurements

Up to now, most investigations on the TDM of TKIs therapy have focused on the measurement of concentrations of the parent TKI drug in plasma. However, TKIs act intracellularly and their concentrations in cell cytoplasm, besides being determined by circulating blood levels, are also controlled by various transmembrane drug transporters influencing cellular uptake and release. TKIs' pharmacological activity in cells is modulated by complex mutual biological, genetic, and environmental influences, which remain poorly known, and efforts have been recently made to study their cellular disposition, i.e., at determining their "cellular" concentration that would closely reflect the intracellular environment of the therapeutic target, than the systemic blood concentrations that are currently measured.

In that context, Klawitter et al. [107] developed and validated an LC-MS/MS method using a turbo ion spray coupled to TSQ mass spectrometer for the quantification of imatinib in human leukemia cells, using a first step of protein precipitation followed by column switching.

A method has been proposed for the determination of cellular levels in lung cancer cell lines of the TKIs dasatinib and lapatinib [141]. Cellular samples were extracted with a mixture of *tert*-butyl methyl ether/ACN/ammonium formate pH 3.5 (6:2:1, v/v/v). The organic layer was subjected to evaporation and the samples were reconstituted in acetonitrile, followed by chromatographic separation on a C18 column. Dasatinib and lapatinib were monitored by tandem MS equipped with a positive electrospray ionization interface in positive ion mode. These cellular experiments showed that lapatinib is not actively expelled from P-gp over-expressing cancer cells, while P-gp activity significantly decreases cellular levels of dasatinib [141].

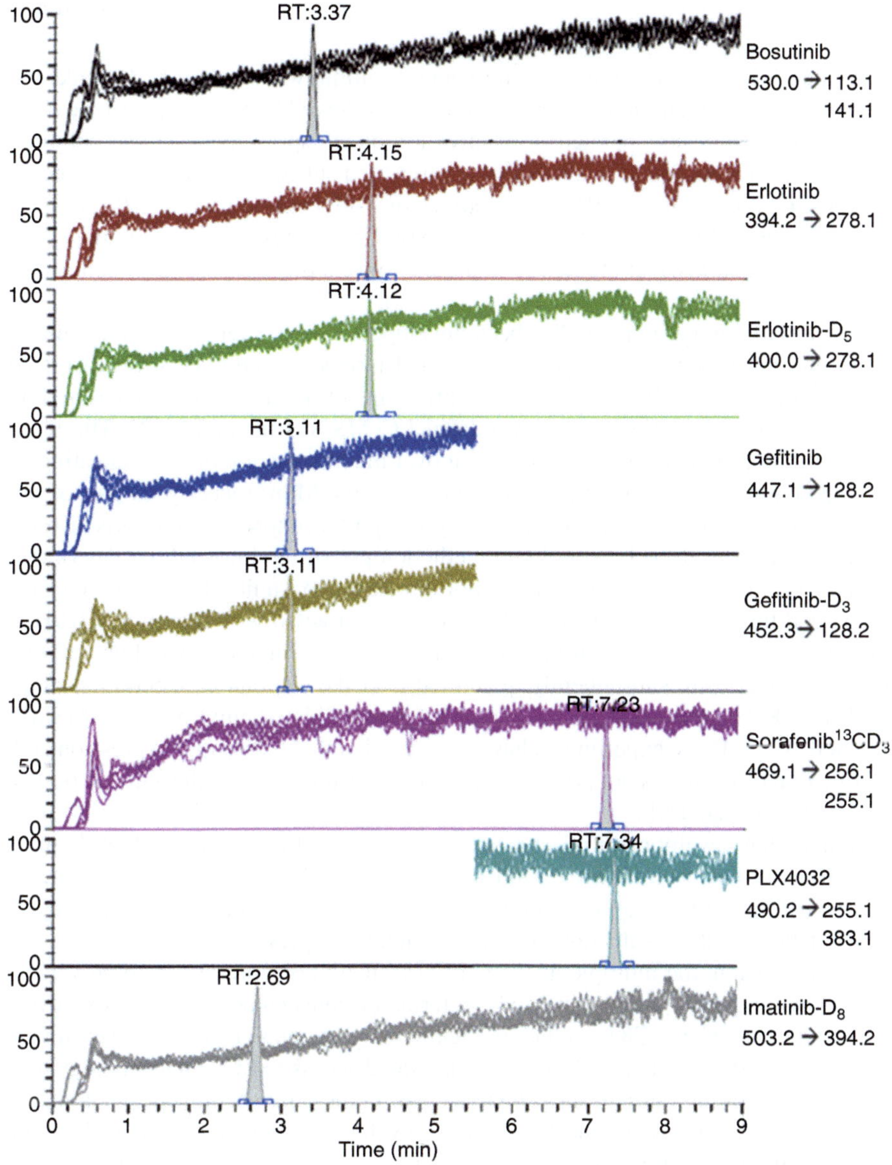

Fig. 1 A multiplex LC tandem MS for the quantification of the targeted anticancer agents bosutinib, gefitinib, erlotinib, and vemurafenib (formerly PLX4032). The chromatographic profile of a QC control containing 37.5 ng/ml of bosutinib, 375 ng/ml of gefitinib and erlotinib, and 3,000 ng/ml of vemurafenib is shown. Chromatographic separations were performed on a column Waters XTerra MS C18 2.1 × 50 mm. Solvent A consisted of 10 mM ammonium acetate containing 1.5 % formic acid (pH 2.3). Solvent B was acetonitrile with 1 % formic acid. The mobile phase was delivered at 0.3 ml/min according to the gradient elution program: 0–9 min, solvent B 5 % → 85 % B, followed by a re-equilibration step. Quantifications were done using the internal standards erlotinib-D_6 and gefitinib-D_3. The internal standards (I.S.) imatinib-D_8 and sorafenib-^{13}C-D_3 have similar retention times as bosutinib and verumafenib, respectively, and were used as I.S. for these latter drugs, because of the lack of labeled standards at the time the assay was developed. On the same chromatographic profiles are shown in offset the superimposed ionization traces of the selected transitions during the analysis of six blank plasma extracts with post-column infusion of a solution containing the four TKIs (1 μg/ml)

Quantification of nilotinib, dasatinib, and bosutinib was done in K562 leukemia cells by protein precipitation using acetonitrile, followed by triple quadrupole mass spectrometer analysis operated in positive ion electrospray mode [142]. Nilotinib and dasatinib were found to act both as transported substrates and, at high concentrations, inhibitors of *ABCB1* (gene coding for P-glycoprotein) and *ABCG2* (gene coding for BCRP) [142]. Whereas neither ABCB1 nor ABCG2 could confer bosutinib resistance; this TKI efficiently inhibited both transporters at higher concentrations [142].

In our laboratory, an assay has been developed for the determination of cellular concentration of imatinib in peripheral blood monocytes cells (PBMCs) isolated from patients [42]. Intracellular concentrations of imatinib were measured in PBMCs from five patients using validated LC MS/MS methods [42]. The intra/extracellular ratio appeared to be constant over the observation period indicating an average eightfold accumulation of imatinib in cells. More recently, as part of our in vitro studies on the consequence of drug transporters expression on TKIs disposition, we have developed a simplified methodology for the intracellular determination of several major TKIs (imatinib, nilotinib, dasatinb, sunitinib, and sorafenib) in K562 cell lines [143]. Incubated cells were first extracted with 0.5 ml MeOH/H_2O 50:50 by vortex mixing, ultrasonication, and centrifugation, yielding cellular extracts. TKIs were subsequently quantified over the relevant concentration range of 0.1–5,000 ng/ml with an adaptation of our validated multiplex LC-MS/MS method [122]. These experiments have revealed that the differential expression and/or function of P-gp was not affecting the cellular disposition of nilotinib, in contrast to the other tested TKIs.

Lately, the development of an assay by LC coupled to single quadrupole mass spectrometry has been recently published for the determination of cellular concentrations of imatinib, dasatinib, and nilotinib in PBMCs, but the authors did not give much details on the results obtained with patients samples [144].

In conclusion, the multiple-analytes LC-MS methods represent an improvement over previous single-analyte methods in terms of convenience (a single extraction procedure for several TKIs, reducing significantly the analytical time), sensitivity, selectivity, and throughput. The current facilitated access to LC-MS technology may contribute to filling our current knowledge gap in the pharmacokinetics–pharmacodynamics relationships of the latest TKIs developed following imatinib. It might better define therapeutic ranges of TKIs in various patient populations prior to the evaluation of a systematic TDM-guided dose adjustment of these anticancer drugs.

3 Tamoxifen as the First Targeted Anticancer Agent

Introduced into the clinic some 30 years ago, tamoxifen selectively modulates estrogen receptors and thus can be considered as one of the first examples of "targeted" anticancer therapy, years before the era of TKIs described in Section 2. This largely justifies that a section of the present review is devoted to tamoxifen, especially in the light of recent findings —some of them made possible by the advent of new powerful mass spectrometry techniques— suggesting that tamoxifen pharmacological activity and clinical outcomes do not rely on the parent drug only, but also depends on the presence of several tamoxifen metabolites produced in patients via complex metabolic pathways. A comprehensive review of mass spectrometry methods for tamoxifen and its metabolites is therefore presented in the context of the current growing interest for monitoring tamoxifen metabolites as a potentially clinically useful tool to monitor tamoxifen treatment in breast cancer patients.

3.1 Clinical Rational for a TDM and Metabolites Profiling of Tamoxifen [145]

The non-steroidal selective estrogen receptor modulator (SERM), tamoxifen, was the first molecularly targeted cancer therapy approved by the U.S. Food and Drug Administration (FDA) since 1977 and 1998, respectively, for the treatment and prevention of estrogen-sensitive breast cancer (BC) [146–148]. Tamoxifen—the *Z* geometric isomer of a triphenylethylene derivative—has been for more than three decades the most widely used antihormonal therapy for premenopausal and postmenopausal women with metastatic breast cancer, for adjuvant and neo-adjuvant treatment of primary breast cancer and as a preventive agent for women at high risk of developing the disease [149–157].

Selective estrogen receptor modulators, such as tamoxifen, display tissue-selective estrogen agonist or antagonist effects. In breast tissues, tamoxifen exerts an anti-estrogenic activity mediated by the competitive inhibition of 17beta-estradiol (E2) binding to estrogen receptors alpha and beta (ERα and ERβ), resulting in the suppression of ERα transcriptional activity and inhibition of estrogen-dependent growth and proliferation of malignant breast epithelial cells [158–161]. However, several lines of evidence indicate that the overall anti-proliferative effects of tamoxifen depend on the formation of the pharmacologically active metabolites 4-hydroxy-tamoxifen and notably 4-hydroxy-*N*-desmethytamoxifen (endoxifen) which have up to 100-fold greater affinity to ERs and 30 to 100-fold greater potency in suppressing breast cancer cell proliferation as compared to the parent drug [10, 162–164].

Of these active metabolites, endoxifen is suggested to be the primary active metabolite responsible for the majority of tamoxifen clinical efficacy, as endoxifen plasma concentrations are about five to tenfold higher than those of 4-hydroxy-tamoxifen [11, 165]. Endoxifen may have additional mechanisms of action than 4-hydroxy-tamoxifen by targeting Erα for degradation by proteasome [166] and through the promotion of ERα/ERβ heterodimerization, blocking ERα transcriptional activity [167].

Tamoxifen could thus be considered a quasi-prodrug that requires metabolic bio-activation to exert its effects. The metabolism of tamoxifen is complex and undergoes extensive phase I and phase II transformation (Fig. 2). Various potentially

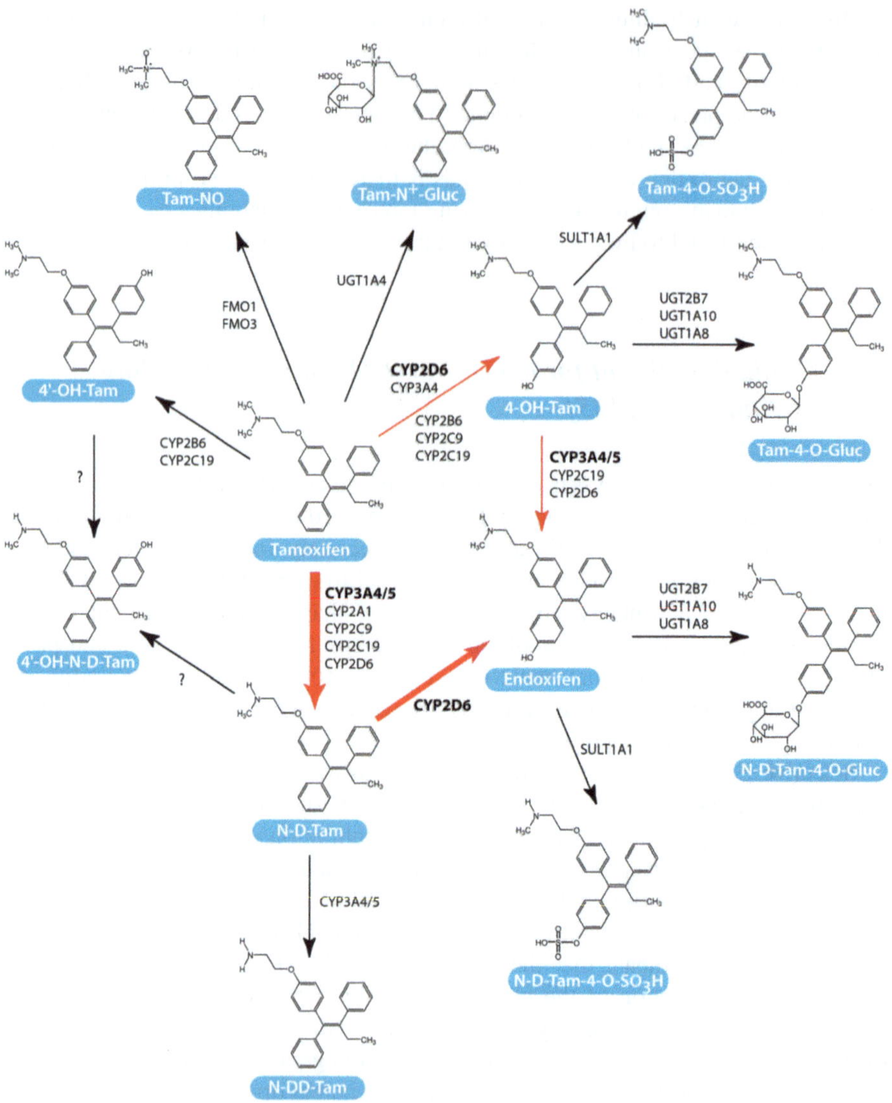

Fig. 2 Principal tamoxifen metabolic pathways of clinical interest. Abbreviations: Tam (Tamoxifen), *N*-D-Tam (*N*-desmethyl-tamoxifen), *N*-DD-Tam (*N*,*N*-didesmethyl-tamoxifen), 4-OH-Tam (4-Hydroxy-Tamoxifen), 4′-OH-Tam (4′-Hydroxy-tamoxifen), Tam-NO (Tamoxifen-*N*-oxide), Endoxifen (4-Hydroxy-*N*-desmethyl-tamoxifen), 4′-OH-*N*-D-Tam (4′-Hydroxy-*N*-desmethyl-tamoxifen), Tam-*N*+-Gluc (Tamoxifen-*N*+-glucuronide), Tam-4-*O*-Gluc (Tamoxifen-4-*O*-glucuronide), *N*-D-Tam-4-*O*-Gluc (*N*-desmethyl-tamoxifen-4-*O*-glucuronide), Tam-4-*O*-SO_3H (Tamoxifen-4-*O*-sulfate), *N*-D-Tam-4-*O*-SO_3H (*N*-desmethyl-tamoxifen-4-*O*-sulfate)

polymorphic cytochrome P450 (CYP) enzymes including CYP3A4, 3A5, 1A2, 2B6, 2 C9, 2 C19 and 2D6 catalyze, to different extents, the hepatic biotransformation of tamoxifen into active and inactive primary and secondary metabolites [168–171].

Briefly, tamoxifen is primarily oxidized to *N*-desmethyl-tamoxifen (the most abundant metabolite in human plasma) and 4-hydroxy-tamoxifen predominantly by CYP3A4/5 and CYP2D6, respectively, followed by endoxifen formation from *N*-desmethyl-tamoxifen, exclusively catalyzed by CYP2D6 and from 4-hydroxy-tamoxifen by CYP3A4/5 (Fig. 2).

N-desmethyl-tamoxifen is quantitatively the major metabolite found in patients' plasma and serum. It accounts approximately for 92 % of primary tamoxifen oxidation [171]. In women receiving a daily dose of 20 mg tamoxifen, steady-state plasma concentrations of *N*-desmethyl-tamoxifen are 1.5 to 2-fold higher than those of tamoxifen. Plasma levels of 4-hydroxy-tamoxifen are five to tenfold lower than those of endoxifen [11, 165, 172, 173]. Steady-state plasma concentration of tamoxifen is achieved after 1 month with terminal elimination half-life of about 5–7 days. *N*-desmethyl-tamoxifen (elimination half-life of about 10–14 days), 4-hydroxy-tamoxifen, endoxifen have longer elimination half-life than tamoxifen and their steady-state plasma concentrations are achieved 3–4 months after treatment initiation [11, 172].

Tamoxifen and its metabolites undergo further glucuronidation and sulfation. Different hepatic and extra-hepatic UDP-glucuronosyltransferases (UGTs) exhibited in-vitro glucuronidation activities towards tamoxifen and its metabolites leading to inactive metabolites [174]. The hepatic enzyme UGT1A4 is considered the major UDP-glucuronosyltransferase responsible – in vitro – for the *N*-glucuronidation of tamoxifen and 4-hydroxy-tamoxifen [175–178]. Hydroxylated active tamoxifen metabolites (i.e., *Z*-4-hydroxy-tamoxifen and *Z*-endoxifen) equally go through *O*-glucuronidation involving mainly UGT2B7 and the extra-hepatic glucuronidating enzymes UGT1A10 and 1A8 [179, 180]. Typical chromatographic profiles of tamoxifen phase I and phase II glucuronidated metabolites, observed in a plasma sample from a BC patient, are depicted in Fig. 3. Sulfotransferase (SULT) 1A1 is the major phase II metabolizing enzyme involved in the sulfation of 4-hydroxy-tamoxifen and endoxifen [181–184]. These sulfated and glucuronidated metabolites are further eliminated in urine and bile and undergo enterohepatic circulation (EHC) [185–187].

As an adjuvant therapy, in pre- and post-menopausal women with ER-positive BC, a standard 5 years treatment with tamoxifen has been demonstrated to almost half (43 %) the rate of disease recurrence and reduce the annual breast cancer death by a third (31 %). In the preventive setting, tamoxifen also reduces the risk of developing a new breast cancer by nearly one-half [150, 151, 154, 157].

Despite the obvious benefits of this drug in the different treatment settings, the clinical outcomes of tamoxifen treatment in terms of efficacy and side effects are incomplete and inconstant, and almost 30–50 % of patients either fail to respond or become resistant to tamoxifen [188]. One of the proposed mechanisms that may account for the impaired response to tamoxifen therapy is an altered bioactivation of the parent drug into endoxifen, either by genetic or environmental factors [188, 189].

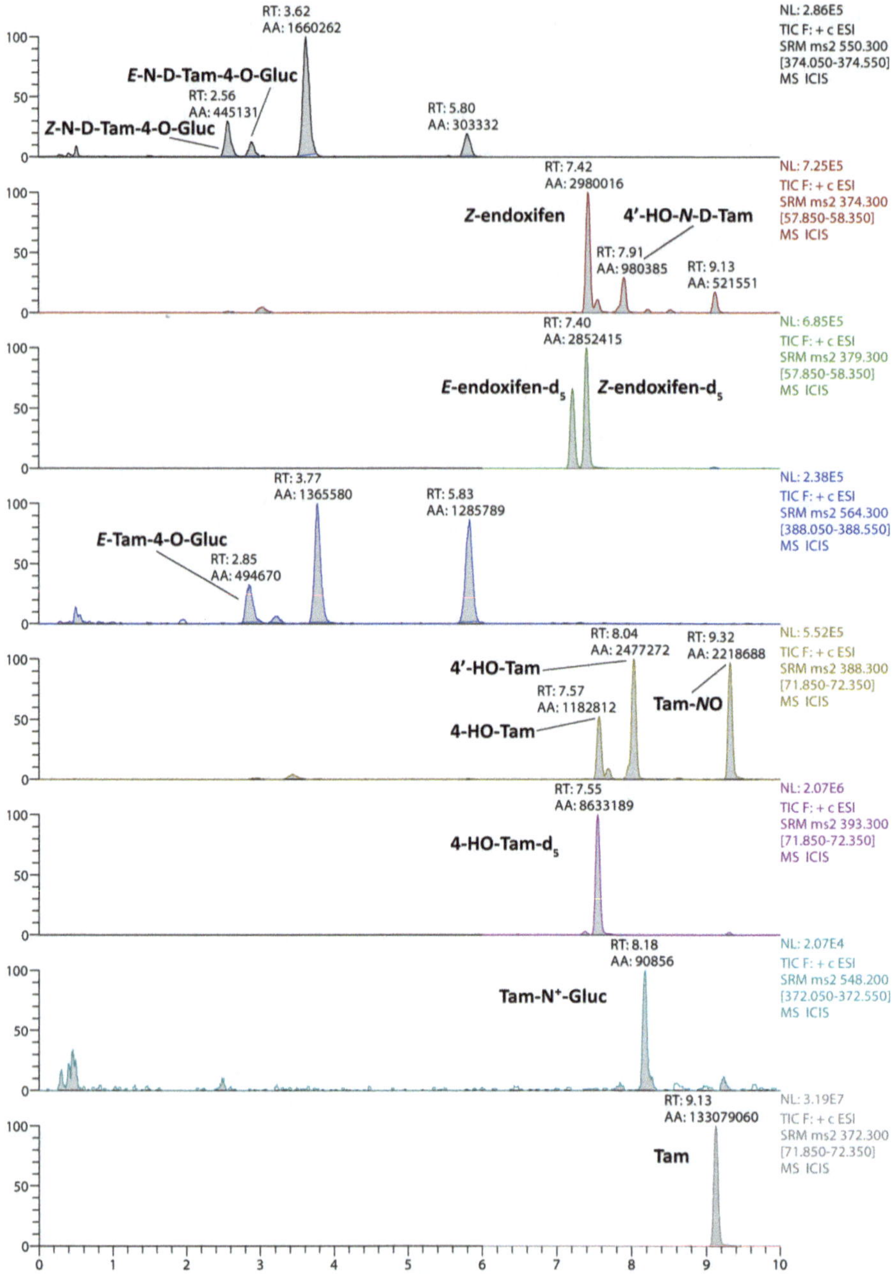

Fig. 3 Chromatographic profiles of main tamoxifen phase I metabolites and some of its identified glucuronidated metabolites in a plasma sample from a breast cancer patient receiving tamoxifen 20 mg BID (modified elution gradient from reference [145]). See legends in Fig. 2 for abbreviations

It appears that CYP2D6 is the key enzyme responsible for the generation of endoxifen [171]. The metabolizing activity of this enzyme is highly polymorphic and varies considerably within a population and between ethnic groups. This large variability is partly determined by genetic polymorphisms in the CYP2D6 gene, with over 100 allelic variants identified to date, resulting in different phenotypic patterns [190, 191]. Currently, on the basis of CYP2D6 activity, the population is usually categorized into four phenotypes including ultrarapid metabolizers (UMs), extensive metabolizers (EMs), intermediate metabolizers (IMs), and poor metabolizers (PMs).

Actually, CY2D6 gene polymorphisms, associated with null or reduced enzyme activity, have been reported to negatively influence (in a gene-dose manner) the blood level of endoxifen in numerous prospective pharmacokinetic studies [11, 172, 192–196]. Some retrospective and prospective studies have shown that CYP2D6 polymorphism was associated with worse clinical outcomes in PMs and IMs patients in terms of recurrence and disease free survival or BC development in the chemoprevention setting [192, 194, 195, 197–205].

This has prompted the consideration of a potential role for CYP2D6 genotype testing in patients' management and choice of alternative adjuvant therapy. Whether genotype-guided tamoxifen administration is a valuable and useful option to optimize antihormonal adjuvant therapy remains, however, controversial, and no clear consensus has yet been reached regarding the insufficient and somewhat conflicting retrospective clinical data relating CYP2D6 genotype to tamoxifen efficacy [206–210]. Moreover, large interpatient variability in endoxifen levels still subsists even after correcting for CYP2D6 status. This remaining variability may depend on one hand on the activity of other cytochromes (CYP3A4/5, 2 C9, 2 C19, 2B6), phase II conjugation enzymes (SULT1A1, UGT1A4, 2B7, 1A10, 1A8, 2B15) some of them known to be polymorphic [11, 173, 182, 196, 206, 207, 211], as well as transporters (other than P-glycoprotein or multidrug resistance-associated protein 2 [MRP2] that seems to have no or limited impact on tamoxifen and metabolites systemic exposure) [195, 212–214]. On the other hand, environmental factors such as treatment adherence [215–220] and particularly, interacting co-medications do modulate drug exposure independent of genetic traits [10, 11, 172]. In fact, it is estimated that 20–30 % of patients under tamoxifen therapy are also taking antidepressants. Of importance are some selective serotonin reuptake inhibitors (SSRIs) with strong CYP2D6-inhibiting activity, such as paroxetine and fluoxetine, that can be prescribed to treat depression or to alleviate tamoxifen-induced hot flushes. The latter drugs are known to reduce endoxifen plasma concentration and may therefore be associated with poorer tamoxifen efficacy [221–225].

The monitoring of plasma concentration of tamoxifen-active metabolites (mainly endoxifen) may therefore constitute a better predicting tool for tamoxifen efficacy than genotype testing. In fact, endoxifen levels correspond to the final phenotypic trait of patients' drug exposure, accounting for the combined effects of all genetic polymorphisms, physiological and environmental factors that may affect drug disposition and bioactivation.

However, whether the monitoring of endoxifen plasma concentrations in breast cancer patients would constitute a valid approach to optimize individual dosage and improve treatment effectiveness remains to be demonstrated. So far, only one study has been recently published that sought for the association between endoxifen concentrations and breast cancer outcomes [173]. This pioneering study suggested a probable nonlinear dose–response relationship for tamoxifen effect and identified a threshold concentration for endoxifen (of about 6 ng/ml) above which approximately 30 % reduction in disease recurrence rate was observed.

Early attempts that examined the feasibility and usefulness of tamoxifen dose-adjustment strategy were based exclusively on CYP2D6 genotype. Genotype-guided dose-adjustment studies have shown that tamoxifen dose increase to 30 mg or 40 mg/day significantly increases 4-hydroxy-tamoxifen and endoxifen concentrations in IM and even in PM patients carrying two null alleles (reflecting metabolism by other enzymes), without any significant difference in adverse effects. However, an important variability is still observed in 4-hydroxy-tamoxifen and endoxifen levels between the genotypic groups [226, 227] and this would be a strong argument for considering TDM of tamoxifen and its active metabolites levels as a valuable strategy for tamoxifen dose adjustment further reducing the residual variability within CYP2D6 genotype groups.

Barginear et al. [228]. investigated in another prospective study the effect of tamoxifen dose increase on the concentrations of tamoxifen, 4-hydroxy-tamoxifen, endoxifen, and their position isomers (4′-hydroxylated) and proposed an "antiestrogenic activity score" (AAS) based on the concentrations of these metabolites and their respective antiestrogenic activities. According to Barginear et al. this AAS score would constitute a better approach to estimate the biologic effectiveness of tamoxifen and, therefore, to guide future tamoxifen dose optimization. However, this approach has yet to be validated by larger studies.

3.2 *Tamoxifen and Metabolites Identification and Quantification*

To date, several quantitative analytical methods have been developed for the monitoring of tamoxifen and some of its metabolites in human biological fluids and tissues, including conventional [229–231] and micellar [232] liquid chromatography (LC) methods coupled to fluorescence detection, capillary electrophoresis-mass spectrometry (CE-MS) [233], gas chromatography–mass spectrometry (GC-MS) [234], as well as liquid chromatography methods hyphenated with mass spectrometry (LC-MS) [195] and tandem mass spectrometry (LC-MS/MS) [145, 173, 193, 194, 196, 218, 226, 227, 235–247]. Reports have also been published describing liquid chromatography method coupled to mass spectrometry or fluorescence detection for the study of tamoxifen metabolism in vitro and in vivo in animal models [185, 248–254]. Most of these qualitative and quantitative LC, GC, and CE methods have already been reviewed by Teunissen et al. [255].

Various hyphenated LC-MS-based assays, using either the electrospray ionization (ESI) or the atmospheric pressure chemical ionization (APCI) interface, have been developed and applied in the clinical setting in order to support pharmacokinetic (PK), pharmacogenetic-pharmacokinetic (PG-PK), and pharmacokinetic–pharmacodynamic (PK-PD) studies in BC patients under tamoxifen therapy (Table 3).

Table 3 Overview of LC-MS and LC-MS/MS methods for the quantification of tamoxifen and its metabolites in human blood samples

Matrix (volume)	Analytes[e]	Internal standard	LOQ (ng/ml)	Sample preparation	Column (particle size, dimensions)	Ionization and detection mode	Reference
Plasma[a] (100 µl)	Tam 4-OH-Tam Others SERMS (raloxifene, nafoxidine, idoxifene)	Idoxifene-d5	5 –	LLE (hexane/ isoamyl-alcohol (96:4 v/v)) Dilution (DMSO) Evaporation (hexane layer) Dilution (H20)	Luna C18 (3 µm, 30×1 mm)	ESI-TSQ	[235]
Plasma (250 µl)	Tam N-D-Tam 4-OH-Tam	Toremifene	–	PP (ACN) Dilution (0.5 M ammonium acetate)	Hypersil BDS C18 (3 µm, 150×2.1 mm)	ESI-Q-TOF (MS mode)	[236, 237]
Serum[b] (75 µl)	Tam 4-OH-Tam N-D-Tam N-D-D-Tam Tam-NO Endoxifen[c]	Tam-d5	0.25 0.25 0.25 1.0 1.0 –	PP (ACN) On-line SPE (Oasis HLB, 50×1 mm)	Chromolith Performance, RP-18e (100×4.6 mm)	ESI-LTQ	[193, 238]
Serum[b] (100 µl)	Tam N-D-Tam 4-OH-Tam Endoxifen[d] Soy isoflavone (genistein, daidzein, equol)	Tam-$^{13}C_2$,^{15}N N-D-Tam-d5 4-OH-Tam-d5 Endoxifen-d5[d] (genistein-d4, daidzein-d3, equol-d4)	0.4 0.4 0.2 1.1	PP (ACN) Hydrolysis (glucuronidase/ sulfatase) SPE (SPEC 96-WELL PLATE C18) Evaporation Reconstitution (MeOH/H2O (1:1 v/v))	Luna C18 (3 µm, 150×2 mm)	ESI-TSQ	[239, 240]

(continued)

Table 3 (continued)

Matrix (volume)	Analytes[e]	Internal standard	LOQ (ng/ml)	Sample preparation	Column (particle size, dimensions)	Ionization and detection mode	Reference
Serum (1 ml)	Tam N-D-Tam 4-OH-Tam	– – –	5 5 0.5	LLE (*n*-hexane/isoamyl alcohol (98:2)) Evaporation Reconstitution (MeOH)	Beckman C8 (5 μm, 50×4.6 mm)	ESI-LTQ	[241]
Serum	Tam 4-OH-Tam	Propranolol	10 1	LLE (hexane/isopropanol (95:5 v/v))	HiQ-Sil C18 (5 μm, 150×2.1 mm)	ESI-TSQ	[194]
Serum[b] (50 μl)	Tam Tam-NO[c] N-D-Tam 4-OH-Tam Endoxifen[d] Soy isoflavone (genistein, daidzein, glycitein)	Tam-d5 N-D-Tam-d5 4-OH-Tam-d5 Endoxifen-d5[d] (genistein-d4, daidzein-d6)	6.76 6.19 6.72 1.13 2.69	PP (ACN) Dilution (3.5 mM ammonium formate buffer, pH 3.5)	Synergi Hydro-RP (4 μm, 150×2 mm)	ESI-TSQ	[242]
Plasma (100 μl)	Tam N-D-Tam 4-OH-Tam Endoxifen	Imipramine	20 20 1 3.75	PP (ACN) SPE (BOND ELUTE-C18 cartridges, 100 mg/1 ml)	XBridge C18 (3.5 μm, 150×3 mm)	ESI-TOF	[195]
Plasma[b] (1 ml)	Tam AIs (Anastrozole, letrozole)	Bunitrolol	25	PP (2 % aqueous phosphoric acid) Polymer-based mixed-mode SPE (Strata X-C, 200 mg/3 ml)	Eurosphere Si-C18 (5 μm, 200×0.5 mm)	ESI-LTQ	[218]

Plasma (100 µl)	Endoxifen	Anastrozole	–	PP (ACN)	Kromasil 100 C8 (5 µm, 150×4.6 mm)	ESI-TSQ	[243, 244]
Plasma[b] (100 µl)	Tam N-D-Tam 4-OH-Tam 4′-OH-Tam[c] Endoxifen[d] 4′-OH-N-D-Tam[c]	Tam-d5 N-D-Tam-d5 4-OH-Tam-d5 Endoxifen-d5[d] 	1 2 0.4 0.4 1 1	FP (ACN) Evaporation Reconstitution (MeOH/20 mM ammonium formate buffer, pH 2.9 (1:1 v/v)) Centrifugation	Acquity UPLC BEH C18 (1.7 µm, 30×2.1 mm)	ESI-TSQ	[145]
Plasma[b] (100 µl)	Tam Tam-NO N-D-Tam 4-OH-Tam[d] 4′-OH-Tam Endoxifen[d] 4′-OH-N-D-Tam	Tam-$^{13}C_2$,^{15}N N-D-Tam-d5 4-OH-Tam-d5 Endoxifen-d5[d] 	1.1 – 0.5 0.2 0.2 0.2 0.2	PP (MeOH) Filtration Dilution (H_2O)	Zobrax SB-C18 (1.8 µm, 50×2.1 mm)	ESI-TSQ	[245]
Serum (200 µl)	Tam N-D-Tam 4-OH-Tam Endoxifen	Tam-d5 N-D-Tam-d5 Endoxifen-d5	– – – –	Polymer-based mixed-mode SPE (Oasis MCX 1 ml cartridges)	XTerra MS C18 (3.5 µm, 100×2.1 mm)	ESI-LTQ	[173]

(continued)

Table 3 (continued)

Matrix (volume)	Analytes[e]	Internal standard	LOQ (ng/ml)	Sample preparation	Column (particle size, dimensions)	Ionization and detection mode	Reference
Plasma (50 μl)	Tam	Tam-d3	0.5	PP (ACN + 1 % acetic acid)	Zobrax Eclipse plus C18 (1.8 μm, 100 × 2 mm)	ESI-TSQ	[196]
	Tam-NO		0.2	Dilution (H_2O + 1 % acetic acid)			
	N-D-Tam	N-D-Tam-d5	1				
	N-D-D-Tam		0.2				
	4-OH-Tam[d]	4-OH-Tam-d5[d]	0.1				
	3-OH-Tam		0.1				
	4′-OH-Tam[d]		0.1				
	α-OH-Tam		0.02				
	Endoxifen[d]	Endoxifen-d5[d]	0.1				
	3-OH-N-D-Tam[c]		–				
	4′-OH-N-D-Tam[d]		0.05				
	α- OH-N-D-Tam[c]		–				
	Tam-N⁺-Gluc	Tam-4-O-Gluc-d5[d]	0.05				
	Tam-3-O-Gluc[c]		–				
	Tam-4-*O*-Gluc[d]		0.1				
	N-D-Tam-3-O-Gluc[c]		–				
	N-D-Tam-4-O-Gluc[d]		0.05				
Serum[b] (50 μl)	Tam	Tam-d5	5	PP (ACN)	Kinetex C18 (2.6 μm, 150 × 2.1 mm)	APCI-TSQ	[246]
	N-D-Tam	N-D-Tam-d5	5	Evaporation			
	4-OH-Tam	4-OH-Tam-d5	0.4	Reconstitution (ACN/4 mM ammonium formate buffer, pH 3.5, (3:7 v/v))			
	4′-OH-Tam		0.2				
	Endoxifen[d]	Endoxifen-d5[d]	1				
	4′-OH-N-D-Tam		1				
Plasma (250 μl)	Tam	Diphenhydramine	–	LLE (ethyl acetate under pH 11.3)	Luna C18 (3 μm, 100 × 2 mm)	ESI-TSQ	[226]
	N-D-Tam		–	Evaporation			
	4-OH-Tam		–	Reconstitution (mobile phase)			
	Endoxifen		–				

Plasma[b] (200 μl)	Tam N-D-Tam 4-OH-Tam Endoxifen	Tam-d5 N-D-Tam-d5 4-OH-Tam-d5 Endoxifen-d5[d]	1.86 1.78 0.194 0.187	PP (ACN/acetone (1:1)) LLE (n-hexane/isopropanol (95:5)) Evaporation Reconstitution (ACN/H_2O/formic acid (40:60:0.1 v/v/v)) Centrifugation	Acquity UPLC BEH C18 (1.7 μm, 100×2.1 mm)	ESI-TSQ	[247]
Plasma (100 μl)	Tam N-D-Tam 4-OH-Tam Endoxifen	Imipramine	20 40 1 4	PP (ACN)	Acquity UPLC BEH C18 (1.7 μm, 100×2.1 mm)	ESI-TSQ	[227]

[a]Non-clinical samples
[b]Fully validated method
[c]Method has not been validated for the quantification of the analyte
[d]Reported E/Z isomers chromatographic resolution
[e]Abbreviations (see Fig. 2 and Table 4)

Among these, both LC-MS and LC-MS/MS approaches have been described using different mass analyzers operating in the positive ion mode scan such as triple stage quadrupole (TSQ) mass spectrometers [145, 194, 226, 227, 235, 239, 240, 242–247] and hybrid quadrupole-linear ion trap (LTQ) [173, 193, 218, 238, 241] mass spectrometers working in SRM mode as well as time-of-flight (TOF) [195] and hybrid quadrupole-TOF (Q-TOF) [236, 237] mass spectrometers working in the MS mode.

3.3 Chromatographic Conditions and Tamoxifen Metabolites Separation

Since the introduction of ionization sources working at atmospheric pressure such as ESI interface, LC–MS has become the gold standard in the field of quantitative bioanalysis due mainly to the selectivity, sensitivity, and high-throughput detection in LC-MS systems. However, LC-MS features depend not only on the ionization technique and mass spectrometer unrivaled inherent selectivity, sensitivity, and speed acquisition but are also challenged, notably in drug metabolism studies, by the availability of stable isotope labeled (SIL) version of metabolites (see below) and the need of efficient and adequate chromatographic resolution of multiple analytes from interfering metabolites or endogenous biological components in a minimum time frame.

Reversed-phase LC (RPLC) methods using conventional, microbore [218], narrow-bore [194, 242], and short [241] HPLC columns have been used for the separation of tamoxifen/metabolites either under isocratic or gradient elution conditions. Narrow-bore columns present the advantages of being solvent saving and by the need of low sample injection (or loading) volumes. These advantages were illustrated by Beer et al. [218], who developed an analytical method for the separation of tamoxifen, anastrozole, and letrozole under gradient of 30 μl/min of acetone in aqueous heptafluorobutyric acid solution and volumes as low as 2 μl, from the processed samples, were injected into the system. Furlanut et al. [241] used a short analytical column for the separation of tamoxifen and two of its metabolites within almost 8 min under isocratic conditions at flow rate of 1 mL/min. Although, the use of conventional short columns is a simple method for shortening analytical run times, these columns suffer from a loss in efficiency and resolution.

For enhanced throughput, fast RPLC methods using monolithic silica columns [238], small size particles (3 μm) packed columns [173, 195, 226, 235–237, 239, 240], ultra high pressure liquid chromatography (UHPLC) columns packed with sub-2 μm particles [145, 196, 227, 245, 247] and 2.6 μm core-shell particles HPLC columns [246] have been proposed for the high-throughput separation and quantification of tamoxifen/metabolites.

Five UHPLC methods have already been described to improve speed, resolution, and sensitivity of HPLC assays for the quantification of tamoxifen phase I as well as phase II metabolites. These methods exclusively enabled, within run times of about 12 min or even less, to reach an excellent overall resolution for all considered metabolites including (*E*/*Z*) endoxifen isomers and position isomers of 4-hydroxy and 4-hydroxy-*N*-desmethyl-tamoxifen. Alternatively, Zweigenbaum J and Henion J [235] developed a high-throughput analysis technique for the separation of tamoxifen, 4-hydroxytamoxifen, and other SERMs within only 30 s using a narrow-bore short analytical column packed with small (3 μm) particles. Separation was performed under isocratic conditions at flow rate of 500 μl/min. Gjerde et al. [238] also described an online SPE-LC-MS/MS procedure where chromatographic resolution of tamoxifen and five of its metabolites was achieved within 6 min using a monolithic silica column (Separation was performed under a gradient program at a flow rate of 500 μl/min). However this method, like other HPLC assays, clearly failed to resolve all the hydroxylated and *N*-desmethyl-hydroxylated tamoxifen metabolites.

Tamoxifen is metabolized to a plethora of *N*-desmethylated, hydroxylated, and their corresponding glucurono- or sulfo-conjugated metabolites. Some of these hydroxylated metabolites are position isomers (such as 4-hydroxy-tamoxifen, 3-hydroxy-tamoxifen, and 4′-hydroxy-tamoxifen; endoxifen; and 4′-hydroxy-*N*-desmethyl-tamoxifen) and have similar molecular mass and fragmentation pattern (Table 4). Besides, *E*/*Z* isomerization (around the ethylenic double bond of tamoxifen and its metabolites) may occur either in biological samples or as contaminants or degradation products in pure standards. Some pure standards are also best synthesized as an equimolar *E*/*Z* mixture. Therefore, the chromatographic resolution of these metabolites and their (*E*/*Z*) geometric isomers is of paramount importance to ensure reliable and accurate bioanalytical methods.

However, of the LC-MS and LC-MS/MS methods developed so far for the comprehensive and quantitative study of levels variability in tamoxifen metabolites, there is limited data with respect to the resolution of both 4-hydroxytamoxifen and endoxifen position isomers (notably 4′-hydroxylated metabolites) and their corresponding (*E*/*Z*) geometric isomers. In fact, apart from the most recently published articles [145, 173, 196, 245–247], no data have been provided regarding this issue. We were the first group that focused on method selectivity and on the effective separation on potentially interfering hydroxylated tamoxifen metabolites. This has allowed us to identify for the first time the occurrence of 4′-hydroxy-tamoxifen and 4′-hydroxy-*N*-desmethyl-tamoxifen and to estimate their plasma levels in a subset of BC patients [145]. Such differences in chromatographic performances, between assays, can thus affect the selectivity, the accuracy, and the reliability of some of the proposed bioanalytical methods, potentially leading to discrepant data (or results) between the PK, PG-PK and PK-PD studies. Actually, Mürdter et al. [196] reported twice or even higher differences in median concentrations of (*Z*)-endoxifen between studies conducted in the United States, Japan and Norway. They also found a plausible explanation for these discrepancies in method selectivity problems. Madlensky et al. [173] compared the performance of their assay to that of another laboratory performing similar measurements of tamoxifen metabolites in human serum. They found discordant results for 4-hydroxy-tamoxifen levels measured in the same serum samples.

Table 4 Molecular masses and SRM transitions for tamoxifen and some of its metabolites of interest

Analytes	Abbreviation	Molecular weight	Precursor ion[a] $[M+H]^+$	Production
Tamoxifen	Tam	371	372	72
N-Desmethyl-tamoxifen	*N*-D-Tam	357	358	58
N,*N*-Didesmethyl-tamoxifen	*N*-D-D-Tam	343	344	44
4-Hydroxy-tamoxifen	4-OH-Tam	387	388	72
3-Hydroxy-tamoxifen	3-OH-Tam	387	388	72
4′-Hydroxy-tamoxifen	4′-OH-Tam	387	388	72
α-Hydroxy-tamoxifen	α-OH-Tam	387	388	72
Tamoxifen-*N*-oxide	Tam-NO	387	388	72
4-Hydroxy-*N*-desmethyl-tamoxifen	Endoxifen	373	374	58
3-Hydroxy-*N*-desmethyl-tamoxifen	3-OH-*N*-D-Tam	373	374	58
4′-Hydroxy-*N*-desmethyl-tamoxifen	4′-OH-*N*-D-Tam	373	374	58
α-Hydroxy-*N*-desmethyl-tamoxifen	α- OH-*N*-D-Tam	373	374	58
Tamoxifen-*N*⁺-glucuronide	Tam-*N*⁺-Gluc	548	548	372
Tamoxifen-4-*O*-glucuronide	Tam-4-*O*-Gluc	563	564	388
Tamoxifen-3-*O*-glucuronide	Tam-3-*O*-Gluc	563	564	388
N-Desmethyl-tamoxifen-4-*O*-glucuronide	*N*-D-Tam-4-*O*-Gluc	549	550	374
N-Desmethyl-tamoxifen-3-*O*-glucuronide	*N*-D-Tam-3-*O*-Gluc	549	550	374

[a]Molecule protonation occurs on the amino group

Another drawback, challenging the routine applicability of some of these LC-MS and LC-MS/MS assays for measuring exposure to tamoxifen and its active metabolites is that for some assays no data have been provided concerning the validation process. Other methods have only been partially validated and have not or limitedly addressed matrix effects (ME) issues.

3.4 Handling Matrix Effects

Matrix effects (ME), caused by co-eluting endogenous and exogenous matrix components, significantly affect the efficiency and reproducibility of the ionization process of target analytes. This phenomenon represents a major concern for LC-MS bioanalytical method precision, accuracy, sensitivity, and robustness. Amongst the atmospheric pressure ionization interfaces used in LC-MS systems, ESI source is more prone to signal alteration (ion suppression or enhancement) due to matrix. Therefore, careful evaluation and correction for ME must be considered particularly with ESI-MS.

The use of stable isotope labeled (SIL) version of the target analyte as an internal standard (IS) is theoretically considered to be the best approach to compensate or correct for matrix effects and minimize their influence on the accuracy and precision of ESI-MS quantitative assays.

With the exception of the LC-MS/MS methods recently published, previous assays were using either no IS [241], structurally related IS [194, 195, 218, 226, 227, 236, 237, 243, 244], or a single SIL-IS [193, 235, 238] as a surrogate IS for the quantification of tamoxifen/metabolites.

Since SIL-ISs are not always available and their use rather expensive, especially in the case of multiple analytes analysis, the use of structurally related compounds or analogue IS with different mass and with close or similar chromatographic behavior to that of the analytes can represent an acceptable alternative. Nevertheless, in these latter instances, ME variability between different sources of plasma (relative matrix effect variability) must be investigated and quantified. From the assays operating with either no IS or a unique analogue IS, only three methods quantitatively assessed for ME variability. Zweigenbaum J and Henion J [235] reported a significant ion suppression which approximately halved 4-hydroxytamoxifen signal. This ion suppression was not corrected by the IS and affected the precision and accuracy of the method that failed to meet the acceptance criteria for 4-hydroxy-tamoxifen quantification. Furlanut et al. [241] monitored Tam, *N*-D-Tam, and 4-OH-Tam in serum and tissue of BC patients, employing external standard calibration and reported no ion suppression problem after quantitative evaluation of ME. Unfortunately, no detailed information was available regarding the extent of matrix effects variability and the number of plasma lots tested. Only the recent method described by Beer et al. [218] thoroughly examined ME variability using the quantitative approach proposed by Matuszewski et al. [256, 257].

It is noteworthy that ME variability should be investigated even when using SIL-ISs. In fact, SIL-IS may not fully correct for matrix effects, obviously when they do not completely co-elute with their corresponding analyte. This phenomenon has been particularly observed with deuterated SIL-IS that were found to be less lipophilic than their corresponding non-deuterated analogues, causing a slightly earlier elution on a reversed-phase column [258].

Although most recent developed assays used SIL-IS, only few methods quantitatively investigated potential ME variability on tamoxifen and its metabolites quantification [145, 245].

In our proposed assay [145], we thoroughly investigated ME both qualitatively using the post-column infusion system proposed by Bonfiglio et al. [259] and quantitatively using the recommendations of Matuszewski et al. [256, 257] and the 2007 Washington workshop/conference report [260]. Although the qualitative examination of ME did not show any signal alteration, probably due to the infusion of high concentration of analytes, quantitative ME examination showed an ion suppression of approximately 40 % for the signal of *N*-D-Tam. We observed a similar extent of ion suppression with the deuterated *N*-D-Tam (*N*-D-Tam-d5) and ascertained that SIL-IS effectively corrected for the absolute and relative ME (or ME effect variability among six different lots of plasma). Therefore, this was a good illustration of the value of SIL-IS use for an efficient control of residual matrix effects.

Besides the use of SIL-IS, another upstream and primordial approach that allows to anticipate and drastically reduce matrix effects is the optimization of sample preparation procedure.

Plasma protein precipitation (PP) with either ACN or methanol (MeOH) was the most frequently used sample cleanup technique in the described bioanalytical methods [145, 196, 236, 237, 242–245]. Of these, ACN was the prevalent precipitant used, as it was considered to be an optimal choice for protein removal than methanol (MeOH) [261–263]. Although PP is a simple and fast way for preparing samples, it does not result in a very clean extract, as it fails to remove endogenous components such as lipids, phospholipids (such as glycerophosphocholines) and fatty acids. However, if necessary, the elimination of most endogenous lipidic compounds from PP extracts can be performed by subjecting the PP extracts to an additional step of evaporation under nitrogen (or, even better, by submitting them to speed-vac technology) followed by the reconstitution of dried residues with medium polarity solvent system (e.g., MeOH-buffer mixture) wherein lipids would not be resolubilized.

Solid phase extraction allows yielding a much cleaner extract than PP, since it significantly lowers phospholipids levels which represent the major endogenous compounds causing significant matrix effects [263–265].

Different reversed phase [195, 239, 240], mixed mode (ion exchange and reversed phase) SPE cartridges [173, 218] and online SPE column [193, 238] have been also reported for samples preparation and extraction. Some of these assays combined both PP and SPE in order to achieve an extensive sample cleanup [193, 195, 238–240]. Likewise SPE, LLE provides cleaner plasma extracts than PP. Nevertheless, LLE procedure does not always provide satisfactory results with regard to extraction recovery and selectivity, especially with polar analytes and particularly in the case of multicomponent analysis such as in drug-metabolism studies, where analytes polarity varies widely. This issue was addressed by Zweigenbaum J and Henion J [235] and extraction solvent optimization, using isoamyl alcohol, to achieve acceptable extraction selectivity and recovery for polar analytes has been discussed.

To sum up, there is a great heterogeneity in the described methods that have so far been developed and, for the great majority of them, used in the clinical setting to support pharmacogenetic-pharmacokinetic–pharmacodynamic (PG-PK-PD) studies. Of these methods, only the most recent fully validated ones that have proven enough accuracy, precision, robustness, and selectivity seems to be reliable and suitable for measuring exposure of tamoxifen and its metabolites in tamoxifen-treated breast cancer patients.

Whether the monitoring of endoxifen plasma concentrations in breast cancer patients would constitute a valid approach to optimize individual dosage and improve treatment efficacy is under scrutiny and remains to be demonstrated. In that purpose large prospective studies relating endoxifen plasma levels to clinical outcomes are as yet needed. In this perspective, it is critical to settle analytical and selectivity discrepancies between methods and laboratories and to ensure reproducible quantification results between laboratories. These concerted harmonization efforts can be carried out within the frame of an international external quality control program, which as yet, remains to be organized.

4 Discussion and Conclusions

An increasing body of evidence is accumulating for legitimating the blood monitoring of targeted anticancer TKIs and tamoxifen metabolites profiles, especially given their high interindividual pharmacokinetic variability, due to influences of co-medications, diet, and comorbidities, in addition to patients' genetic constitution.

Overall, the analytical developments by mass spectrometry have been instrumental (1) for the development of initial population pharmacokinetics–pharmacodynamics models for some targeted anticancer drugs (mostly imatinib), possibly also integrating the underlying patients' pharmacogenetic background and (2) for being able to respond to clinically relevant issues on drug interaction problems with first-generation targeted anticancer agents [63]. At present, however, it must be acknowledged that the information on the relationships existing between the pharmacokinetics, pharmacodynamics, and in some cases pharmacogenetics, is for most TKIs frequently lacking, or supported by a limited number of—often anecdotal—studies. Therapeutic intervals remain therefore to be determined for the majority of TKIs, and PK-PD studies are best suited to that endeavor. Renewed translational efforts integrating population pharmacokinetics analysis and patients' clinical responses should therefore be carried out in the field of targeted anticancer therapy. Once established, they should open the way to randomized clinical trials for formally validating the clinical usefulness of TDM for TKIs dosage adjustment, before being integrated into standard of care. This raises ethical concerns, as once analytical methods have been developed, clinicians are usually reluctant to deny control patients group to TDM service. Alternate study designs should thus be considered, such as comparison of "routine" TDM (i.e., done even in the absence of clinical problems) versus a "rescue" TDM (done in case of unsatisfactory clinical response or adverse events) [266].

Even though not yet validated by RCTs for approved or more recent anticancer-targeted agents, TDM can already be expected to bring clinically useful information for the optimal management of selected cancer patients, e.g., in case of less-than-optimal clinical response, occurrence of adverse side effects, treatment initiation in the presence of interacting agents, or questionable compliance. For example, TDM appears to be presently used to a rather large extent for the first TKI imatinib, based on recommendations about target plasma levels to maintain for optimal clinical response. Nevertheless, results from randomized controlled studies about TDM usefulness are still eagerly expected for this TKI [266]. This is no less the case for all the more recent TKIs as well as for tamoxifen and its metabolites.

The TDM of TKIs is thus likely to become a very rapidly evolving field, with new targeted anticancer agents approved at a regular pace. Further developments for the TDM of several new TKIs are therefore anticipated to occur within the next few years. In that context, not only a facilitated access to powerful mass spectrometry instruments, but also the availability of robust methodologies for TKIs and tamoxifen/metabolites analysis is a necessity for academic hospital centers that provide TDM service for targeted anticancer therapy. In particular, bioanalytical methods cross-validation is a general problem that should be prioritized amongst the clinical pharmacology community working in the field of targeted anticancer therapy. Initially, external quality control program have been organized for imatinib at the Bordeaux University Hospital (France) within the frame European Treatment and Outcome Study (EUTOS) of the European Leukemia Net [267]. Some private laboratories currently provide external quality control samples for imatinib, nilotinib and dasatinib [268]. Given the growing armamentarium anticipated for targeted anticancer therapy in the next decade, a reinforced analytical collaboration must be deployed between laboratories for harmonizing the assays for current and new TKIs to come, as well as for tamoxifen and metabolites, and possibly also for other anticancer endocrine agents administered chronically (i.e., aromatase inhibitors) as well as the m-TOR inhibitor everolimus, increasingly used in oncology. Beyond working out analytical issues, collaborative research efforts should also be devoted to structuring the collection of data internationally, so that translational research aimed at understanding pharmacokinetics, pharmacodynamics, pharmacogenetics of newest anticancer-targeted therapy will allow without delay the return to clinicians of relevant measurements and their validated interpretations. The systematic and efficient collection of accurate clinical information along with TDM samples indeed represents no less challenging issues that the measurement of those samples.

In complement to the diagnostic tests already approved for selecting patients who are more likely to benefit from a given anticancer treatment [5], individualization of TKIs drug dosage by TDM represents the next step towards a further refinement for targeted anticancer therapies, aiming at administering "the right dose of the right drug to the right patient." In this emerging field of personalized medicine, the development of TDM for patient-tailored dose adjustment should allow to maximize both the therapeutic benefit and the tolerability of these new drugs. These issues are certainly relevant both to individual patients, given the frequency of suboptimal clinical responses, toxicities, intolerance, and treatment discontinuations, and to the society, given the elevated costs of TKIs treatments and of their shortcomings.

Acknowledgments This chapter has been realized within the frame of the research project "Integrative cellular pharmacokinetics/pharmacodynamics/pharmacoproteomics studies of anticancer TKIs in leukemia" (SNF grant no. 310030_138097/1 to LAD) supported by the Swiss National Science Foundation (SNF, Switzerland). It also benefited from the support of the SNF-funded initiative Nano-Tera (ISyPeM project [269] to TB).

References

1. Jemal A et al (2010) Cancer statistics, 2010. CA Cancer J Clin 60:277–300
2. Chabner BA et al (2005) Timeline: chemotherapy and the war on cancer. Nat Rev Cancer 5:65–72
3. Giamas G et al (2010) Kinases as targets in the treatment of solid tumors. Cell Signal 22:984–1002

4. Hughes TP, et al. Monitoring disease response to tyrosine kinase inhibitor therapy in CML. Hematology Am Soc Hematol Educ Program. 2009:477–87
5. Table of pharmacogenomic biomarkers in drug labels. Last accessed 5 Jan 2012
6. Schilsky RL (2010) Personalized medicine in oncology: the future is now. Nat Rev Drug Discov 9:363–366
7. McDermott U et al (2009) Personalized cancer therapy with selective kinase inhibitors: an emerging paradigm in medical oncology. J Clin Oncol 27:5650–5659
8. Cortes JE et al (2009) Pharmacokinetic/pharmacodynamic correlation and blood-level testing in imatinib therapy for chronic myeloid leukemia. Leukemia 23:1537–1544
9. von Mehren M et al (2011) Correlations between imatinib pharmacokinetics, pharmacodynamics, adherence, and clinical response in advanced metastatic gastrointestinal stromal tumor (GIST): an emerging role for drug blood level testing? Cancer Treat Rev 37:291–299
10. Stearns V et al (2003) Active tamoxifen metabolite plasma concentrations after coadministration of tamoxifen and the selective serotonin reuptake inhibitor paroxetine. J Natl Cancer Inst 95:1758–1764
11. Borges S et al (2006) Quantitative effect of CYP2D6 genotype and inhibitors on tamoxifen metabolism: implication for optimization of breast cancer treatment. Clin Pharmacol Ther 80:61–74
12. Widmer N et al (2004) Determination of imatinib (gleevec) in human plasma by solid-phase extraction-liquid chromatography-ultraviolet absorbance detection. J Chromatogr B Analyt Technol Biomed Life Sci 803:285–292
13. Krause DS et al (2005) Tyrosine kinases as targets for cancer therapy. N Engl J Med 353:172–187
14. Oliff A et al (1996) New molecular targets for cancer therapy. Sci Am 275:144–149
15. Larson RA et al (2008) Imatinib pharmacokinetics and its correlation with response and safety in chronic-phase chronic myeloid leukemia: a subanalysis of the IRIS study. Blood 111:4022–4028
16. Picard S et al (2007) Trough imatinib plasma levels are associated with both cytogenetic and molecular responses to standard-dose imatinib in chronic myeloid leukemia. Blood 109: 3496–3499
17. Demetri GD et al (2009) Imatinib plasma levels are correlated with clinical benefit in patients with unresectable/metastatic gastrointestinal stromal tumors. J Clin Oncol 27(19):3141–3147
18. Peng B et al (2005) Clinical pharmacokinetics of imatinib. Clin Pharmacokinet 44:879–894
19. Widmer N et al (2008) Relationship of imatinib-free plasma levels and target genotype with efficacy and tolerability. Br J Cancer 98:1633–1640
20. Widmer N et al (2010) Imatinib plasma levels: correlation with clinical benefit in GIST patients. Br J Cancer 102:1198–1199
21. Badalamenti G et al (2007) Gastrointestinal stromal tumors (GISTs): focus on histopathological diagnosis and biomolecular features. Ann Oncol 18(Suppl 6):vi136–vi140
22. Geyer CE et al (2006) Lapatinib plus capecitabine for HER2-positive advanced breast cancer. N Engl J Med 355:2733–2743
23. Le TC et al (2008) New developments in multitargeted therapy for patients with solid tumours. Cancer Treat Rev 34:37–48
24. Mitsudomi T et al (2005) Mutations of the epidermal growth factor receptor gene predict prolonged survival after gefitinib treatment in patients with non-small-cell lung cancer with postoperative recurrence. J Clin Oncol 23:2513–2520
25. Cadranel J et al (2011) Genetic profiling and epidermal growth factor receptor-directed therapy in nonsmall cell lung cancer. Eur Respir J 37:183–193
26. Commander H et al (2011) Vandetanib: first global approval. Drugs 71:1355–1365
27. Flaherty KT et al (2010) Inhibition of mutated, activated BRAF in metastatic melanoma. N Engl J Med 363:809–819
28. Sasaki T et al (2011) New strategies for treatment of ALK-rearranged non-small cell lung cancers. Clin Cancer Res 17:7213–7218

29. Widmer N et al (2008) Principles of therapeutic drug monitoring. Rev Med Suisse 4:1644–1648
30. Widmer N et al (2008) Therapeutic drug monitoring: the clinical practice. Rev Med Suisse 4:1649–1660
31. Decosterd LA et al (2010) Multiplex ultra-performance liquid chromatography-tandem mass spectrometry method for simultaneous quantification in human plasma of fluconazole, itraconazole, hydroxyitraconazole, posaconazole, voriconazole, voriconazole-N-oxide, anidulafungin, and caspofungin. Antimicrob Agents Chemother 54:5303–5315
32. Fayet A et al (2009) A LC-tandem MS assay for the simultaneous measurement of new antiretroviral agents: raltegravir, maraviroc, darunavir, and etravirine. J Chromatogr B Analyt Technol Biomed Life Sci 877:1057–1069
33. Fayet Mello A et al (2011) Successful efavirenz dose reduction guided by therapeutic drug monitoring. Antivir Ther 16:189–197
34. Gotta V et al (2010) Suivi thérapeutique de l'imatinib. Forum Med Suisse 10:403–406
35. Gervasini G et al (2010) Pharmacogenetic testing and therapeutic drug monitoring are complementary tools for optimal individualization of drug therapy. Eur J Clin Pharmacol 66:755–774
36. McMahon G et al (2009) Therapeutic drug monitoring in oncology: does it have a future? Bioanalysis 1:507–511
37. Lennard L (2001) Therapeutic drug monitoring of cytotoxic drugs. Br J Clin Pharmacol 52(Suppl 1):75S–87S
38. Kamath AV et al (2008) Preclinical pharmacokinetics and in vitro metabolism of dasatinib (BMS-354825): a potent oral multi-targeted kinase inhibitor against SRC and BCR-ABL. Cancer Chemother Pharmacol 61:365–376
39. Lathia C et al (2006) Lack of effect of ketoconazole-mediated CYP3A inhibition on sorafenib clinical pharmacokinetics. Cancer Chemother Pharmacol 57:685–692
40. O'Brien SG et al (2003) Effects of imatinib mesylate (STI571, glivec) on the pharmacokinetics of simvastatin, a cytochrome p450 3A4 substrate, in patients with chronic myeloid leukaemia. Br J Cancer 89:1855–1859
41. White DL et al (2006) OCT-1-mediated influx is a key determinant of the intracellular uptake of imatinib but not nilotinib (AMN107): reduced OCT-1 activity is the cause of low in vitro sensitivity to imatinib. Blood 108:697–704
42. Widmer N et al (2006) Population pharmacokinetics of imatinib and the role of alpha-acid glycoprotein. Br J Clin Pharmacol 62:97–112
43. Haouala A et al. (2011) Prediction of free imatinib concentrations based on total plasma levels in GIST patients. Br J Clin Pharmacol, in press
44. Hazarika M et al (2008) Tasigna for chronic and accelerated phase Philadelphia chromosome–positive chronic myelogenous leukemia resistant to or intolerant of imatinib. Clin Cancer Res 14:5325–5331
45. Koch KM et al (2009) Effects of food on the relative bioavailability of lapatinib in cancer patients. J Clin Oncol 27:1191–1196
46. Dai G et al (2008) Importance of characterizing determinants of variability in exposure: application to dasatinib in subjects with chronic myeloid leukemia. J Clin Pharmacol 48(11):1254–1269
47. Houk BE et al (2010) Relationship between exposure to sunitinib and efficacy and tolerability endpoints in patients with cancer: results of a pharmacokinetic/pharmacodynamic meta-analysis. Cancer Chemother Pharmacol 66:357–371
48. Klumpen HJ et al (2011) Moving towards dose individualization of tyrosine kinase inhibitors. Cancer Treat Rev 37(4):251–260
49. Moore M et al (2005) Phase I study to determine the safety and pharmacokinetics of the novel Raf kinase and VEGFR inhibitor BAY 43-9006, administered for 28 days on/7 days off in patients with advanced, refractory solid tumors. Ann Oncol 16:1688–1694
50. Peng B et al (2004) Pharmacokinetics and pharmacodynamics of imatinib in a phase I trial with chronic myeloid leukemia patients. J Clin Oncol 22:935–942

51. Strumberg D et al (2005) Phase I clinical and pharmacokinetic study of the novel Raf kinase and vascular endothelial growth factor receptor inhibitor BAY 43-9006 in patients with advanced refractory solid tumors. J Clin Oncol 23:965–972
52. Tan AR et al (2004) Evaluation of biologic end points and pharmacokinetics in patients with metastatic breast cancer after treatment with erlotinib, an epidermal growth factor receptor tyrosine kinase inhibitor. J Clin Oncol 22:3080–3090
53. Burris HA III et al (2005) Phase I safety, pharmacokinetics, and clinical activity study of lapatinib (GW572016), a reversible dual inhibitor of epidermal growth factor receptor tyrosine kinases, in heavily pretreated patients with metastatic carcinomas. J Clin Oncol 23:5305–5313
54. Kantarjian H et al (2010) Dasatinib versus imatinib in newly diagnosed chronic-phase chronic myeloid leukemia. N Engl J Med 362:2260–2270
55. Soulieres D et al (2004) Multicenter phase II study of erlotinib, an oral epidermal growth factor receptor tyrosine kinase inhibitor, in patients with recurrent or metastatic squamous cell cancer of the head and neck. J Clin Oncol 22:77–85
56. Buclin T et al (2011) Who is in charge of assessing therapeutic drug monitoring? The case of imatinib. Lancet Oncol 12:9–11
57. van Erp NP et al (2009) Clinical pharmacokinetics of tyrosine kinase inhibitors. Cancer Treat Rev 35:692–706
58. Bouchet S et al (2010) Therapeutic drug monitoring of tyrosine-kinase inhibitors in the treatment of chronic myelogenous leukaemia: interests and limits. Therapie 65:213–218
59. Gambacorti-Passerini C et al (2000) Role of alpha1 acid glycoprotein in the in vivo resistance of human BCR-ABL(+) leukemic cells to the abl inhibitor STI571. J Natl Cancer Inst 92:1641–1650
60. le CP et al (2004) Pharmacokinetics and cellular uptake of imatinib and its main metabolite CGP74588. Cancer Chemother Pharmacol 53:313–323
61. Wang L et al (2008) Expression of the uptake drug transporter hOCT1 is an important clinical determinant of the response to imatinib in chronic myeloid leukemia. Clin Pharmacol Ther 83:258–264
62. Haouala A et al (2010) Cardiovascular drug interactions with tyrosine kinase inhibitors. Cardiovasc Med 13:147–154
63. Haouala A et al (2011) Drug interactions with the tyrosine kinase inhibitors imatinib, dasatinib, and nilotinib. Blood 117:e75–e87
64. Gambillara E et al (2005) Severe pustular eruption associated with imatinib and voriconazole in a patient with chronic myeloid leukemia. Dermatology 211:363–365
65. Druker BJ et al (2006) Five-year follow-up of patients receiving imatinib for chronic myeloid leukemia. N Engl J Med 355:2408–2417
66. Takahashi N et al (2010) Correlation between imatinib pharmacokinetics and clinical response in Japanese patients with chronic-phase chronic myeloid leukemia. Clin Pharmacol Ther 88:809–813
67. Singh N et al (2009) Drug monitoring of imatinib levels in patients undergoing therapy for chronic myeloid leukaemia: comparing plasma levels of responders and non-responders. Eur J Clin Pharmacol 65:545–549
68. Awidi A et al (2010) Relationship of serum imatinib trough level and response in CML patients: long term follow-up. Leuk Res 34:1573–1575
69. Faber E et al (2010) Imatinib dose escalation in two patients with chronic myeloid leukemia, with low trough imatinib plasma levels measured at various intervals from the beginning of therapy and with suboptimal treatment response, leads to the achievement of higher plasma levels and major molecular response. Int J Hematol 91:897–902
70. Larson RA (2009) Therapeutic monitoring of drug plasma concentrations and improved clinical outcomes in CML. Clin Adv Hematol Oncol 7:S3–S5
71. Kantarjian H et al (2006) Nilotinib in imatinib-resistant CML and Philadelphia chromosome-positive ALL. N Engl J Med 354:2542–2551
72. Saglio G et al (2010) Nilotinib versus imatinib for newly diagnosed chronic myeloid leukemia. N Engl J Med 362:2251–2259

73. Milojkovic D et al (2009) Mechanisms of resistance to imatinib and second-generation tyrosine inhibitors in chronic myeloid leukemia. Clin Cancer Res 15:7519–7527
74. Wang X et al. (2008) Dasatinib pharmacokinetics and exposure-response (E-R): relationship to safety and efficacy in patients (pts) with chronic myeloid leukemia (CML). J Clin Oncol. 26 (No 15 S) (May 20 Suppl, Abstract 3590)
75. Mendel DB et al (2003) In vivo antitumor activity of SU11248, a novel tyrosine kinase inhibitor targeting vascular endothelial growth factor and platelet-derived growth factor receptors: determination of a pharmacokinetic/pharmacodynamic relationship. Clin Cancer Res 9:327–337
76. Faivre S et al (2006) Safety, pharmacokinetic, and antitumor activity of SU11248, a novel oral multitarget tyrosine kinase inhibitor, in patients with cancer. J Clin Oncol 24:25–35
77. Billemont B, et al. (2009) Correlation of sorafenib plasma concentrations and clinical toxicity: a prospective population pharmacodynamic and pharmacokinetic study. J Clin Oncol 27 (No 15 S) (May 20 Suppl, e14585)
78. Ratain MJ et al (2007) The value meal: how to save $1,700 per month or more on lapatinib. J Clin Oncol 25:3397–3398
79. Li J et al (2006) CYP3A Phenotyping approach to predict systemic exposure to EGFR tyrosine kinase inhibitors. J Natl Cancer Inst 98:1714–1723
80. Martin P et al (2012) Pharmacokinetics of vandetanib: three phase I studies in healthy subjects. Clin Ther 34(1):221–237
81. Zhang L et al (2011) Pharmacokinetics and tolerability of vandetanib in Chinese patients with solid, malignant tumors: an open-label, phase I, rising multiple-dose study. Clin Ther 33:315–327
82. Martin P et al (2011) Pharmacokinetic drug interactions with vandetanib during coadministration with rifampicin or itraconazole. Drugs R&D 11:37–51
83. Weil A et al (2010) Pharmacokinetics of vandetanib in subjects with renal or hepatic impairment. Clin Pharmacokinet 49:607–618
84. Weisberg E et al (2007) Second generation inhibitors of BCR-ABL for the treatment of imatinib-resistant chronic myeloid leukaemia. Nat Rev Cancer 7:345–356
85. Santos FP et al (2010) Bafetinib, a dual Bcr-Abl/Lyn tyrosine kinase inhibitor for the potential treatment of leukemia. Curr Opin Investig Drugs 11:1450–1465
86. Keisner SV et al (2011) Pazopanib: the newest tyrosine kinase inhibitor for the treatment of advanced or metastatic renal cell carcinoma. Drugs 71:443–454
87. Abbas R et al (2011) Pharmacokinetics of oral neratinib during co-administration of ketoconazole in healthy subjects. Br J Clin Pharmacol 71:522–527
88. Abbas R et al (2011) Effect of ketoconazole on the pharmacokinetics of oral bosutinib in healthy subjects. J Clin Pharmacol 51:1721–1727
89. Jost LM et al (2006) Metabolism and disposition of vatalanib (PTK787/ZK-222584) in cancer patients. Drug Metab Dispos 34:1817–1828
90. Chiorean EG et al (2010) A phase I dose escalation and pharmacokinetic study of vatalanib (PTK787/ZK 222584) in combination with paclitaxel in patients with advanced solid tumors. Cancer Chemother Pharmacol 66:441–448
91. Reardon DA et al (2009) Phase I pharmacokinetic study of the vascular endothelial growth factor receptor tyrosine kinase inhibitor vatalanib (PTK787) plus imatinib and hydroxyurea for malignant glioma. Cancer 115:2188–2198
92. Fox E et al (2010) A phase 1 trial and pharmacokinetic study of cediranib, an orally bioavailable pan-vascular endothelial growth factor receptor inhibitor, in children and adolescents with refractory solid tumors. J Clin Oncol 28:5174–5181
93. Yamamoto N et al (2009) Phase I, dose escalation and pharmacokinetic study of cediranib (RECENTIN), a highly potent and selective VEGFR signaling inhibitor, in Japanese patients with advanced solid tumors. Cancer Chemother Pharmacol 64:1165–1172
94. Goss G et al (2009) A phase I and pharmacokinetic study of daily oral cediranib, an inhibitor of vascular endothelial growth factor tyrosine kinases, in combination with cisplatin and gemcitabine in patients with advanced non-small cell lung cancer: a study of the national cancer institute of Canada clinical trials group. Eur J Cancer 45:782–788

95. Li C et al (2009) In vitro metabolism of the novel, highly selective oral angiogenesis inhibitor motesanib diphosphate in preclinical species and in humans. Drug Metab Dispos 37:1378–1394
96. TARCEVA®: Scientific Discussion page 11. European Medicines Agency 2005. www.ema.europa.eu, last accessed 12 Dec 2011
97. Costa DB et al (2011) CSF concentration of the anaplastic lymphoma kinase inhibitor crizotinib. J Clin Oncol 29:e443–e445
98. Tan W, et al. (2010) Pharmacokinetics (PK) of PF-02341066. a dula ALK/MET inhibitor after multiple oral dose to advanced cancer patients. J Clin Oncol 28 (228 s) (supplement, abstract 2596)
99. Hidalgo M et al (2003) Pharmacokinetics and pharmacodynamics: maximizing the clinical potential of erlotinib (tarceva). Semin Oncol 30:25–33
100. Lu JF et al (2006) Clinical pharmacokinetics of erlotinib in patients with solid tumors and exposure-safety relationship in patients with non-small cell lung cancer. Clin Pharmacol Ther 80:136–145
101. Mohamed MK et al (2005) Skin rash and good performance status predict improved survival with gefitinib in patients with advanced non-small cell lung cancer. Ann Oncol 16:780–785
102. Eskens FA et al (2006) The clinical toxicity profile of vascular endothelial growth factor (VEGF) and vascular endothelial growth factor receptor (VEGFR) targeting angiogenesis inhibitors; a review. Eur J Cancer 42:3127–3139
103. Awidi A et al (2010) Determination of imatinib plasma levels in patients with chronic myeloid leukemia by high performance liquid chromatography-ultraviolet detection and liquid chromatography-tandem mass spectrometry: methods' comparison. Leuk Res 34:714–717
104. Bakhtiar R et al (2002) High-throughput quantification of the anti-leukemia drug STI571 (gleevec) and its main metabolite (CGP 74588) in human plasma using liquid chromatography-tandem mass spectrometry. J Chromatogr B Analyt Technol Biomed Life Sci 768: 325–340
105. Boddy AV et al (2007) Pharmacokinetic investigation of imatinib using accelerator mass spectrometry in patients with chronic myeloid leukemia. Clin Cancer Res 13:4164–4169
106. Guetens G et al (2003) Quantification of the anticancer agent STI-571 in erythrocytes and plasma by measurement of sediment technology and liquid chromatography-tandem mass spectrometry. J Chromatogr A 1020:27–34
107. Klawitter J et al (2009) Development and validation of a sensitive assay for the quantification of imatinib using LC/LC-MS/MS in human whole blood and cell culture. Biomed Chromatogr 23:1251–1258
108. Parise RA et al (2003) Liquid chromatographic-mass spectrometric assay for quantitation of imatinib and its main metabolite (CGP 74588) in plasma. J Chromatogr B Analyt Technol Biomed Life Sci 791:39–44
109. Rochat B et al (2008) Imatinib metabolite profiling in parallel to imatinib quantification in plasma of treated patients using liquid chromatography-mass spectrometry. J Mass Spectrom 43:736–752
110. Titier K et al (2005) Quantification of imatinib in human plasma by high-performance liquid chromatography-tandem mass spectrometry. Ther Drug Monit 27:634–640
111. Parise RA et al (2009) A high-performance liquid chromatography-mass spectrometry assay for quantitation of the tyrosine kinase inhibitor nilotinib in human plasma and serum. J Chromatogr B Analyt Technol Biomed Life Sci 877:1894–1900
112. Tanaka C et al (2010) Clinical pharmacokinetics of the BCR-ABL tyrosine kinase inhibitor nilotinib. Clin Pharmacol Ther 87:197–203
113. de Bruijn P et al (2010) Bioanalytical method for the quantification of sunitinib and its n-desethyl metabolite SU12662 in human plasma by ultra performance liquid chromatography/tandem triple-quadrupole mass spectrometry. J Pharm Biomed Anal 51:934–941
114. Minkin P et al (2008) Quantification of sunitinib in human plasma by high-performance liquid chromatography-tandem mass spectrometry. J Chromatogr B Analyt Technol Biomed Life Sci 874:84–88

115. Jain L et al (2008) Development of a rapid and sensitive LC-MS/MS assay for the determination of sorafenib in human plasma. J Pharm Biomed Anal 46:362–367
116. Zhao M et al (2007) A rapid and sensitive method for determination of sorafenib in human plasma using a liquid chromatography/tandem mass spectrometry assay. J Chromatogr B Analyt Technol Biomed Life Sci 846:1–7
117. Bai F et al (2006) Determination of lapatinib (GW572016) in human plasma by liquid chromatography electrospray tandem mass spectrometry (LC-ESI-MS/MS). J Chromatogr B Analyt Technol Biomed Life Sci 831:169–175
118. Bai F et al (2011) Determination of vandetanib in human plasma and cerebrospinal fluid by liquid chromatography electrospray ionization tandem mass spectrometry (LC-ESI-MS/MS). J Chromatogr B Analyt Technol Biomed Life Sci 879:2561–2566
119. Lankheet AG et al (2009) A validated assay for the quantitative analysis of vatalanib in human EDTA plasma by liquid chromatography coupled with electrospray ionization tandem mass spectrometry. J Chromatogr B Analyt Technol Biomed Life Sci 877:3625–3630
120. Sparidans RW et al (2009) Liquid chromatography-tandem mass spectrometric assay for the light sensitive tyrosine kinase inhibitor axitinib in human plasma. J Chromatogr B Analyt Technol Biomed Life Sci 877:4090–4096
121. Novakova L et al (2006) Advantages of application of UPLC in pharmaceutical analysis. Talanta 68:908–918
122. Haouala A et al (2009) Therapeutic drug monitoring of the new targeted anticancer agents imatinib, nilotinib, dasatinib, sunitinib, sorafenib and lapatinib by LC tandem mass spectrometry. J Chromatogr B Analyt Technol Biomed Life Sci 877:1982–1996
123. Baratte S et al (2004) Quantitation of SU1 1248, an oral multi-target tyrosine kinase inhibitor, and its metabolite in monkey tissues by liquid chromatograph with tandem mass spectrometry following semi-automated liquid-liquid extraction. J Chromatogr A 1024:87–94
124. Anja Goedl P. Pfizer, Personal communication. 12 Dec 2007
125. Yamazaki S et al (2011) Prediction of oral pharmacokinetics of cMet kinase inhibitors in humans: physiologically based pharmacokinetic model versus traditional one-compartment model. Drug Metab Dispos 39:383–393
126. Christopher LJ et al (2008) Metabolism and disposition of dasatinib after oral administration to humans. Drug Metab Dispos 36:1357–1364
127. Furlong MT et al (2012) A validated LC-MS/MS assay for the simultaneous determination of the anti-leukemic agent dasatinib and two pharmacologically active metabolites in human plasma: application to a clinical pharmacokinetic study. J Pharm Biomed Anal 58: 130–135
128. He K et al (2008) N-in-1 dosing pharmacokinetics in drug discovery: experience, theoretical and practical considerations. J Pharm Sci 97:2568–2580
129. Wu JT et al (2000) Direct plasma sample injection in multiple-component LC-MS-MS assays for high-throughput pharmacokinetic screening. Anal Chem 72:61–67
130. Hsieh Y et al (2002) Direct cocktail analysis of drug discovery compounds in pooled plasma samples using liquid chromatography-tandem mass spectrometry. J Chromatogr B Analyt Technol Biomed Life Sci 767:353–362
131. Kuo BS et al (1998) Sample pooling to expedite bioanalysis and pharmacokinetic research. J Pharm Biomed Anal 16:837–846
132. Kummar S et al (2010) Utilizing targeted cancer therapeutic agents in combination: novel approaches and urgent requirements. Nat Rev Drug Discov 9:843–856
133. De Francia S et al (2009) New HPLC-MS method for the simultaneous quantification of the antileukemia drugs imatinib, dasatinib, and nilotinib in human plasma. J Chromatogr B Analyt Technol Biomed Life Sci 877:1721–1726
134. Demetri GD et al (2009) Phase I dose-escalation and pharmacokinetic study of dasatinib in patients with advanced solid tumors. Clin Cancer Res 15:6232–6240
135. Chahbouni A et al (2009) Simultaneous quantification of erlotinib, gefitinib, and imatinib in human plasma by liquid chromatography tandem mass spectrometry. Ther Drug Monit 31: 683–687

136. Honeywell R et al (2010) Simple and selective method for the determination of various tyrosine kinase inhibitors used in the clinical setting by liquid chromatography tandem mass spectrometry. J Chromatogr B Analyt Technol Biomed Life Sci 878:1059–1068
137. Gotze L et al (2012) Development and clinical application of a LC-MS/MS method for simultaneous determination of various tyrosine kinase inhibitors in human plasma. Clin Chim Acta 413:143–149
138. Bouchet S et al (2011) Simultaneous determination of nine tyrosine kinase inhibitors by 96-well solid-phase extraction and ultra performance LC/MS-MS. Clin Chim Acta 412:1060–1067
139. Neeman M et al. (2011) A LC-MS/MS for the assay of verumafenib, bosutinib, gefitinib and erlotinib in plasma from cancer patients, Master thesis, School of Pharmaceutical Sciences, University of Geneva and Lausanne, Switzerland
140. Rudin CM et al (2008) Pharmacogenomic and pharmacokinetic determinants of erlotinib toxicity. J Clin Oncol 26:1119–1127
141. Roche S et al (2009) Development of a high-performance liquid chromatographic-mass spectrometric method for the determination of cellular levels of the tyrosine kinase inhibitors lapatinib and dasatinib. J Chromatogr B Analyt Technol Biomed Life Sci 877:3982–3990
142. Hegedus C et al (2009) Interaction of nilotinib, dasatinib and bosutinib with ABCB1 and ABCG2: implications for altered anti-cancer effects and pharmacological properties. Br J Pharmacol 158:1153–1164
143. Haouala A et al (2010) SiRNA-mediated knock-down of P-glycoprotein expression reveals distinct cellular disposition of anticancer tyrosine kinases inhibitors. Drug Metab Lett 4:114–119
144. D'Avolio A et al (2012) HPLC-MS method for the simultaneous quantification of the antileukemia drugs imatinib, dasatinib and nilotinib in human peripheral blood mononuclear cell (PBMC). J Pharm Biomed Anal 59:109–116
145. Dahmane E et al (2010) An ultra performance liquid chromatography-tandem MS assay for tamoxifen metabolites profiling in plasma: first evidence of 4′-hydroxylated metabolites in breast cancer patients. J Chromatogr B Analyt Technol Biomed Life Sci 878:3402–3414
146. Jordan VC (2003) Tamoxifen: a most unlikely pioneering medicine. Nat Rev Drug Discov 2:205–213
147. Jordan VC (2006) Tamoxifen (ICI46,474) as a targeted therapy to treat and prevent breast cancer. Br J Pharmacol 147(Suppl 1):S269–S276
148. Jordan VC (2007) New insights into the metabolism of tamoxifen and its role in the treatment and prevention of breast cancer. Steroids 72:829–842
149. Osborne CK (1998) Tamoxifen in the treatment of breast cancer. N Engl J Med 339:1609–1618
150. Tamoxifen for early breast cancer: an overview of the randomised trials. Early Breast Cancer Trialists' Collaborative Group (1998). Lancet 351:1451–1467
151. Early Breast Cancer Trialists' Collaborative Group (EBCTCG) (2005) Effects of chemotherapy and hormonal therapy for early breast cancer on recurrence and 15-year survival: an overview of the randomised trials. Lancet 365:1687–1717
152. Goldhirsch A et al (2007) Progress and promise: highlights of the international expert consensus on the primary therapy of early breast cancer 2007. Ann Oncol 18:1133–1144
153. Goldhirsch A et al (2009) Thresholds for therapies: highlights of the St gallen international expert consensus on the primary therapy of early breast cancer 2009. Ann Oncol 20:1319–1329
154. Visvanathan K et al (2009) American society of clinical oncology clinical practice guideline update on the use of pharmacologic interventions including tamoxifen, raloxifene, and aromatase inhibition for breast cancer risk reduction. J Clin Oncol 27:3235–3258
155. Mathew J et al (2009) Neoadjuvant endocrine treatment in primary breast cancer - review of literature. Breast 18:339–344
156. Chia YH et al (2010) Neoadjuvant endocrine therapy in primary breast cancer: indications and use as a research tool. Br J Cancer 103:759–764

157. Cuzick J et al (2011) Preventive therapy for breast cancer: a consensus statement. Lancet Oncol 12(5):496–503
158. Jensen EV et al (2003) The estrogen receptor: a model for molecular medicine. Clin Cancer Res 9:1980–1989
159. Riggs BL et al (2003) Selective estrogen-receptor modulators – mechanisms of action and application to clinical practice. N Engl J Med 348:618–629
160. Nilsson S et al (2005) Oestrogen receptors and selective oestrogen receptor modulators: molecular and cellular pharmacology. Basic Clin Pharmacol Toxicol 96:15–25
161. Orlando L et al (2010) Molecularly targeted endocrine therapies for breast cancer. Cancer Treat Rev 36(Suppl 3):S67–S71
162. Johnson MD et al (2004) Pharmacological characterization of 4-hydroxy-N-desmethyl tamoxifen, a novel active metabolite of tamoxifen. Breast Cancer Res Treat 85:151–159
163. Lim YC et al (2005) Endoxifen (4-hydroxy-N-desmethyl-tamoxifen) has anti-estrogenic effects in breast cancer cells with potency similar to 4-hydroxy-tamoxifen. Cancer Chemother Pharmacol 55:471–478
164. Lim YC et al (2006) Endoxifen, a secondary metabolite of tamoxifen, and 4-OH-tamoxifen induce similar changes in global gene expression patterns in MCF-7 breast cancer cells. J Pharmacol Exp Ther 318:503–512
165. Goetz MP et al (2008) Tamoxifen pharmacogenomics: the role of CYP2D6 as a predictor of drug response. Clin Pharmacol Ther 83:160–166
166. Wu X et al (2009) The tamoxifen metabolite, endoxifen, is a potent antiestrogen that targets estrogen receptor alpha for degradation in breast cancer cells. Cancer Res 69:1722–1727
167. Wu X et al (2011) Estrogen receptor-beta sensitizes breast cancer cells to the anti-estrogenic actions of endoxifen. Breast Cancer Res 13:R27
168. Crewe HK et al (1997) Variable contribution of cytochromes P450 2D6, 2 C9 and 3A4 to the 4-hydroxylation of tamoxifen by human liver microsomes. Biochem Pharmacol 53:171–178
169. Coller JK et al (2002) The influence of CYP2B6, CYP2C9 and CYP2D6 genotypes on the formation of the potent antioestrogen Z-4-hydroxy-tamoxifen in human liver. Br J Clin Pharmacol 54:157–167
170. Crewe HK et al (2002) Metabolism of tamoxifen by recombinant human cytochrome P450 enzymes: formation of the 4-hydroxy, 4′-hydroxy and N-desmethyl metabolites and isomerization of trans-4-hydroxytamoxifen. Drug Metab Dispos 30:869–874
171. Desta Z et al (2004) Comprehensive evaluation of tamoxifen sequential biotransformation by the human cytochrome P450 system in vitro: prominent roles for CYP3A and CYP2D6. J Pharmacol Exp Ther 310:1062–1075
172. Jin Y et al (2005) CYP2D6 Genotype, antidepressant use, and tamoxifen metabolism during adjuvant breast cancer treatment. J Natl Cancer Inst 97:30–39
173. Madlensky L et al (2011) Tamoxifen metabolite concentrations, CYP2D6 genotype, and breast cancer outcomes. Clin Pharmacol Ther 89:718–725
174. Zheng Y et al (2007) Elimination of antiestrogenic effects of active tamoxifen metabolites by glucuronidation. Drug Metab Dispos 35:1942–1948
175. Kaku T et al (2004) Quaternary ammonium-linked glucuronidation of tamoxifen by human liver microsomes and UDP-glucuronosyltransferase 1A4. Biochem Pharmacol 67:2093–2102
176. Ogura K et al (2006) Quaternary ammonium-linked glucuronidation of trans-4-hydroxytamoxifen, an active metabolite of tamoxifen, by human liver microsomes and UDP-glucuronosyltransferase 1A4. Biochem Pharmacol 71:1358–1369
177. Sun D et al (2006) Characterization of tamoxifen and 4-hydroxytamoxifen glucuronidation by human UGT1A4 variants. Breast Cancer Res 8:R50
178. Benoit-Biancamano MO, et al. (2009) A pharmacogenetics study of the human glucuronosyltransferase UGT1A4. Pharmacogenet Genomics [Epub ahead of print]
179. Sun D et al (2007) Glucuronidation of active tamoxifen metabolites by the human UDP glucuronosyltransferases. Drug Metab Dispos 35:2006–2014
180. Blevins-Primeau AS et al (2009) Functional significance of UDP-glucuronosyltransferase variants in the metabolism of active tamoxifen metabolites. Cancer Res 69:1892–1900

181. Chen G et al (2002) 4-Hydroxytamoxifen sulfation metabolism. J Biochem Mol Toxicol 16:279–285
182. Nishiyama T et al (2002) Reverse geometrical selectivity in glucuronidation and sulfation of cis- and trans-4-hydroxytamoxifens by human liver UDP-glucuronosyltransferases and sulfotransferases. Biochem Pharmacol 63:1817–1830
183. Falany JL et al (2006) Sulfation of raloxifene and 4-hydroxytamoxifen by human cytosolic sulfotransferases. Drug Metab Dispos 34:361–368
184. Nowell S et al (2006) Pharmacogenetics of human cytosolic sulfotransferases. Oncogene 25:1673–1678
185. Lien EA et al (1988) Identification of 4-hydroxy-N-desmethyltamoxifen as a metabolite of tamoxifen in human bile. Cancer Res 48:2304–2308
186. Lien EA et al (1989) Distribution of 4-hydroxy-N-desmethyltamoxifen and other tamoxifen metabolites in human biological fluids during tamoxifen treatment. Cancer Res 49:2175–2183
187. Kisanga ER et al (2005) Excretion of hydroxylated metabolites of tamoxifen in human bile and urine. Anticancer Res 25:4487–4492
188. Ring A et al (2004) Mechanisms of tamoxifen resistance. Endocr Relat Cancer 11:643–658
189. Hoskins J et al (2009) CYP2D6 And tamoxifen: DNA matters in breast cancer. Nat Rev Cancer 9:576–586
190. Zhou SF (2009) Polymorphism of human cytochrome P450 2D6 and its clinical significance: part I. Clin Pharmacokinet 48:689–723
191. Zhou SF (2009) Polymorphism of human cytochrome P450 2D6 and its clinical significance: part II. Clin Pharmacokinet 48:761–804
192. Lim HS et al (2007) Clinical implications of CYP2D6 genotypes predictive of tamoxifen pharmacokinetics in metastatic breast cancer. J Clin Oncol 25:3837–3845
193. Gjerde J et al (2008) Effects of CYP2D6 and SULT1A1 genotypes including SULT1A1 gene copy number on tamoxifen metabolism. Ann Oncol 19:56–61
194. Xu Y et al (2008) Association between CYP2D6 *10 genotype and survival of breast cancer patients receiving tamoxifen treatment. Ann of Oncol Off J Eur Soc Med Oncol/ESMO 19:1423–1429
195. Kiyotani K et al (2010) Significant effect of polymorphisms in CYP2D6 and ABCC2 on clinical outcomes of adjuvant tamoxifen therapy for breast cancer patients. J Clin Oncol 28:1287–1293
196. Murdter TE et al (2011) Activity levels of tamoxifen metabolites at the estrogen receptor and the impact of genetic polymorphisms of phase I and II enzymes on their concentration levels in plasma. Clin Pharmacol Ther 89:708–717
197. Goetz MP et al (2005) Pharmacogenetics of tamoxifen biotransformation is associated with clinical outcomes of efficacy and hot flashes. J Clin Oncol 23:9312–9318
198. Goetz MP et al (2007) The impact of cytochrome P450 2D6 metabolism in women receiving adjuvant tamoxifen. Breast Cancer Res Treat 101:113–121
199. Schroth W et al (2007) Breast cancer treatment outcome with adjuvant tamoxifen relative to patient CYP2D6 and CYP2C19 genotypes. J Clin Oncol 25:5187–5193
200. Kiyotani K et al (2008) Impact of CYP2D6*10 on recurrence-free survival in breast cancer patients receiving adjuvant tamoxifen therapy. Cancer Sci 99:995–999
201. Newman WG et al (2008) Impaired tamoxifen metabolism reduces survival in familial breast cancer patients. Clin Cancer Res 14:5913–5918
202. Schroth W et al (2009) Association between CYP2D6 polymorphisms and outcomes among women with early stage breast cancer treated with tamoxifen. JAMA 302:1429–1436
203. Bijl MJ et al (2009) The CYP2D6*4 polymorphism affects breast cancer survival in tamoxifen users. Breast Cancer Res Treat 118:125–130
204. Ramon y Cajal T et al (2010) Impact of CYP2D6 polymorphisms in tamoxifen adjuvant breast cancer treatment. Breast Cancer Res Treat 119:33–38
205. Serrano D et al (2011) Efficacy of tamoxifen based on cytochrome P450 2D6, CYP2C19 and SULT1A1 genotype in the Italian tamoxifen prevention trial. Pharmacogenomics J 11:100–107

206. Nowell S et al (2005) Association of genetic variation in tamoxifen-metabolizing enzymes with overall survival and recurrence of disease in breast cancer patients. Breast Cancer Res Treat 91:249–258
207. Wegman P et al (2005) Genotype of metabolic enzymes and the benefit of tamoxifen in postmenopausal breast cancer patients. Breast Cancer Res 7:R284–R290
208. Wegman P et al (2007) Genetic variants of CYP3A5, CYP2D6, SULT1A1, UGT2B15 and tamoxifen response in postmenopausal patients with breast cancer. Breast Cancer Res 9:R7
209. Okishiro M et al (2009) Genetic polymorphisms of CYP2D6 10 and CYP2C19 2, 3 are not associated with prognosis, endometrial thickness, or bone mineral density in Japanese breast cancer patients treated with adjuvant tamoxifen. Cancer 115:952–961
210. Lash T et al (2009) Genotype-guided tamoxifen therapy: time to pause for reflection? Lancet Oncol 10:825–833
211. Nowell S et al (2002) Association between sulfotransferase 1A1 genotype and survival of breast cancer patients receiving tamoxifen therapy. J Natl Cancer Inst 94:1635–1640
212. Teft WA et al (2011) Endoxifen, the active metabolite of tamoxifen, is a substrate of the efflux transporter P-glycoprotein (multidrug resistance 1). Drug Metab Dispos 39:558–562
213. Iusuf D et al (2011) P-glycoprotein (ABCB1) transports the primary active tamoxifen metabolites endoxifen and 4-hydroxytamoxifen and restricts their brain penetration. J Pharmacol Exp Ther 337:710–717
214. Kiyotani K et al (2012) Pharmacogenomics of tamoxifen: roles of drug metabolizing enzymes and transporters. Drug Metab Pharmacokinet 27(1):122–131
215. Partridge AH et al (2003) Nonadherence to adjuvant tamoxifen therapy in women with primary breast cancer. J Clin Oncol 21:602–606
216. McCowan C et al (2008) Cohort study examining tamoxifen adherence and its relationship to mortality in women with breast cancer. Br J Cancer 99:1763–1768
217. Dezentje VO et al (2010) Effect of concomitant CYP2D6 inhibitor use and tamoxifen adherence on breast cancer recurrence in early-stage breast cancer. J Clin Oncol 28:2423–2429
218. Beer B et al (2010) Development and validation of a liquid chromatography-tandem mass spectrometry method for the simultaneous quantification of tamoxifen, anastrozole, and letrozole in human plasma and its application to a clinical study. Anal Bioanal Chem 398:1791–1800
219. Hershman DL et al (2010) Early discontinuation and nonadherence to adjuvant hormonal therapy in a cohort of 8,769 early-stage breast cancer patients. J Clin Oncol 28:4120–4128
220. Hershman DL et al (2011) Early discontinuation and non-adherence to adjuvant hormonal therapy are associated with increased mortality in women with breast cancer. Breast Cancer Res Treat 126:529–537
221. Kelly C et al (2010) Selective serotonin reuptake inhibitors and breast cancer mortality in women receiving tamoxifen: a population based cohort study. BMJ 340:c693
222. Lammers LA et al (2010) The impact of CYP2D6-predicted phenotype on tamoxifen treatment outcome in patients with metastatic breast cancer. Br J Cancer 103:765–771
223. Ahern TP et al (2009) No increase in breast cancer recurrence with concurrent use of tamoxifen and some CYP2D6-inhibiting medications. Cancer Epidemiol Biomarkers Prev 18:2562–2564
224. Cronin-Fenton D et al (2010) Selective serotonin reuptake inhibitors and adjuvant tamoxifen therapy: risk of breast cancer recurrence and mortality. Future Oncol 6:877–880
225. Lash T et al (2010) Breast cancer recurrence risk related to concurrent use of SSRI antidepressants and tamoxifen. Acta Oncol 49(3):305–312
226. Irvin WJ Jr et al (2011) Genotype-guided tamoxifen dosing increases active metabolite exposure in women with reduced CYP2D6 metabolism: a multicenter study. J Clin Oncol 29:3232–3239
227. Kiyotani K et al (2012) Dose-adjustment study of tamoxifen based on CYP2D6 genotypes in Japanese breast cancer patients. Breast Cancer Res Treat 131(1):137–145
228. Barginear MF et al (2011) Increasing tamoxifen dose in breast cancer patients based on CYP2D6 genotypes and endoxifen levels: effect on active metabolite isomers and the antiestrogenic activity score. Clin Pharmacol Ther 90:605–611

229. Lien EA et al (1987) Determination of tamoxifen and four metabolites in serum by low-dispersion liquid chromatography. Clin Chem 33:1608–1614
230. Lee KH et al (2003) Quantification of tamoxifen and three metabolites in plasma by high-performance liquid chromatography with fluorescence detection: application to a clinical trial. J Chromatogr B Analyt Technol Biomed Life Sci 791:245–253
231. Zhu YB et al (2008) Optimizing high-performance liquid chromatography method with fluorescence detection for quantification of tamoxifen and two metabolites in human plasma: application to a clinical study. J Pharm Biomed Anal 46:349–355
232. Esteve-Romero J et al (2010) Tamoxifen monitoring studies in breast cancer patients by micellar liquid chromatography. Anal Bioanal Chem 397:1557–1561
233. Carter SJ et al (2001) Biomonitoring of urinary tamoxifen and its metabolites from breast cancer patients using nonaqueous capillary electrophoresis with electrospray mass spectrometry. Electrophoresis 22:2730–2736
234. Mihailescu R et al (2000) Identification of tamoxifen and metabolites in human male urine by GC/MS. Biomed Chromatogr 14:180–183
235. Zweigenbaum J et al (2000) Bioanalytical high-throughput selected reaction monitoring-LC/MS determination of selected estrogen receptor modulators in human plasma: 2000 samples/day. Anal Chem 72:2446–2454
236. Sheth HR et al (2003) Aging may be associated with concentrations of tamoxifen and its metabolites in breast cancer patients. J Womens Health 12:799–808
237. Tucker AN et al (2005) Polymorphisms in cytochrome P4503A5 (CYP3A5) may be associated with race and tumor characteristics, but not metabolism and side effects of tamoxifen in breast cancer patients. Cancer Lett 217:61–72
238. Gjerde J et al (2005) Identification and quantification of tamoxifen and four metabolites in serum by liquid chromatography-tandem mass spectrometry. J Chromatogr A 1082:6–14
239. Williams LD et al (2006) Quantification of tamoxifen and metabolites and soy isoflavones in human plasma using liquid chromatography with electrospray ionization tandem mass spectrometry. J AOAC Int 89:1168–1173
240. Wu AH et al (2007) Tamoxifen, soy, and lifestyle factors in Asian American women with breast cancer. J Clin Oncol 25:3024–3030
241. Furlanut M et al (2007) Tamoxifen and its main metabolites serum and tissue concentrations in breast cancer women. Ther Drug Monit 29:349–352
242. Teunissen SF et al (2009) Development and validation of a quantitative assay for the analysis of tamoxifen with its four main metabolites and the flavonoids daidzein, genistein and glycitein in human serum using liquid chromatography coupled with tandem mass spectrometry. J Chromatogr B Analyt Technol Biomed Life Sci 877:2519–2529
243. Ahmad A et al (2010) Orally administered endoxifen is a new therapeutic agent for breast cancer. Breast Cancer Res Treat 122:579–584
244. Ahmad A et al (2010) Endoxifen, a new cornerstone of breast cancer therapy: demonstration of safety, tolerability, and systemic bioavailability in healthy human subjects. Clin Pharmacol Ther 88:814–817
245. Jaremko M et al (2010) Tamoxifen metabolite isomer separation and quantification by liquid chromatography-tandem mass spectrometry. Anal Chem 82:10186–10193
246. Teunissen SF et al (2011) Development and validation of a quantitative assay for the determination of tamoxifen and its five main phase I metabolites in human serum using liquid chromatography coupled with tandem mass spectrometry. J Chromatogr B Analyt Technol Biomed Life Sci 879:1677–1685
247. Binkhorst L et al (2011) Quantification of tamoxifen and three of its phase-I metabolites in human plasma by liquid chromatography/triple-quadrupole mass spectrometry. J Pharm Biomed Anal 56:1016–1023
248. Poon GK et al (1993) Analysis of phase I and phase II metabolites of tamoxifen in breast cancer patients. Drug Metab Dispos 21:1119–1124
249. Poon GK et al (1995) Identification of tamoxifen metabolites in human Hep G2 cell line, human liver homogenate, and patients on long-term therapy for breast cancer. Drug Metab Dispos 23:377–382

250. Jones RM et al (1996) On-line high-performance liquid chromatographic-electrospray ionization mass spectrometric method for the study of tamoxifen metabolism. J Chromatogr A 722:249–255
251. Lim CK et al (1997) Identification and mechanism of formation of potentially genotoxic metabolites of tamoxifen: study by LC-MS/MS. J Pharm Biomed Anal 15:1335–1342
252. Boocock DJ et al (2002) Identification of human CYP forms involved in the activation of tamoxifen and irreversible binding to DNA. Carcinogenesis 23:1897–1901
253. Mazzarino M et al (2008) A mass spectrometric approach for the study of the metabolism of clomiphene, tamoxifen and toremifene by liquid chromatography time-of-flight spectroscopy. Eur J Mass Spectrom 14:171–180
254. Mazzarino M et al (2010) Mass spectrometric characterization of tamoxifene metabolites in human urine utilizing different scan parameters on liquid chromatography/tandem mass spectrometry. Rapid Commun Mass Spectrom 24:749–760
255. Teunissen SF et al (2010) Bioanalytical methods for determination of tamoxifen and its phase I metabolites: a review. Anal Chim Acta 683:21–37
256. Matuszewski BK et al (2003) Strategies for the assessment of matrix effect in quantitative bioanalytical methods based on HPLC-MS/MS. Anal Chem 75:3019–3030
257. Matuszewski BK (2006) Standard line slopes as a measure of a relative matrix effect in quantitative HPLC-MS bioanalysis. J Chromatogr B Analyt Technol Biomed Life Sci 830: 293–300
258. Lindegardh N et al (2008) Development and validation of a liquid chromatographic-tandem mass spectrometric method for determination of piperaquine in plasma stable isotope labeled internal standard does not always compensate for matrix effects. J Chromatogr B Analyt Technol Biomed Life Sci 862:227–236
259. Bonfiglio R et al (1999) The effects of sample preparation methods on the variability of the electrospray ionization response for model drug compounds. Rapid Commun Mass Spectrom 13:1175–1185
260. Viswanathan CT et al (2007) Quantitative bioanalytical methods validation and implementation: best practices for chromatographic and ligand binding assays. Pharm Res 24:1962–1973
261. Polson C et al (2003) Optimization of protein precipitation based upon effectiveness of protein removal and ionization effect in liquid chromatography-tandem mass spectrometry. J Chromatogr B Analyt Technol Biomed Life Sci 785:263–275
262. Souverain S et al (2004) Protein precipitation for the analysis of a drug cocktail in plasma by LC-ESI-MS. J Pharm Biomed Anal 35:913–920
263. Ismaiel OA et al (2010) Investigation of endogenous blood plasma phospholipids, cholesterol and glycerides that contribute to matrix effects in bioanalysis by liquid chromatography/mass spectrometry. J Chromatogr B Analyt Technol Biomed Life Sci 878:3303–3316
264. Van Eeckhaut A et al (2009) Validation of bioanalytical LC-MS/MS assays: evaluation of matrix effects. J Chromatogr B Analyt Technol Biomed Life Sci 877:2198–2207
265. Hewavitharana AK (2011) Matrix matching in liquid chromatography-mass spectrometry with stable isotope labelled internal standards–is it necessary? J Chromatogr A 1218: 359–361
266. Imatinib Concentration Monitoring Evaluation (I-COME). http://www.controlled-trials.com/ISRCTN 31181395
267. European Treatment and Outcome Study (EUTOS). http://www.eutos.org
268. ASQUALAB, Paris, France. http://www.asqualab.com/
269. Guiducci C et al. ISyPeM: Intelligent Integrated Systems for Personalised Medicine. http://www.nano-tera.ch/projects/405.php

Applications of Mass Spectrometry in Analyses of Steroid Hormones

Zimeng Yan, Chang Cheng, and Shaorong Liu

Abstract Steroid hormones are endogenous chemicals controlling many endocrinology functions. Mass spectrometry technologies have been applied for analyses of steroid hormones as biomarkers in endocrinology and pathology diagnoses, doping drugs in athletes and racing horses, residuals in food safety concerns, and environmental pollutants in water and sediments. Both liquid chromatography mass spectrometry (LC-MS or LC-MS/MS) and gas chromatography mass spectrometry (GC-MS or GC-MS/MS) are broadly used in research, clinical, pharmaceutical industry, competition sports, food safety, and environmental testing laboratories. Sample preparation techniques, such as deconjugation and extraction, are critical procedures for isolating steroid hormone from sample matrices, including biological fluids, tissues, environmental water and sediments. Chemical derivatization modifies the physicochemical properties of steroid hormone molecules to improve their chromatographic performances and to enhance their sensitivities to mass detection. Chromatographic techniques such as HPLC, UPLC, and GC have direct impact on separation of analytes, MS interface, and analysis throughput. The method sensitivity and specificity of LC-MS and GC-MS depend largely on the analyte status, i.e., easiness of ionization, derivatization, sample matrix, and MS detection mode, e.g., ESI, APCI, APPI, MAILDI, or EI. LC-MS and GC-MS methodologies should be developed and validated following scientific and regulatory guidelines, and the steroid hormones analyses should be standardized.

Z. Yan, Ph.D. (✉)
Research Pharmaceutical Services, Inc., 520 Virginia Drive, Fort Washington, PA 19034, USA
e-mail: zyan@rpsweb.com

C. Cheng, Ph.D.
Analytical Development, Albany Molecular Research, Inc. 33 Riverside Avenue, Rensselaer, NY 12144, USA
e-mail: Chang.Cheng@amriglobal.com

S. Liu, Ph.D.
Department of Chemistry and Biochemistry, University of Oklahoma, Norman, OK 73019, USA
e-mail: shaorong.liu@ou.edu

Q. Alan Xu and Timothy L. Madden (eds.), *LC-MS in Drug Bioanalysis*,
DOI 10.1007/978-1-4614-3828-1_10,

1 Introduction

1.1 Steroid Hormones: Chemical Structures and Functions

Steroid hormones are composed of 18–21 carbon atoms, which are bonded together by four fused rings: three cyclohexane rings (designated as rings A, B, and C) and one cyclopentane ring (as ring D), with an exception of estrogens' ring A as a benzene ring. Hydroxyl and/or ketone groups are attached at C_3, C_{11}, C_{17}, C_{20}, or C_{21} positions, and methyl group(s) is attached at C_{18} and/or C_{19} positions of the steroid skeletons, as shown in Fig. 1. These molecules are neutral, unionized, nonvolatile, lipid soluble, and stable during extraction and analysis by mass spectrometry. Originated from cholesterol, the biosynthetic pathways of steroid hormones are well established [1–3]. The major biotransformation reactions for steroid hormones include oxidation, reduction, conjugations with glucuronic acid, sulfate, and glutathione [4, 5]. As most of the biotransformation pathways and metabolites of steroid hormones are clearly defined, structure elucidation or identification is not a major concern for steroid analysis [6]. However, since a large portion of endogenous steroid hormones exists as their glucuronide and sulfate conjugates, especially in urine, deconjugation of the steroid hormones may be needed before LC-MS/MS and GC-MS analysis [7, 8].

According to bonding receptors and biological functions, steroid hormones are classified as progestagens (or progestogens, pregnancy hormones), mineralocorticoids (mineral retention hormones), glucocorticoids (glucose metabolism and inflammation hormones), androgens (male hormones), and estrogens (female hormones). Endogenous steroid hormones are chemical messengers regulating many life functions, such as controlling pregnancy, salt and water balance, metabolism, inflammation, immune functions, and development of sexual characteristics. Many serious health problems, diseases, and clinical syndromes are indicated by steroid hormones deviations from normal levels. Steroid hormone analysis plays important roles in health care, including disease diagnosis, food safety, and environmental protection. For example, endogenous steroid hormones are critical biomarkers for clinical diagnoses in endocrinology, physiology, and pathology; and steroid hormones are also broadly used as medicines for treatment of varieties of diseases, such as inflammation, malfunctions in immune systems, underdevelopment syndromes, and disorders in endocrine systems [2, 9]. On the other hand, steroid hormones are abused by some athletes [10] and racing horse owners [11] to enhance their sport performances. Steroid hormones are also found as hazardous residuals in eatable meats [12] and are monitored as environmental pollutants in water and sediments [13]. Therefore, development and application of specific, sensitive, and robust analytical methods and methodologies for steroid hormone analyses have significant impacts on clinical analyses, disease diagnoses and treatments, antidoping drug screening, food safety examination, and environmental quality monitoring. A large number of articles have been published on analyses of steroid hormones by mass spectrometry

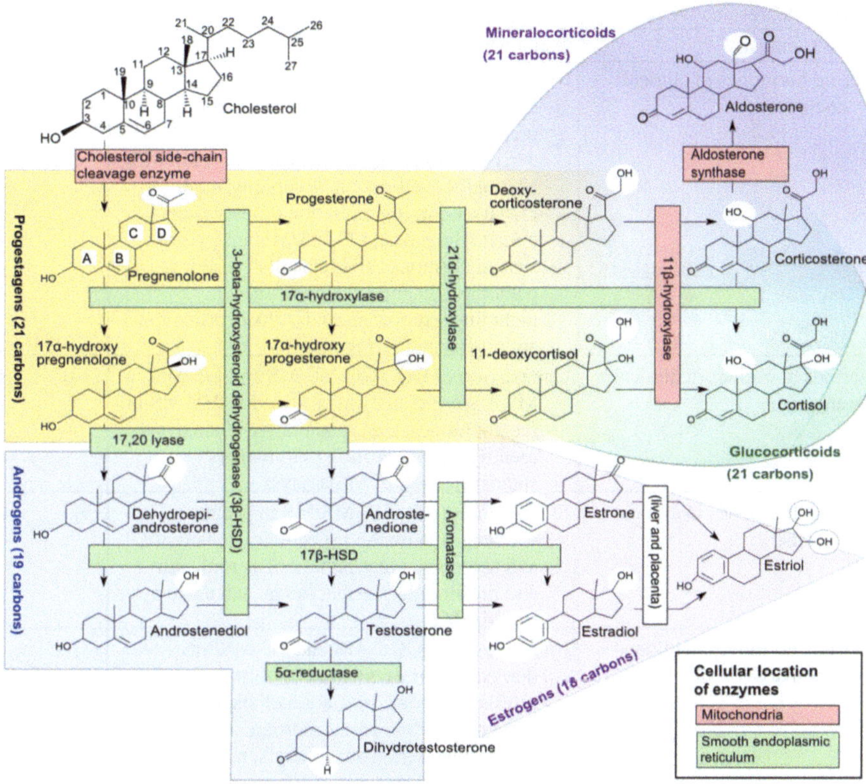

Fig. 1 The major classes of steroid hormones: progestagens, mineralocorticoids, glucocorticoids, androgens, and estrogens. Enzymes, their cellular location, substrates, and products in human steroidogenesis. Boron WF, Boulpaep EL (2003) Medical Physiology: a cellular and molecular approach, p 1300, Elsevier/Saunders (reprinted with permission from Elsevier/Saunders)

and other technologies, and progress and challenges of MS technologies and applications in steroid hormone analyses have been briefly reviewed (see Table 1).

1.2 Challenges to Analytical/Bioanalytical Technologies and Methodologies

Immunoassays (IA) and radioimmunoassays (RIA) are widely accepted and routinely applied for clinical and environmental analyses of steroid hormones, but they have limitations in specificity and accuracy. Therefore, more selective, accurate, and sensitive technologies such as LC-MS/MS and GC-MS/MS methodologies are needed to improve the IA and RIA testing techniques for steroid hormones and

Table 1 Representative reviews on steroids analyses by LC-MS/MS and GC-MS

Steroid analysis	Major topics	Reference
Steroid hormones in clinical chemistry	Role of MS in clinical diagnostic testing and endocrine biomarker analysis; sample preparation, e.g., LLE vs. SPE, and derivatization to enhance sensitivity; LC-MS/MS method validation, specificity, quality management, interpretation of test results for clinical diagnosis; LC-MS/MS analysis of steroid hormones and related physiology/pathology diagnostic significances, including: adrenal steroids, glucocorticoids, androgens and estrogens.	[2]
Steroids hormones in clinical chemistry	Comparison of RIA, GC-MS/MS and LC-MS/MS in steroid analysis; role of LC-MS/MS in metabolomics, e.g., quantitative bioanalysis, identifying and profiling biomarkers; standardization of MS assays.	[14]
Steroids hormones in clinical chemistry	Steroid analysis by LC-MS/MS in pediatrics—challenges in method specificity, sensitivity, and test results interpretation; steroid assays and profiling in plasma, saliva, and urine by LC-MS/MS.	[9]
Steroids hormones in clinical chemistry	Immunoassay vs. GC-MS and LC-MS/MS; derivatization vs. nonderivatization; steroid profiles for newborns, adrenal insufficiency, prostatitis/pelvic pain syndrome, premature adrenarche, sera from smokers, metabolic diseases, diabetes, water contaminant, athletes doping.	[3]
Estrogens in clinical chemistry	RIA, GC-MS/MS, and LC-MS/MS analyses of estrogens in serum and plasma; isotope internal standard; sample derivatization; ionization modes and sensitivities of GC-MS/MS and LC-MS/MS.	[4]
Estrogens in clinical chemistry	Measurement of endogenous estrogens by immunoassays, LC-MS (ESI, APCI, APPI) and GC-MS.	[15]
Endogenous conjugated androgens in clinical chemistry	GC-GS and LC-MS/MS analyses of conjugated androgens—deconjugation, derivatization, and associated issues.	[7]
Steroids hormone residuals in meat safety concerns	GC-MS^n and LC-MS^n analyses of steroids in edible matrices, e.g., meat, liver, kidney, kidney fat, and milk.	[12]
Steroids hormones as environmental pollutants	Analysis of steroids as environmental endocrine disrupting compounds; sample preparation, e.g., LLE vs. SPE; immunoassay vs. GC-MS/MS and LC-MS/MS analyses; sensitivities, e.g., LOD at pg–ng/mL level.	[13]

(continued)

Table 1 (continued)

Steroid analysis	Major topics	Reference
Steroid hormones chemical derivatization, reagent selection and optimization	Derivatization of steroids to enhance LC/MS detection, e.g., $ESI^{+/-}$/MS, $APCI^{+/-}$/MS, and $APPI^{+/-}$/MS.	[16, 17]
Steroid hormones by GC/MS and LC/MS	GC/MS and LC/MS in determination of androgens, corticoids, estrogens, cholesterol and related, bile acids, vitamin D and metabolites, phytosteroids, etc.	[1]
Steroid hormones assay standardization	Immunoassay vs. GC-MS and LC-MS/MS, validation of assay methodologies, establishing standard pools of steroid hormones in women and man, utilizing the pools for cross comparison of various methodologies.	[18]

metabolites [4, 12, 14, 19]. LC-MS/MS and GC-MS methods are increasingly used for analyzing steroid hormones due to their high selectivity and sensitivity. The typical LC-MS/MS methods by direct injection [20, 21] and GC-MS methods [6] have limit of quantitation (LOQ) at ng/mL level, and a number of LC-MS/MS [2, 8, 22] and GC-MS/MS [23] methods are able to achieve LOQ at pg/mL levels when the steroid hormone samples are chemically derivatized before injection. Nevertheless, in many cases, clinical diagnostic tests need to determine steroid hormones at low pg/mL, and even fg/mL levels [4, 15, 18]. In order to improve the MS method sensitivity and sample preparation efficiency, many studies on sample preparation have been carried out, including deconjugation, extraction, and derivatization. Different chromatographic techniques and MS detection modes have also been investigated.

On the other hand, many LC-MS/MS and GC-MS methodologies and their applications to steroid hormone analyses have been developed for the purpose of scientific research or specific studies only, while they might not have been validated according to regulatory guidelines, and their results might not correlate with those from the widely accepted bioanalytical techniques, e.g., IA and RIA. Therefore, standardization of the LC-MS/MS and GC-MS technologies and methodologies, including facilities, instruments, reference standards, procedures, data system, etc., plays a critical role in transferring the LC-MS/MS and GC-MS technologies to daily testing procedures for clinical diagnosis, sports antidoping screening, food safety control, and analysis of environmental contaminations in water and sediments. Only until their specificity, accuracy, precision, calibration mode, sensitivity, and robustness are validated according to regulatory guidelines, and the test results are comparable or consistent with those obtained from the existing techniques such as IA and RIA, the LC-MS/MS and GC-MS/MS techniques may not be accepted as reliable clinical diagnostic methodologies for steroid hormone determination [18].

2 Sample Preparation

The major sample preparation procedures of steroid hormone analyses include extraction, deconjugation, and derivatization, as shown in Fig. 2. A number of examples of steroid hormone sample preparation are summarized in Table 2. Application of stable isotope labeled (e.g., deuterated or ^{13}carbon labeled) steroid hormones as the internal standard or isotope dilution is a standard of practice during quantitative steroid hormone analyses by LC-MS/MS and GC-MS whenever possible. These isotope labeled internal standards are added into the samples before deconjugation or extraction procedures. They undergo the same deconjugation and/or extraction procedures and LC-MS/MS or GC-MS analysis as the steroid samples do. The efficiency of deconjugation and/or extraction, assay accuracy and precision are calibrated and calculated with the isotope internal standards, leading to more accurate, precise, and robust methods and results.

2.1 *Deconjugation of Steroid Hormones*

A large percentage of steroid hormones exist as glucuronide, sulfate, and glutathione conjugates in body fluids and tissues. These steroid conjugates may be analyzed directly by LC-MS/MS using either electrospray ionization (ESI) [5, 40–43] or atmospheric pressure chemical ionization (APCI) [44] mode. However, in many cases, the presence of glucuronide and sulfate conjugates in samples may reduce MS detection sensitivity and the total amount of a hormone (both unconjugated and conjugated) that can be determined by LC-MS/MS [8] or GC-MS [6] after the

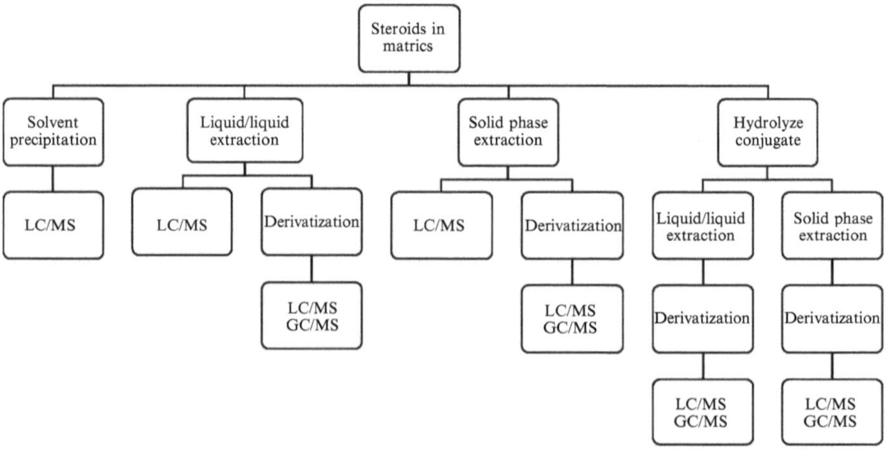

Fig. 2 Steroid hormone sample preparation procedures: extraction, deconjugation, and derivatization

Table 2 Typical steroid hormone sample preparation techniques

Steroids in matrix	Hydrolyzing conjugates	Extraction	Derivatization	Internal standard	MS mode	Reference
Estrogens in serum Steroids in testicular fluid	NA	Solvent precipitation vs. LE LLE	NA	Deuterated isotopes	LC-APPI$^{-/+}$/MS/MS LC-ESI^{+}/MS/MS	[22] [24]
Estrogens in breast tissue, plasma, brain	NA	Solvent precipitation/LLE LLE	Hydroxylamine Dansyl chloride	Deuterated isotopes	LC-TIS^{+}/MS/MS LC-ESI^{+}/MS/MS LC-ESI^{+}/MS/MS	[25] [26] [27]
Steroids in urine Estrogens in urine, serum	β-Glucuronidase/ H_2SO_4 β-Glucuronidase/ sulfatase	LLE LLE	NA Dansyl chloride	Deuterated isotopes	LC-ESI^{+}/MS/MS	[20] [8, 28]
Estrones in cell culture medium Steroids in water	NA	SPE (96-well C_{18}) Magnetic-mediated SPE	NA	Deuterated isotopes NA	LC-APCI$^{+/-}$/MS/MS vs. LC-ESI$^{+/-}$/MS/MS LC-ESI^{-}/MS/MS	[29] [30]
Androgens in prostatic tissue Androgens in plasma Estrogens in sediment and sewage sludge Estrogens in water	NA	SPE (96-well Oasis HLB) Sep-Pak C_{18} Sep-Pak C_{18} Supelco Discovery C_{18}	2-Fluoro-1-methypyridinium *p*-toluenesulfonate (FPMTS) Pentafluoropropionic anhydride BSTFA MSTFA	Deuterated isotopes	LC-ESI^{+}/MS/MS GC-EI^{+}/MS LC-ESI^{+}/MS and GC-EI^{+}/MS GC-EI^{+}/MS	[31] [32] [33] [34]
Estrones in urine	β-Glucuronidase/ sulfatase	SPE (Bobd Elut LRC $C_{18,}$ Extrlut QE)	*p*-Toluenesulfonhydrazide	Deuterated isotopes	LC-ESI^{+}/MS and LC-APCI^{+}/MS, LC-EI^{+}/MS^{n}	[35, 36]
Estrogens in urine	β-Glucuronidase	SPE (Sep-pak C18, Sephadex A25, Oasis HLB)	MSTFA	Deuterated isotopes	GC-EI^{+}/MS	[37, 38]
Steroids in urine	β-Glucuronidase/ arylsulfatase	βCD (Oasis HLB)			GC-EI^{+}/MS	[6, 39]

NA, not applicable

deconjugation and derivatization of the steroid hormones. Thus deconjugation of steroid hormone conjugates become a critical sample preparation procedure in LC-MS/MS and GC-MS analyses.

Both estrogen glucuronide and sulfate conjugates may be deconjugated with hydrochloric acid in methanol [45], but acid solvolysis is not a selective reaction, and the harsh condition may cause side reactions and by-products. An alternative deconjugation procedure applies β-glucuronidase from *Escherichia coli* to hydrolyze steroid glucuronide conjugates first, and then hydrolyzes steroid sulfate conjugates with sulfuric acid [20]. Steroid hormone glucuronide conjugates may be hydrolyzed with β-glucuronidase from different sources, e.g., *Escherichia coli*. [38, 46], limpets [47] or *Helix pomatia* [39]. An attention should be paid to the fact that β-glucuronidase from *Escherichia coli* does not affect the chemical structures of steroid hormones during deconjugation, while β-glucuronidase from *Helix pomatia* can convert 3β-hydroxy-5-ene steroids into 3-oxo-4-ene steroids, and change 3β-hydroxy-5α-reduced and 3β-hydroxy-5β-reduced steroids into 3-oxo-5α-reduced and 3-oxo-5β-reduced steroids, respectively, because *Helix pomatia* also contains other enzymes, including cholesterol oxidase, 3β-hydroxysteroid oxidoreductase/3-oxosteroid-4,5-eneisomerase, and 6-hydroxylase. When glucuronide conjugates of steroid hormones with 3-hydroxy-5-ene structure, e.g., progestagens and androgens, are deconjugated using β-glucuronidase from *Helix pomatia*, the hormone quantitation may not be accurate or representative [48]. However, β-glucuronidase/sulfatase from *Helix pomatia* are commonly used for hydrolyzing both glucuronide and sulfate conjugates simultaneously, especially for estrogens and metabolites, because the phenolic ring A of estrogens is not affected by those enzymes in *Helix pomatia*. The deconjugation of steroid glucuronide and sulfate conjugates is simplified as one step incubation of β-glucuronidase/sulfatase with steroid samples at 37 °C for 20 h or at 55 °C for 3 h [8, 28, 39].

2.2 Extraction of Steroid Hormones from Biological Matrices, Environmental Water and Sediments

The techniques used for extracting steroid hormones and metabolites include protein precipitation (PP) with an organic solvent, liquid–liquid extraction (LLE), and solid phase extraction (SPE). As shown in Fig. 2, selection of a sample extraction technique is based on the steroid hormone sample (unconjugated or conjugated), quantity, matrix and the objective of an analytical method or test. For example, if a method or test needs to analyze only the unconjugated steroid hormones in a small volume of serum, e.g., <1 mL, a small volume of acetonitrile can precipitate the proteins and extract steroids from the sample [25]. The procedure is very simple, but the acetonitrile extract contains more nonrelated components than the extract from LLE [22]. When a small volume of biological fluid is extracted by LLE with a solvent, whether the sample undergoes deconjugation and/or derivatization or not, the LLE extract is cleaner than PP extract, but may not be as clean as an extract from

SPE. However, LLE may reduce the sample loss, experiment procedures and errors, and save time in comparison to SPE [8, 25, 28]. In contrast, if a larger amount of sample is available, e.g., 2–10 mL of urine or 1–10 L of environmental water, SPE is a better choice, because it concentrates the sample and minimizes the interferences from other materials, leading to a higher sensitivity and selectivity of the method or test [34, 37, 39]. The sample extraction throughput can be significantly enhanced by using automated 96-well SPE plates [31, 49].

2.3 Derivatization

Chemical derivatization is a standard procedure for GC-MS analysis of steroid hormones, because steroid hormones are not volatile to go through GC column [1, 6]. The major concerns of derivatization reagents for GC-MS analysis of steroid hormones are the completeness of the derivatization reaction and the volatility of hormone derivatives. The typical derivatization reagents for GC-MS samples are silylation reagents, e.g., *N*-methyl-*N*-trifluorotrimethyl acetamide (MSTFA, [37, 39]) and *N*,*O*-bis(trimethylsilyl) trifluoroacetamide (BSTFA, [33]), which react with both alcoholic and phenolic hydroxy groups on steroid hormone molecules.

One of the major advantages of LC-MS/MS over GC-MS or GC-MS/MS is that steroid hormones may be analyzed directly by LC-MS/MS without derivatization procedures, which are time-consuming and tedious [22, 50–53]. However, a number of studies demonstrated that the chemically derivatized steroid hormones were significantly more sensitive to LC-MS/MS detection than the underivatized hormones, because the neutral molecules of estrogens and metabolites might not be effectively ionized under electrospray ionization (ESI) or atmospheric pressure chemical ionization (APCI) modes [4, 21, 25, 54, 55]. In order to enhance the steroid hormone molecules sensitivity for LC-MS/MS analysis at pg/mL level, chemical derivatization is an effective technique for analysis of steroid hormones and metabolites. A list of derivatization reagents and application examples for steroid hormone analyses by LC-MS/MS and GC-MS are presented in Table 3.

An ideal derivatization reagent is able to react with steroid hormones and metabolites selectively and quantitatively under mild conditions within a short time, and those hormone derivatives are stable and easily ionized during LC-MS/MS analysis. Based on their functional groups, the derivatization reagents used for LC-MS/MS analyses of steroid hormones and metabolites may be classified into seven major classes:

1. Hydrazide, e.g., (carboxymethyl)trimethylammonium chloride hydrazide (Girard T reagent) [4, 56], and *p*-toluenesulfonhydrazide [36]; and hydroxylamine [25]
2. Benzyl bromide, e.g., pentafluorobenzyl bromide [55, 59] and 4-nitrobenzyl bromide [60]
3. Fluorobenzene or fluoropyridine, e.g., 2,4-dinitro-5-fluorobenzene analogues [54] and 2-fluoro-1-methyl-2-pyridinium *p*-toluenesulfonate [61]

Table 3 Typical derivatization reagents and target chemical groups of steroid hormones

Derivatizing reagent	Typical compound and reference	Reaction target	MS mode
Silylation agent	*N*-Methyl-*N*-tryfluorotrimethyl acetamide [6] Pentafluoropropionic anhydride [32]	Both phenolic OH and alcoholic OH	GC-EI+/MS
Hydrazide and hydroxylamine	(Carboxymethyl)trimethylammonium chloride hydrazide (Girard T reagent, [4, 56]); 2-hydrazinopyridine [57]; *p*-toluenesulfonhydrazide [36] Girard P reagent [58] Hydroxylamine [25]	Ketolic C=O	LC-ESI+/MS/MS MALDI/MS/MS LC-ESI+/MS/MS
Benzyl bromide	Pentafluorobenzyl bromide [55, 59] 4-Nitrobenzyl bromide [60]	Phenolic OH	LC-APCI−/MS/MS
Fluorobenzene and fluoropyridine	2,4-Dinitro-5-fluorobenzene analogues [54] 2-Fluoro-1-methyl-2-pyridinium *p*-toluenesulfonate [61]	Phenolic OH	LC-ESI+/MS/MS
Sulfonyl chloride	Dansyl chloride, 1,2 -dimethylimidazole-4-chloride, pyridine-3-sulfonyl chloride 4-(1-H-pyrazol-1-yl)benzenesulfonyl chloride [62] 10-Ethyl-acridine-2-sulfonyl chloride [63]	Phenolic OH	LC-ESI+/MS/MS LC-APCI+/MS/MS
Carboxylic acid *N*-hydroxy-succinimide ester	*N*-Methyl-nicotinic acid *N*-hydroxysuccinimide ester [21]	Phenolic OH	LC-ESI+/MS/MS
Carbonyl chloride or acetic anhydride	Picolinoyl chloride [64] 4-Nitrobenzoyl chloride [60] Pentafluorobenzoyl chloride [65] Acetic anhydride [66]	Both phenolic OH and alcoholic OH	LC-ESI+/MS/MS LC-APCI−/MS/MS GC-EI+/MS LC-APCI+/MS^2
o-Phenylenediamine	*o*-Phenylenediamine [67]	Estrogen *o*-quinones	LC-ESI+/MS/MS

4. Sulfonyl chloride, e.g., dansyl chloride, 1,2-dimethylimidazole-4-chloride, and pyridine-3-sulfonyl chloride; 4-(1-H-pyrazol-1-yl)benzenesulfonyl chloride [62]
5. Carboxylic acid *N*-hydroxysuccinimide ester, e.g., *N*-methyl-nicotinic acid *N*-hydroxysuccinimide ester [21]
6. Carbonyl chloride and acetic anhydride, e.g., picolinoyl chloride [64]
7. *o*-Phenylenediamine [67]

The first class of derivatization reagents, hydrazide reagents and hydroxylamine, react with most of the ketolic steroid hormones and metabolites (ketone group on C_3, C_{17}, or C_{20}), i.e., androgens, progestagens, corticoids, and ketolic estrogens [4, 25, 36, 56]. However, they are not suitable for steroid hormones without ketolic group(s), such as estradiol, estriol and their related metabolites. Hydroxylamine is a typical derivatization reagent for steroid hormone profiling, because it can react with all the ketolic hormones, and provide unique mass fragments for each steroid moiety during LC-MS/MS analysis. When these unique mass fragments or daughter ions are used for multiple reaction monitoring (MRM) quantitation, the method selectivity is much higher than those using a daughter ion from a derivatization reagent, e.g., *m/z* 171 for dansyl ion [25].

The second to seventh classes of reagents react with hydroxyl steroid hormones and metabolites, especially estrogens. Pentafluorobenzyl bromide estrogen derivatives, belonging to the second class, are sensitive to both ESI^+ [1] and $APCI^-$ [55, 61] modes, and usually have lower limits of quantitation (LOQ) values under $APCI^-$ mode than those of derivatives of dansyl chloride and 2-fluoro-1-methyl-pyridinium *p*-toluenesulfonate under ESI^+ mode, because there were little interference from analogue compounds and the matrix background under $APCI^-$ mode. Nevertheless, the derivatization reaction of estrogens with pentafluorobenzyl bromide was ten times longer than the derivatization reaction with dansyl chloride (30 min vs. 3 min at 60 °C) [61]. A study by Higashi et al. indicated that the derivatization reactions of estrogens with 4-nitrobenzyl bromide (the second class), 2,4-dinitrofluorobenzene (the third class) and 4-nitrobenzoyl chloride (the sixth class) were not as complete as the reaction with 4-nitrobenzene sulfonyl chloride (the fourth class) [60].

The fifth class, carboxylic acid *N*-hydroxysuccinimide ester, is also not as reactive as sulfonyl chloride, and its derivative is not as sensitive to LC-MS/MS either [21]. The sixth class of reagents, carbonyl chloride and carboxylic acid anhydride, can react with both phenolic and alcoholic hydroxyl groups of steroids. However, the selectivity, speed, and completeness of derivatization reactions of these reagents with steroids are not as good as those of sulfonyl chloride derivatization reagents [60, 64, 66]. Since the other four classes of derivatization reagents, sulfonyl chloride, benzyl bromide, carboxylic acid *N*-hydroxysuccinimide ester, and fluorobenzene, are able to selectively react with phenolic hydroxyl group of estrogens and metabolites, the carbonyl chloride and carboxylic acid anhydride reagents become less preferable for derivatizing estrogens and metabolites. The seventh class, *o*-phenylenediamine, is a specific derivatization reagent for estrogen *o*-quinones, potential carcinogens, such as estrone-2,3-quinone, estrone-3,4-quinone, estradiol-2,3-quinone, estradiol-3,4-quinone [67].

Works published so far suggest that sulfonyl chloride is a preferred reagent for derivatizing estrogens and their metabolites, due to its reaction completeness and selectivity. In addition, a sulfonyl chloride reagent containing a basic or preionized nitrogen atom, e.g., on dansyl, pyridine, imidazole, pyrazole, or piperizine ring, could significantly enhance the ionization of estrogen derivatives under ESI^+ mode, and increase the detection sensitivity [4, 54, 62]. Dansyl chloride is a typical sulfonyl chloride reagent used for derivatizing estrogens and metabolites from varieties of matrices, such as river water [50, 61], mouse plasma and brain [27], human urine [28], breast tissue [26], and serum [8]. However, dansyl derivatives have a disadvantage that the common dansyl fragment *m/z* 171 from background of the derivatization reagent and from those derivatives may have negative impacts on the method selectivity and accuracy during LC-MS/MS analysis, especially for those isomers with the same molecular ions and fragments, because the elevated dansyl fragment background noise may reduce the analyte signal–noise ratio, and cross interfere quantitation of other analytes with the same fragments [8, 25, 28].

3 Comparison of LC-MS/MS, GC-MS, and Immunoassays

3.1 LC-MS/MS and UHPLC-MS/MS vs. GC-MS and GC-MS/MS

The bioanalytical methods developed in recent years focused more on LC-MS/MS and GC-MS/MS techniques, because the earlier studies demonstrated that LC-MS/MS and GC-MS/MS are significantly more sensitive in analyzing estrogens and metabolites than LC-MS and GC-MS [18, 68]. Analyzing both unconjugated and conjugated steroid hormones directly is the major advantage of LC-MS and LC-MS/MS over GC-MS and GC-MS/MS, because the sample preparation procedures of deconjugation and derivatization can be avoided [5, 7, 40–42, 44]. In complex samples, separation of steroid hormones and metabolites by LC or GC is still one of the major concerns, because many steroid hormones and metabolites have the same molecular weights and MS fragments, and they may interfere with each other if not separated. For example, 2-hydroxyestrone and 4-hydroxyestrone have same molecular weight, even after they are derivatized with MSTFA for GC-MS analysis [48] or with dansyl chloride or *p*-toluenesulfonhydrazide for LC-MS/MS analysis [8, 35]. If the derivatives are not separated, GC-MS or LC-MS/MS is unable to distinguish the derivatives of 2-hydroxyestrone from those of 4-hydroxyestrone, whether the silylated, dansylated, or hydrozone steroid fragments are used for quantitation.

An LC column with a length of 150 mm can separate up to 23 steroid hormones and metabolites [20]. A typical LC-MS/MS method developed by Xu et al. was able to separate 15 estrogens and metabolites using a 150×2 mm, 4 μm LC column, which had a run time of 100 min [8, 28]. The separation efficiency can be improved by using smaller particle size LC columns, e.g., 100×2.0 mm, 2.5 μm column [69] or 50×2.1 mm, 1.8 μm column [21], also leading to a significantly reduced run time (e.g., less than 30 min). Similarly, ultra high performance liquid chromatography (UHPLC) is able to significantly improve separation efficiency and to reduce run time [40, 70–72]. Two-dimensional (2D) LC-MS/MS with column-switching technique has been used for determination of unconjugated and conjugated estrogens in river water [51] and sediment [73], and this technique can significantly reduce the analysis time, and increase method separation capability and detection sensitivity. When estrogens and their derivatives of dansyl chloride and pentafluorobenzyl bromide are analyzed by HPLC, 2D-LC and UHPLC with ESI, APCI, APPI, and APCI/APPI MS modes, the UHPLC-ESI$^-$-MS/MS significantly enhances sensitivities of the derivatives over the native estrogens, in the order UHPLC > 2D-LC > LC, and ESI > APPI > APCI = APCI/APPI [74]. On the other hand, the cleaner and more efficient supercritical fluid chromatography (SFC)-mass spectrometry was also used for analysis of estrogens and metabolites [75], but the SFC technology was less versatile and robust than LC, UHPLC, and GC.

Since the neutral molecules of steroid hormones and metabolites are not easily ionized under either APCI$^{+/-}$ or ESI$^{+/-}$ modes, LC-MS/MS is less sensitive when used directly in either APCI$^{+/-}$ or ESI$^{+/-}$ modes, with LOQs at ng/mL level as shown in Table 4 [20, 21, 52]. It has been observed that estrone, 16α-hydroxyestrone, 2-methoxyestrone, 4-methoxyestrone, and 2-hydroxy-3-methoxyestrone are sensitive to APCI$^+$ mode, while 2-hydroxyestrone and 4-hydroxyestrone are sensitive to APCI$^-$ mode, and even more sensitive to ESI$^-$ mode [29]. Estrone, estradiol, estradiol, and estriol are sensitive to ESI$^-$ mode, and testosterone is sensitive to ESI$^+$ mode [76, 77]. Similarly, estrone and estradiol are sensitive to APPI$^-$ mode, and testosterone is sensitive to APPI$^+$ mode with LOQs in a range of 1.5–10 pg/mL [22], which are comparable with those steroid hormones and metabolites derivatized with hydroxylamine or dansyl chloride, and detected under ESI$^+$ mode [2, 8].

As most steroid hormones and metabolites are already identified, high resolution MS technologies with higher selectivity, e.g., time of flight mass spectrometry (TOF-MS), Fourier transform ion cyclotron resonance mass spectrometry (FTICR-MS), Orbitrap-MS, MALDI, etc., may not have major advantages over triple quadrupole MS in quantitative analysis, because the triple quadrupole MS takes shorter data acquisition time, leading to a higher sensitivity. For example, LC-ESI$^-$-MS/MS is 4–6 times more sensitive than LC-ESI$^-$-TOF-MS in analysis of estrogens [76]. On the other hand, MS interface also plays an important role in steroid hormone analysis. When analyzed by nanospray ES-Q-TOF-MS, Girard P derivative of testosterone at 0.5 pg/μg (0.1 pg of sample) can be detected, while 10 pg of sample is needed for matrix-assisted laser desorption ionization (MALDI)-Q-TOF-MS analysis, because of significant background interferences from the MALDI matrix [58].

GC-MS is a matured technology in analyses of steroid hormones, because GC-MS interface and electron impact ionization (EI) MS mode are stable and easily standardized [1]. GC-MS is a very powerful tool for profiling steroid hormones in biological matrices, such as that more than 70 steroid hormones and metabolites can be separated and quantitated by a single GC-MS run, with LOQ in range of 0.1–10 ng/mL [6, 48]. The derivatization procedure may not be a challenge, because there are many derivatization reagents available, e.g., BSTFA and MSTFA, and the silylation reaction is straightforward and quantitative. In addition, the sensitivity of GC-MS can be improved from low ng/mL to 0.6 pg/mL LOQ level by GC-MS/MS technology [23]. That is why GC-MS is still broadly used today for analyses of

Table 4 Comparison of GC-MS, GC-MS/MS, and LC-MS-MS in separation and sensitivity

			Method sensitivity, LOQ (S/N≥5)								
			GC/MS	GC/NCI/ MS/MS	LC/APCI⁺/ MS/MS	LC/APCI⁻/ MS/MS	LC/ESI⁻/ MS/MS	LC/ESI⁺/ MS/MS	LC/ESI⁺/ MS/MS	LC/APPI⁺/ MS/MS	LC/APPI⁻/ MS/MS
			Urine, 2 mL	Serum, 1.0 mL	Prostatic tissue	Serum, 0.5 mL	Milk, 10 mL	Urine, 0.2 mL	Serum	Serum, 0.2 mL	Serum, 0.2 mL
Compound	Abbreviation	FW	MSTFA derivatives (ng/mL)	PFB, MSTFA (pg/mL)	carbonyl derivative (ng/g) (LOD)	Underivatized (ng/L) (LOD)	Underivatized (pg/L)	Underivatized (ng/mL)	NH_2OH derivative (pg/mL)	Underivatized (pg/mL)	Underivatized (pg/mL)
Androgens											
Androstenedione	A-dione	286	2.0								
Dehydroepiandrosterone	DHEA	288	50.0					1			
Androstanedione	5α-dione	288	5.0								
Epitestosterone	Epi-T	288	0.1					0.3	10		
Testosterone	T	288	0.1					0.3		10	
5β-Dihydrotestosterone	5β-DHT	290	0.2							50*	
Androsterone	An	290	0.5					1			
Etiocholanolone	Etio	290	0.5					1			
Epietiocholanolone	Epi-Etio	290						1			
Epiandrosterone	Epi-An	290	5.0					1			
Androstenediol	A-diol	290	2.0		<0.35						
5α-Dihydrotestosterone	5αDHT	290	0.2					0.5		50*	
5β-Androstan-3α, 17α-diol	βαα-diol	292	1.0							200	
5β-Androstan-3β, 17α-diol	ββα-diol	292	0.5							200	
5α-Androstan-3α, 17α-diol	ααα-diol	292									

5β-Androstan-3β, 17β-diol	βββ-diol	292							
5α-Androstan-3α, 17β-diol	ααβ-diol	292	1.0				0.5		
5β-Androstan-3α, 17β-diol	βαβ-diol	292	0.5						
5α-Androstan-3β, 17α-diol	αβα-diol	292							
Epidihydrotestosterone	Epi-DHT	292	0.2						
5α-Androstan-3β, 17β-diol	αββ-diol	292	20.0						
5α-Androstan-3, 17-dione	5α-Adione	288					1		
16α-Hydroxyandros-tenedione	16α-OH-Adione	302							
11-Keto-androsterone	11-keto-An	304	0.5						
11-Keto-etiocholanolone	11-keto-Etio	304	0.5				0.5		
16α-Hydroxy-DHEA	16α-OH-DHEA	304	0.2						
11β-Hydroxyandroster-one	11β-OH-An	306	0.5						
11β-Hydroxyetiocho-lanolone	11β-OH-Etio	306	0.5				1		
Testosterone-17β-glucuronide	T-Gluc	464					0.3		
Estrogens									
Estrone	E1	270	0.2		37,000	20	1	<0.5 mL>	1.5
17α-Estradiol	17α-E2	272	0.1					1 <8>	
17β-Estradiol	17β-E2	272	0.1	0.6	143	50	0.3	1 <8>	2.5

(continued)

Table 4 (continued)

Compound	Abbreviation	FW	Method sensitivity, LOQ (S/N≥5)								
			GC/MS	GC/NCI/ MS/MS	LC/APCI+/ MS/MS	LC/APCI−/ MS/MS	LC/ESI−/ MS/MS	LC/ESI+/ MS/MS	LC/ESI+/ MS/MS	LC/APPI+/ MS/MS	LC/APPI−/ MS/MS
			Urine, 2 mL	Serum, 1.0 mL	Prostatic tissue	Serum, 0.5 mL	Milk, 10 mL	Urine, 0.2 mL	Serum	Serum, 0.2 mL	Serum, 0.2 mL
			MSTFA derivatives (ng/mL)	PFB, MSTFA (pg/mL)	carbonyl derivative (ng/g) (LOD)	Underivatized (ng/L) (LOD)	Underivatized (pg/L)	Underivatized (ng/mL)	NH_2OH derivative (pg/mL)	Underivatized (pg/mL)	Underivatized (pg/mL)
2-Hydroxyestrone	2-OH-E1	286	0.2			14			<8>		
4-Hydroxyestrone	4-OH-E1	286	0.2			50			<8>		
16-Keto-17β-estradiol	16-keto-E2	286	0.2			4,200			<8>		
16α-Hydroxyestrone	16α-OH-E1	286	0.2			4,800			<8>		
2-Hydroxy-17β-estradiol	2-OH-E2	288	0.2			14			<8>		
4-Hydroxy-17β-estradiol	4-OH-E2	288	0.2			30					
17-Epiestriol	17-epi-E3	288	1.0						<8>		
Estriol	E3	288	0.5			15,700	50	3	1 (8)		
16-Epiestriol	16-epi-E3	288	1.0			6,600			<8>		
4-Methoxyestrone	4-MeO-E1	300				8,700			<8>		
4-Methoxy-17β-estradiol	4-MeO-E2	302	0.5			19,200			<8>		
2-Methoxyestrone	2-MeO-E1	300	0.1			32,400			<8>		
3-Methoxyestrone	3-MeO-E1	300							<8>		
2-Methoxy-17β-estradiol	2-MeO-E2	302	0.1			61,500			<8>		
2-Hydroxyestriol	2-OH-E3	304	0.5								
Progestagens											
Progesterone	Prog	314	1.0					0.3	50		
5α-Dihydroprogesterone	5α-DHP	316	0.5								
5β-Dihydroprogesterone	5β-DHP	316	0.2								

20α-Hydroprogesterone	20α-DHP	316	0.5			
Pregnenolone	Preg	316	2.0			
Pregnanolone	P-one	318	2.0			
Allopregnanolone	Allo-P-one	318	0.5			
Epipregnanolone	Epi-P-one	318	2.0			
Isopregnanolone	Iso-P-one	318				
Pregnanediol	P-diol	320	1.0	0.3	50	
17α-Hydroxyprogesterone	17α-OH-Prog	330	0.5			
11β-Hydroxyprogesterone	11β-OH-Prog	330	5.0			250
17α-Hydroxypregnenolone	17α-OH-Preg	332	10.0			
21-Hydroxypregnenolone	21-OH-Preg	332				
Pregnanetriol	P-triol	336	0.5			
Pregnanediol glucuronids	P-diol-Gluc	496		0.5		
Corticoids						
11-Deoxycorticosterone	11-deoxyB	330				
5β-Dihydrodeoxycorticosterone	5βDHDOC	332				
Tetrahydrodeoxycorticosterone	THDOC	334				
11-Dehydrocorticosterone	11-dehydroB	344				
11-Deoxycortisol	11-deoxyF	346				50
21-Deoxycortisol	21-deoxyF	346				
Corticosterone B		346				
Allodihydrocorticosterone	Allo-DHB	348				

(continued)

Table 4 (continued)

Compound	Abbreviation	FW	Method sensitivity, LOQ (S/N≥5)								
			GC/MS	GC/NCI/ MS/MS	LC/APCI$^+$/ MS/MS	LC/APCI$^-$/ MS/MS	LC/ESI$^-$/ MS/MS	LC/ESI$^+$/ MS/MS	LC/ESI$^+$/ MS/MS	LC/APPI$^+$/ MS/MS	LC/APPI$^-$/ MS/MS
			Urine, 2 mL	Serum, 1.0 mL	Prostatic tissue	Serum, 0.5 mL	Milk, 10 mL	Urine, 0.2 mL	Serum	Serum, 0.2 mL	Serum, 0.2 mL
			MSTFA derivatives (ng/mL)	PFB, MSTFA (pg/mL)	carbonyl derivative (ng/g) (LOD)	Underivatized (ng/L) (LOD)	Underivatized (pg/L)	Underivatized (ng/mL)	NH_2OH derivative (pg/mL)	Underivatized (pg/mL)	Underivatized (pg/mL)
Tetrahydrocorticosterone	THB	350									
Tetrahydrodeoxycortisol	THS	350							5,000		
Cortisone E		360							5,000		
Cortisol F		362						0.3			
20α-Dihydrocortisone	20α-DHE	362									
20α-Dihydrocortisol	20α-DHF	364									
Allodihydrocortisol	Allo-DHF	364									
Tetrahydrocortisone	THE	364									
Tetra-11-dehydrocorticosterone	THA	364									
α-Cortolone		366									
β-Cortolone		366									
Tetrahydrocortisol	THF	366						1			
Allotetrahydrocortisol	Allo-THF	366						3			
α-Cortol		368						3			
β-Cortol		368									
Sterols											
Cholestenone		384									
Desmolesterol		384									

Cholesterol	Chol	386	2.0							
20α-Hydroxycholesterol	20α-OH-Chol	402								
24S-Hydroxycholesterol	24S-OH-Chol	402	2.0							
Lanosterol		428								
Reference			[6]	[23]	[16]	[21]	[52]	[20]	[2]	[22]

* <8> is 8 pg/mL in serum and derivatized with densyl chloride [8].

steroid hormones in human plasma [32], urine [7, 37, 39], and environmental water [34, 45, 78, 79]. If both LC-MS/MS and GC-MS/MS methodologies require similar sample preparation procedures, e.g., deconjugation, extraction, and derivatization, and they provide the same or comparable sensitivity in steroid hormone analyses, the choice of an LC-MS/MS or a GC-MS/MS methodology may depend on the instrument availability and the cost of the testing.

3.2 LC-MS/MS and GC-MS vs. Immunoassays

IA and RIA are broadly used technologies in steroid hormone analyses by both clinical laboratories and environmental agencies, because they are sensitive, convenient, simple, rapid, and inexpensive. Their disadvantage is lack of selectivity or specificity [4, 14, 15]. Since selectivity or specificity of bioanalytical methodologies is essential for clinical diagnosis testing, a number of studies have been conducted to compare LC-MS/MS and GC-MS methodologies with immunoassays in steroid hormone analysis. A study for determining nine androgens and three estrogens in blood showed that LC-MS/MS, GC-MS, and RIA provided similar results, however, RIA resulted in significant higher values for all related hormones than LC-MS/MS and GC-MS, due to its cross interferences from analogue compounds [80]. Another study also indicated that direct RIA overestimated estrogen sulfate in plasma than GC-MS and LC-MS/MS [65]. Similar results were observed in analysis of estrogens by IA and RIA against LC-MS/MS as well [19]. All these studies suggest that LC-MS/MS and GC-MS are more selective or specific and accurate than RIA. Nevertheless, LC-MS/MS and GC-MS instruments require more sophisticated expertise to perform the bioanalytical testing, and to provide reliable interpretations of the results for clinical diagnoses.

4 Standardization of Analytical Procedures

Development of LC-MS/MS or GC-MS analytical methods is just the first step in bioanalysis. A number of additional procedures should also be established to make the whole bioanalytical platform including reference standards, methodologies and data systems in compliance with scientific and regulatory guidelines or requirements. Otherwise, the results from various techniques, e.g., IA, RIA, LC-MS/MS, and GC-MS, may not be comparable and acceptable for clinical diagnosis. As there are a large number of scientific publications and clinical applications on steroid hormone analyses, organizations like the Center for Disease Control and Prevention, Division of Laboratory Science of National Center for Environmental Health have attempted to standardize the clinical laboratory practice, analytical methodology and data management in steroid hormone analyses (CDC/NCEH/DLS, [81–83]). The National Institute of Standards and Technology (NIST) has made efforts on development of standard methods and reference materials for the determination of hormones in human serum [84–86]. A number of researchers also proposed a series of standardization procedures, including method development, validation, data interpretation and correlation between different bioanalytical techniques, and establishing international databases of human hormones [2, 14, 18]. In contrast to the clinical laboratories, EPA considers steroid hormones as environmental pollutants in water and sediments, and it has established an LC-MS/MS Method #539 following EPA Chemical QC Guidelines [77].

4.1 Guidelines for GLP, Reference Standards, and Analytical Methodology

Current Good Laboratory Practice from Food and Drug Administration (FDA GLP, 21CFR58) is a general guideline for pharmaceutical industry, and Good Laboratory Practice from Environmental Protection Agency (EPA GLP, 40CFR792) is a guideline for agriculture chemical industry, and environmental protection agencies and organizations. Both guidelines emphasize on compliance in personnel, facilities, articles, and documentation for animal and analytical experiments. In order to standardize the analytical methodology of steroid hormone analysis, the clinical and the environmental laboratories and facilities should follow FDA or EPA GLP guidelines. These guidelines include the following major requirements: (1) to train related analysts, (2) to calibrate and maintain instruments, (3) to characterize reference standard, (4) to conduct analytical testing following standard operation procedures (SOPs), (5) to record analytical procedures, protocols, reports, deviations, investigation, etc. Besides GLPs, guidelines for reference standards and analytical methodology are also important for steroid hormone analyses, as summarized in Table 5.

4.2 Reference Standards

The primary reference standards, including internal standards, of steroid hormones used for LC-MS/MS or GC-MS analyses in GLP laboratories should be obtained from authentic sources, e.g., US Pharmacopeia (USP) and NIST. Relative inexpensive secondary or working standards for daily testing may be used as alternatives of the primary standards, and they may be obtained from commercial sources. These working standards should be characterized against the primary reference standards using compendia or validated methods before being applied for GLP testing purposes. A reference standard program should be established in each laboratory, institute, or company to monitor the specification, quality, characterization, stability, storage, inventory, and replacement of those reference standards. A certificate of analysis should be issued after characterization of each standard batch, and it should include the information of manufacturer or source, date of manufacture, date of

Table 5 Guidelines for cGLP, reference standards, and analytical methodology

Guideline	Major topics
FDA GLP (21CFR58)	Ensures the quality and integrity of test data, personnel, facilities, articles, and documentation in pharmaceutical industry
EPA GLP (40CFR792)	Ensures the quality and integrity of test data, personnel, facilities, articles, and documentation in agriculture chemical industry and environmental protection organizations
USP34-NF29 S1 <11>	*General Requirements*/<11> USP Reference Standards—definition and applications
ISO guides for reference materials	ISO Guide 30:1992/Amd 1:2008 Revision of definitions for reference material and certified reference material ISO Guide 31:2000 Reference materials—Contents of certificates and labels ISO Guide 32:1997 Calibration in analytical chemistry and use of certified reference materials ISO Guide 33:2000 Uses of certified reference materials ISO Guide 34:2009 General requirements for the competence of reference material producers ISO Guide 35:2006 Reference materials—General and statistical principles for certification
USP34-NF29 S1 <1225>	Validation of compendial procedures
ICH Q2(1)	Validation of analytical procedures
IUPAC method validation	International Union of Pure and Applied Chemistry, Harmonized guidelines for single laboratory validation of method of analysis
NIST Tech Note 1297	Guidelines for evaluating and expressing the uncertainty of NIST measurement results
ISO 15193—Bioanalytical procedures	ISO 15193:2009 In vitro diagnostic medical devices—Measurement of quantities in samples of biological origin—Requirements for content and presentation of reference measurement procedures
ISO 5725—Accuracy (trueness and precision) of measurement methods and results	ISO 5725-1:1994 General principles and definitions ISO 5725-2:1994 Basic method for the determination of repeatability and reproducibility of a standard measurement method ISO 5725-3:1994 Intermediate measures of the precision of a standard measurement method ISO 5725-4:1994 Basic methods for the determination of the trueness of a standard measurement method ISO 5725-5:1998 Alternative methods for the determination of the precision of a standard measurement method ISO 5725-6:1994 Use in practice of accuracy values
FDA bioanalytical method validation	Guidance for Industry, Bioanalytical Method Validation
NIST methods and Reference Materials	Development of reference methods and reference materials for the determination of hormones in human serum
EPA	EPA Chemical QC Guidelines EPA method 539: Determination of hormones in drinking water by solid phase extraction (SPE) and liquid chromatography electrospray ionization tandem mass spectrometry (LC-MS/MS) [77]

analysis, testing results (purity or strength, moisture, impurities, etc.), storage conditions, and expiration or retesting date. More detailed information on reference standard can be found in USP34-NF29 and International Organization for Standardization (ISO) Guides 30-35.

4.3 Standardization of Methodology

In general, the analytical methodology evaluates and defines bioanalytical/analytical procedures using a number of parameters, including selectivity or specificity, accuracy, precision, matrix effect, recovery, calibration model (linearity) and range, sensitivity (limit of quantitation—LOQ, and limit of detection—LOD), sample stability, ruggedness and robustness [87]. The USP and ICH guidelines listed in Table 4 focus on the role of analytical methodology in quality control and compliance of drug substances and drug products in pharmaceutical industry. The IUPAC, ISO, and NIST 1297 guidelines emphasize on the definition and evaluation of trueness and uncertainty in analytical methodology. The FDA guideline is the most direct guidance on bioanalysis for human and nonhuman (animal and biological) studies in pharmacology, toxicology, pharmacokinetics, and drug metabolism. Typical bioanalytical methodology guidelines and LC-MS/MS methods for steroid hormone analyses are shown in Table 6.

To standardize the steroid hormone analyses, standard operation procedures (SOPs) should be generated for all laboratory functions, e.g., facilities, instrument, reference standards, samples, procedures, data collection and processing, documentation, etc., following GLP, reference standard, and the methodology guidelines. All the laboratory activities should follow these SOPs. Furthermore, the experimental procedures or techniques, e.g., deconjugation, extraction (LLE or SPE), derivatization, isotope dilution, instrument setting (e.g., LC-MS/MS modes), etc., should also be standardized. It is a challenge to ask different laboratories meet the same criteria, e.g., sensitivity and precision, using different models of instruments, and to make the results comparable and reliable for clinical diagnosis, due to the significant differences in LC-MS/MS hardware (configuration, ionization modes, and parameter setting) and software (data acquisition speed and processing) from different vendors and different models.

Standardization of GC-MS methodology for steroid hormones is relatively straightforward, because most of the unconjugated steroid hormones have a hydroxyl group(s) at C_3, C_{11}, C_{17}, or C_{21} position, which may be easily derivatized with a trimethylsilylation agent, e.g., MSTFA, and derivatives may be well separated by GC, and detected by MS using EI^+ mode [6]. Similarly, standardization of LC-MS/MS methodology for estrogens and metabolites is also not very complicated, because all unconjugated estrogens and metabolites have phenolic hydroxyl group(s) at C_3, and C_2 or C_4 positions, which may be derivatized with a sulfonyl chloride, e.g., dansyl chloride, and analyzed LC-MS/MS using ESI^+ [8, 84, 88]. The other derivatization agents, e.g., hydrazide and hydroxylamine, may react only with those ketolic

Table 6 Typical bioanalytical methodology guidelines and LC-MS/MS methods for steroid hormone analyses

Parameter	FDA Guidance for bioanalytical method validation (http://fda.gov/edcr/guidance/index.htm)	EPA Method 539—7 hormones in drinking water by LC/ESI$^{+/-}$-MS/MS (http://water.epa.gov/drink/)	LC/ESI^{+}-MS/MS methods for hormones in serum following NIST 297 and ISO15193 guidelines [84, 85]	An LC/ESI^{+}-MS/MS method for estrogens in plasma following IUPAC guideline [88]
Specificity	No interference from biological matrix endogenous compounds and metabolites	Interference from reagent blank ≤ 1/3MRL	Endogenous hormones were subtracted from the blank	>95 % certainty of no analyte in blank (n = 10)
Accuracy	Nominal ±15 % Nominal ±20 % at LLOQ	Nominal ±30 % at >2× MRL Nominal ±50 % at ≤2× MRL	100.7–101.8 % for estradiol 100.0–100.3 % for testosterone	93–105 %
Precision	%RSD ≤ 15 % at ≥ 3 levels n ≥ 5 at each level %RSD ≤ 20 % at LLOQ	%RSD ≤ 20 % n = 4–7	CV ≤ 2.2 % Uncertainty evaluation: Coefficient of variation (CV) 1 % Uncertainty of volumetric error 0.3 % Uncertainty of reference standard 0.1 % Uncertainty of weighing 0.5 % Uncertainty of other systematic errors Combined standard uncertainty Coverage factor Expanded uncertainty Relative expanded uncertainty (%)	Intraassay CV = 2–9 % Interassay CV = 4–12 %
Calibration	≥6 (6–8) standard points spiked in matrix including LLOQ A function of analyte vs. response within the calibration range	≥5 standard points spiked in 50%MeOH/reagent water The lowest at or below MRL Linear or quadratic regression	six points calibration curve	Seven points calibration curve
QC samples	Replicates at three levels, e.g., LQC (3× LOQ), MQC and HQC 67 % (4 out of 6) within ± 15 % of nominal value	Single run of LQC, MQC, … and HQC LQC within ± 50 % of nominal value, and others within ± 30 %	Single run of working standards at two levels	NA

Sensitivity	LLOQ=S/N≥5 Accuracy: 80–120 %	MRL is confirmed by: Upper PIR=(Mean+HR_{PIR})/fortified concentration×100≤150 % Lower PIR=(Mean−HR_{PIR})/fortified concentration×100≥50 %	LOD=S/N=3	LOQ=the lowest measureable concentration of actual sample with interassay CV<20 %
Sample stability	3× Freeze–thaw cycles for samples between −70 °C and room temperature; 4–24 h at room temperature; long-term storage; sample stock and postpreparation stability.	5–10 ng/L×6 at≤6 °C for 28 days	NA	3× Freeze–thaw cycles for samples between −20 °C and room temperature; Ambient and 4 °C for 7 days
Internal standard	Isotope labeled analyte with optimized binding assessment, and verified standard curve	IS peak area within ±50 % of each analyte $^{13}C_6$-Estradiol, 16α-Hydroxyestradiol-d_2, Testosterone-d_3, $^{13}C_2$-Etheylnylestradiol	IS peak area within±50 % of each analyte Testosterone-d_3, Estradiol-d_3	Estron-d_4, Estradiol-d_5
QC samples	Replicates at three levels, e.g., LQC (3× LOQ), MQC and HQC 67 % (4 out of 6) within±15 % of nominal value	Single run of LQC, MQC, … and HQC LQC within±50 % of nominal value, and others within±30 %	Single run of working standards at two levels	NA
Sensitivity	LLOQ=S/N≥5 Accuracy: 80–120 % Precision: ≤20 %	MRL is confirmed by: Upper PIR=(Mean+HR_{PIR})/fortified concentration×100≤150 % Lower PIR=(Mean−HR_{PIR})/fortified concentration×100≥50 % HR_{PIR}=3.963 S	LOD=S/N=3	LOQ=the lowest measureable concentration of actual sample with interassay CV<20 %

(continued)

Table 6 (continued)

Parameter	FDA Guidance for bioanalytical method validation (http://fda.gov/edcr/guidance/index.htm)	EPA Method 539—7 hormones in drinking water by LC/ESI$^{+/-}$-MS/MS (http://water.epa.gov/drink/)	LC/ESI^{+}-MS/MS methods for hormones in serum following NIST 297 and ISO15193 guidelines [84, 85]	An LC/ESI^{+}-MS/MS method for estrogens in plasma following IUPAC guideline [88]
Sample stability	3× Freeze–thaw cycles for samples between −70 °C and room temperature; 4–24 h at room temperature; long-term storage; sample stock and postpreparation stability.	5–10 ng/L × 6 at ≤ 6 °C for 28 days	NA	3× Freeze–thaw cycles for samples between −20 °C and room temperature; Ambient and 4 °C for 7 days
Internal standard	Isotope labeled analyte with optimized binding assessment, and verified standard curve	IS peak area within ±50 % of each analyte $^{13}C_6$-Estradiol, 16α-Hydroxyestradiol-d_2, Testosterone-d_3, $^{13}C_2$-Etheylnylestradiol	IS peak area within ± 50 % of each analyte Testosterone-d_3, Estradiol-d_3	Estron-d_4, Estradiol-d_5
Sample preparation	Extraction recovery (at three levels) of analyte and internal standard should be consistent, precise, and reproducible.	SPE, without derivatization	SPE + LLE (≥80 %), no derivatization for testosterone, Estrogens derivatized with dansyl chloride	LLE (100 %), derivatized with dansyl chloride

steroids, but will miss those nonketolic steroids [4, 25, 56]. If steroid hormones are analyzed without derivatization, they may need to be analyzed by LC-MS/MS using both positive and negative modes for different hormones, and the LOQ values of steroid hormones fall in a very broad range, because of the sensitivity variations of steroid hormones to different MS modes [22, 29, 77].

5 Applications of Mass Spectrometry in Steroid Hormones Analyses

5.1 Clinical Chemistry and Food Safety

Varieties of LC-MS/MS and GC-MS technologies and methodologies have been developed and applied for clinical chemistry diagnostic testing. For example, adrenal steroids, glucocorticoids, androgens, and estrogens are biomarker of many endocrinology diseases [2], pediatric development syndromes [9], and congenital adrenal hyperplasia [89]. Endogenous steroids are also related to prevalent cardiovascular disease in old men and women [90], and other age-related diseases [60]. Elevated estrogens and metabolites in plasma and serum of postmenopausal women are used as biomarkers for risk assessment of breast, ovary, and thyroid cancers, bone homeosis and osteoporosis in postmenopausal women [4, 15, 39, 91]. Androgens may be related to prostate cancer progression and treatment [31], and hyperandrogenism may cause polycystic ovary syndrome and androgen-secreting tumors [32].

In addition, steroid hormones in many kinds of biological fluids and tissues have been determined by LC-MS/MS and GC-MS technologies and methodologies. For example, LC-MS/MS has been utilized for monitoring (1) plasma corticosteroids and metabolites to evaluate their therapeutic and side effects as clinically used medicines [92], (2) estrogens in human cerebrospinal fluid [93] and peritoneal fluid [94], (3) urinary endogenous estrogen metabolites [95, 96], (4) estrogens in breast tissue [26], and (5) steroid hormones as residuals in edible matrices [12]. Polyphenol phytoestrogens in foods and human biological fluids are also measured by mass spectrometry technologies [97].

5.2 Antidoping Steroid Screen for Athletes and Racing Horses

Corticosteroids are used by some athletes and racing horses to enhance their performances. In order to prohibit drug doping, many sport organizations, e.g., International Olympic Committee, attempt to monitor the corticosteroids in urine of athletes. An example of earlier GC-MS method in steroid screen consists of procedures of deconjugation of glucuronide and sulfate, derivatization with MSTFA and GC-MS analysis [98]. The more recent LC-MS/MS methods separate the corticosteroids by

LC, and analyze them with ESI-MS (either positive or negative mode) without derivatization [10, 99]. Similarly, corticosteroids in racing horse urine are monitored by LC-ESI-MS/MS as well [11].

5.3 *Determination of Steroid Hormones as Environmental Pollutants*

Steroid hormones are found as pollutants in drinking water, waste water, river and sediments. The major concerns of analytical methodologies for monitoring steroid hormones from environmental samples are extraction techniques from aqueous or solid matrices. Since sample volume or amount is not an issue in most cases, SPE is the method of choice. Both LC-MS/MS and GC-MS technologies are broadly applied for steroid analyses of environmental samples, such as LC-MS/MS analyses of steroid hormones in effluents of wastewater treatment plants [100] and estrogens in water [101, 102], and GC-MS analyses of steroid hormones in environmental water [34, 45, 78, 79]. A study by Grover and colleagues showed that GC-MS was the simplest technique in determination of steroid hormones in environmental water samples, but lack of sensitivity; LC-MS/MS was more sensitive than GC-MS, but susceptible to matrix interferences; and GC-MS/MS was the recommended technique, because it was more selective and sensitive than GC/MS and LC-MS/MS [103].

6 Summary

LC-MS/MS has similar sensitivities in analysis of steroid hormones as IA, RIA, and GC-MS/MS, but LC-MS/MS and GC-MS/MS have much higher selectivity or specificity. The steroid hormones from different sources or matrices should be extracted with LLE or SPE, depending on the sample volume and matrices. Isotope (as internal standard) dilution is a standard procedure for quantitative analysis of steroid hormone and metabolites. Deconjugation is required in order to determine the total steroid hormones, because the very low levels of unconjugated and conjugated steroid hormones in biomatrices may not be feasible to be analyzed at the same time by LC-MS/MS, GC-MS, or GC/MS/MS. Derivatization can enhance MS detection sensitivity for many steroid hormones and metabolites, while the derivatization reagents and procedures should be selected based on the techniques of LC-MS/MS, GC-MS, or GC/MS/MS, and their ionization modes. Depending on the objectives of the method applications, bioanalytical procedures should be developed and validated according to the scientific and regulatory guidelines, e.g., FDA, EPA, ISO, NIST, IUPAC, and then these procedures may be standardized following CDC/NCEH/DLS practices. LC-MS/MS is increasingly applied for steroid analysis in research, clinical, pharmaceutical, and food industries, sports and environmental testing, due to its selectivity, sensitivity, and versatility.

References

1. Shimada K, Mitamura K, Higashi T (2001) Gas chromatography and high-performance liquid chromatography of natural steroids. J Chromatogr A 935:141–172
2. Kushnir MM, Rockwood AL, Bergquist J (2010) Liquid chromatography–tandem mass spectrometry applications in endocrinology. Mass Spectrom Rev 29:480–502
3. Soldin SJ, Soldin OP (2009) Steroid hormone analysis by tandem mass spectrometry. Clin Chem 55:1061–1066
4. Blair IA (2010) Analysis of estrogens in serum from postmenopausal women: past, present and future. Steroids 75:297–306
5. Rathahao E, Page A, Jouanin I, Paris A, Debrauwer L (2004) Liquid chromatography coupled to negative electrospray/ion trap mass spectrometry for the identification of isomeric glutathione conjugates of catechol estrogens. Int J Mass spectrom 231:119–129
6. Moon J, Jung H, Moon M, Chung B, Choi M (2009) Heat-map visualization of gas chromatography-mass spectrometry based quantitative signatures on second steroids metabolism. J Am Soc Mass Spectrom 20:1626–1637
7. Gomes RL, Meredith W, Snape CE, Sephton MA (2009) Analysis of conjugated steroid androgens: deconjugation, derivatization and associated issues. J Pharm Biomed Anal 49:1133–1140
8. Xu X, Roman JM, Issaq HJ, Keefer LK, Veenstra TD, Ziegler RG (2007) Quantitative measurement of endogenous estrogens and estrogen metabolites in human serum by liquid chromatography-tandem mass spectrometry. Anal Chem 79:7813–7821
9. Rauh M (2010) Steroid measurement with LC–MS/MS. Application examples in pediatrics. J Steroid Biochem Mol Biol 121:520–527
10. Deventer K, Delbeke FT (2003) Validation of a quantitative screening method for corticosteroids by liquid chromatography tandem mass spectrometry. In: Shanzer W, Geyer H, Gotzmann A, Mareck U (eds) Recent advances in doping analysis, vol 11, Spert and Buch Straub, Köln, pp 23–31
11. Leung GNW, Chung EW, Ho ENM, Kwok WH, Leung DKK, Tang FPW, Wan TSM, Yu NH (2005) High throughput screening of corticosteroids and basic drugs in horse urine by liquid chromatography-tandem mass spectrometry. J Chromatogr B 825:47–56
12. Noppe H, Le Bizec B, Verheyden K, De Brabander HF (2008) Novel analytical methods for the determination of steroid hormones in edible matrices. Anal Chim Acta 611:1–16
13. Pacakova V, Loukotkova L, Bosakova Z, Stulik K (2009) Analysis for estrogens as environmental pollutants—a review. J Sep Sci 32:867–882
14. Stanczyk FZ, Clarke NJ (2010) Advantages and challenges of mass spectrometry assays for steroid hormones. J Steroid Biochem Mol Biol 121:491–495
15. Giese RW (2003) Measurement of endogenous estrogens: analytical challenges and recent advances. J Chromatogr A 1000:401–412
16. Higashi T (2006) Trace determination of steroids causing age-related diseases using LC/MS combined with detection-oriented derivatization. Chem Pharm Bull 54:1479–1485
17. Higashi T, Shimada K (2004) Derivatization of neutral steroids to enhance their detection characteristics in liquid chromatography–mass spectrometry. Anal Bioanal Chem 378: 875–882
18. Stanczyk FZ, Lee JS, Santen RJ (2007) Standardization of steroid hormone assays: why, how, and when? Cancer Epidemiol Biomarkers Prev 16:1713–1719
19. Ziegler RG, Faupel-Badger JM, Sue L, Fuhrman BJ, Falk RT, Boyd-Morin J, Henderson MK, Hoover RN, Veenstra TD, Keefer LK, Xu X (2010) A new approach to measure estrogen exposure and metabolism in epidemiologic studies. J Steroid Biochem Mol Biol 121:538–545
20. Hauser B, Deschner T, Boesch C (2008) Development of a liquid chromatography–tandem mass spectrometry method for the determination of 23 endogenous steroids in small quantities of primate urine. J Chromatogr B 862:100–112

21. Yang W, Regnierb FE, Slivac D, Adame J (2008) Stable isotope-coded quaternization for comparative quantification of estrogen metabolites by high-performance liquid chromatography–electrospray ionization mass spectrometry. J Chromatogr B 870:233–240
22. Harwood DT, Handelsman DJ (2009) Development and validation of a sensitive liquid chromatography–tandem mass spectrometry assay to simultaneously measure androgens and estrogens in serum without derivatization. Clin Chim Acta 409:78–84
23. Santen RJ, Demers L, Ohorodnik S, Settlage J, Langecker P, Blanchettd D, Gosse PE, Wang S (2007) Superiority of gas chromatography/tandem mass spectrometry assay (GC/MS/MS) for estradiol for monitoring of aromatase inhibitor therapy. Steroids 72:666–671
24. Zhao M, Bakera SD, Yan X, Zhao Y, Wright WW, Zirkinb BR, Jarow JP (2004) Simultaneous determination of steroid composition of human testicular fluid using liquid chromatography tandem mass spectrometry. Steroids 69:721–726
25. Regal P, Vázquez BI, Franco CM, Cepeda A, Fente C (2009) Quantitative LC–MS/MS method for the sensitive and simultaneous determination of natural hormones in bovine serum. J Chromatogr B 877:2457–2464
26. Taioli E, Im A, Xu X, Veenstra TD, Ahrendt G, Garte S (2010) Comparison of estrogens and estrogen metabolites in human breast tissue and urine. Reprod Biol Endocrinol 8:93–99
27. Xia Y, Chang SW, Patel S, Bakhtiar R, Karanam B, Evans DC (2004) Trace level quantitation of deuterated 17b-estradiol and estrone in ovarietomized mouse plasma and brain using liquid chromatography/tandem mass spectrometry following dansylation reaction. Rapid Commun Mass Spectrom 18:1621–1628
28. Xu X, Veenstra TD, Fox SD, Roman JM, Issaq HJ, Falk R, Saavedra JE, Keefer LK, Ziegler RG (2005) Measuring fifteen endogenous estrogens simultaneously in human urine by high-performance liquid chromatography-mass spectrometry. Anal Chem 77:6646–6654
29. Hsu J, Chang Y, Chen T, Lin L, Liao P (2007) Evaluation of electrospray ionization and atmospheric pressure chemical ionization for simultaneous detection of estrone and its metabolites using high-performance liquid chromatography/tandem mass spectrometry. J Chromatogr B 860:49–56
30. Li Q, Lam MHW, Wu RSS, Jiang B (2010) Rapid magnetic-mediated solid-phase extraction and pre-concentration of selected endocrine disrupting chemicals in natural waters by poly(divinylbenzene-co-methacrylic acid) coated Fe_3O_4 core-shell magnetite microspheres for their liquid chromatography–tandem mass spectrometry determination. J Chromatogr A 1217:1219–1226
31. O'Brien Z, Post N, Brown M, Madan A, Coon T, Luo R, Kohout TA (2009) Validation and application of a liquid chromatography–tandem mass spectrometric method for the simultaneous determination of testosterone and dihydrotestosterone in rat prostatic tissue using a 96-well format. J Chromatogr B 877:3515–3521
32. Yokokawa A, Yamamoto K, Omori Y, Shibasaki H, Shinohara Y, Kasuya Y, Furuta T (2009) Simultaneous determination of androstenedione, 11b-hydroxyandrostenedione, and testosterone in human plasma by stable isotope dilution mass spectrometry. J Chromatogr B 877:621–626
33. Gomes RL, Avcioglu E, Scrimshaw MD, Lester JN (2004) Steroid estrogen determination in sediment and sewage sludge: a critique of sample preparation and chromatographic/mass spectrometry considerations, incorporating a case study in method development. Trends Anal Chem 23:737–744
34. Zuo Y, Zhang K, Lin Y (2007) Microwave-accelerated derivatization for the simultaneous gas chromatographic–mass spectrometric analysis of natural and synthetic estrogenic steroids. J Chromatogr A 1148:211–218
35. Xu X, Ziegler RG, Waterhouseb DJ, Saavedrac JE, Keefer LK (2002) Stable isotope dilution high-performance liquid chromatography–electrospray ionization mass spectrometry method for endogenous 2- and 4-hydroxyestrones in human urine. J Chromatogr B 780:315–330
36. Xu X, Keefer LK, Waterhouse DJ, Saavedra JE, Veenstra TD, Ziegler RG (2004) Measuring seven endogenous ketolic estrogens simultaneously in human urine by high-performance liquid chromatography-mass spectrometry. Anal Chem 76:5829–5836

37. Knust U, Strowitzki T, Spiegelhalder B, Bartsch H, Owen RW (2007) Optimization of an isotope dilution gas chromatography/mass spectrometry method for the detection of endogenous estrogen metabolites in urine samples. Rapid Commun Mass Spectrom 21:2245–2254
38. Moon J, Jung H, Moon M, Chung B, Choi M (2008) Inclusion complex-based solid-phase extraction of steroidal compounds with entrapped b-cycodextrin polymer. Steroids 73:1090–1097
39. Moon J, Kim K, Moon M, Chung B, Choi M (2011) A novel GC-MS method in urine estrogen analysis from postmenopausal women with osteoporosis. J Lipid Res 52:1595–1603
40. Kumar V, Nakada N, Yasojima M, Yamashita N, Johnson AC, Tanaka H (2009) Rapid determination of free and conjugated estrogen in different water matrices by liquid chromatography–tandem mass spectrometry. Chemosphere 77:1440–1446
41. Nguyen HP, Yang SH, Wigginton JG, Simpkins JW, Schug KA (2010) Retention behavior of estrogen metabolites on hydrophilic interaction chromatography stationary phases. J Sep Sci 33:793–802
42. Ramanathan R, Cao K, Cavalieri E, Gross ML (1998) Mass spectrometric methods for distinguishing structural isomers of glutathione conjugates of estrone and estradiol. J Am Soc Mass Spectrom 9:612–619
43. Reddy S, Iden CR, Brownawell BJ (2005) Analysis of steroid conjugates in sewage influent and effluent by liquid chromatography-tandem mass spectrometry. Anal Chem 77:7032–7038
44. Schlusener MP, Bester K (2005) Determination of steroid hormones, hormone conjugates and macrolide antibiotics in influents and effluents of sewage treatment plants utilizing high-performance liquid chromatography/tandem mass spectrometry with electrospray and atmospheric pressure chemical ionization. Rapid Commun Mass Spectrom 19:3269–3278
45. Liu Z, Kanjo Y, Mizutani S (2011) Removal of natural free estrogens and their conjugates in a municipal wastewater treatment plant. Clean – Soil, Air, Water 39:128–135
46. Hauser B, Mugisha L, Preis A, Deschner T (2011) LC–MS analysis of androgen metabolites in serum and urine from east African chimpanzees (Pan troglodytes schweinfurthii). Gen Comp Endocrinol 170:92–98
47. Tang PW, Law WC, Wan TSM (2001) Analysis of corticosteroids in equine urine by liquid chromatography-mass spectrometry. J Chromatogr B 754:229–244
48. Moon J, Ha Y, Moon M, Chung B, Choi M (2010) Systematic error in gas chromatography–mass spectrometry based quantitation of hydrolyzed urinary steroids. Cancer Epidemiol Biomarkers Prev 19:388–397
49. Zhang H, Henion J (1999) Quantitative and qualitative determination of estrogen sulfates in human urine by liquid chromatography/tandem mass spectrometry using 96-well technology. Anal Chem 71:3955–3964
50. Qin F, Zhao Y, Sawyer MB, Li X (2008) Hydrophilic interaction liquid chromatography-tandem mass spectrometry determination of estrogens conjugates in human urine. Anal Chem 80:3404–3411
51. Qin F, Zhao Y, Sawyer MB, Li X (2008) Column-switching reversed phase–hydrophilic interaction liquid chromatography/tandem mass spectrometry method for determination of free estrogens and their conjugates in river water. Anal Chim Acta 627:91–98
52. Tso J, Aga DS (2010) A systematic investigation to optimize simultaneous extraction and liquid chromatography tandem mass spectrometry analysis of estrogens and their conjugated metabolites in milk. J Chromatogr A 1217:4784–4795
53. Yan W, Zhao L, Feng Q, Wei Y, Lin J (2009) Simultaneous determination of ten estrogens and their metabolites in waters by improved two-step SPE followed by LC–MS. Chromatographia 69:621–628
54. Nishio T, Higashi T, Funaishi A, Tanaka J, Shimada K (2007) Develpment and application of electrospray-active derivatization reagents for hydroxysteroids. J Pharm Biomed Anal 44:786–795

55. Penning TM, Lee S, Jin Y, Gutierrez A, Blair IA (2010) Liquid chromatography-mass spectrometry (LC-MS) of teroid hormone metabolites and its applications. J Steroid Biochem Mol Biol 121:546–555
56. Johnson DW (2005) Ketosteroid profiling using Girard T derivatives and electrospray ionization tandem mass spectrometry: direct plasma analysis of androstenedione, 17-hydroxyprogesterone and cortisol. Rapid Commun Mass Spectrom 19:193–200
57. Hala D, Overturf MD, Petersen LH, Huggett DB (2011) Quantification of 2-hydrazinopyridine derivatized steroid hormones in fathead minnow (pimephales promelas) blood plasma using LC-ESI+/MS/MS. J Chromatogr B 879:591–598
58. Griffiths WJ, Liu S, Alvelius G, Sjovall J (2003) Derivatization for the characterisation of neutral oxosteroids by electrospray and matrix-assisted laser desporption/ionization tandem mass spectrometry: the Girard P derivatives. Rapid Commun Mass Spectrom 17:924–935
59. Arai S, Miyashiro Y, Shibata Y, Kashiwagi B, Tomaru Y, Kobayashi M, Watanabe Y, Honma S, Suzuki K (2010) New quantification method for estradioal in the prostatic tissues of nenign prostatic hyperplasia using liquid chromatogrophy-dandem mass spectromentry. Steroids 75:13–19
60. Higashi T, Takayama N, Kyutoku M, Shimada K, Kohb E, Namiki M (2006) Liquid chromatography–mass spectrometric assay of androstenediol in prostatic tissue: influence of androgen deprivation therapy on its level. Steroids 71:1007–1013
61. Lin Y, Chen C, Wang G (2007) Analysis of steroid estrogens in water using liquid chromatography/tandem mass spectrometry with chemical derivatizations. Rapid Commun Mass Spectrom 21:1973–1983
62. Xu L, Spink DC (2008) Analysis of steroidal estrogens as pyridine-3-sulfonyl derivatives by liquid chromatography electrospray tandem mass spectrometry. Anal Biochem 375:105–114
63. You J, Zhao H, Sun Z, Suo Y, Chen G (2009) 10-Ethyl-acridine-2-sulfonyl chloride: a new derivatization agent for enhancement of atmospheric pressure chemical ionization of estrogens in urine. Chromatographia 70:45–55
64. Yamashita K, Okuyama M, Watanabe Y, Honma S, Kobayashi S, Numazawa M (2007) Highly sensitive determination of estrone and estradiol in human serum by liquid chromatography-electrospray ionization tandem mass spectrometry. Steroids 72:819–827
65. Giton F, Caron P, Bérubé R, Bélanger A, Barbier O, Fiet J (2010) Plasma estrone sulfate assay in men: comparison of radioimmunoassay, mass spectrometry coupled to gas chromatography (GC–MS), and liquid chromatography-tandem mass spectrometry (LC-MS/MS). Clin Chim Acta 411:1208–1213
66. Mitamura K, Yatera M, Shimada K (2000) Studies on neurosteroids Part XIII. Characterization of catechol estrogens in rat brains using liquid chromatography-mass spectrometry-mass spectrometry. Analyst 125:811–814
67. Yamashita K, Masuda A, Hoshino Y, Komatsu S, Numazawa M (2010) Assay of labile estrogen *o*-quinones, potent carcinogenic molecular species, by high performance liquid chromatography–electrospray ionization tandem mass spectrometry with phenazine derivatization. J Steroid Biochem Mol Biol 119:141–148
68. Díaz-Cruz MS, López de Alda MJ, López R, Barceló D (2003) Determination of estrogens and progestogens by mass spectrometry techniques (GC/MS, LC/MS and LC/MS/MS). J Mass Spectrom 38:917–923
69. Cheng C, Hou J, Wang S, Xu B, Liu S, Yan Z (2011) Development and validation of an LC-MS/MS method for determination of fifteen estrogens and metabolites in human serum. In: Pittsburgh conference, Atlanta, GA, 18 March 2011
70. Nordstrom A, O'Maille G, Qin C, Siuzdak G (2006) Nonlinear data alignment for UPLC-MS and HPLC-MS based metabolomics: quantitative analysis of endogenous and exogenous metabolites in human serum. Anal Chem 78:3289–3295
71. Novakoca L, Matysova L, Solich P (2006) Advantages of application of UPLC in pharmaceutical analysis. Talanta 68:908–918

72. Zelena E, Dunn WB, Broadhurst D, Francis-McIntyre S, Carroll KM, Begley P, O'Hagan S, Knowles JD, Halsall A, Consortium H, Wilson ID, Kell DB (2009) Development of a robust and repeatable UPLC-MS method for the long-term metabolomic study of human serum. Anal Chem 81:1357–1364
73. Matějíček D (2011) On-line two-dimensional liquid chromatography–tandem mass spectrometric determination of estrogens in sediments. J Chromatogr A 1218:2292–2300
74. Lien G, Chen C, Wang G (2009) Comparison of electrospray ionization, atmospheric pressure chemical ionization and atmospheric pressure photoionization for determining estrogenic chemicals in water by liquid chromatography tandem mass spectrometry with chemical derivatizations. J Chromatogr A 1216:956–966
75. Xu X, Roman JM, Veenstra TD, Van Anda J, Ziegler RG, Issaq HJ (2006) Analysis of fifteen estrogen metabolites using packed column supercritical fluid chromatography-mass spectrometry. Anal Chem 78:1553–1558
76. Labadie P, Hill EM (2007) Analysis of estrogens in river sediments by liquid chromatography–electrospray ionisation mass spectrometry—comparison of tandem mass spectrometry and time-of-flight mass spectrometry. J Chromatogr A 1141:174–181
77. U.S. Environmental protection agency method 539: determination of hormones in drinking water by solid phase extraction (SPE) and liquid chromatography electrospray ionization tandem mass spectrometry (LC-MS/MS). EPA document No. 815-B-10-001, November 2010. http://www.epa.gov/drink/
78. Mouatassim-Souali A, Tamisier-Karolak S, Perdiz D, Cargouet M, Levi Y (2003) Validation of a quantitative assay using GC/MS for trace determination of free and conjugated estrogens in environmental water samples. J Sep Sci 26:105–111
79. Zhou Y, Zhou J, Xu Y, Zha J, Ma M, Wang Z (2009) An alternative method for the determination of estrogens in surface water and wastewater treatment plant effluent using pre-column trimethylsilyl derivatization and gas chromatography/mass spectrometry. Environ Monit Assess 158:35–49
80. Hsing AW, Stanczyk FZ, Bélanger A, Schroeder P, Chang L, Falk RT, Fears TR (2007) Reproducibility of serum sex steroid assays in men by RIA and mass spectrometry. Cancer Epidemiol Biomarkers Prev 16:1004–1008
81. Myers GL (2008) Introduction to standardization of laboratory results. Steroids 73: 1293–1296
82. Rosner W, Vesper H (2008) CDC workshop report improving steroid hormone measurements in patient care and translation research. Steroids 7:1285
83. Vesper HW, Botelho JC, Shacklady C, Smith A, Myers GL (2008) CDC project on standardizing steroid hormone measurements. Steroids 73:1286–1292
84. Tai SS-C, Welch MJ (2005) Development and evaluation of a reference measurement procedure for the determination of estradiol-17b in human serum using ID-LC/MS/MS. Anal Chem 77:6359–6363
85. Tai SS-C, Xu B, Welch MJ, Phinney KW (2007) Development and evaluation of a candidate reference measurement procedure for the determination of testosterone in human serum using isotope dilution liquid chromatography/tandem mass spectrometry. Anal Bioanal Chem 388:1087–1094
86. Tai SS-C, Xu B, Welch MJ (2006) Development and evaluation of a candidate reference measurement procedure for the determination of progesterone in human serum using ID-LC/MS/MS. Anal Chem 78:6628–6633
87. Peters FT, Maurer HH (2002) Bioanalytical method validation and its implications for forensic and clinical toxicology—a review. Accred Qual Assur 7:441–449
88. Nelson RE, Grebe SK, O'Kane DJ, Singh RJ (2004) Liquid chromatography mass spectrometry assay for simultaneous measurement of estradiol and estrone in human plasma. Clin Chem 50:373–384
89. Lai C, Tsai C, Tsai F, Wu J, Lin W, Lee C (2002) Monitoring of congenital adrenal hyperplasia by microbore HPLC–electrospray ionization tandem mass spectrometry of dried blood spots. Clin Chem 48:354–356

90. Naessen T, Sjogren U, Bergquist J, Larsson M, Lind L, Kushnir MM (2010) Endogenous steroids measured by high-specificity liquid chromatography-tandem mass spectrometry and prevalent cardiovascular disease in 70-year-old men and women. J Clin Endocrinol Metab 95:1889–1897
91. Santen RJ, Lee JS, Wang S, Demers LM, Mauras N, Wang H et al (2008) Potential role of ultra-sensitive estradiol assays in estimating the risk of breast cancer and fractures. Steroids 73:1318–1321
92. Ionita IA, Akhlaghi F (2010) Quantification of unbound prednisolone, prednisone, cortisol and cortisone in human plasma by ultrafiltration and direct injection into liquid chromatography tandem mass spectrometry. Ann Clin Biochem 47:350–357
93. Nguyen HP, Li L, Gatson JW, Maass D, Wigginton JG, Simpkins JW, Schug KA (2011) Simultaneous quantification of four native estrogen hormones at trace levels in human cerebrospinal fluid using liquid chromatography–tandem mass spectrometry. J Pharm Biomed Anal 54:830–837
94. Xu X, Othman ER, Issaq HJ, Hornung D, Al-Hendy A, Veenstra TD (2008) Multiplexed quantitation of endogenous estrogens and estrogen metabolites in human peritoneal fluid. Electrophoresis 29:2706–2713
95. Nielen MWF, van Bennekom EO, Heskamp HH, van Rhijn JA, Bovee TFH, Hoogenboom LAP (2004) Bioassay-directed identification of estrogen residues in urine by liquid chromatography electrospray quadrupole time-of-flight mass spectrometry. Anal Chem 76:6600–6608
96. Xu X, Keefer LK, Ziegler RG, Veenstra TD (2007) A liquid chromatography–mass spectrometry method for the quantitative analysis of urinary endogenous estrogen metabolites. Nat Protoc 2:1350–1355
97. Wilkinson AP, Wahala K, Williamson G (2002) Identification and quantitation of polyphenol phytoestrogens in foods and human biological fluids. J Chromatogr B 777:93–109
98. Bowers LD (1997) Analytical advances in detection of performance-enhancing compounds. Clin Chem 43:1299–1304
99. Peng L, Farcase T, McGinley Identification of steroids in urine and plasma by LC/MS/MS using strata X and Gemini C18. Application note: TN-1026. www.phonomenex.com
100. Ingrand V, Herry G, Beausse J, de Roubin M (2003) Analysis of steroid hormones in effluents of wastewater treatment plants by liquid chromatography-tandem mass spectrometry. J Chromatogr A 1020:99–104
101. Habauzit D, Armentgaud J, Roig B, Chopineau J (2008) Determination of estrogen presence in water by SPR using estrogen receptor dimerization. Anal Bioanal Chem 390:873–883
102. Miége C, Bados P, Brosse C, Coquery M (2009) Method validation for the analysis of estrogens (including conjugated compounds) in aqueous matrices. Trends Anal Chem 28:237–244
103. Grover DP, Zhanga ZL, Readman JW, Zhou JL (2009) A comparison of three analytical techniques for the measurement of steroidal estrogens in environmental water samples. Talanta 78:1204–1210

Guidelines

International Conference on Harmonisation of Technical Requirements for Registrations of Pharmaceuticals for Human Use, ICH Harmonised Tripartite Guideline, Validation of Analytical Procedures: Text and Methodology, Q2(R1), Current Step 4 version, 2005, 13 pp. http://www.ich.org/products/guidelines/quality/article/quality-guidelines.html
U.S. Food and Drug Administration CFR—Code of Federal Regulations Title 21, Part 58—Good laboratory practice for nonclinical laboratory studies. http://www.accessdata.fda.gov/scripts/cdrh/cfdocs/cfCFR/CFRSearch.cfm?CFRPart=58

U.S. Food and Drug Administration, Center for Drug Evaluation and Research (CDER), Center for Veterinary Medicine (CVM), May 2001, Guidance for industry—bioanalytical method validation. http://www.fda.gov/Drugs/GuidanceComplianceRegulatoryInformation/Guidances/ucm064964.htm

The United States Pharmacopeia–National Formulary (USP–NF), USP34-NF29 S1. *General Requirements/*<11>USP Reference Standards, pp 38–40. Pharmacopeial Forum 35(6):1507

The United States Pharmacopeia–National Formulary (USP–NF), USP34-NF29 S1. *General Information/*<1225>Validation of Compendial Procedures, pp 779–782. Pharmacopeial Forum 35(2):444

U.S. Environmental protection agency CFR—Code of Federal Regulations Title 40—Protection of Environment PART 792—Good laboratory practice standards. http://www.access.gpo.gov/nara/cfr/waisidx_07/40cfr792_07.html

U.S. Environmental protection agency, Chemical QC guidelines. http://www.epa.gov/sam/qc-chem.htm

The National Institute of Standards and Technology, Analytical chemistry division, Development of Reference Methods and Reference Materials for the Determination of Hormones in Human Serum. http://www.nist.gov/mml/analytical/organic/hormonesinserum.cfm

The National Institute of Standards and Technology, Taylor BN, Kuyatt CE (1994) Guidelines for evaluating and expressing the uncertainty of NIST measurement results. NIST Technical Note 1297, pp 1–20. http://www.nist.gov/pml/pubs/tn1297/index.cfm

Thomson M, Ellison SLR, Wood R (2002) International union of pure and applied chemistry, harmonized guidelines for single laboratory validation of method of analysis. Pure Appl Chem**74**:835–855. http://www.iupac.org/objID/Article/pac7405x0835

International Organization of Standardization, Guide:

ISO Guide 30:1992 Terms and definitions used in connection with reference materials

ISO Guide 30:1992/Amd 1:2008 Revision of definitions for reference material and certified reference material

ISO Guide 31:2000 Reference materials—contents of certificates and labels

ISO Guide 32:1997 Calibration in analytical chemistry and use of certified reference materials

ISO Guide 33:2000 Uses of certified reference materials

ISO Guide 34:2009 General requirements for the competence of reference material producers

ISO Guide 35:2006 Reference materials—general and statistical principles for certification. http://www.iso.org/iso/search.htm?qt=guide+35&searchSubmit=Search&sort=rel&type=simple&published=on

International Organization of Standardization, Published:

ISO 5725-1:1994 Accuracy (trueness and precision) of measurement methods and results—Part 1: general principles and definitions

ISO 5725-2:1994 Accuracy (trueness and precision) of measurement methods and results—Part 2: basic method for the determination of repeatability and reproducibility of a standard measurement method

ISO 5725-3:1994 Accuracy (trueness and precision) of measurement methods and results—Part 3: intermediate measures of the precision of a standard measurement method

ISO 5725-4:1994 Accuracy (trueness and precision) of measurement methods and results—Part 4: basic methods for the determination of the trueness of a standard measurement method

ISO 5725-5:1998 Accuracy (trueness and precision) of measurement methods and results—Part 5: alternative methods for the determination of the precision of a standard measurement method
ISO 5725-6:1994 Accuracy (trueness and precision) of measurement methods and results—Part 6: use in practice of accuracy values. http://www.iso.org/iso/search.htm?qt=5725&published=on&active_tab=standards

Liquid Chromatography-Mass Spectrometric Analysis of Tropane Alkaloids in Mammalian Samples: Techniques and Applications

Harald John

Abstract Tropane alkaloids (TA) are bioactive small molecules containing an esterified bicyclic tropane moiety. TA represent natural plant poisons and (semi) synthetic drugs, most of them (e.g. atropine, benzatropine, *N*-butyl scopolamine, cimetropium, homatropine, ipratropium, *N*-methyl scopolamine, scopolamine, tiotropium, trospium) antagonizing acetylcholine at muscarinic receptors (MR) and some of them (e.g. bemesetron, granisetron, scopolamine, tropisetron) antagonizing serotonin at the 5-HT_3 receptor (5-HT_3R). Therapeutic effects on MR include mydriasis, spasmolysis of the gastrointestinal tract, overactive bladder and of the respiratory system. Due to their binding to the 5-HT_3R, TA are used as antiemetic to treat vomiting and nausea. In addition, a few TA interact with α1-adrenoreceptors thus being used to improve blood flow for the treatment of septic shock (e.g. anisodamine, anisodine). Furthermore, ingestion of plants containing natural TA such as hyoscyamine and scopolamine may cause fatal intoxications.

Determination of TA is required for pharmacokinetic and distribution studies, to elucidate biotransformation in vivo and in vitro as well as to identify and quantify any poison in toxicological and forensic samples. For this purpose LC-MS-based methods are often applied. The present chapter comprehensively introduces and discusses diverse LC-MS procedures for 19 natural and synthetic TA that are of relevance as drug and poison. Individual compounds are briefly introduced providing basic pharmacological information. Their physico-chemical properties are addressed exemplifying the impact on sample preparation (e.g. precipitation, liquid–liquid extraction, solid-phase extraction) and chromatographic separation. Mass analyzers and scan modes used following electrospray and atmospheric pressure chemical ionization are commented and statistical evaluation of their frequency of use is illustrated. MS/MS-based metabolite identification strategies applied to TA

H. John (✉)
Bundeswehr Institute of Pharmacology and Toxicology,
Neuherbergstrasse 11, 80937 Munich, Germany
e-mail: HaraldJohn@bundeswehr.org

Q. Alan Xu and Timothy L. Madden (eds.), *LC-MS in Drug Bioanalysis*,
DOI 10.1007/978-1-4614-3828-1_11,

analysis are pointed out. At least diverse fields of applications are categorized to review several examples of PK, distribution, and biotransformation studies as well as toxicological analysis. Concluding remarks point out potential future trends and possibilities of LC-MS in clinical pharmacology.

Abbreviations

5-HT_3R	5-Hydroxytryptamine (serotonin) receptor
ACN	Acetonitrile
Ada	Anisodamine
Adi	Anisodine
AGP	Alpha-glycoprotein
APCI	Atmospheric pressure chemical ionization
API	Active pharmaceutical ingredient
Atr	Atropine
AtrE	Atropinesterase
BBB	Blood–brain barrier
Becg	Benzoylecgonine
Beme	Bemesetron
Benz	Benztropine
BuS	*N*-butyl-scopolamine
CID	Collision-induced dissociation
Cim	Cimetropium
CNS	Central nervous system
Coc	Cocaine
COPD	Chronic obstructive pulmonary disease
Da	Dalton
EBQ_1Q_2	Double focusing sector field mass spectrometer
EE	Ethylacetate
ESI	Electrospray ionization
Et_2O	Diethylether
FA	Formic acid
FAB	Fast atom bombardment
GIT	Gastrointestinal tract
glucu	Glucuronide conjugate
Gran	Granisetron
Hep-SA	Heptanesulfonic acid
HPLC	High-performance liquid chromatography
hyo	Hyoscyamine
IBS	Irritable bowel syndrome
Ipra	Ipratropium
iso	Isocratic

IT	Ion trap
lin range	Linear range
LLE	Liquid–liquid extraction
log P	Logarithm of octanol/water partition coefficient
MeOH	Methanol
Me-O-^{t}Bu	Methyl-^{tert}butyl ether
MeS	Methylscopolamine
MRM	Multiple reaction monitoring
MS	Full scan mass spectrometry
MS/MS	Tandem mass spectrometry, product ion scan
MW	Mono-isotopic molecular weight
n.s.	Not specified
OAc	Acetate
OP	Organophosphorus compound
QqQ	Triple quadrupole mass spectrometer
QTA	Quaternary tropane alkaloid
ref	Reference
Sat	Satropane
Scp	Scopolamine
SIM	Selected ion monitoring
Solv	HPLC solvent
SPE	Solid-phase extraction
SQ	Single quadrupole mass spectrometer
T	Temperature
TA	Tropane alkaloid
Tio	Tiotropium
Trop	Tropisetron
Tros	Trospium
TSP	Thermospray
TTA	Tertiary tropane alkaloids
WHO	World Health Organisation
α1-AR	α1-Adrenoreceptor
α7-nAChR	α7-Nicotinic receptor

1 Introduction

The present book chapter presents an actual summary on the use of LC-MS-based procedures applied to mammalian samples for qualitative and quantitative analysis of tropane alkaloids (TA). Relevant literature obtained from PubMed database search and references cited therein is summarized and commented to provide an overview on various natural TA structures and their related pharmaceutical derivatives. A comprehensive unique set of TA was compiled that is categorized by diverse fields of analytical applications.

The first main part of this chapter opens with general aspects of chemical TA structures and the resulting physico-chemical properties, which determine the adequate choice of sample preparation, separation and detection techniques. Thereafter, pharmacological and toxicological basics of TA are outlined in general, followed by individual introduction of those TA, that are referred to in detail in this chapter.

The second main part of this contribution is addressed to LC-MS-based analytical procedures. Initially, we discuss important methods for sample preparation. Afterwards characteristics of chromatographic separation are presented followed by important issues of mass spectrometric detection including type of ionization, kind of mass analyzers and typical scan modes for MS operation.

Subsequently, different fields of LC-MS applications are presented covering (a) PK studies, (b) investigations of in vivo and in vitro distribution, (c) biotransformation processes in living organisms and in isolated in vitro systems, and (d) toxicological screening and methods to diagnose drug abuse.

At last, the concluding remarks will point out most important facts and future perspectives.

Cocaine was not discussed in detail as numerous reviews we refer to provide good overviews on this drug and relevant LC-MS/MS methods.

2 Tropane Alkaloids

TA belong to a class of small molecules characterized by a derivatized tropane ring skeleton (8-methyl-8-azabicyclo[3.2.1]octane). Hydroxylated tropane (tropine) is esterified with diverse organic acids often representing structural analogues of, e.g. tropic, benzoic or mandelic acid (Fig. 1). Manifold variations are found either in natural sources or as (semi-) synthetic active pharmaceutical ingredients (API) with individual optimized properties.

TA have been used as magic potions, recreational, shamanic and arrow poison, knockout drops, sedatives, hallucinogens, and medicine since ancient times [1–3]. TA are accessible from plants especially of the *solanaceae* family including *Atropa belladonna* (deadly nightshade), *Hyoscyamus niger* (henbane), *Scopolia carniolica* (scopola), *Datura stramonium* (thornapple), *Duboisia myoporoides* (corkwood) and *Brugmansia arborea* (angel's trumpets) [1]. In addition, some TA (e.g. 6β-acetoxy-2β-hydroxy-nortropane, anisodine, anisodamine, scopolamine) were identified in herbs used in Chinese traditional medicine, e.g. *Erycibe obtusifolia* (Benth) [4] and *Anisodus tanguticus* (Maxim.) Pascher [5–7].

Isolation and chemical characterization of TA started in the early nineteenth century (e.g. hyoscyamine in 1833; scopolamine in 1881; cocaine in 1862 [1]) thus representing early examples of pharmacognosy, which denominates the discovery of drugs from medicinal plants, fungi, bacteria and marine organisms [1, 8].

Today, more than 200 different natural TA are known, but only a very few are of clinical or toxicological importance [9]. Most common representatives of these

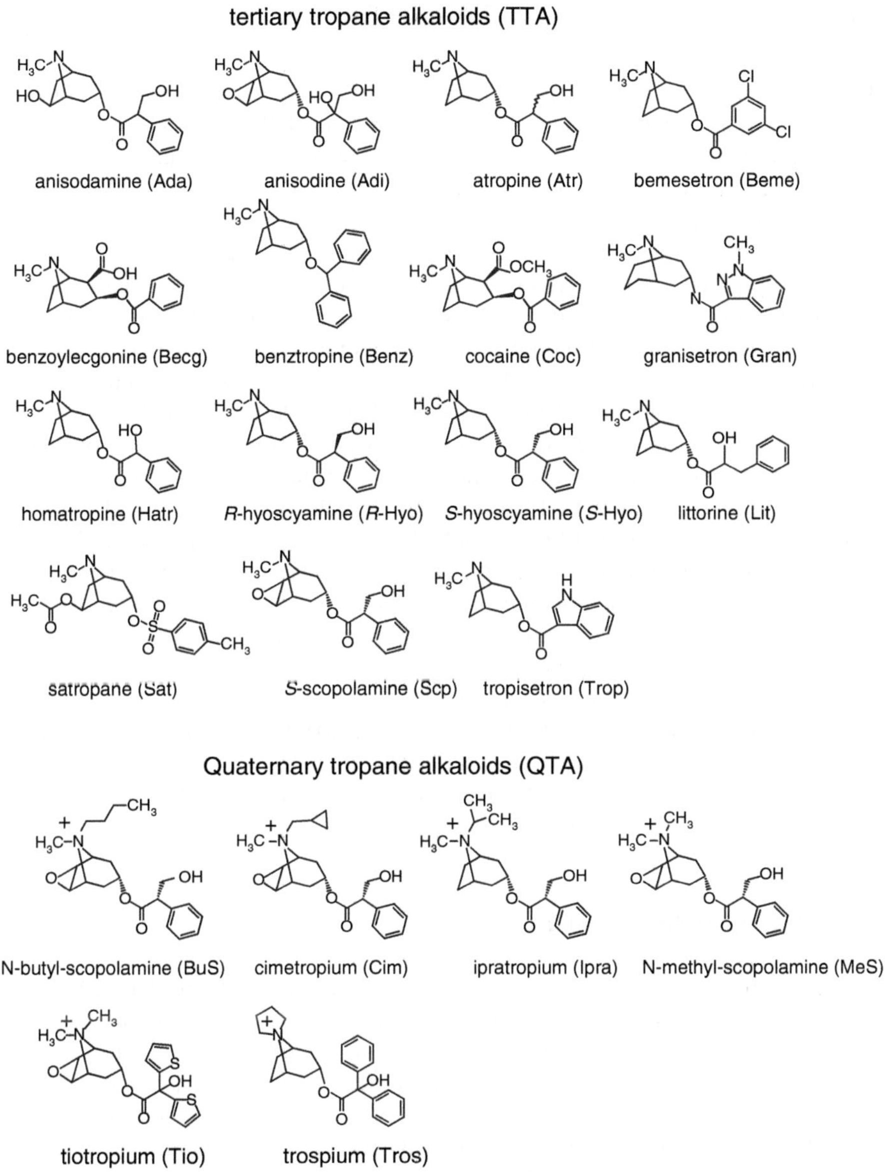

Fig. 1 Chemical structures of selected tropane alkaloids

plant secondary metabolites are hyoscyamine (atropine), scopolamine (hyoscine) and cocaine, the latter one originated from *Erythroxylaceae coca* (Fig. 1).

The following section will discuss chemical structures of TA and the resulting implications for the design of analytical approaches.

2.1 Chemical Structures and Properties

Biosynthesis of TA is a stereoselective multi-step process generating secondary metabolites, e.g. *S*-scopolamine and *S*-hyoscyamine, the latter one is synonymous for L-hyoscyamine or (−)-hyoscyamine (Fig. 1). Therefore, *S*-hyoscyamine is a chiral molecule of esterified 3α-hydroxytropane and *S*-tropic acid (Fig. 1) [1]. While alkaline extraction of plant material for pharmaceutical purposes, *S*-hyoscyamine, easily undergoes racemization thus producing *R*-hyoscyamine (Fig. 1) to yield a mixture named atropine [10]. However, sometimes atropine is misleadingly used to assign the natural *S*-hyoscyamine content of plants especially in cases of intoxication [11–15]. Both hyoscyamine and scopolamine serve as raw material for next generation drugs representing derivatized analogues being mostly N-alkylated thus yielding positively charged quaternary TA (QTA), e.g. *N*-butyl scopolamine and cimetropium from scopolamine and ipratropium from hyoscyamine (Fig. 1). Apart from those QTA a large number TA has been synthesized exchanging the natural tropic acid moiety against, e.g. paramethyl-benzenesulfonic acid (satropane, Fig. 1), dichlorobenzoic acid (bemesetron, Fig. 1), indole-3-carboxylic acid (tropisetron, Fig. 1) or mandelic acid (homatropine, Fig. 1), thus regulating pharmacological activity.

In contrast to TA of the *Solanaceae* family cocaine from *Erythroxylaceae coca* contains a substituted 3β-hydroxytropane (3β-tropine, pseudotropine) skeleton (ecgonine methyl ester) (Fig. 1).

For elucidation of biosynthesis and TA identification in plants mass spectrometric procedures including LC-MS [16, 17], LC-MS/MS [18] and DART MS (direct analysis in real time MS) [19] were successfully applied.

A review on biosynthesis of TA is given by Humphrey and O´Hagan [20] and derived TA drugs are reviewed by Grynkiewicz and Gadzikowska [1].

2.1.1 Charge Properties of TA

Due to the presence of the tertiary amine in the tropane moiety, tertiary TA (TTA) behave as weak bases allowing protonation of the N-atom. Table 1 summarizes calculated pK_b-values of TTA being in the range from 4.9 to 6.8 indicating this character. For comparison, NH_3 has a pK_b of 4.7. Calculations were based on SMILES notation of the relevant TTA structures whose usefulness is briefly addressed below (*SMILES notation*). Even though the listed theoretical pK_b calculations, which are based on the SPARC software [21], do not represent experimental data, dimensions and trends of pK_b will have a satisfying prognostic value. For comparison, the pK_b of atropine and scopolamine were experimentally determined to be 4.57 and 6.25, respectively [22], and calculation resulted in sufficient congruency with 5.08 and 6.78, respectively (Table 1).

The pH influences the ratio of protonated to non-protonated TTA, which has impact on pharmacological behaviour (see Sect. 2.2) as well as on sample preparation procedures (see Sect. 3.1). Especially, when extracting by liquid–liquid extraction (LLE), the partition between an aqueous and an organic non-polar phase is

Table 1 Log *P* and SMILES code of tropane alkaloids used in LC-MS-based analysis

Log *P*[a]	Analyte	pK_b[b]	SMILES code[a]
−2.93	MeS	n.c.	C[N+]2(C)C4CC(OC(=O)C(CO)c1ccccc1)CC2C3OC34
−2.06	Cim	n.c.	C[N+]3(CC1CC1)C5CC(OC(=O)C(CO)c2ccccc2)CC3C4OC45
−1.94	Tio	n.c.	C[N+]3(C)C5CC(OC(=O)C(O)(c1cccs1)c2cccs2)CC3C4OC45
−1.49	BuS	n.c.	CCCC[N+]2(C)C4CC(OC(=O)C(CO)c1ccccc1)CC2C3OC34
−1.47	Ipra	n.c.	CC(C)[N+]1(C)C3CCC1CC(OC(=O)C(CO)c2ccccc2)C3
−0.61	Tros	n.c.	OC(C(=O)OC1CC2CCC(C1)[N+]23CCCC3)(c4ccccc4)c5ccccc5
0.64	Adi	6.82	CN2C3CC(OC(=O)C(O)(CO)c1ccccc1)CC2C4OC34
0.85	Ada	6.08	CN1C3CC(O)C1CC(OC(=O)C(CO)c2ccccc2)C3
1.05	Scp	6.78	CN2C3CC(OC(=O)C(CO)c1ccccc1)CC2C4OC34
1.77	Atr	5.08	CN1C3CCC1CC(OC(=O)C(CO)c2ccccc2)C3
1.77	Hyo	5.08	CN1C3CCC1CC(OC(=O)C(CO)c2ccccc2)C3
1.79	Hatr	5.10	CN1C3CCC1CC(OC(=O)C(O)c2ccccc2)C3
2.13	Lit	5.08	CN1C3CCC1CC(OC(=O)C(O)Cc2ccccc2)C3
2.24	Sat	6.80	CN2C3CC(OS(=O)(=O)c1ccc(C)cc1)CC2C(OC(C)=O)C3
2.43	Gran	4.72	CN1C4CCCC1CC(NC(=O)c2nn(C)c3ccccc23)C4
2.72	Trop	4.76	CN1C4CCC1CC(OC(=O)c2c[nH]c3ccccc23)C4
2.87	Coc	6.49	COC(=O)C2C(OC(=O)c1ccccc1)CC3CCC2N3C
4.19	Beme	5.03	CN1C3CCC1CC(OC(=O)c2cc(Cl)cc(Cl)c2)C3
4.27	Benz	4.88	CN1C4CCC1CC(OC(c2ccccc2)c3ccccc3)C4

For abbreviation of analyte names see Sect. "Abbreviations". *log P* logarithm of octanol/water partition coefficient; *n.c.* not calculable as N-atom of tropane moiety is present as alkylated quaternary amine, pK_b negative decadal logarithm of base constant, *SMILES* simplified molecular input line entry system

[a]Determined and calculated by the molinspiration software after drawing the chemical structure [25]

[b]$pK_b = 14 - pK_a$, pK_a was calculated by SPARC software based on SMILES code for hydrolysis of protonated N-atom of the tropane moiety [26]

influenced by the acid–base equilibrium and thus by pH. Due to optimum extractability from alkaline media, TA were historically classified as alkaloids.

In contrast to TTA, QTA possess a fourfold alkylated N-atom within the tropane moiety and therefore a permanent positive charge (ammonium cation), e.g. *N*-methyl scopolamine and *N*-butyl scopolamine, tiotropium, cimetropium, ipratropium and trospium (Fig 1). Accordingly, pK_b values cannot be calculated. Again, the charge influences pharmacological properties like distribution in vivo thus preventing the passage of the blood–cerebrospinal fluid or blood–brain barrier. Furthermore, the permanent charge of QTA is independent of pH and causes highly polar compounds with minimum tendency to be transferred into organic non-polar solvents. Therefore, special attention has been paid to the development of appropriate LLE procedures for QTA leading to the modification of ion-pair LLE as presented in the Sect. 3.1 [23, 24].

The positive charge of QTA and the explicit proton affinity of TTA make these cationic compounds ideal candidates for mass spectrometric detection after positive electrospray ionization (ESI) and atmospheric pressure chemical ionization (APCI).

2.1.2 Lipophilicity of TA

Despite the presence of polar moieties including, e.g. tertiary amine, ester-, hydroxyl- and epoxy-groups, non-protonated uncharged TTA are quite lipophilic. Lipophilicity is mainly caused by the heterocyclic tropane structure and aromatic (substituted) phenyl rings (Fig. 1).

The octanol-water partition coefficient (K_{OW}), characterizing distribution of a non-ionized compound between an octanol (o) and an immiscible aqueous (w) layer, may function as a measure of lipophilicity, that is often listed as its logarithmic value (log *P*). Therefore, log *P* may be used as a predictor of extractability in LLE. Table 1 presents log *P* values of TA calculated by the *Molinspiration* software [25] using SMILES notation for chemical structures. The SMILES concept is addressed in the next section [26]. Conformity of calculated (calc.) and experimental (exp.) log *P* values is satisfying, as exemplarily shown for atropine (1.83 exp. [27]; 1.77 calc.) and scopolamine (0.98 exp. [22]; 1.05 calc.).

The most lipophilic TA are benztropine (log *P* 4.27) (Fig. 1) and bemesetron (log *P* 4.19) (Fig. 1), possessing either two non-substituted phenyl rings (benztropine) or one dichlorinated phenyl ring (bemesetron). In contrast, the most hydrophilic compounds are QTA due to the presence of the permanent positive charge. However, as determination of K_{ow} requires non-ionized molecular species, calculated log *P* values might be inaccurate for QTA but surely demonstrate the weak tendency of these substances to be transferred into an organic non-polar phase (Table 1).

Consideration of log *P* can also be useful to assess extractability by solid-phase extraction (SPE) using reversed-phase (RP) material as well as retention properties on HPLC columns of such material [28].

2.1.3 SMILES Notation

SMILES is the abbreviation for simplified molecular input line entry system. This system allows to describe two-dimensional chemical strcutures in a simple linear text form as shown for diverse TA in Table 1. To give an example, atropine (Fig. 1) is described by the sequence: CN1C3CCC1CC(OC(=O)C(CO)c2ccccc2)C3, which was directly generated by the *Molinspiration* software after presetting the chemical structure. As obvious hydrogen bonds are not symbolized. The pattern (OC(=O) C(CO) for example represents the ester moiety (OC(=O)), that is bound to a methylene carbon atom, C, that carries the hydroxymethylene group (CO) (Fig. 1). The phenyl ring is coded by c2ccccc2, describing six aromatic carbon atoms (six times lower case C) spanning a second (2) cyclic structure in the molecule (besides the tropane bicyclus). For more details on construction rules of the SMILES code the reader is referred to an online tutorial offered by, e.g. the U.S. Environmental Protection Agency (EPA) [26].

SMILES provides an-easy-to-handle-format useful for entries in any software application. Therefore, based on the SMILES code pK_b and log *P* values can be predicted beneath other chemical and biological properties (Table 1).

2.2 *Pharmacology and Toxicology*

Based on the chemical structure, TA can be subdivided into TTA and QTA depending on the grade of N-alkylation in the tropane moiety. These structural differences have major influence on bioavailability and distribution in vivo and should briefly be outlined below.

2.2.1 Tertiary TA and Quaternary TA

Due to their lipophilicity TTA are rapidly and completely absorbed in the GIT and cross the blood–brain barrier allowing drug administration as tablets [29]. Accordingly, severe intoxication may be caused after ingestion of toxic plants and berries.

The permanent positive charge of QTA influences distribution in vivo and prevents passage of blood–brain barrier and blood–cerebrospinal fluid barrier [30, 31]. Myolitic QTA are muscarinic receptor antagonists but allow a better therapeutic index as they are insoluble in lipids and thus poorly systemically absorbed (e.g. bioavailability of *N*-butyl-scopolamine after oral intake: <1 % [30]). Therefore, spasmolytic activity in the GIT (by, e.g. cimetropium, butropium or *N*-butyl-scopolamine, Fig. 1), respiratory tract (ipratropium, Fig. 1) and overactive bladder (trospium, Fig. 1) appears as the primary local effect whereas systemic side effects are markedly minimized or absent [32–34].

More common reviews on TA drugs and their synthetic derivatives are given by Christen [9] and Grynkiewicz and Gadzikowska [1].

2.2.2 Mechanism of Action

In general, diverse TA are known to bind to (a) muscarinic receptors (MR), (b) 5-hydroxytryptamine (serotonin) receptor 3 (5-HT_3R), (c) α1-adrenoreceptors (α1-AR), or (d) α7-nicotinic receptors (α7-nAChR) thus causing different physiological effects. The corresponding mechanisms of action should briefly be addressed below.

2.2.3 Binding to MR

MR are present in, e.g. the central nervous system (CNS, for respiratory and cardiovascular activity, cognition and stress processing), peripheral nervous system (PNS, for smooth muscle contraction, control of heart rate, vasodilatation), as well as the sympathetic and parasympathetic ganglion cells [1]. Five metabotropic cholinergic MR subtypes (M1-M5) were identified [1], but selectivity of TA is merely apparent [9] except for tiotropium and ipratropium [31].

Most TA referred herein (e.g. hyoscyamine, scopolamine, *N*-butyl-scopolamine, Fig. 1) primarily act as competitive antagonists of acetylcholine (anticholinergics) in the synaptic cleft (parasympatholytic) [29, 35]. Thus therapeutic effects may include mydriasis, spasmolysis of the respiratory system, GIT and overactive bladder, anaesthesia and analgesia [33].

Nevertheless, certain TA are known to exhibit agonistic activity (cholinergics), e.g. (−)-satropane (Fig. 1) or baogongteng [4]. These compounds cause, e.g. smooth muscle contraction and may exhibit protective and trophic effects on retinal ganglion cells thus being considered for the treatment of glaucoma [4].

2.2.4 Binding to 5-HT$_3$R

The 5-HT$_3$R receptor is expressed in areas involved in the emetic reflex in the CNS and GIT. Serotonin (5-hydroxytryptamine) functions as its physiological agonist mediating neuronal excitations involving a ligand-gated cation channel [36]. Due to chemotherapy, serotonin release is increased thus causing vomiting and nausea. Regulation by antagonization is recommended for therapy.

Some synthetic TA primarily act as selective serotonin antagonist and are thus considered and used as antiemetics, including bemesetron, granisetron, tropisetron and scopolamine (Fig. 1) [36, 37].

2.2.5 Binding to a1-AR and a7-nAChR

α1-AR plays an important role in blood pressure regulation as well as contraction and growth of smooth and cardiac muscles. Anisodamine and anisodine (Fig. 1) are muscarinic cholinergic antagonists and they also exhibit blocking activity on α1-AR. This mechanism supports the clinical use to treat septic shock by improvement of blood flow (vasodilation of the microcirculation) [38].

Tropisetron (a selective 5-HT$_3$R antagonist, Fig. 1) has been found to be also a potent and selective but partial agonist of α7-nAChR thus potentially effecting learning and memory processes [39]. Clinical implications and use are still not elaborated.

2.2.6 TA Analysed by LC-MS Methods

In the following, 19 individual TA are listed in alphabetical order that represent analytes of referred LC-MS procedures. The short summaries provide important keywords for characterizing important background information.

Anisodamine (Ada; $C_{17}H_{23}NO_4$; MW 305.16 g/mol; CAS-No; 55869-99-3; [(1S,3S,5S,7S)-7-hydroxy-8-methyl-8-azabicyclo[3.2.1]octane-3-yl](2S)-3-hydroxy-2-phenyl-propanoate).

Anisodamine is a natural derivative of hyoscyamine mono-hydroxylated at the tropane skeleton (Fig. 1). The compound was extracted from traditional Chinese medicine *Anisodus tanguticus* evoking typical non-specific effects of cholinergic antagonists (spasmolysis, anaesthesia, mydriasis, analgesia) in combination with α1-AR blocking properties. In China it is used for therapy of, e.g. circulatory disorders, gastrointestinal colic, nephritis, hepatitis and eclampsia but primarily against septic shock and severe acute respiratory syndrome (SARS). Anisodamine is less toxic than atropine and of less CNS toxicity than scopolamine [6, 38].

Biotransformation of anisodamine in an in vivo rat model as well as in liver homogenates was analysed by LC-ESI MS/MS and is referred later on [6].

Anisodine (Adi; $C_{17}H_{21}NO_5$; MW 319.14 g/mol; CAS-No. 52646-92-1; 9-methyl-3-oxa-9-azatricyclo[3.2.1.0^{2,4}]nonane-7-yl α-hydroxy-α-hydroxy-methyl-benzeneacetate).

Anisodine is a natural TTA that represents a derivative of the scopolamine structure mono-hydroxylated at the tropic acid moiety (Fig. 1). Similar to anisodamine it was extracted from Chinese herb *Anisodus tanguticus* (Maxim.) Pascher and also exhibits α1-AR blocking properties and non-specific anticholinergic effects. Accordingly, in China anisodine is used for the therapy of the same indications as described for anisodamine, most often to treat transmissible shock. Toxicity and side effects of anisodine are smaller than those for atropine, scopolamine and anisodamine [5].

This book chapter refers to rat in vivo biotransformation studies performed by LC-ESI MS/MS [5, 40].

Atropine (Atr; $C_{17}H_{23}NO_3$; MW 289.17 g/mol; CAS-No. 51-55-8; 8-methyl-8-azabicyclo[3.2.1]octane'-3-yl 3-hydroxy-2-phenylpropanoate).

This TTA is an injectable core medicine listed in the World Health Organization's (WHO) "Model list of essential medicines" [41]. Atropine (Fig. 1) acts as a competitive MR antagonists used clinically as, e.g. parasympatholytic for pre-anaesthesia medication, ophthalmologic procedures and as antidote for the therapy of anticholinesterase poisoning [42, 43]. A corresponding PK study in man monitoring atropine as antidote by LC-MS/MS is referred in this chapter [44].

Atropine is the racemic mixture of *R*- and *S*-hyoscyamine produced during the pharmaceutical plant extraction process. *R*-hyoscyamine is nearly inactive on MR (distomer) whereas S-hyoscyamine exhibits high affinity (eutomer). Nevertheless, due to economic reasons atropine is typically administered even though only half of the applied dose (*S*-hyoscyamine) is pharmacologically active on MR. Surprisingly, there is still little information about different pharmacokinetic behaviour of both enantiomers anyhow [46, 47].

LC-MS-based procedures for chiral and enantioselective analysis of mammalian samples are discussed in the Sects. 3.2 and 3.2.2 [47–50]. Detailed data on biotransformation in vivo especially in man are quite rare. LC-ESI MS/MS procedures are referred that allowed metabolite identification in animals in in vivo and in vitro models [51, 52].

Intoxications with higher concentrations will cause tachycardia, mydriasis, CNS excitations and hallucinations, coma and ultimately death [42]. Incorporation of atropine (more correctly *S*-hyoscyamine) is the predominant reason for TA intoxication after ingestion of *Datura* plants.

LC-MS methods to investigate atropine intoxications by analysis of plasma, serum or whole blood [11, 53, 54], urine [12–14, 55], hair [56, 57] and viscera [15] are presented in this chapter.

Bemesetron (Beme; MDL 72,222; $C_{15}H_{17}NO_2Cl_2$; MW 313.06 g/mol; CAS-No; 40796-97-2; [(1S,5R)-8-methyl-8-azabicyclo[3.2.1]octane-3-yl] 3,5-dichlorobenzoate).

Bemesetron is a synthetic ester of tropine and dichlorobenzoylic acid (Fig. 1). In contrast to TA introduced above, bemesetron primarily acts as a selective serotonin antagonist on the 5-HT_3R in the CNS and GIT. Therefore, it was tested as an injectable antiemetic drug for cancer chemotherapy to treat vomiting and nausea but is not in clinical use [36, 37, 58].

Biotransformation products present in urine were identified by LC-APCI MS [37].

Benztropine (Benz; benzatropine; $C_{21}H_{25}NO$; MW 307.19 g/mol; CAS-No; 86-13-5; (3-endo)-3-(diphenylmethoxy)-8-methyl-8-azabicyclo[3.2.1]octane).

Benztropine is a synthetic tropine derivative (Fig. 1) that acts as anticholinergic agent with similar pharmacological properties as atropine but exhibiting much stronger antihistamine and dopamine reuptake inhibitor activity. It is clinically used as an orally administered second-line drug to treat Parkinson's disease improving tremor and rigidity [59, 60].

Studies on in vivo biotransformation of benztropine in rats were supported by LC-ESI MS analysis and are referred herein [59].

Cimetropium (Cim; $C_{21}H_{28}NO_4$; MW 358.20 g/mol; CAS-No. 51598-60-8; 9-cyclopropylmethyl-7-(3-hydroxy-2-phenyl-propionyloxy)-9-methyl-3-oxa-9-azonia-tricyclo [3.3.1.0^{2,4}]nonane).

Cimetropium is a QTA representing an N-alkylated derivative of scopolamine (Fig. 1). It exhibits antispasmodic activity on smooth muscle cells in the GIT. Therefore, it is clinically used for oral long-term treatment of irritable bowel syndrome with minimal effects on vascular receptors [32, 61].

We refer to LC-MS-based methods for a PK study [61] as well as in vitro biotransformation studies with liver microsomes [23, 62].

Cocaine (Coc, benzoylmethylecgonine, $C_{17}H_{21}NO_4$; MW 303.15 g/mol; CAS-No. 50-36-2; methyl (1R,2R,3S,5S)-3-(benzoyloxy)-8-methyl-8-azabicyclo[3.2.1]octane-2-carboxylate).

In contrast to other TA, cocaine derives from *Erythroxylaceae coca* and may be considered as a derivatized tropine structure (Fig. 1). Cocaine is a serotonin–norepinephrine–dopamine reuptake inhibitor that stimulates CNS, suppresses appetite and may be applied as topical local anaesthetic being notorious for causing devastating addiction [1]. Furthermore, cocaine induces euphoria, decreased fatigue together with alterations in pulse and respiration. Due to the completely different pharmacology of cocaine when compared to other TA and the huge number of scientific publications on LC-MS analysis of cocaine in biological fluids, cocaine will not be considered in this summary. Instead we like to refer the reader to excellent reviews on this topic providing relevant and complementary information [63–69].

Granisetron (Gran; $C_{18}H_{24}N_4O$; MW 312.19 g/mol; CAS-No. 109889-09-0; endo-1-methyl-*N*-[9-methyl-9-azabicyclo-[3,3,1]nonane-3-yl]-1H-imidazole-3-carboxamide).

In contrast to TA introduced above, granisetron does not possess an azabicyclo-octane moiety (tropane skeleton) but instead an azabicyclo-nonane system representing a methylene-extended bicyclic compound (Fig. 1). Nevertheless, similarities of chemical structure and pharmacological activity justify its integration into the group of TA. Similar to bemesetron, granisetron acts as a selective serotonin antagonist on 5-HT_3 receptors. It is clinically used as daily antiemetic especially administered orally or intravenously to treat post-operative as well as chemotherapy-associated vomiting and nausea [70, 71].

We address LC-MS/MS procedures for granisetron that were applied for PK studies in rats [72] and man [71, 73].

Homatropine (Hatr; $C_{16}H_{21}NO_3$; MW 275.15 g/mol; CAS-No; 87-00-3; 8-methyl-8-azabicyclo[3.2.1]octane-3-yl 2-hydroxy-2-phenylacetate).

Homatropine is a synthetic parasympatholytic representing the ester of tropine and mandelic acid (Fig. 1). This drug antagonizes acetylcholine on MR and is thus used as mydriatic with less detrimental paralytic effects on ciliary eye muscle than atropine but showing faster and shorter action [1, 33]. Therefore, it belongs to the WHO list of essential medicines as mydriatic eye drops representing an alternative to atropine [41].

We refer to a LC-ESI MS/MS method applied to characterize hydrolytic stability of homatropine in plasma [50].

Hyoscyamine (Hyo; $C_{17}H_{23}NO_3$; MW 289.17 g/mol; CAS-No; 101-31-5; (8-methyl-8-azabicyclo[3.2.1]octane-3-yl) 3-hydroxy-2-phenylpropanoate).

Hyoscyamine is the natural TTA (*S*-hyoscyamine) isolated from plants of the *solanaceae* family, that represents the lead structure for all other TA-derived drugs (Fig. 1). Despite its chemical polarity hyoscyamine exhibits strong lipophilic properties (especially under basic, non-protonating conditions) that enable rapid and nearly complete absorption in the GIT and passage of the blood–cerebrospinal fluid barrier. This alkaloid acts as competitive non-selective antagonist of acetylcholine on MR. Hyoscyamine is given orally or sublingually to relief visceral spasm of the GIT as well as spasm of the biliary and genito-urinary tract. If administered as pure *S*-hyoscyamine (eutomer) the drug has twice the potency of racemic atropine [33].

LC-MS approaches were introduced to investigate biotransformation in vitro [49, 50] and PK behaviour in man [48, 49].

Besides scopolamine, *S*-hyoscyamine represents the most toxic TA in *Datura* plants causing severe to fatal intoxications. Evidence of *Datura* intoxication was enabled by LC-MS procedures presented in this chapter [11–15, 53, 55, 57].

Ipratropium (Ipra; $C_{20}H_{30}NO_3$; MW 332.22 g/mol; CAS-No. 60205-81-4; [8-methyl-8-(1-methylethyl)-8-azoniabicyclo[3.2.1] octane-3-yl] 3-hydroxy-2-phenylpropanoate).

Ipratropium is a QTA produced by N-alkylation (iso-propyl) of hyoscyamine (Fig. 1). When inhaled as aerosolized agent it is a short-acting (3–6 h) antimuscarinic especially on MR subtypes M1–3 present in the lung. Therefore, it is clinically used as bronchodilator, antiasthmatic and drug for chronic obstructive pulmonary disease (COPD) [31, 33]. Ipratropium is part of the WHO list of essential medicines [41].

We refer to LC-MS/MS methods that were used to investigate diffusion through bronchial tissue in vitro [74] and for toxicological screening [54]. In addition, the misuse of ipratropium in equine sports was investigated by LC-MS [24].

Littorine (Lit; $C_{17}H_{23}NO_3$; MW 289.17 g/mol; CAS-No. 21956-47-8; 8-methyl-8-azabicyclo[3.2.1]octane-3-yl) (2R)-2-hydroxy-3-phenylpropanoate).

Littorine is the natural plant precursor for biosynthesis of *S*-hyoscyamine representing its isobaric isomer (Fig. 1). Even though littorine exhibits high affinity to MR similar to hyoscyamine [35] it is not used as drug, most reasonably due to its low concentration in plants.

A LC-ESI MS/MS method is referred for analysis of in vitro stability in human and rabbit serum [50].

N-methyl scopolamine (MeS; methscopolamine; $C_{18}H_{24}NO_4$; MW 318.17 g/mol; CAS-No; 13265-10-6; (1R,2R,4S,7S)-7-{[(2S)-3-hydroxy-2-phenylpropanol]oxy}-9,9-dimethyl-3-oxa-9-azoniatricyclo[3.3.1.0^{2,4}]nonane).

Methylscopolamine is a QTA obtained by N-methylation of scopolamine (Fig. 1). It exhibits non-selective antimuscarinic activity and is used as adjunctive therapy for the treatment of peptic ulcer, to treat nausea and vomiting due to motion sickness and for the management of irritable bowel syndrome [75]. The tritiated form of methylscopolamine is often used in in vitro MR binding assays to investigate receptor affinity and competition by other agents [76].

Here we refer to a study on TA CNS penetration analysed by LC-ESI MS/MS [77].

N-butyl scopolamine (BuS; $C_{21}H_{30}NO_4$; MW 360.22 g/mol; CAS.-No; 149-64-4; [7(S)-(1α,2β,4β,5α,7β)]-9-butyl-7-(3-hydroxy-1-oxo-2-phenylpropoxy)-9-methyl-3-oxa-9-azonitricyclo[3.3.1.0^{2,4}]nonane).

N-butyl-scopolamine is a QTA representing the N-butylated derivative of scopolamine (Fig. 1). *N*-butyl-scopolamine acts as an anticholinergic and since mid 1950s is widely used as antispasmodic inducing smooth-muscle relaxing to treat acute colics, abdominal pain and cramps. *N*-butyl-scopolamine also binds to nicotinic receptors thus inducing ganglion-blocking [30, 32].

We refer to LC-MS techniques that were used to investigate hydrolytic stability in plasma [50], prove drug abuse in horses [78], and analyse PK behaviour in man [79].

Tiotropium (Tio; $C_{19}H_{22}NO_4S_2$; MW 392.10 g/mol; CAS-No; 186691-13-4; (1α,2β,4β,7β)-7-[(hydroxydi-2-thienylacetyl)oxy]-9,9-dimethyl-3-oxa-9-azoniatricyclo[3.3.1.0^{2,4}]nonane).

Tiotropium is a QTA representing a synthetic ester of N-methylated scopine (Fig. 1). This drug is clinically used as anticholinergic long-acting bronchodilator applied once daily as aerosolized solution to treat airway diseases such as COPD. Tiotropium is considered as a selective inhibitor of MR subtype M1 and M3 thus not enhancing mucus production [31].

PK studies in man analysed by LC-MS and LC-MS/MS are addressed [80, 81].

Tropisetron (Trop; $C_{17}H_{20}N_2O_2$; MW 284.15 g/mol; CAS-No; 89565-68-4; (1R,5S)-8-methyl-8-azabicyclo[3.2.1]octane-3-yl 1H-indole-3-carboxylate).

Tropisetron is a synthetic TTA containing an esterified tropine moiety (Fig. 1). This drug is a potent antiemetic selective 5-HT_3 receptor antagonist to treat post-operative and chemotherapy-induced vomiting and nausea by daily intravenous or oral administration [70]. Furthermore, tropisetron is a potent and selective partial agonist of α_7-nAChR [39].

We herein refer to an in vitro biotransformation study in human liver microsomes [82] and a PK study in man [83] both analysed by LC-MS-based methods.

Trospium (Tros; $C_{25}H_{30}NO_3$; MW 392.22 g/mol; CAS-No. 10405-02-4; spiro[8-azoniabicyclo-[3,2,1]octane-8,1′-pyrrolidinium]-3-(hydroxydiphenyl-acetyl)-oxy).

Trospium is a QTA in which the tropine N-atom is part of two alkyl ring structures (Fig. 1). This drug acts as an antimuscarinic agent that is clinically used as orally administered first-line urinary antispasmodic to treat overactive bladder (OAB) [34]. Due to the absence of significant biotransformation, trospium is excreted mainly unchanged into urine allowing to exhibit primarily local activity [84].

We introduce a PK study in man [84] and an in vivo distribution study in rat to elucidate CNS penetration of trospium [77] both analysed by LC-MS/MS.

Satropane (Sat; $C_{17}H_{20}NO_5S$; MW 535.13 g/mol; [6β-acetoxy-8-methyl-8-azabicyclo[3.2.1]octane-3-yl] paramethyl-benzenesulfonate).

Satropane is a synthetic derivative of tropine (Fig. 1). The structure represents an analogue of baogongteng, a TTA isolated from Chinese herb *Erycibe obtusifolia Benth* [4]. In contrast to other TA, (−)-satropane exhibits agonistic activity on MR. Due to activation of iris MR subtype M3 present in the iris-ciliary body, hypertensive intraocular pressure (IOP) is reduced. Therefore, (−)-satropane is considered as future drug to treat high IOP and glaucoma, caused by, e.g. degeneration of optic nerve axons [4, 85].

A LC-ESI MS/MS method to study distribution of satropane in rabbit is referred [85].

Scopolamine (Scp; hyoscine; $C_{17}H_{21}NO_4$; MW 303.14 g/mol; CAS-No; 51-34-3; (−)-(S)-3-hydroxy-2-phenylpropionic acid (1R,2R,4S,7S,9S)-9-methyl-3-oxa-9-azatricyclo [3.3.1.$0^{2,4}$] non-7-yl ester).

S-scopolamine is a natural TTA biosynthesised from hyoscyamine present in plants of the *solanaceae* family (Fig. 1). Scopolamine acts as a non-selective antimuscarinic drug but exhibits sedative depressant effects on the CNS. Typically it is applied to prevent and control motion sickness and visceral spasm but also as mydriatic, cycloplegic and antiemetic [29]. Scopolamine, similar to atropine, inhibits secretions in the respiratory tract, stomach and of saliva and may cause psychotic reactions in higher doses [33]. Administration may be oral, via injection or as transdermal patch. The latter variant has also been elaborated for the treatment of OP poisoning in dogs [86].

LC-MS/MS procedures to investigate biotransformation in vivo and in vitro [7, 87], to monitor PK profiles [88, 89], and to screen for intoxications [11, 13–15, 55–57] are referred in this chapter.

Hints on diverse sample origins and fields of LC-MS application given above point up a broad spectrum of sample matrices with various complexity as well as different requirements on qualitative and quantitative analysis of TA. Therefore, the following main part of this chapter is addressed to corresponding sample preparation procedures and specific characteristics depending on the aim of TA measurement.

3 Analysis of Biological Samples

Selective and sensitive analysis of TA requires adequate sample preparation to reduce sample matrix complexity and concentrate the analyte. Subsequent chromatography and MS detection should also account for potential matrix interferences allowing rapid measurement and a sufficient limit of quantification. The next section describes most important and frequent techniques for sample preparation.

3.1 Sample Preparation

In principle, sample preparation is performed (a) to clean the analyte by separating from matrix compounds that might deteriorate selectivity and sensitivity in subsequent LC-MS analysis and (b) to concentrate the target molecules to achieve optimum limits of quantification. Several conventional, less selective techniques are well established including simple protein precipitation, LLE and SPE. The appropriate choice of the procedure depends on the purpose and requirements of analysis. Non-selective protein precipitation appears useful when a larger number of analytes with differing polarity (e.g. the drug and its polar and conjugated biotransformation products) has to be measured simultaneously. For this purpose LLE might be less adequate as charged and highly polar derivatives exhibit minor tendency to be transferred into a non-polar environment. In contrast, modern SPE columns provide solid materials that allow interaction with analytes exhibiting a broad range of polarity.

The following paragraphs present these techniques for sample preparation applied to TTA and QTA LC-MS analysis.

3.1.1 Protein Precipitation

Protein precipitation represents one of the most simple and quite frequent preparation methods for protein-containing fluids allowing rapid performance to reduce costs. The addition of organic solvents, e.g. acetonitrile (ACN), methanol (MeOH) or $ZnSO_4$, causes deproteinization allowing separation of particulate matter from clear analyte containing supernatant with good recovery. Elimination of high molecular weight proteins is an important step for subsequent chromatographic separation to minimize the risk of clotting, column obstruction and shifting retention properties caused by irreversible protein adsorption to the stationary phase. Nevertheless, protein precipitation lacks selectivity and produces a still quite complex matrix of salts and other ionic, hydrophilic or slight hydrophobic components that might cause interferences in MS detection. Therefore, analyte concentrations have to be effectively high or additional preparation steps will be necessary. Furthermore, co-precipitation of analytes bound to precipitated proteins might reduce recovery. However, such effects have not been discussed in the literature for TA analysis.

The use of organic solvents may constitute a matrix compatible to subsequent liquid chromatography, thus not requiring any concentration or evaporation step. However, protein precipitation seems to be inappropriate for automation and thus requires a manual workflow.

Table 2 summarizes reports describing sample precipitation for TA analysis.

John et al. added twice the volume of ACN to plasma samples thus generating a supernatant containing 66 % (v/v) ACN. Merely dilution with aqueous HPLC solvent A [0.1 % formic acid (FA)] was necessary prior to injection allowing quantitative analysis of *R*- and *S*-hyoscyamine [49] as well as atropine, cocaine, homatropine, ipratropium, littorine, *N*-butyl-scopolamine and scopolamine, simultaneously [50]. Recoveries were greater than 85 % (Table 2). Despite this crude preparation procedure effects on ionization were negligible.

Chen et al. used MeOH for precipitation of rat plasma in a 3:1 volume ratio to investigate a broad spectrum of biotransformation products generated from anisodine [5], anisodamine [6], atropine [52] and scopolamine [87]. Biotransformation products covered a broad spectrum of polarity including sulfo- and glucuronide conjugates, oxidized, hydroxylated, methoxylated and demethylated metabolites of the parent drug as well as its hydrolysis products. Unfortunately, recoveries were not reported (Table 2).

Kajbaf et al. obtained 100 % recovery for cimetropium when adding aqueous 1 %(w/v) $ZnSO_4$ solution to buffered liver microsomes incubation mixtures. Subsequently a SPE step was carried out to clean and concentrate the cimetropium containing fraction [23, 62].

In comparison to LLE and SPE, protein precipitation was applied less frequently especially for quantitative measurement of TTA or QTA.

3.1.2 Liquid–Liquid Extraction

For LLE the liquid sample is mixed with a larger volume of a non-polar organic solvent to induce a partition equilibrium of the analyte between the aqueous and organic layer. Typically, the latter one contains the major fraction of analytes and is further processed for LC-MS analysis. In contrast to protein precipitation, LLE thus requires a subsequent drying step by evaporation or with a gentle stream of nitrogen

Table 2 Preparation of mammalian samples by precipitation prior to LC-MS analysis

Precipitation solvent[a]	Sample matrix	Analyte	Sample:solvent volume ratio	Comment	Recovery (%)	Reference
ACN	Hum plasma	Atr	1:1.5	Solvent containing IS	38–41	[44]
ACN	Hum plasma	Atr, Coc, Hatr, Ipra, Lit, BuS, Scp	1:2	+IS a.p.	93, 94, 86, 90, 93, 92, 88	[50]
ACN	Hum plasma	*R*/*S*-Hyo	1:2	+AtrE p.p. +IS a.p.[b]	93	[49]
ACN	Porcine plasma	*R*/*S*-Hyo	1:2	+AtrE p.p. +IS a.p.[b]	93	[47]
ACN	Rabbit plasma	*R*/*S*-Hyo	1:2		93	[49]
ACN	Rat plasma	Gran	1:1.1	+IS p.p.	> 100	[72]
ACN/MeOH 5:1	Hum serum	Atr	1:7.5	Solvent containing IS	n.s.	[53]
Chlorobutane	Rabbit plasma	Scp	1:2		n.s.	[89]
MeOH	Rat plasma	Adi, metab.	1:3		n.s.	[5]
MeOH	Rat plasma	Ada, metab.	1:3		n.s.	[6]
MeOH	Rat plasma	Atr, metab.	1:3		n.s.	[52]
MeOH	Rat plasma	Scp, metab.	1:3		n.s.	[87]
MeOH	Hum liver microsomes	Trop, metab.	1:1		n.s.	[82]
MeOH	Rat plasma	Sat	1:2.7	+IS p.p.	76–85	[85]
$ZnSO_4$ aq 1 % w/v	Liver microsomes[c]	Cim, metab.	1:0.3	+IS p.p., SPE a.p.	100	[23, 62]

For abbreviation of analyte names see Sect. "Abbreviations". *ACN* acetonitrile, *a.p.* after precipitation, *AtrE* atropinesterase from rabbit serum, *hum* human, *IS* internal standard, *metab.* metabolites (biotransformation products produced in vivo), *MeOH* methanol, *n.s.* not specified, *p.p.* prior to precipitation

[a]Numbers give volume ratio v/v

[b]AtrE was added to hydrolyze *S*-hyoscyamine selectively for enantioselective analysis

[c]Microsomes from rat, mouse, hamster, guinea pig

to remove the organic layer that is inapplicable for direct injection to reversed-phase chromatography. Therefore, LLE is more laborious and time consuming but often enables efficient extraction, good separation from deteriorating matrix ingredients and effective analyte concentration with satisfying recovery (Table 3).

The partition equilibrium depends on the solubility of the analyte in the organic solvent but more critical on the analyte's polarity that is determined by the functional groups and charge. As discussed above (*Lipophilicity of TA*) the log *P* value is a measure of extractability into an organic phase. The higher the log *P* value the more lipophilic is the compound (Table 1).

Besides substance lipophilicity, the polarity of the organic solvent also determines extraction efficiency premising sufficient solubility. Polarity of liquids can be characterized by the Snyder polarity index (P') sorting solvents from smallest polarity (pentane, $P'0$), over small (methyl-*tert*-butyl ether, MTBE, $P'2.5$; diethylether, Et_2O $P'2.8$; dichloromethane $P'3.1$), and medium (chloroform $P'4.1$; ethylacetate, EE $P'4.4$) to high polarity (dimethylsulfoxide, DMSO $P'7.2$; water $P'10.2$) [91].

Therefore, diverse organic solvents were used for LLE of TA including pentane, chloroform, very often dichloromethane, most frequently EE, sometimes Et_2O and seldom MTBE as listed in Table 3.

In case of TTA extraction efficiency requires to avoid acidic pH to minimize the grade of protonation of the weak basic N-atom of the tropane moiety. Resulting uncharged TTA are predominantly transferred into the organic layer. The charged protonated form exhibits better solubility in aqueous media and often causes insolubility in non-polar aprotic solvents, e.g. Et_2O, chloroform. Therefore, a slight basic 0.001 % Na_2CO_3 solution was added to rat faeces prior to LLE of anisodamine [6], anisodine [5], atropine [52] and scopolamine [87] and their biotransformation products using ethylacetate (Table 3). Xu et al. added 1 M NH_4OH to human plasma prior to extraction of atropine with dichloromethane (recovery 110 %) [90] and Ahmed et al. also alkalized human plasma for LLE of scopolamine with MTBE (recovery not specified) [88] (Table 3). A recovery of 98 % for granisetron was obtained after addition of 0.1 M NaOH to plasma before LLE with EE [71].

Intending to improve extract purity of atropine and scopolamine from horse urine, Gerber et al. performed an excessive sequence of consecutive LLE steps [14]. Initially, urine was incubated overnight at pH 5.0 with β-glucuronidase from *E. coli* to hydrolyze glucuronide-conjugates of both TTA and thus enhance the concentration of structurally unaltered compounds (Table 3). Afterwards, the incubation mixture was alkalized and extracted for the first time with dichloromethane/iso-propanol 4:1 (v/v) for 15 min. The liquid organic layer containing the analytes was separated from the frozen aqueous phase after freezing in an ice/alcohol bath. Subsequently, the organic layer was acidified with HCl and extraction was performed a second time for 15 min transferring protonated TTA back into the aqueous layer. After separating both layers the aqueous solution was extracted with EE to remove any kind of undesired non-polar matrix components. Subsequent to discarding the EE phase, the aqueous layer was alkalized again (pH 9.0) by addition of NH_4OH followed by LLE with dichloromethane for 15 min thus transferring TA into the organic

Table 3 Preparation of mammalian samples by liquid–liquid extraction (LLE) prior to LC-MS analysis

Organic solvent[a]	Sample matrix	Analyte	Sample:solvent volume ratio[b]	Steps prior to LLE	Recovery (%)	Reference
$C_5H_{12}/H_2CCl_2/^{iso}prop$ 46:49:5	Rat urine, bile	Benz	n.s. (3×)	Lyophilization, +MeOH +PB, pH 7.2[c]	n.s.	[59]
$HCCl_3$	Liver microsomes[d]	Cim	1:1.4 (2×)	+IS +NaCl +hepSA, pH 3.7	75	[23]
H_2CCl_2	Hum plasma	*S*-Hyo	1:4.3	+IS +1 M NH_4OH	110	[90]
H_2CCl_2	Hum plasma	Cim	1:11.8	+IS	64	[61]
H_2CCl_2	Hum plasma	BuS	1:8.6	+IS	69	[79]
H_2CCl_2	Rat, mouse, hamster, guinea pig liver microsomes	Cim, metab.	1:1.4 (2×)	+IS +NaCl +hepSA, pH 3.7	62	[23]
H_2CCl_2	Hum plasma	Tio	1:2	+IS +ACN (precipitation)[e]	76–81	[80]
$H_2CCl_2/^{iso}prop$ 4:1	Equine urine	Atr, Scp	1:0.73	+1 M NaOAc pH 5.0 +β-glucu +25 % NH_4OH[f]	n.s.	[14]
EE	Liver microsomes[d]	Cim	1:1.4 (2×)	+IS +NaCl +hepSA, pH 3.7	32	[23]
EE	Rat faeces	Ada, metab.	1:2	+0.001 % Na_2CO_3	n.s.	[6]
EE	Rat liver homogenate	Ada, metab.	1:2		n.s.	[6]
EE	Rat faeces	Adi, metab.	1:2	+0.001 % Na_2CO_3	n.s.	[5]
EE	Rat liver homogenate	Adi, metab.	1:2		n.s.	[5]
EE	Rat faeces	Atr, metab.	1:2	+0.001 % Na_2CO_3	n.s.	[52]
EE	Rat faeces	Scp	1:2	+0.001 % Na_2CO_3	n.s.	[87]
EE	Equine urine	Ipra	1:1	+IS +borax buffer +5 M NaOH, pH 10.0[g]	82	[24]
EE	Hum plasma	Gran	1:6.3	+IS +0.1 M NaOH	98	[71]
EE/C_6H_{14} 75:25	Dog, monkey urine	Beme (MDL 72,222), metab.	n.s. (2×)	Lyophilization, +acid, pH 2.0 or +base, pH 10.0	n.s.	[37]
Et_2O	Liver microsomes[d]	Cim	1:1.4 (2×)	+IS +NaCl +hepSA, pH 3.7	54	[23]

Et_2O/H_2CCl_2 2:1	Hum plasma	Trop	1:3.4	+IS +PB, pH 10.0	72–77	[83]
Et_2O/H_2CCl_2 7:3	Hum plasma	Gran	1:9.5	+IS	63	[73]
Me-O-^{t}Bu	Hum plasma	Scp	n.s.	+IS +alkalized	n.s.	[88]

For abbreviation of analyte names see Sect. "Abbreviations". *ACN* acetonitrile, *β-glucu* solution of β-glucuronidase (*E. coli*), *EE* ethyl acetate, *hep-SA* heptanesulfonic acid, *^{iso}prop* iso-propanol, *IS* internal standard, *MeOH* methanol, *metab.* metabolites, *Me-O-^{t}Bu*, methyl-^{t}butyl ether; *n.s.* not specified, *OAc* acetate, *PB* phosphate buffer

[a]Numbers give volume ratio v/v

[b]Numbers in parentheses give the number of repeated LLE cycles

[c]A second LLE step followed by adding 0.1 M PB, pH 7.4 to the organic layer (ratio n.s.)

[d]Microsomes from rat, mouse, hamster, guinea pig

[e]A second LLE step followed by adding 0.5 M HCl to the organic layer (1:4)

[f]After the first LLE step a number of additional extraction steps followed: the organic layer of first LLE was mixed with 0.1 M HCl (1:0.38) for second LLE. EE was added to that acidic aqueous layer (1:3) for third LLE. That aqueous layer was alkalized (pH 9.0) with 25 % NH_4OH and mixed with H_2CCL_2 (1:2.7) for fourth LLE. That organic phase was evaporated to dryness, re-dissolved and analysed by LC-APCI MS/MS

[g]After the first LLE step an additional step followed: potassium iodide, glycine, NaOH, and NaCl were added to the aqueous layer prior to second and third LLE with H_2CCl_2 (1:0.7). The latter organic layers were evaporated to dryness and analysed by LC-ESI MS.

solvent. After evaporating the methylene chloride fraction to dryness, extracts were dissolved in MeOH for final LC-MS/MS analysis.

According to the limited tendency of charged molecules to be transferred into an organic phase, only few examples can be found extracting QTA from plasma by simple LLE. Dichloromethane was used for cimetropium [61] and *N*-butyl-scopolamine [79] requiring a large relative excess of organic solvent (1:11.8 for cimetropium; 1:8.6 for *N*-butyl-scopolamine) but yielding only in moderate recoveries of less than 70 % (Table 3).

To improve extraction efficiency, a modification of LLE—named ion pair LLE—was established and applied to QTA extraction [23, 24]. Incubation mixtures containing liver microsomes and cimetropium were mixed with NaCl and heptane sulfonic acid, which is a typical ion pairing reagent also used as HPLC solvent additive in ion pair chromatography [92, 93]. Extraction was performed twice by adding the 1.4-fold volume of an organic solvent. Diverse solvents were tested yielding recoveries of 75 % for chloroform, 62 % for dichloromethane, 54 % for Et_2O and 32 % for EE [23] (Table 3). It is noteworthy that chromatographic separation was also carried out as ion pair chromatography and mass spectrometric detection was done offline by fast atom bombardment (FAB).

An additional example for ion pair extraction of QTA (ipratropium) was described by Tang et al. mixing equine urine with alkaline saturated borax buffer prior to extraction with EE allowing to remove more lipophilic compounds. Subsequently, the EE layer was discarded and an alkaline potassium iodide-glycine solution was added to the aqueous phase. Afterwards, the ipratropium–ion pair complex was extracted twice with dichloromethane yielding in a recovery of 82 % [24] (Table 3).

However, ion pair extraction for QTA did not achieve broad acceptance.

The addition of organic solvents to biological samples was also carried out to extract TA from human hair. For forensic analysis hair represents a versatile specimen allowing to analyse incorporation of a compound [94, 95]. Segment-wise analysis of hairs further allows to estimate the duration and time point of exposure. Kintz et al. successfully analysed hair by LC-MS/MS to prove the abuse of scopolamine and atropine [56, 57]. For sample preparation hair was initially washed (decontaminated) twice with dichlormethane for 2 min. Subsequently, hair was segmented and each segment was cut into small pieces (<1 mm). The chopped hair was incubated for several hours in phosphate buffer (pH 8.4) containing deuterated atropine as an internal standard (IS) prior to TTA extraction with organic solvent (dichloromethane/iso-propanol/*n*-heptane 50:17:33). Extraction recovery was found to be 67 % for scopolamine and 82 % for atropine [57]. Limits of quantification were satisfying at 5 pg/mg hair.

In the following section SPE will be discussed as a third wide-spread method for sample extraction.

3.1.3 Solid-Phase Extraction

SPE is a frequently used procedure for extraction and clean-up of analytes from fluids by prepacked cartridges. Perculating the sample solution through the preconditioned sorbent enables analyte adsorption to the solid phase and elution of unbound matrix components. Subsequent washing steps may be performed by means of reduced pressure to remove interfering compounds prior to analyte elution with an appropriate mixture of organic solvents. The eluate requires evaporation to remove the organic solvent that might cause fatal matrix effects in chromatography. At last, the dried residue has to be re-dissolved for LC-MS analysis. Functionalized sorbents are commercially available in high quality providing excellent reproducibility. Typical materials include diverse hydrophobic reversed-phase (RP) sorbents (C_4–C_{18}) but also cation exchanger or mixed-mode solids that will be addressed below. In comparison to LLE, SPE avoids disposal of larger volumes of organic solvents and prevents from incomplete phase separation and thus optimizes recovery. Therefore, SPE represents a modern, but multi-step sample preparation procedure that is capable for automation to moderate costs.

For TTA and QTA extraction from urine, plasma, incubation mixtures and organ homogenates, SPE has been successfully applied as summarized in Table 4. Extraction of anisodamine [6], anisodine [5], atropine [51, 52] and scopolamine [7, 87] and their biotransformation products was done with AccuBOND^{II} C18 cartridges (Agilent). This sorbent consists of irregular silica particles that were derivatized with octadecyl chains to obtain a non-polar, non-end-capped, reversed-phase material well suited for non-polar, rather weak hydrophobic analytes. Accordingly, end-capped Sep-PAK C18 (Waters) was also used for atropine extraction [96].

With respect to Ingelse et al. compounds with a log $P>0$ are considered as more or less polar and water soluble but can be extracted by RP-SPE in contrast to molecules with log $P<0$ [28]. Log P values of TTA mentioned above (Table 1) are between 0.64 (anisodine) and 1.77 (atropine) and thus confirm the assumption of extractability. Surprisingly, C_{18}-SPE was also applied to extract the charged polar QTA compounds cimetropium [23, 62] and tiotropium [81] from biological samples with excellent recoveries of 100 and 92 %, although corresponding log P values were calculated to be negative (−2.06 and −1.94, respectively, Table 1). Kajbaf et al. investigated optimum extraction procedures for cimetropium from liver microsome incubation mixtures by comparing precipitation with ACN, LLE with diverse solvents and RP-SPE [23]. By far the highest recovery was obtained after protein precipitation (100 %) and SPE (100 %), whereas LLE yielded in much smaller rates (e.g. 32 % for EE to 75 % for chloroform).

Besides RP-sorbents, cation exchange material was also applied appropriate for extraction of positively charged protonated TTA or QTA. Atropine and scopolamine were extracted from human plasma at pH 3.0 using a C8-SCX (strong cation exchange) mixed mode sorbent (Bond Elute certify, Agilent/Varian) [11]. This solid phase is an irregular silica-based, non-end-capped sorbent combining reversed-phase C8 and SCX (benzoylsulfonic acid) properties. In addition, the QTA Ipra was separated from human whole blood [54] and equine urine [78] by the use of CBA

Table 4 Preparation of mammalian samples by solid-phase extraction (SPE) prior to LC-MS analysis

SPE material	Sample matrix	Analyte	Previous step	Elution solvent[a]	Recovery (%)	Reference
C18 (AccuBOND[II])	Rat urine	Ada, metab.	None	MeOH	n.s.	[6]
C18 (AccuBOND[II])	Rat urine	Adi, metab.	None	MeOH	n.s.	[5]
C18 (AccuBOND[II])	Rat urine	Atr, metab.	Hydrolysis[b], adjust pH 8.0	MeOH	n.s.	[51, 52]
C18 (AccuBOND[II])	Rat urine	Scp, metab.	Hydrolysis[b], adjust pH 8.0	MeOH	n.s.	[7, 87]
C18 (Sep-Pak)	Dog plasma	Atr	None	MeOH/0.1 M NH_4OAc 3:1	107	[96]
C18 (Sep-Pak)	Rat, mouse, hamster, guinea pig liver microsomes	Cim, metab.	+1 % $ZnSO_4$ (precipitation)	MeOH	100	[23, 62]
C18 (Supelclean LC)	Hum plasma	Tio	+IS	MeOH/Et_3N/HOAc 99:0.5:0.5	92	[81]
C8-SCX mixed mode (Bond Elute certify)	Hum plasma	Atr, Scp	Adjust pH 3.0, +IS	MeOH/NH_4OH 98:2	61–84	[11]
CBA Bond Elut LRC	Hum blood[c]	Ipra	+PB, pH 8.0; +IS after SPE	MeOH/1 M HCl 30:70	96–103	[54]
CBA ISOLUTE	Equine urine	Ipra, BuS	+0.01 M $(NH_4)_2CO_3$, pH 9.3	1 % FA in MeOH	n.s.	[78]
HLB Oasis	Hum viscera	Atr, Scp	Leaching with PBS	MeOH	n.s.	[15]
HLB Oasis	Hum urine	Atr	+IS	MeOH	n.s.	[12]
HLB Oasis	Hum plasma	Tros	+IS +H_3PO_4	2 mM NH_4OAc/ACN 20:80	73	[84]
HLB Oasis[d]	Hum serum	Scp	+IS +PB	MeOH	51	[97–99]
HLB Oasis[e]	Rat plasma, brain	Tros, Scp, MeS	+PB, pH 8.0, +IS	MeOH	n.s.	[77]
MCX Oasis	Hum plasma	*R/S*-Hyo	+IS, +2 % FA	2.25 % NH_4OH in MeOH	81–83	[48]

For abbreviation of analyte names see Sect. "Abbreviations". *ACN* acetonitrile, *CBA* carboxylic acid, *FA* formic acid, *HLB* hydrophilic–lipophilic balance, *hum* human, *IS* internal standard, *LRC* large reservoir capacity, *MCX* mixed-mode cation exchange, *metab.* metabolites (biotransformation products produced in vivo), *MeOH* methanol, *n.s.* not specified, *OAc* acetate, *PB* phosphate buffer, *PBS* phosphate-buffered saline, *SCX* strong cation exchange

[a]Ratio given as v/v

[b]Hydrolysis of phase II biotransformation products (conjugates) of TA was either done enzymatically (β-glucuronidase) or non-enzymatically under acidic conditions

[c]Human whole blood

[d]Automated performance

[e]In 96-well plate format

Bond Elut (Agilent) and CBA ISOLUTE (International Sorbent technology) both representing an irregular end-capped mixed-mode phase of a weak cation exchanger (carboxylic acid) and a mid-polarity sorbent. *N*-butyl-scopolamine was extracted accordingly from equine urine [78]. Elution of analytes was achieved by the use of a methanolic solution acidified either with 1 M HCl [54] or 1 % formic acid [78] allowing competition between QTA and hydronium ions. Recoveries were between 96 and 103 % [54] (Table 4).

In contrast, alkalized solvents were used to elute hyoscyamine [48] and atropine and scopolamine [11] from cation exchange adsorbent by inducing deprotonation of protonated TTA thus neutralizing the analytes and preventing from attractive ionic interaction.

In addition to the silica-based materials mentioned above, modern polymers are widely used for TTA and QTA sample preparation allowing SPE not impaired by undesirable silanol activities. HLB Oasis (Waters) is the tradename for a hydrophilic–lipophilic balance reversed-phase sorbent enabling lipophilic interaction to benzene moieties and hydrophilic interactions to pyrrolidone groups as present in the macroporous copolymer of poly(divinylbenzene-co-*N*-vinylpyrrolidone). Elution of analytes is often performed with solvents containing MeOH or ACN. Applying this adsorbent TA such as atropine and scopolamine were extracted from human viscera [15], human serum [97–99], human urine [12] as well as from rat plasma and brain microdialysate [77]. Furthermore, this hydrophilic–lipophilic balance phase was also suitable for extraction of the QTA trospium from human and rat plasma [77, 84] and methyl scopolamie from rat plasma [77] (Table 4).

More conventional SPE procedures make use of single cartridges or tubes that may be processed in a multiple-port vacuum manifold for simultaneous preparation of a reasonable number of samples. Further improvement of sample throughput and reduced manual working steps was realized by an automated design originally presented by Oertel et al. for scopolamine analysis in a PK study using atropine as IS [97]. An ASPEC XL sample processor (Gilson) operated with Oasis HLB cartridges was used for microanalytical preparation of just about 200 μl serum for each sample. Any preconditioning, sample loading, washing and elution step was carried out automatically. Liquid flow was established by high air pressure at the top of cartridge in contrast to typical vacuum applied at the bottom of the device. At last, evaporation of eluates required an additional manual working step. However, recovery of scopolamine was found to be only 51 % independent of the concentration yielding an appropriate LOQ of 20 pg/ml [97] (Table 4). Interday precision was adequate (5.9–14.1 %).

The same instrumental design was also applied by Stetina et al. to quantify scopolamine in human serum for a PK study [98] (Table 4).

Automation for improved throughput was also established by the use of 96-well plates allowing the extraction of smallest sample volumes in the microliter range. Callegari et al. recently published a method to extract the QTA trospium from 50 μl plasma and brain homogenate samples using HLB Oasis sorbent [77] (Table 4). A satisfying LOQ (0.1 ng/ml) was obtained when adding atropine as IS thus allowing investigation of CNS penetration in a rat in vivo study. Unfortunately, recovery was not reported.

Whereas examples described above are characterized by separated steps of sample extraction and subsequent separation by the HPLC system, Boppana et al. presented a quick online automated sample enrichment design that allowed direct injection of plasma samples with immediate consecutive chromatographic separation for MRM analysis of granisetron (Fig. 1) and its biotransformation products [100]. Following this approach plasma was directly flushed onto a guard column packed with restricted access media (RAM) characterized by an internal surface reversed-phase (ISRP). RAM is a special porous extraction sorbent with dual properties caused by its non-adsorptive outer hydrophilic surface and its adsorptive internal surface. It exhibits both a sieving effect (size exclusion mechanism) thus preventing large proteins from entering the small pores of the sorbent and an adsorption effect (often reversed-phase) to small molecules that have entered the pores. Therefore, plasma proteins and other large matrix components immediately pass through the guard column without clogging whereas small molecule analytes are retained on the inner surface. Switching the flow direction of an appropriate solvent (reverse to injection) allows elution of the analytes, which are directly transferred onto the separation column for analysis.

A similar approach with multiple columns operating as automated multi-dimensional LC has also been reported by Machtejevas et al. for peptides from human blood filtrate [101].

All examples for sample preparation procedures discussed above were followed by LC-MS or LC-MS/MS analysis of TTA and QTA. The next sections are addressed to the typical performance of sample separation and analyte detection.

3.2 *Chromatographic Separation*

Reliable chromatographic methods for LC-MS have to achieve efficient separation of multiple analytes or interfering matrix components by using solvents that allow compatibility to MS detection. Therefore, HPLC solvents should not contain inorganic salts as typical for ion pair or ion exchange chromatography as they might cause clotting of the MS entrance. Optimum prerequisites for coupling to MS are provided by reversed-phase chromatography separating compounds on a hydrophobic stationary phase (C_8–C_{18}, phenyl) in combination with an aqueous mobile phase containing an appropriate ratio of organic modifier (e.g. ACN, MeOH) that elutes the analytes.

3.2.1 Reversed-Phase Chromatography

Nearly without exception TTA and QTA were separated on RP-material as summarized in Tables 5–8. In practice, the choice of a specific RP-column often depends

Table 5 Tropane alkaloid analysis in plasma or serum by LC-MS and LC-MS/MS for pharmacokinetic studies

Analyte	Column (mm × mm)/flow (ml/min)/*T* (°C)	Mobile phase[a] (run mode)	Lin range (ng/ml)[b]	Ionization/mass analyser/scan mode	Reference
Atr	X-Terra MS C8 (100 × 2.1)/0.2/n.s.	Solv A: ACN Solv B: 2 mM NH_4FA, pH 3.0 (gradient)	0.25–50	+ESI/QqQ/MRM	[44]
Atr, *R*-/*S*-Hyo	Atlantis T3 C18 (150 × 4.6)/1.0/30	Solv A: 0.1 % v/v FA Solv B: ACN/0.1 % FA 80:20 (gradient)	0.1–50	+ESI/QqQ/MRM	[47, 49]
BuS	C18 n.s. (50 × 4.6)/0.45/40	Solv. A: ACN Solv. B: 5 mM NH_4OAc, 0.1 % FA, 60:40 (iso)	0.1–40	+ESI/QqQ/MRM	[79]
Cim	YMC Hydrosphere C18 (50 × 2)/0.2/35	MeOH/10 mM NH_4OAc, pH 4.0, 81:19 (iso)	0.2–100	+ESI/QqQ/MRM	[61]
Gran	Lichrospher C18 (50 × 4.6)/1.0/30	10 mM NH_4OAc, 0.5 % HOAc in ACN/H_2O 40:60 (iso)	0.02–20	+ESI/QqQ/MRM	[71]
Gran	Xterra MS (150 × 2.1)/0.2/40	H_2O/ACN/10 mM NH_4OAc, pH 3.5, 27:23:50 (iso)	5–1,000	+ESI/SQ/SIM	[72]
Gran	Symmetry C18 (150 × 4.6)/1.0/30	0.03 % FA/ACN 30:70 (iso)	0.1–20	+ESI/QqQ/MRM	[73]
Ipra[c]	RP DB-8 (75 × 3)/0.7/25	Solv A: 10 mM NH_4OAc, pH 3.0 Solv B: ACN (gradient)	10–100	+ESI/IT/MS	[24]
R-/*S*-Hyo	Chirobiotic V (250 × 4.6)/1.0/30	MeOH/HOAc/Et_3N 100:0.05:0.04 (gradient)	0.5–50	+APCI/SQ/SIM	[48]
Scp[d]	Purospher STAR RP-18e (55 × 2)/0.3 or 0.4/n.s.	Solv A: ACN/2 mM NH_4OAc/FA 5:95:0.2 Solv B: ACN/2 mM NH_4OAc/FA 95:5:0.2 (gradient)	Serum: 0.02–5 Dial.: 0.05–10	+ESI/QqQ/MRM	[97–99]
Scp	n.s./n.s./n.s.	n.s.	0.02–0.5	n.s.	[88]
Scp	Sunfire C18 (250 × 4.6)/0.3/n.s.	10 mM NH_4OAc, pH 4.0/MeOH/ACN 30:40:30 (iso)	10–2,000	+ESI/SQ/SIM	[89]
Tio	Zorbax Extend C18 (150 × 4.6)/0.8/30	ACN/10 mM NH_4OAc, 1 % FA 40:60 (iso)	0.0005–0.05	+ESI/QqQ/MRM	[80]

(continued)

Table 5 (continued)

Analyte	Column (mm × mm)/flow (ml/min)/*T* (°C)	Mobile phase[a] (run mode)	Lin range (ng/ml)[b]	Ionization/mass analyser/scan mode	Reference
Tio	SB-C18 Zorbax (250 × 4.6)/0.6/22	40 mM NH_4OAc/MeOH 56:44 (iso)	0.0015–0.03	+ESI/SQ/SIM	[81]
Trop	Diamonsil C_{18} (150 × 4.6)/0.5/25	MeOH/H_2O 80:20, 0.2 % FA (iso)	0.1–100	+APCI/QqQ/MRM	[83]
Tros	HyPurity C_{18} (50 × 4.6)/0.5/24	ACN/2 mM NH_4OAc 80:20 (iso)	0.05–10	+ESI/QqQ/MRM	[84]

For abbreviation of analyte names see Sect. "Abbreviations". *ACN* acetonitrile, *APCI* atmospheric pressure chemical ionization, *dial.* microdialysis samples, *ESI* electrospray ionization, *FA* formic acid, *iso* isocratic, *IT* ion trap, *lin range* linear range, *MeOH* methanol, *MRM* multiple reaction monitoring, *MS* full scan mass spectrometry, *n.s.* not specified, *OAc* acetate, *QqQ* triple quadrupole mass spectrometer, *SIM* selected ion monitoring, *Solv* HPLC solvent, *SQ* single quadrupole mass spectrometer, *T* temperature

[a]Ratios given as v/v

[b]Lower limit of linear range corresponds to lower limit of quantification

[c]Analysis of equine urine

[d]Analysis of human serum and microdialysate

Table 6 Tropane alkaloid analysis by LC-MS and LC-MS/MS for distribution studies in vivo and in vitro

Analyte	Column (mm × mm)/flow (ml/min)/T (°C)	Mobile phase[a] (run mode)	Lin range (ng/ml)[b]	Ionization/mass analyser/scan mode	Reference
		Distribution study			
Ipra[c]	SB-C18 Zorbax (150 × 2.1)/0.25/40	Solv A: 0.1 % FA Solv B: 0.1 % FA in MeOH (gradient)	n.s.	+ESI/QqQ/MRM	[74]
Sat[d]	Zorbax SB-C18 (50 × 2.1)/0.25/n.s.	0.02 % FA in ACN/0.02 % FA in H_2O 32:68 (iso)	Dial.: 2–500 Plasma: 5–1,000	+ESI/QqQ/MRM	[85]
Tros, Scp, MeS[e]	Luna C18 (50 × 2.1)/0.3/n.s.	Solv A: ACN Solv B: 5 mM NH_4Oac, 0.1 % FA (gradient)	0.1–100	+ESI/QqQ/MRM	[77]
		No real life sample cpplication described			
Atr[f]	C18 Hitachi gel 3056 (150 × 4.6)/1.0/RT	MeOH/0.1 M NH_4OAc 3:2 (iso)	2.5–2,500	+APCI/SQ/SIM	[96]
Gran[f]	Inertsil C8 (50 × 4.6) with guard column opti-guard C8 (15 × 1)/1.0/n.s.	Loading: H_2O/ACN 95:5 v/v Elution: 0.05 M NH_4OAc, pH 5.0/ACN 73:27 (iso)	0.05–50	+APCI/QqQ/MRM	[100]
S-Hyo[g]	BDS C18 (50 × 3)/0.5/RT	ACN/MeOH/10 mM NH_4OAc; 625:325:150 (iso)	0.02–0.5	+APCI/QqQ/MRM	[90]

For abbreviation of analyte names see Sect. "Abbreviations". *ACN* acetonitrile, *APCI* atmospheric pressure chemical ionization, *dial.* microdialysate, *ESI* electrospray ionization, *FA* formic acid, *iso* isocratic, *lin range* linear range, *MeOH* methanol, *MRM* multiple reaction monitoring, *n.s.* not specified, *OAc* acetate, *QqQ* triple quadrupole mass spectrometer, *SIM* selected ion monitoring, *Solv* HPLC solvent, *SQ* single quadrupole mass spectrometer, *T* temperature

[a] Ratios given as v/v

[b] Lower limit of linear range corresponds to lower limit of quantification

[c] In vitro: passage through porcine bronchial tissue

[d] In vivo: distribution in rabbit plasma and aqueous humour (dialysate)

[e] In vivo: distribution in rat plasma, brain and cerebrospinal fluid

[f] Matrix: dog plasma

[g] Human plasma

Table 7 Tropane alkaloid analysis by LC-MS and LC-MS/MS for biotransformation studies in vivo and in vitro

Analyte	Column (mm × mm)/flow (ml/min)/*T* (°C)	Mobile phase[a] (run mode)	Lin range (ng/ml)[b]	Ionization/mass analyser/scan mode	Reference
Ada[c]	AICHROM ReliAsil C18 (150 × 2)/0.2/40	MeOH/0.01 % Et_3N, pH 3.5, 60:40 (iso)	n.s.	±ESI/IT/MS/MS	[6]
Adi[c]	AICHROM ReliAsil C18 (150 × 2)/0.2/40	MeOH/0.01 % Et_3N, pH 3.5, 60:40 (iso)	n.s.	±ESI/IT/MS/MS	[5]
Atr[c]	Zorbax extend C18 (100 × 3)/0.2/40	MeOH/2 mM NH_4OAc, pH 3.5 70:30 (iso)	n.s.	+ESI/IT/MS/MS	[51, 52]
Beme[d]	Spherisorb CN (250 × 4.6) or Spherex C-18 (250 × 4.6)/1.0/n.s.	Solv A: 0.05 M NH_4OAc, pH 4.4 Solv B: 0.05 M NH_4OAc/ACN 55:45 Solv C: 0.05 M NH_4OAc/ACN 40:60 (gradient)	n.s.	+TSP/SQ/n.s.	[37]
Benz[e]	CSC spherisorb cyano (100 × 25)/0.006[j]/n.s.	H_2O/ACN 80:20, 0.1 % HOAc (iso)	n.s.	+ESI/SQ/MS	[59]
Cim[f]	Spherisorb ODS-2 (250 × 4.6)/1.0/n.s.	20 or 35 mM hep-SA, 100 mM Et_3N in MeOH/ACN/H_2O 20:20:55 or 5:10:55 , pH 3.7 (iso)	30–150	+FAB (Xe)/EBQ_1Q_2/MS/MS	[23, 62]
Scp[g]	Zorbax extend C18 (100 × 3)/0.2/40	MeOH/2 mM NH_4OAc, pH 3.5 70:30 (iso)	n.s.	+ESI/IT/MS/MS	[7, 87]
Trop[h]	Nucleosil C4 (250 × 4.6)/0.5/35	Solv A: ACN/20 mM NH_4OAc 10:90, pH 6.0 Solv B: ACN/20 mM NH_4OAc 60:40, pH 6.0 (gradient)	70–1,425	+ESI/SQ/SIM	[82]
Atr, BuS, Coc, Hatr, Ipra, Lit, Scp[i]	Atlantis T3 C18 (150 × 4.6)/1.0/30	Solv A: 0.1 % v/v FA Solv B: ACN/0.1 % FA 80:20 v/v (gradient)	Atr, Coc, Ipra, Lit, Scp: 0.05–50 Hatr: 0.1–50 BuS: 0.8–50	+ESI/QqQ/MRM	[50]
R-/*S*-Hyo[i]	AGP (2x 150 × 2)/0.3/40	Solv A: 10 mM NH_4FA, pH 8.0 Solv B: 10 mM NH_4FA in 25 % ACN, pH 8.0; 85:15 (iso)	1–500	+ESI/QqQ/MRM	[49]

Atr, *R*-/*S*-Hyo[i]	Atlantis T3 C18 (150×4.6)/1.0/30	Solv A: 0.1 % v/v FA Solv B: ACN/0.1 % FA 80:20 (gradient)	0.1–50	+ESI/QqQ/MRM	[49]

For abbreviation of analyte names see Sect. "Abbreviations". *ACN* acetonitrile, EBQ_1Q_2 double focusing sector field mass spectrometer, *ESI* electrospray ionization, *FA* formic acid, *FAB* fast atom bombardment, *hep-SA* heptanesulfonic acid, *iso* isocratic, *IT* ion trap, *lin range linear range*, *MeOH* methanol, *MRM* multiple reaction monitoring, *MS* full scan mass spectrometry, *MS/MS* tandem mass spectrometry, *n.s. not specified*, *OAc* acetate, *QqQ* triple quadrupole mass spectrometer, *SIM* selected ion monitoring, *Solv* HPLC solvent, *SQ* single quadrupole mass spectrometer, *T* temperature, *TSP* thermospray

[a]Ratios given as v/v

[b]Lower limit of linear range corresponds to lower limit of quantification

[c]In vivo: rat urine, plasma, faeces, in vitro: rat liver homogenates

[d]In vivo: dog and monkey urine and plasma

[e]In vivo: rat urine and bile

[f]In vitro: liver microsomes from rat, mouse, hamster, and guinea pig

[g]In vivo: rat urine

[h]Human liver microsomes

[i]In vitro: human and rabbit serum

[j]The low flow published (6 µl/min) seems to be erroneous

Table 8 Tropane alkaloid analysis by LC-MS and LC-MS/MS applied to intoxication and cases of drug abuse

Analyte	Column (mm × mm)/flow [ml/min]/*T* (°C)	Mobile phase[a] (run mode)	Lin range (ng/ml)[b]	Ionization/mass analyser/scan mode	Reference
		Analysis of plasma, serum or whole blood			
Atr	C18 Hypurity Aquastar Javelin Express (50 × 2.1)/0.2/n.s.	0.13 M NH_4OAc +30 mM TFA/H_2O/ACN (gradient)	5–100	+ESI/QqQ/MRM	[53]
Atr, Scp	Superspher60 RP Select B (125 × 2)/04–0.6/25	ACN/50 mM NH_4FA, pH 3.5 (gradient)	*APCI* Atr: 5–100 Scp: 5–100 *ESI* Atr: 0.1–100 Scp: 0.1–100	+APCI/SQ/SIM +ESI/QqQ/MRM	[11]
Ipra	Atlantis dC18 (100 × 2.1)/0.2/30	Solv A: 15 mM HFBA, 20 mM NH_4FA, pH 3.3 Solv B: MeOH (gradient)	19.5–200	+ESI/IT/MRM	[54]
		Analysis of urine			
Atr	MB Nucleodur C18 (150 × 1)/0.05/n.s.	Solv A: 2 mM NH_4FA, pH 3.0 Solv B: ACN/solv A 90:10 (gradient)	n.s.	+ESI/QqQ/MRM	[12]
Atr, Scp	Hypersil GOLD (100 × 2.1)/0.2/n.s.	Solv A: ACN/10 mM FA 1:99 Solv B: ACN/10 mM FA 60:40 (gradient)	Atr: 5–5,000 Scp: 10–5,000	+ESI/QqQ/MRM	[13, 55]
Atr, Scp	Luna C18(2) (150 × 2)/0.25/40	Solv. A: 10 mM NH_4FA in water Solv. B: 10 mM NH_4FA in MeOH (gradient)	n.s.	+APCI/IT/MRM	[14]
Ipra, BuS	Suplecosil LC-8-DB (100 × 2.1)/0.2/30	Solv A: 10 mM NH_4FA, pH 3.0 Solv B: ACN (gradient)	n.s.	+ESI/IT/MRM	[78]
		Analysis of hair			
Scp	ACQUITY C18 (50 × 2.1, 1.7 μm)/0.3/n.s.	Solv. A: ACN Solv. B: 2 mM NH_4FA, pH 3.0 (gradient)	0.2–20 pg/mg	+ESI/QqQ/MRM	[56]

Atr, Scp	XTerra MS C18 (100×2.1)/0.2/n.s.	ACN/formate buffer (gradient)	Atr: 5–100 pg/mg Scp: 5–100 pg/mg	+ESI/QqQ/MRM	[57]
		Analysis of hum viscera			
Atr, Scp	Xterra Phenyl (150×2.1)/0.2/n.s.	ACN/10 mM NH_4OAc, pH 10.5 (gradient)	Atr: 100–10,000 Scp: 100–10,000	+ESI/SQ/SIM	[15]

For abbreviation of analyte names see Sect. "Abbreviations". *ACN* acetonitrile, *APCI* atmospheric pressure chemical ionization, *ESI* electrospray ionization, *FA* formic acid, *HFBA* heptafluorobutyric acid, *IT* ion trap, *lin range* linear range, *MeOH* methanol, *MRM* multiple reaction monitoring, *n.s.* not specified, *OAc* acetate, *QqQ* triple quadrupole mass spectrometer, *SIM* selected ion monitoring, *Solv* HPLC solvent, *SQ* single quadrupole mass spectrometer, *T* temperature, *TFA* trifluoroacetic acid

[a]Ratios given as v/v

[b]Lower limit of linear range corresponds to lower limit of quantification

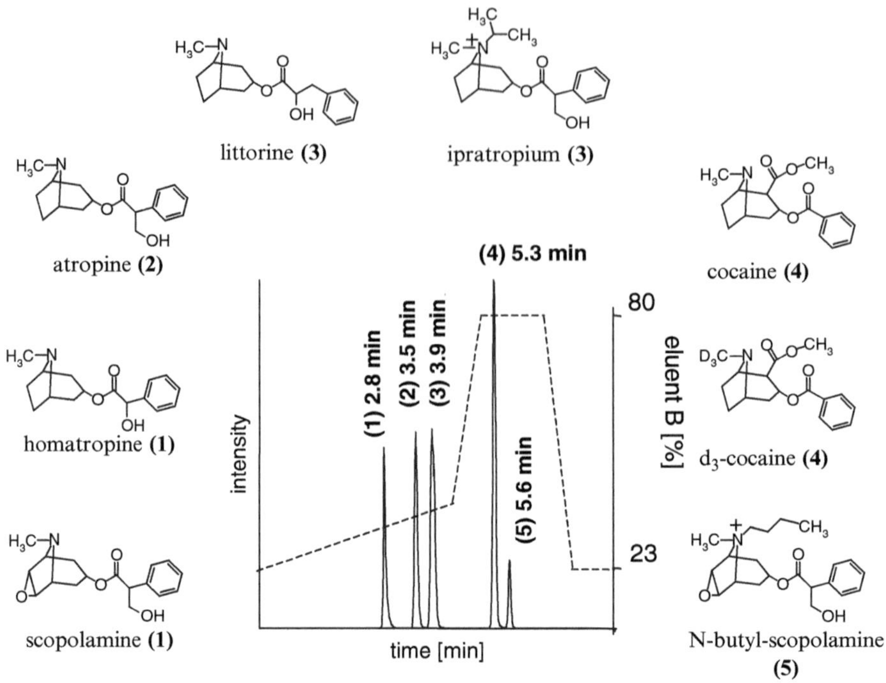

Fig. 2 Simultaneous LC-MS/MS analysis of diverse tropane alkaloids. Analysis of plasma samples was performed according to John et al. [50] using an Atlantis T3 C18 column (150 mm×4.6 mm I.D., 5 μm) in gradient mode (as indicated) with solvent A (0.1 % v/v FA) and solvent B (ACN/water 80:20 v/v 0.1 % v/v FA) at 1 ml/min and 30 °C. Detection was done by positive ESI MS/MS in MRM mode

simply on properties of columns, availability of columns in the laboratory, experience made with other applications and relationship to the providers and their special offers. Therefore, a large variety of diverse RP-columns has been applied not breeding any favourite. A representative chromatogram of a generic LC-ESI MS/MS method for simultaneous quantification of 7 TA in plasma according to John et al. [50] is shown in Fig. 2.

Merely, the choice of column dimension demonstrates the awareness of cost reduction for organic solvents as well as analytical run time and thus of sample throughput and economic efficiency. Only 2 % of published reports make use of a column with an inner diameter (I.D.) as small as 1 mm [12] corresponding to a flow of 50 μl/min. Columns with an I.D. of 2.0–2.1 mm are used in 48 % of reports determining a flow between 200 and 500 μl/min and half of all applications is performed with a column I.D. between 3.0 and 4.6 mm and a flow ranging from 500 to 1,000 ml/min (Tables 5–8).

Typically used solvents contain ammonium acetate (NH_4OAc) or ammonium formate (NH_4FA) in low millimolar concentrations to improve separation quality and ionization efficiency in the positive ion mode. In 64 % of the reports referred to, acetonitrile was used as organic modifier whereas methanol was applied in 36 % of all applications (Tables 5–8). As discussed above, the extent of sample clean-up depends on the preparation procedure. Crude methods such as protein precipitation cause a complex mixture of compounds with a large variety of polarity. This fact has to be taken into account for subsequent chromatographic separation. An increasing gradient of the organic modifier appears necessary to elute lipophilic phospholipids that might cause severe ion suppression in electrospray MS. In contrast to crude protein precipitation, LLE or even SPE allows better sample clean-up. Therefore, about half the separations were performed isocratically and the other half in gradient mode (Table 5–8).

In addition to reversed-phase separations, two examples were found for analysis on non-end-capped cyano-modified columns (Spherisorb CN, Waters) separating TA of highest hydrophobicity, namely bemesetron (log *P* 4.19, Table 1, Fig. 1) [37] and benztropine (log *P* 4.27, Table 1, Fig. 1) [59]. Cyano-phases are of low hydrophobicity, allow dipolar and ionic interactions and are thus in principle suitable for normal- and reversed-phase applications.

Beside column dimension the size of stationary phase particles is a matter of recent progress. More traditional columns are packed with 3.0–5 μm particles enabling satisfying resolution and reasonable column back pressure of solvent suitable to be processed by conventional HPLC pumps. In contrast, sub 2-μm particles (e.g. 1.7 and 1.8 μm) as applied in rapid or fast LC or ultra high-performance LC (UHPLC) allow better resolved separations in shorter run times. Column back pressure (>12,000 psi) is remarkably high demanding more robust solvent pumps.

Excellent column efficiency delivered by a large number of theoretical plates (low plate height of about 0.005 mm) is theoretically deduced from the Van Deemter equation which predicts that use of sub 2.5 μm particles does not diminish efficiency at increased flow rates (increased speed of analysis). For more details the reader is referred to general textbooks on chromatography. Therefore, shorter columns (15–50 and 2.1 mm I.D.) provide sufficient resolution and improved sensitivity within run times of only a few minutes or less.

Improvement of sample throughput was also achieved by using a dual-column design as described by Oertel et al. for quantification of scopolamine from serum [99]. A primary conventional HPLC system was accessorily equipped with a second identical separation column (C18, 55 mm×2 mm I.D., 5 μm) connected to an isocratic auxiliary pump via a 10-port valve. The entire system was operated in a way that enabled sample separation on one column and equilibration of the other column in parallel. Whereas equilibration was done with the auxiliary pump, separation was carried out in gradient mode with the primary pump system, thus increasing sample throughput by about factor 2. A comparable design was used by John et al. for quantitative determination of a biologically active peptide from plasma [102].

In contrast to all other applications referred to in this article, that make use of 3.0–5 μm particles, Kintz et al. chose an ACQUITY C18 column (Waters) of 50 mm length and 2.1 mm I.D. packed with 1.7 μm particles. They analysed scopolamine from hair samples by LC-ESI-MS/MS with a flow of 0.3 ml/min [56]. The retention time of scopolamine was as short as 1.13 min and the LOQ was excellent at 0.2 pg/mg hair. This LOQ was 25-times lower than that obtained with conventional 3.5 μm material as reported before by the same research group (Table 8) [57].

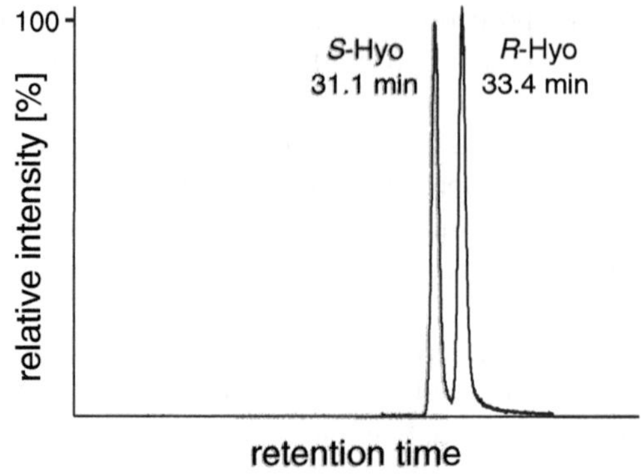

Fig. 3 Chiral separation of atropine (racemic mixture of *S*- and *R*-hyoscyamine). Analysis of buffered serum dilution was performed according to John et al. [49] using two consecutively coupled AGP columns (150 mm × 2.0 mm I.D., 5 μm) in isocratic mode with solvent A (0.01 M NH_4FA, pH 8.0) and solvent B (0.01 M NH_4FA in 25 % v/v ACN, pH 8.0) in 85:15 ratio at 300 μl/min and 40 °C. Detection was done by positive ESI MS/MS in MRM mode

3.2.2 Chiral Separation and Enantioselective Determination

RP-separations discussed above do not differentiate between enantiomers of analytes and thus do not allow chiral analysis. However, stereochemistry of TA agents is a critical issue with respect to pharmacological properties. As discussed above (Sect. 2.1) natural TTA are biosynthesized stereospecifically occurring as *S*-variants of, e.g. hyoscyamine and scopolamine (Fig. 1). Atropine is a racemic mixture of both hyoscyamine antipodes that is administered clinically even though only *S*-hyoscyamine exhibits pharmacological activity on MR (eutomer) [33]. Therefore, analytical distinction of eutomer and distomer appears recommendable when characterizing PK behaviour and correlating clinical effects with circulating drug concentrations. Only three procedures were introduced in the literature to date, that allow quantitative distinction of both hyoscyamine enantiomers in mammalian specimens by LC-MS techniques [47–49]. Two of them represent chiral separations using HPLC columns for selective interaction between the stationary phase and the analytes.

Siluk et al. applied a Chirobiotic V column (250 mm × 4.6 mm I.D., 5 μm, Astec) for separation of *R*- and *S*-hyoscyamine extracted from human plasma for a PK study [48]. The stationary phase consists of spherical silica gel with the covalently attached high-purity macrocyclic glycopeptide vancomycin operated in reversed-phase mode by a gradient mobile phase of methanol, acetic acid, and triethylamine. Coupling to an APCI interface and a single quadrupole mass analyser established a broad linear range of two orders of magnitude (0.5–50 ng/ml) and a sufficient LOQ of 0.5 ng/ml (Table 5). This chiral method revealed that *S*-hyoscyamine concentrations were slightly lower than of *R*-hyoscyamine after i.v. bolus injection of atropine sulphate and subsequent infusion in man [48].

The second chiral method was presented by John et al. [49]. Distinction of hyoscyamine variants from rabbit serum dilutions was enabled on two consecutive AGP columns (each 150 mm × 2 mm I.D., 5 μm, ChromTech) coupled to +ESI-MS/

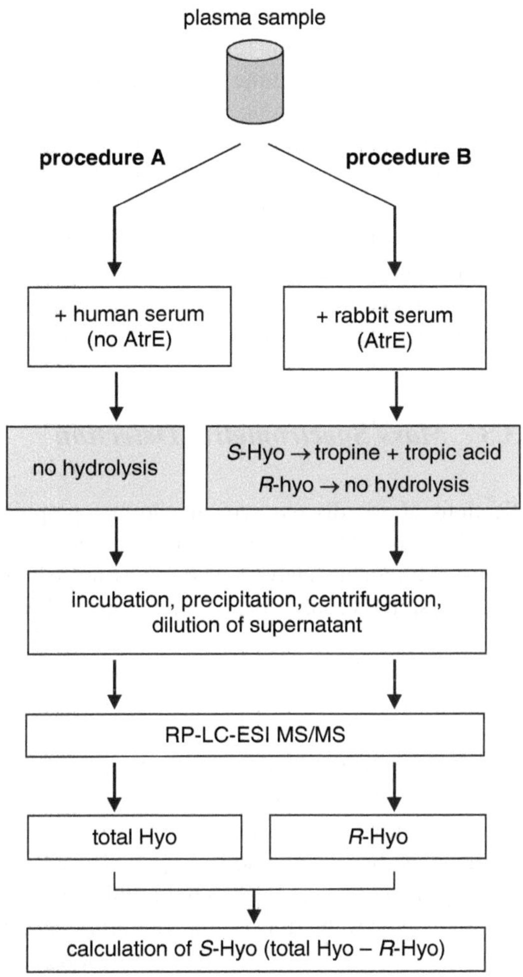

Fig. 4 Scheme of enzymatic plasma sample preparation for enantioselective analysis of *S*- and *R*-hyoscyamine. Atropine-containing plasma samples are mixed either with human serum (*procedure A*) or with rabbit serum (*procedure B*). The latter one contains atropinesterase (AtrE) whereas human serum does not. AtrE catalyses the stereoselective hydrolysis of *S*-Hyo, whereas *R*-Hyo remains unaffected. Following *procedure A* the concentration of total Hyo was determined and following *procedure B* only the remaining *R*-Hyo portion was measured. *S*-Hyo was calculated by the difference of both concentrations. Enantioselective analysis was performed according to John et al. [49]

MS. AGP (α-glycoprotein) is protein of 183 amino acids exhibiting a high carbohydrate content of 45 % (w/w) that is bound to spherical silica particles. Enantioselectivity is mainly driven by the type of organic modifier and most important by the appropriate choice of pH determining the net charge of the chiral selector. As depicted in Fig. 3 separation of *S*- and *R*-hyoscyamine variants (selectivity factor 1.07) was achieved isocratically at 40 °C using 0.01 M ammonium formate (pH 8.0) containing 3.75 % (v/v) acetonitrile as solvent. This method was used to follow enzymatic hydrolysis of atropine by means of atropinesterase from rabbit serum demonstrating enantioselectivity for *S*-hyoscyamine [49].

The third approach for quantification of hyoscyamine enantiomers was carried out as a non-chiral RP-LC-ESI MS/MS procedure combined with an enantioselective sample preparation step [47, 49]. As illustrated in Fig. 4 atropine-containing plasma samples were mixed either with human serum (procedure A) or with rabbit serum (procedure B). The latter one contains atropinesterase (AtrE) whereas human serum does not. Atropinesterase catalyses the stereoselective hydrolysis of *S*-hyoscyamine, whereas *R*-hyoscyamine remains unaffected. Following procedure A the concentration of total hyoscyamine was determined and following procedure B only the remaining *R*-hyoscyamine portion was measured. *S*-hyoscyamine was calculated by the difference of both concentrations. Prepared plasma samples were separated on an Atlantis T3 C18 column (150 mm × 4.6 mm I.D., 5 μm) applying an ACN gradient [49]. This application was based on a LC-ESI MS/MS method previously described by John et al. for the simultaneous measurement of seven TA [50]. So far, the enantioselective procedure was successfully applied to analyse hyoscyamine concentrations in plasma of (a) a pesticide-poisoned patient under atropine therapy [49] and (b) swine treated i.v. with atropine for a PK study [47].

3.3 *Mass Spectrometric Detection*

Sections above discuss typical chromatographic methods applied to TA samples. Beside sufficient separation, sensitive and selective detection is also required for robust and reliable analysis. The following section is addressed to different ionization interfaces, mass analysers and scan modes allowing valuable LC-MS method design.

3.3.1 Ionization

Most often positive ESI and only to a small extent positive atmospheric pressure chemical ionization (APCI) were used as ion source (interface) to generate desolvated free ions of TTA or QTA suitable for MS or MS/MS detection. TTA were detected as their proton adducts $[M+H]^+$, whereas QTA were simply monitored as the original cations $[M]^+$ thus not requiring adduct formation (Table 9).

Statistical evaluation of the literature referred to in this chapter (Table 5–8) revealed a frequency of 2 % for thermospray (TSP), 4 % for offline FAB, 14 % for APCI and 80 % for ESI (Fig. 5a).

Beyer et al. compared the method characteristics of LC-APCI MS with LC-ESI MS/MS for analysis of toxic alkaloids including atropine, scopolamine and deuterated benzoylecgonine as IS in human plasma [11]. Following the same sample preparation (SPE on mixed-mode C8) and gradient chromatographic separation (acetonitrile/ammonium formate, pH 3.5) on a C8 base select column (Superspher 60 RP select B, Merck) eluates were subjected either to APCI or ESI prior to MS detection. ESI MS/MS operating in the multiple reaction monitoring mode (MRM, three transitions) was found to be more sensitive than APCI MS operated in the selected ion monitoring (SIM) mode. This was most presumably due to the better signal-to-noise ratio obtained in the MRM mode. Nevertheless, slightly higher matrix effects causing suppression of atropine and scopolamine ionization were obvious for ESI. Selectivity, accuracy and precision were comparable and satisfying for both methods.

Table 9 Mass spectrometric detection of tropane alkaloids used in LC-MS and LC-MS/MS

		Precursor ion			
Analyte	MW (Da)[a]	Species	m/z[b]	Product ions (m/z)[b]	Reference
Ada	305.16	$[M+H]^+$	306	288; 276; 140; 122	[6]
Adi	319.14	$[M+H]^+$	320	156; 138; 110	[5]
Atr	289.17	$[M+H]^+$	290	260; 142[d]; 124[c]; 93[d]; 91[d]	[11, 12, 44, 50–53, 55, 57, 77, 90, 97]
Atr-d3	292.17	$[M+H]^+$	293	127[c]; 93[d]; 92[d]	[23, 56, 57]
Becg-d3	292.13	$[M+H]^+$	293	171[c]; 105[d]; 77[d]	[11]
Beme	313.06	$[M+H]^+$	314/316	No fragmentation done	[37]
Benz	307.19	$[M+H]^+$	308	No fragmentation done	[59]
BuS	360.22	$[M]^+$	360	212; 194[c]; 156; 138; 121[c]; 110; 103[c]	[50, 61, 78, 79]
Cim	358.20	$[M]^+$	358	103[c]; 138; 121; 156	[61]
Coc	303.15	$[M+H]^+$	304	182[c]; 150; 122; 105; 82	[50]
Coc-d3	306.15	$[M+H]^+$	307	185[c]; 153; 125; 105; 85	[50]
Gran	312.19	$[M+H]^+$	313	282; 176; 159; 138[c]; 96; 80	[71, 73, 100]
Hatr	275.15	$[M+H]^+$	276	142; 124[c]; 93	[50]
Ipra	332.22	$[M]^+$	332	290; 168; 166[c]; 142; 124[c]; 93	[50; 54, 74, 78]
Lit	289.17	$[M+H]^+$	290	142; 124[c]; 93	[50]
R-Hyo	289.17	$[M+H]^+$	290	124[c]; 93	[49, 50]
Sat	353.13	$[M+H]^+$	354	294; 182[c]; 140; 122; 91	[85]
Scp	303.14	$[M+H]^+$	304	156[d]; 138[c]; 121; 110; 103[d]	[7, 11, 50, 55–57, 87, 90, 97, 99]
Tio	392.10	$[M]^+$	392	170[d]; 152[c]	[80]
Trop	284.15	$[M+H]^+$	285	144; 124[c]; 93; 91	[83]
Tros	392.22	$[M]^+$	392	182[c]; 164[c]	[77, 84]
Tros-d8	400.22	$[M]^+$	400	190; 172[c]	[84]

For abbreviation of analyte names see Sect. "Abbreviations". *APCI* atmospheric pressure chemical ionization, *ESI* electrospray ionization

[a]Monoisotopic mass

[b]Due to slight variations within cited literature m/z values of precursor and product ions are given as whole numbers

[c]Typically used as quantifying ion

[d]Typically used as qualifying ion

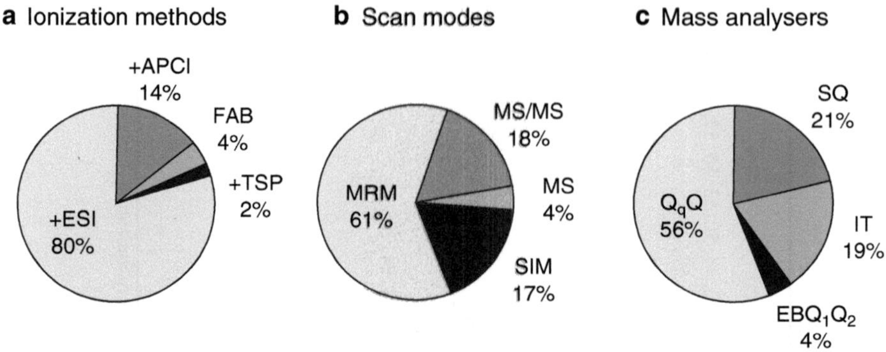

Fig. 5 Statistical evaluation of LC-MS-based methods for tropane alkaloids referred in this chapter. (**a**) Relative frequency of ionization methods. *+APCI* positive atmospheric pressure chemical ionization, *+ESI* positive electrospray ionization, *FAB* fast atom bombardment, *+TSP* positive thermospray. (**b**) Relative frequency of scan modes used. *MS* full scan MS, *MS/MS* tandem mass spectrometry (product ion scan), *MRM* multiple reaction monitoring, *SIM* selected ion monitoring. (**c**) Relative frequency of mass analysers used. EBQ_1Q_2 double focusing sector field mass spectrometer, *IT* ion trap, *QqQ* triple quadrupole, *SQ* single quadrupole. Considered publications were found by PubMed data-based search and references cited in these articles

With respect to matrix effects Björnstad et al. observed maximum reduction in MS detector signals at the void time of chromatographic separation when analysing urine containing atropine and scopolamine relinquishing any sample preparation step [55]. This was due to salts, amino acids and many other highly polar urine components that were not retained on RP material. However, such effects could be compensated by using deuterated IS being ionized under identical conditions as the analytes and were obsolete for atropine and scopolamine that showed remarkable retention.

Suppressing matrix effects on ionization of ipratropium (21 % suppression) were described by Ariffin and Anderson when analysing SPE extracts of whole blood by gradient RP-LC-ESI MS/MS [54]. In contrast, only slight ion suppression was described by John et al. for simultaneous measuring of cocaine (4 %), homatropine (4 %), ipratropium (4 %), *S*-hyoscyamine (3 %), littorine (1 %), *N*-butyl-scopolamine (1 %), and scopolamine (10 %) by gradient RP-LC-ESI MS/MS [50]. In this study plasma samples were precipitated with acetonitrile and the supernatant was further diluted prior to injection.

Besides TTA and QTA analysis, high popularity and wide spread applicability of ESI is also found for numerous additional substances, e.g. peptides [103] and organophosphorus compounds [104].

3.3.2 Mass Analysers

Whereas the interface produces analyte ions the mass analyser identifies the corresponding masses or may allow fragmentation for tandem-mass spectrometry (MS/MS or MS^2) or multiple fragmentation (MS^n). Single quadrupole (SQ), triple quadrupole (QqQ), and ion trap (IT) systems belong to the most frequent mass analysers also applied to TA analysis (Fig. 5c). Statistical evaluation of the literature referred to in this chapter (Table 5–8) revealed a frequency of 4 % for offline double focusing sector field mass spectrometer (EBQ_1Q_2), 19 % for IT, 21 % for SQ and 56 % for QqQ mass analysers (Fig. 5c).

Early mass spectrometers were simply equipped with a SQ mass analyser merely suitable for full scan MS mode and selected ion monitoring. Even though SQ spectrometers are highly superior to UV-detectors with respect to selectivity, they still bear the risk of deterioration by matrix compounds of similar *m/z* values. Especially, peptides and small proteins may cause a series of diverse *m/z* values due to their multiple charge states after ES ionization potentially interfering with the analytes or IS [103, 105].

In contrast to SQ spectrometers, QqQ systems possess at least three quadrupoles that allow full scan MS and SIM but additionally enable, e.g. product ion scan (MS/MS) and multiple reaction monitoring (MRM) modes. Typically, four quadrupoles are assembled in QqQ instruments. The first quadrupole (Q_0) is part of the ion optics that enables focusing of the ion beam originated from the ion source and is not considered or counted as part of the mass analyser. The second quadrupole (Q_1) functions as mass analyser (mass filter) that allows mass scanning for full scan MS mode or selection of ions for SIM. Additionally, that assembly also enables selection of precursor ions for subsequent fragmentation in the following third quadrupole (collision cell, Q_2). Fragmentation by collision-induced dissociation (CID, collision with N_2 or noble gases) generates product ions valuable for structure elucidation and selective quantification. The fourth and last quadrupole (Q_3) again functions either as mass analyser for the entire set of product ions (product ion scan) or as mass filter to select a subset of product ions prior to detection. Monitoring preselected precursor–product ion transitions is performed for MRM providing optimum selectivity and best signal-to-noise ratios that enable optimum sensitivity. Detection of all product ions is well suited for compound identification considering the complete composition of MS/MS signals.

Ion trap machines allow to perform the same set of scan modes as described for QqQ but additionally provide the opportunity for higher grade fragmentation of product ions (MS^n). This option expands the analytical possibilities for structure elucidation considerably [106].

In contrast to QqQ systems, IT mass spectrometers perform sequential ion trapping, isolation of precursor ions and subsequent mass scanning in a single small assembly (ion trap) making this system robust and valuable.

At last, the double focusing sector field mass spectrometer (EBQ_1Q_2) should briefly be addressed. This mass analyser represents a highly sophisticated early design that has rarely been used for routine analysis especially for quantification. For detailed information on the functional principle the reader is referred to respective textbooks on mass spectrometry. We mention this technique as Kajbaf et al. used EBQ_1Q_2 for detection and identification of the QTA cimetropium and its biotransformation products from liver microsomal mixtures in offline analysis of HPLC fractions after FAB ionization [23, 62] (Table 7).

However, most modern highly accurate time-of-flight (TOF), Fourier-transform ion cyclotron resonance (FT-ICR) and Orbitrap mass analysers [107] have not been reported for TA analysis so far and are thus not discussed in this chapter.

Depending on the instrumental set up and study purpose different scan modes were selected, which are addressed below (Fig. 5b).

3.3.3 Scan Modes

The scan mode determines the extent and quality of mass spectrometric data and has thus to be chosen with respect to the analytical requirements. Detection of unknown compounds, identification of unknown structures and confirmation of known molecules as well as quantification of distinct target analytes require different scan modes for reliable optimum analytical acuity.

3.3.3.1 Full Scan (MS)

The most simple, least sensitive and less selective scan mode is the ordinary MS full scan covering a pre-adjusted *m/z* range. This procedure is well suited for the simultaneous detection of known and unknown intact compounds that may have been produced by, e.g. chemical synthesis or biotransformation processes in vitro and in vivo. For retroactive data processing *m/z* ratios of interest are extracted to generate extracted ion chromatograms that illustrate only the selected mass traces and thus depict the analytes' chromatographic behaviour. No need for predefinition of analyte masses represents a major advantage of this scan option thus avoiding to overlook compounds of interest.

A quite early LC-ESI MS approach was presented by He et al. in 1995 using a SQ analyser for the detection of benztropine and its biotransformation products in urine and bile after oral administration of the drug to rat [59] (Table 7). However, the lack of additional structure elucidation by mass spectrometric fragmentation restricted the analytical significance of the SQ instrument.

A similar full scan MS approach was presented by Tang et al. for screening of quarternary ammonium compounds and quantification of ipratropium in horse urine [24] (Table 5). However, they used an IT mass spectrometer that additionally allowed confirmatory analysis of the drug by MS/MS experiments for verification of HPLC peak-drug assignment.

3.3.3.2 Selected Ion Monitoring

In contrast to the full scan MS mode discussed above, SIM provides improved sensitivity but requires preselection of analyte ions thus preventing from posterior search for unknown substances.

Nevertheless, reliable methods using the SIM mode were introduced for TA analysis. The SIM mode was used for analysis of atropine and scopolamine from human viscera [15] and human plasma [11] (Table 8), of tropisetron from liver microsomal incubation mixtures [82] (Table 7), of granisetron from rat plasma [72], of hyoscyamine enantiomers from human plasma [48], of scopolamine from rabbit plasma [89], of tiotropium from human plasma [81] (Table 5) and of atropine from dog plasma [96] (Table 6). The corresponding values for the monoisotopic molecular weight (MW) and the corresponding *m/z*-values of protonated TTA and pure QTA are summarized in Table 9.

However, the SIM technology was only applied with minor frequency of all scan modes (17 %) (Fig. 5b).

3.3.3.3 Product Ion Scan

The product ion scan provides the entire information on product signals and thus allows structural assignment by diagnostic ions or by comparison to spectral data bases [108]. This scan mode was a highly valuable tool for confirmative identification of novel in vivo biotransformation products from diverse TTA as elaborated by Chen et al. for anisodamine [6], anisodine [5], atropine [51, 52], and scopolamine [7, 87]. The strategy and principle of these studies including "neutral loss" and "precursor ion scans" will be discussed below in Sects. 3.4.3. Table 9 summarizes *m/z*-values of the most dominant product ion signals of TTA and QTA referred to in this chapter.

No examples for quantification in the product ion scan mode were found in the literature even though data processing would allow extraction of selected ions, integration of related signal areas, and summation for quantification. This procedure has been used by John et al. for the determination of the human haemoglobin derived peptide hHEM-γ 130–146 in plasma [102]. However, quantification especially of small molecule analytes is best performed in the MRM mode that is addressed below.

3.3.3.4 Multiple Reaction Monitoring

MRM is the most frequently used scan mode (61 % of all applications, Fig. 5b) predestinated for most selective and highly sensitive quantification. Typically, QqQ systems were used even though IT spectrometers are also capable for this scan mode. More rapid scan rates and superior linear range are commonly accepted preferences of the QqQ design.

In principle, simple quantification only requires to monitor one transition from the precursor to one product ion (quantifying ion). The resulting selectivity is often very satisfying but should be improved by additional monitoring of at least a second transition from the precursor to a second product ion (qualifying ion).

Table 9 summarizes precursor ions and product ions that were typically used as quantifying and qualifying ions. Detection of both transitions allows comparison of resulting product ion signal intensities. The calculated ratio should be constant (±20 %) under standard operating conditions. Deviation from the tolerance level is indicative for interferences that deteriorate quantitative measurement. Even though such interfering compounds must exhibit (a) the same chromatographic retention time as the analyte or IS, (b) an identical precursor ion, and (c) an identical product ion, it is not that unlikely that MRM determinations are influenced [12, 109]. Therefore, analysis of samples especially of quite unknown origin and composition as present in toxicological screening should be investigated with a maximum degree of selectivity [109–113].

Modern QqQ instruments equipped with high-end fast electronics, and accompanying optimized software allow to follow several hundreds of analytes in one run by MRM thus making this scan mode the method of choice for qualitative and quantitative analysis. Therefore, MRM was commonly applied to TTA and QTA analysis for quantification in PK (Table 5), distribution (Table 6), and biotransformation studies (Table 7) as well as for toxicological screening and evidence of drug abuse (Table 8). Specificities and remarkable characteristics of these fields of application are addressed in the following sections.

3.4 Fields of Application

Quantification of known analytes in PK and distribution studies makes different demands on the analytical procedure than detection of unknown compounds in biotransformation experiments or identification of unknown and postulated molecules in toxicological screening. For example, requirements for quantitative analysis of fixed analytes with optimum sensitivity and selectivity differ from those for qualitative detection of intact molecular weight or diagnostic MS/MS fragments. Selectivity of sample preparation and applicability of diverse scan modes represent relevant critical issues. The following sections address this context.

3.4.1 PK Studies

PK studies aim to monitor the concentration-time profiles of distinct drugs in the circulation by analysing plasma or serum. Therefore, sample preparation, chromatographic separation and mass spectrometric detection are typically optimized for a single analyte. Achieving best selectivity the MRM mode was used most frequently as summarized in Table 5. Only rare examples are found applying the simple SIM mode detection most often due to the use of SQ instruments. However, the LOQs for TA quantification were often quite similar being approximately 0.5 ng/ml (Table 5).

Nevertheless, LOQs for tiotropium being 1,000-fold lower were reported by Ding et al. [81] and Wang et al. [80] obtained by isocratic chromatography coupled either to a SQ mass spectrometer for the SIM mode [81] or to a QqQ instrument for MRM mode detection [80] (Table 5). Even though Ding et al. performed a 17-fold sample concentration after SPE and injected a 1.5 ml plasma equivalent for analysis [81] and Wang et al. concentrated it by a factor of 2.7 after precipitation and LLE followed by injection of a 107 μl plasma equivalent [80], LOQs appeared surprisingly low. Presumably, the permanent charge of the QTA tiotropium (Fig. 1) was primarily responsible for excellent detector response. Mass spectrometric transitions are listed in Table 9. However, peak plasma concentrations of tiotropium after inhalation of a single 18 μg dose were as small as 20 pg/ml [80] and 13 pg/ml [81] in human PK studies.

Much higher drug concentrations are established when administering atropine as antidote for OP poisoning.

Abbara et al. performed simultaneous quantification of different antidotes (diazepam, pralidoxime and atropine) typically co-administered for the therapy of anticholinesterase poisoning (Table 5) [44]. PK data resulting from i.m. drug injection by means of a bi-compartemental auto-injector were calculated from human plasma concentrations measured by LC-ESI MS/MS with MRM settings. Administration of 2 mg atropine sulphate yielded plasma peak concentrations of about 4 ng/ml 15 min after injection.

John et al. analysed concentration-time profiles of total atropine and both the corresponding hyoscyamine isomers after i.v. administration of atropine sulphate (25 mg) to an organophosphorus (OP) pesticide-poisoned human patient yielding maximum concentrations of about 600 ng/ml for total atropine and about 300 ng/ml for each enantiomer found 30 min after injection (Table 5) [49]. Samples were measured after enzymatic pretreatment (incubation with atropinesterase) prior to LC-ESI MS/MS analysis in the MRM mode allowing enantioselective quantification of *R*- and *S*-hyoscyamine [49]. Concentration-time profiles documented that elimination of *S*-hyoscyamine in man appeared to be more rapid than of *R*-hyoscyamine. These findings were similar to data reported by Siluk et al. using a chiral LC-APCI MS method [48] (Table 5) and by Aaltonen et al. [45] and Kentala et al. [46] both using a combination of radio-receptor assay (RRA) and radioimmunoassay (RIA).

The enantioselective procedure of John et al. was also originally applied to monitor concentration-time profiles of atropine and hyoscyamine variants in a PK study in healthy swine (Table 5) [47]. Mass spectrometric characteristics with respect to precursor and product ions of atropine and hyoscyamine are summarized in Table 9. Following single i.v. administration of 100 μg/kg, maximum plasma concentrations were found to be 48 ng/ml for atropine and 24 ng/ml for both enantiomers, the distomer *R*-hyoscyamine and the eutomer *S*-hyoscyamine. In contrast to data in human, no stereoselective preference for elimination was found in swine thus substantiating the assumption that hyoscyamine kinetics in man differ from that in swine.

Similar results were obtained with the same LC-ESI MS/MS method when quantifying atropine and hyoscyamine variants in plasma of OP-poisoned swine under atropine therapy. Swine were topically exposed to the nerve agent VR (302 μg/kg, t_0)

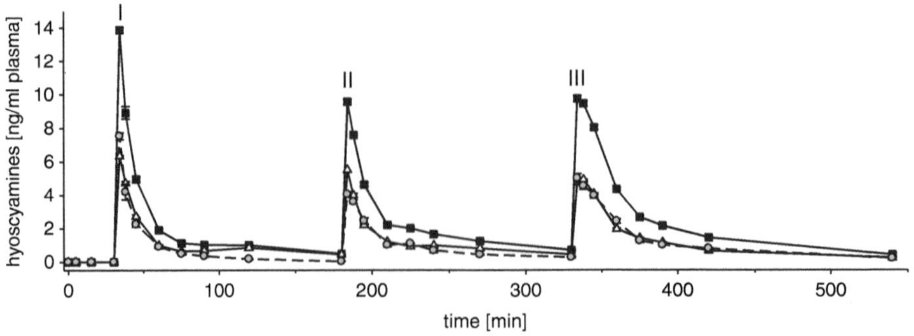

Fig. 6 Concentration-time profile of antidotal atropine and its enantiomers *S*- and *R*-hyoscyamine in plasma of an in vivo swine study. Swine were topically exposed to the nerve agent VR (302 µg/kg, t_0) followed by administration of atropine sulphate (30 µg/kg) and the reactivating oxime HI 6 (12.8 mg/kg) via three i.m. injections into the rear leg at 30 (I), 180 (II) and 330 min (III). Blood samples were collected at distinct time points to generate EDTA plasma. Maximum concentrations were found 4 min after drug administration each. No differences of *S*- and *R*-Hyo concentrations were evident underlining similar elimination kinetics for both enantiomers. Data are mean and SD from duplicate measurement using the enantioselective LC-MS/MS approach of John et al. [47, 49]. *Black circles*, total hyo; *grey circles*, *S*-hyo; *grey triangles*, *R*-hyo

followed by administration of atropine sulphate (30 µg/kg) and the reactivating oxime HI 6 (12.8 mg/kg) via three i.m. injections into the rear leg at 30 (I), 180 (II) and 330 min (III). Blood samples were collected from a catheter placed in a branch of the saphenous artery into EDTA tubes at distinct time points during the 540 min experimental time periods. The animal study was performed by John Mikler from Defence Research and Development Canada-Suffield, Medicine Hat, Alberta, Canada. Subsequent plasma analysis was carried out in our laboratory at the Bundeswehr Institute of Pharmacology and Toxicology, Munich, Germany. Figure 6 illustrates the resulting concentration-time profiles of *R*- and *S*-hyoscyamine and atropine as free base underlining similar elimination kinetics for both enantiomers.

Such enantioselective investigations are necessary to avoid misleading correlations of drug concentrations with clinical parameters based on interaction with muscarinic receptors.

Accordingly, Siluk et al. performed chiral chromatographic APCI-MS analysis of human plasma samples obtained from a PK study after administration of atropine sulphate (10 µg/kg as initial i.v. bolus followed by 30 min infusion of 20 µg/kg) [48]. Peak plasma concentrations of both enantiomers were not reported in detail but were presumably about 5 ng/ml as deduced from the illustrated concentration-time profile. The authors found that *S*-hyoscyamine was eliminated slightly faster than *R*-hyoscyamine.

No further studies were found investigating stereoselective PK behaviour of any TA monitored by LC-MS-based methods.

Similar and considerable lower maximum therapeutic drug concentrations in human plasma were analysed in PK studies with *N*-butyl-scopolamine (about 1 ng/

ml after oral application [79]), cimetropium, granisetron, tropisetron and trospium (about 6 ng/ml after oral dosing [61, 71, 73, 83, 84]), and scopolamine (13 ng/ml after i.v. injection [97], 7 ng/ml after 15 min i.v. infusion [98], 0.25–1.3 ng/ml after nasal administration [88]) (Table 5). Corresponding MS/MS transitions are summarized in Table 9.

Whereas PK studies referred above required analysis of drugs in the circulation, distribution studies discussed next were performed to investigate drug transfer into or through other compartments.

3.4.2 Distribution Studies

Van Zyl et al. reported on the diffusion of ipratropium through porcine bronchial epithelium tissue [74]. In principle, ipratropium is administered via the respiratory tract by inhalation to treat pulmonary diseases associated with bronchoconstriction. Therefore, pulmonary absorption by bronchial tissue determines its local efficacy and was thus investigated in a diffusion cell in vitro. Bronchial epithelium was equilibrated in PBS and discs of 4 mm^2 were mounted on that diffusion cell separating the donor and receiver compartment. The donor compartment contained the drug dissolved in PBS (1 mg/ml) and the receiving chamber was permanently flushed with a low flow (1.5 ml/h) of PBS thus allowing time-resolved fractionation for subsequent direct analysis by LC-ESI MS/MS in MRM mode. Transition to the product ion at *m/z* 124 was monitored for quantification (Table 9). The transfer of ipratropium was characterized by the flux (about 220 ng/cm^2/min) and the permeability coefficient calculated to be 1.6×10^{-8} cm/s.

Whereas ipratropium is inhaled for therapy, the MR agonist (−)-Sat (Fig. 1) is thought to be directly dropped into the eye to treat glaucoma and reduce IOP [85]. Therefore, penetration of satropane into the anterior chamber determines its therapeutic efficacy. Accordingly, Fu et al. analysed this process in an in vivo study using anaesthetized rabbits. A linear probe was implanted in the centre of the anterior chamber of dilated rabbit pupils. Perfusion was performed at 2 μl/min with isotonic PBS. Satropane was instilled into the eyes and microdialysates were collected after distinct periods of time for subsequent direct LC-ESI MS/MS analysis in MRM mode. Transition from the *m/z* 354 precursor ion to the product ion at *m/z* 182 was monitored for measurement as listed in Table 9. Matrix effects of aqueous humour dialysates caused about 18 % suppression of satropane ionization but allowed reliable quantification. Even though the paper of Fu et al. presents careful method validation no physiological data on drug distribution were provided.

Callegari et al. characterized the central nervous system (CNS) penetration of the QTA trospium and methyl scopolamine as well as of the TTA scopolamine [77]. Therefore, they performed in vivo studies administering the drugs to rats subcutaneously. After 1 h animals were euthanized, cerebrospinal fluid (CSF) was taken from cisterna magna and brains were removed and homogenized. Whereas CSF was mixed with IS (atropine) to be directly injected for LC-ESI MS/MS analysis, brain homogenates were extracted by SPE on Oasis HLB material subsequent to IS addition. Dried methanolic eluates were reconstituted in 5 mM ammonium acetate buffer (0.1 % v/v formic acid) prior to mass spectrometric analysis (Table 6). MRM transitions used for trospium are given in Table 9. In addition, Callegari et al. performed in vitro binding studies to characterize the unbound (free) drug fraction in brain homogenates using a 96-well equilibrium dialysis block. Analysis by LC-ESI MS/MS was carried out by the same method. The authors found, that in contrast to the tertiary TA scopolamine, the quarternary TA trospium and methyl scopolamine did not show significant CNS penetration.

The three examples introduced above present applications of LC-MS-based methods to determine specific drug transfer into compartments relevant for systemic or local activity and thus providing basic pharmacological or toxicological data.

Evaluation of biotransformation processes in vivo and in vitro is also an essential issue for drug characterization that is addressed in the next section.

3.4.3 Biotransformation Studies

Biotransformation processes may partly convert a less polar parent drug into activated more polar forms (phase I) and subsequently into highly polar often charged conjugates (phase II) that are excreted via liver and kidney into urine. Most TA undergo such transformation even though not in quantitative yield. Evaluation of drug metabolism in vivo is indispensable for understanding of pharmacokinetic and pharmacological properties. Often laboratory animals were used as adequate models, e.g. rat [5–7, 37, 51, 59, 87], dog and monkey [37]. Despite its obvious limitations several in vitro systems, e.g. organ homogenates (e.g. liver [5, 6, 23, 37, 51, 82]), plasma [49, 50], bacteria cultures [87] as well as pure isolated or engineered enzymes [5, 6, 23, 37, 51, 82] were also used as valuable tools to clarify and describe drug biotransformation.

Table 7 summarizes LC-MS-based methods that were applied to study metabolism of TA and Fig. 7 depicts some selected metabolites identified by mass spectrometry.

Chen et al. impressively presented the possibilities of modern LC-MS equipment to elucidate metabolism of several TA [5–7, 51, 52, 87]. The strategy was based on the following considerations:

1. The general pathways of phase I and II metabolism are effective for TA.
2. Phase I and II metabolism add distinct and predictable chemical moieties to the parent drug that cause defined changes in the molecular mass but keep the basic structure (skeleton) unaltered (e.g. +16 Da for single oxidation, +32 Da for twofold oxidation, +30 Da for methoxylation, −18 Da for dehydration, −14 Da for demethylation, +176 Da for glucuronidation, +80 Da for sulpho-conjugation).
3. The structural skeleton of the parent TA and its modified biotransformation products undergo similar mass spectrometric fragmentation by CID resulting in identical (diagnostic) or specifically shifted signals of product ions (e.g. cleavage of the tropane moiety resulting in fragments at *m/z* 124 and 93, Fig. 8).

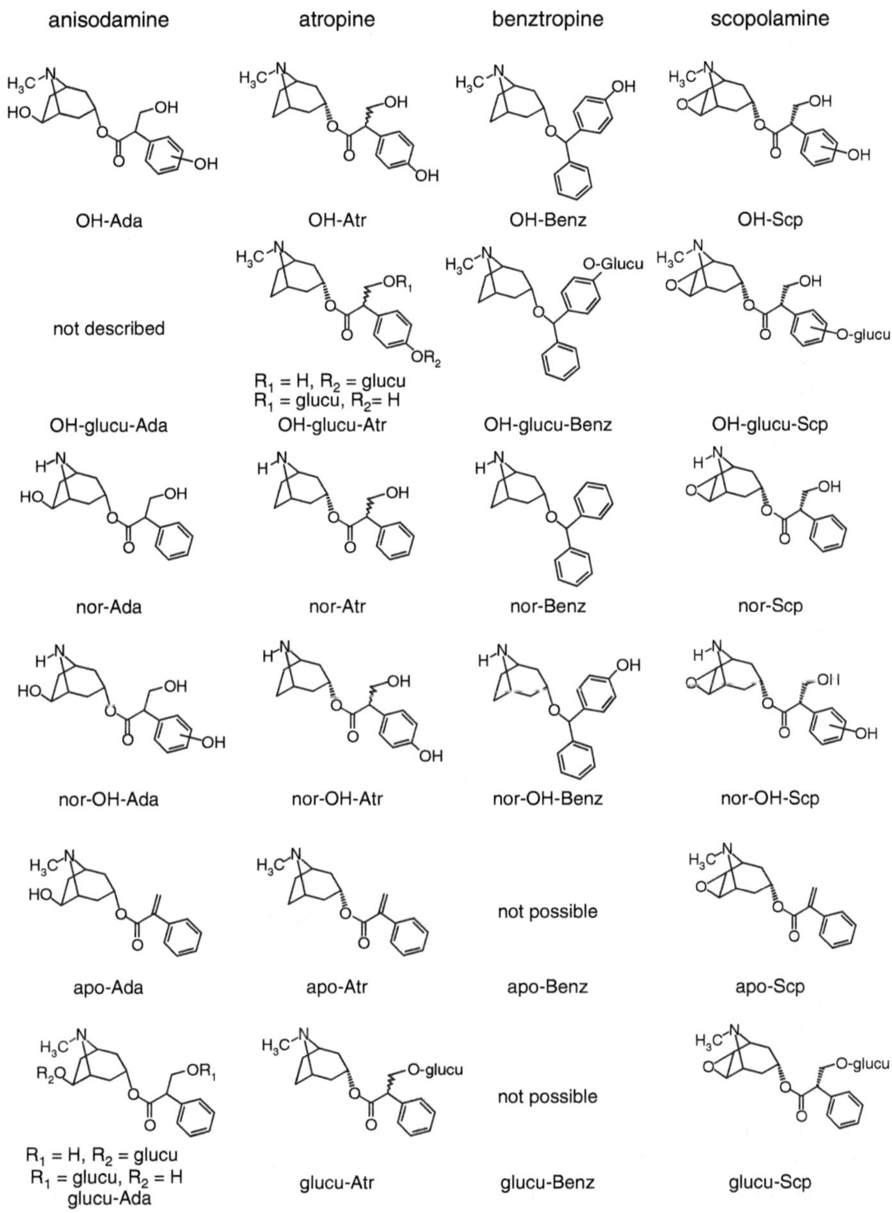

Fig. 7 Selected biotransformation products of diverse tropane alkaloids identified by LC-MS/MS approaches. Depicted metabolites are only an exert of those presented by Chen et al. for anisodamine (Ada) [6], atropine (Atr) [51], and scopolamine (Scp) [7] as well as by He et al. for benztropine (Benz) [59]. Structures were identified from in vivo and in vitro samples following diverse modes of MS/MS analysis as discussed in the section: *Biotransformation studies*

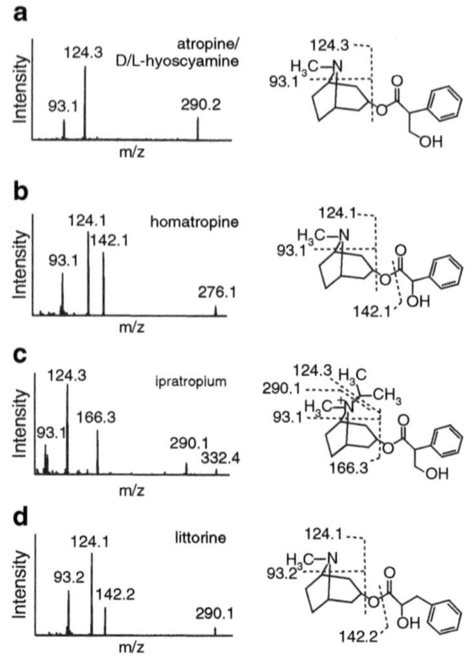

Fig. 8 Product ion spectra of diverse tropane alkaloids after collision-induced dissociation. Spectra were obtained after positive electrospray ionization and collision-induced dissociation (CID) with nitrogen in a QTrap 4000 instrument. Common structural elements cause generation of corresponding diagnostic fragments, e.g. *m/z* 124 and 93. For structural assignment of biotransformation products similar fragmentation pattern were produced and interpreted as shown by Chen et al. [6, 7, 51] and He et al. [59]

4. The parent drug and its metabolites exhibit different retention times in RP-LC separation being predictive for altered polarities.

Accordingly, Chen et al. performed a serious number of charming in vivo and in vitro biotransformation studies for diverse TTA (anisodamine, anisodine, atropine, scopolamine) after oral administration of the drugs to rat [5–7, 51, 52, 87]. Urine was analysed by LC-MS/MS methods that provided the following operation modes:

1. Product ion scan: This type of scan monitors all product ion signals within a pre-adjusted *m/z* range that are produced from a pre-selected precursor ion representing the expected mass of a postulated biotransformation product. The entire set of product ions supports interpretation of diagnostic fragments and thus of structural identification.
2. Precursor ion scan: Following this detection mode all analyte ions are captured that generate a specified product ion during CID. With respect to TA structures the MS/MS signal at *m/z* 124 could be defined as a common product ion (representing the tropane moiety) that allows to detect the protonated masses of any precursor ion that generates this tropane moiety signal. Figure 8 depicts MS/MS spectra of diverse TA that were obtained after positive ESI and CID. The common presence of the signal at *m/z* 124 illustrates that atropine as well as homatropine, ipratropium and littorine would be detectable by a corresponding precursor ion scan.

3. Neutral loss: Neutral loss monitoring allows the detection of any compound that loses a specified mass during CID. This scan type allowed to detect any glucuronides (neutral loss of 176 Da), sulpho-conjugates (loss of 80 Da) and was helpful to identify molecules that lose a tropic acid moiety (loss of 166 Da) or a tropic acid-derived HCOH group (loss of 30 Da) [51, 52]. The occurrence of the latter losses provided helpful hints for the presence of an (partly) unaltered tropic acid moiety in, e.g. atropine.
4. Measurement in negative mode: Basic TTA and positively charged QTA are typically ionized in positive mode (Table 5–8). Nevertheless, charge properties are significantly changed after generation of glucuronide or sulpho-conjugates providing negatively charged groups. Therefore, these biotransformation products can be detected after ionization in the negative mode. Measurement of corresponding MS/MS spectra and comparison to data obtained in positive mode support structural identification. In negative mode Chen et al. detected glucuronides of, e.g. nor-hyoscyamine (*m/z* 450), hyoscyamine (*m/z* 464, Fig. 7) and hydroxylated-hyoscyamine (*m/z* 480, Fig. 7) as well as their corresponding sulpho-conjugates at *m/z* 354, 368 and 384, respectively [51, 52]. Detection after positive ionization provided signals that were 2 Da heavier each. Similar metabolites were also found for anisodine [5], scopolamine [7, 87] (Fig. 7) and anisodamine [6].

In addition to conjugates a large number of diverse phase I and II metabolites were proposed including apo-(dehydrated) and nor-(demethylated) variants as well as hydroxylated and methoxylated forms (Fig. 7). However, HPLC peak-metabolite assignment only based on MS fragmentation data, chromatographic elution order and some enzymatic control studies. Even though assignments sound reasonable, none of the referred reports provided evidence by comparative analysis using chemically synthesized reference substances as typically required in pharmaceutical development.

Similar results were obtained by He et al. analysing biotransformation products of benztropine (Fig. 7) in urine and bile after oral application of the drug to rats using a simple single quadrupole instrument with ESI (Table 7) [59]. Nevertheless, comparison to chemically synthesized standards for *N*-desmethylbenztropine (norbenztropine, Fig. 7) and benztropine *N*-oxide provided important substantiation for proposed metabolite structures.

Accordingly, Gordon et al. used synthetic reference compounds for metabolites of bemesetron (MDL 72,222) to evidence proposed structures (Table 7) [37].

However, in contrast to the IT and SQ analysers Chen et al., He et al. and Gordon et al. used, more modern mass spectrometers providing high-resolution equipment, e.g. Orbitrap, FT-ICR or double time-of-flight (TOF-TOF), would allow more adequate and precise identification by determination of the accurate mass.

Enzymatic degradation of several TTA and QTA by atropinesterase from rabbit serum was investigated by John et al. [50]. It was found that all TTA tested (atropine, *S*-hyoscyamine, littorine, scopolamine, homatropine and cocaine) were hydrolysed (ester cleavage) by atropinesterase with different velocities whereas both QTA (ipratropium and *N*-butyl-scopolamine) remained unaffected. Furthermore, as mentioned above, atropine was degraded to half of its initial concentration due to the enzymatic stereoselective preference for *S*-hyoscyamine (Fig. 4). This enzyme activity is solely found in rabbits and explains the animals' high tolerance towards ingested plants of the *solanaceae* family (e.g. deadly nightshade) that are in general toxic to other mammals [114, 115].

This aspect of poisoning is addressed in the last section introducing LC-MS methods for toxicological and forensic screening to detect toxic TA in body fluids after drug abuse or ingestion of toxic plants.

3.4.4 Toxicological Screening and Evidence of Drug Abuse

In contrast to PK or biotransformation studies that require detection of known or at least postulated compounds, toxicological screenings appear much more sophisticated. In general, the drugs or poisons as well as their concentrations in body fluids are unknown and thus require elaborated and reliable analytical tools for initial unambiguous identification and subsequent quantification. Specifications for post-mortem analysis are even higher [113]. Toxicological judgment and clinical or forensic consulting demands authoritative analysis.

Even though LC-MS/MS procedures promise an optimum of selectivity, careful validation of the entire procedure is mandatory for a highly selective, sensitive, accurate and coherent analysis [116]. Therefore, validation of LC-MS methods should address selectivity, calibration model (linearity), stability (long-term, freeze and thaw, benchtop), accuracy, precision, LOQ, LOD, recovery, reproducibility, ruggedness and matrix effects [117]. Matrix effects represent a highly critical issue in bioanalysis especially for trace analysis when performing ESI that is highly susceptible for these deteriorating phenomena. Ion suppression might be the most frequent impact of co-eluting IS or compounds derived from the matrix. Especially analytes of lowest concentrations might be overseen. To prevent from false results appropriate sample preparation and clean-up have to be worked out.

Very often plasma or serum is used for quantitative purposes allowing correlation to pharmacological or toxicological effects. Even though oral fluid is regarded as alternative to plasma for drug screening it is not capable for TTA and QTA analysis as these compounds are ionized and do thus not correlate to plasma concentrations [116]. Therefore, no LC-MS-based applications were found for oral fluid as summarized in Table 8. Furthermore, atropine intoxication for example causes dry mouth and thus makes sample taking by spitting quite difficult.

In addition, hair might represent a valuable specimen for confirmation of drug abuse [95] even though specific pitfalls should be avoided for reliable measurement [94]. Table 8 refers to two LC-MS/MS methods that were used to detect atropine and scopolamine in hair [56, 57].

Finally, urine is also an important body fluid to detect drug and poisons and their corresponding biotransformation products. However, correlation of concentrations in urine to pharmacological effects is hardly possible. For urine sample preparation simple dilution is often sufficient prior to direct injection onto the HPLC column. Methods used to detect atropine and scopolamine [12–14, 55] as well as ipratropium and *N*-butyl-scopolamine [78] are listed in Table 8.

More conventional drug screening approaches are designed as multi-analyte procedures for detection of a specified and limited number of analytes typically performed by LC-MS/MS in the MRM mode [117]. In contrast, more modern procedures make use of a linear QTrap instrument applying a survey of complex scan modes, e.g., MRM, information dependent acquisition (IDA), and enhanced product ion scan (EPI). Resulting data are compared with library entries for identification as presented by Mueller et al. [118]. This new concept of multi-target screening (MTS) comprises 301 forensically important drugs also including atropine, benztropine, methyl scopolamine, cocaine and scopolamine.

To conclude this section some curious and daunting examples of TA intoxications analysed by LC-MS should be mentioned.

Incidental and accidental intake of atropine and scopolamine, which are the main tropane alkaloids in plants of the *solanecae* family, may provoke poisoning of man and livestock [11, 13–15, 55, 57, 119–122] causing agitation, aggression, hallucinations, dry mouth and skin, mydriasis, loss of consciousness followed by coma combined with tachycardia, hypotension, and hyperthermia [57, 121]. A detailed statistical analysis of paediatric plant exposures in Germany within the years 1998–2004 has been provided by Pietsch et al. [123]. They found that most prevalent victims of accidental plant exposures are children in the age of 1–6 years presumably being misled by the attractive plump berries.

A curious case of hallucinogenic scopolamine intoxication has been reported from a prison inmate who smoked cigarettes spiked with the bronchodilatory antispasmotic buscopan. LC-ESI MS/MS analysis proved that the API of this drug (*N*-butyl-scopolamine, Fig. 1) undergoes pyrolysis during smoking thus producing scopolamine that was directly inhaled with the cigarette smoke and caused systemic effects [124].

An alarming toxidram of atropine-poisoning in combination with cocaine abuse was described by Boermans et al. [53]. A high number of habitual cocaine users were brought to the hospital due to assumed cocaine intoxication but showed untypical signs and symptoms (e.g. restlessness, excitability, hallucinations, abdominal pain, vomiting and muscular spasm). Analysis of serum by LC-ESI MS/MS revealed the presence of atropine and confirmed the suspicion that users unintentionally consumed atropine-adulterated cocaine (25 % w/w Atr_2SO_4) thus provoking acute mortal danger by cardiovascular effects (Table 8). The source and reason for atropine adulteration was not addressed in the paper.

An unsavoury case study of TA poisoning was reported by Kintz et al. proving the systematic and continuous scopolamine intake of children who were forced by their own mother to take 4–10 Feminax tablets per day for months. Evidence of scopolamine was achieved by LC-ESI MS/MS analysis of hair samples (Table 3). The Children survived and their mother had to face a charge of their offence [56].

Moreover, TA were also misused to commit suicide sometimes in combination with alcohol and other drugs [125–127].

4 Conclusions

Many TA are well characterized with respect to their binding properties towards diverse receptors including MR, 5-HT$_3$R, α1-AR and α7-nAChR. Most frequently TA were identified as competitive acetylcholine antagonists on MR thus exhibiting therapeutic benefit to induce mydriasis, spasmolysis of the respiratory system, GIT and overactive bladder, anaesthesia and analgesia (atropine, benztropine, *N*-butyl-scopolamine, cimetropium, homatropine, ipratropium, methyl scopolamine, scopolamine, *S*-hyoscyamine, tiotropium, and trospium). In contrast, certain TA such as satropane show contrary effects such as smooth muscle contraction thus being considered for glaucoma treatment. TA that exhibit strong binding to 5-HT$_3$R allow their use as antiemetic (bemesetron, granisetron, scopolamine, tropisetron). In addition, affinity of some TA towards α1-AR provides activity to treat circulatory disorders and septic shock by improvement of blood flow (anisodamine, anisodine).

Whereas most TA discussed in this chapter are well-established drugs either worldwide or in specific countries, a huge number of synthetic derivatives might be designed to improve efficacy, selectivity and reduce potential adverse side effects.

Such novel APIs are to be characterized with respect to, e.g. biotransformation, distribution, PK behaviour and other pharmacological properties. Elaboration of these data will require sensitive, selective and robust analytical methods most presumably based on LC-MS techniques. With respect to optimized economic feasibility procedures should be preferred that allow both (a) reduced consumption of organic solvents for sample preparation and chromatographic separation and (b) short run-time analysis.

These requirements may be best fulfilled when using automated approaches for sample preparation or implementing in-line sample processing that minimize the extent of manual working steps. Chromatography could be optimized by the use of sub-2 μm particles for improved resolution and solvent reduction. Comparable results might be expected when applying fused core particles (2.7 μm overall diameter) providing a solid, non-porous core (1.7 μm diameter) that is coated with a thin porous shell (0.5 μm thickness). The shell contains the functionalized chromatographic phase, e.g. C18, interacting with the analytes. Therefore, only the thin layer is accessible for diffusion thus improving mass transfer kinetics, reducing axial dispersion of solute and minimizing peak width.

With respect to MS equipment significant performance improvements are promised by the use of most modern mass analysers (e.g. Orbitrap, FT-ICR, TOF-TOF) that provide highest resolution and mass accuracy important for, e.g. in-depth elucidation of biotransformation.

Nevertheless, despite the use of high-end LC and MS systems validation of any method should be carried out carefully to characterize validity and reliability. Special attention should be paid to matrix effects occurring during ionization of analytes that might deteriorate analytical performance.

At least, extended compound/spectra data bases especially for ESI that allow adequate fitting and search algorithms will help to improve identification of unknown analytes relevant for toxicological and forensic samples.

To conclude, TA analysis by LC-MS methods will benefit from general technical progress in the near future allowing to obtain more rapid and exact data for qualitative and quantitative measurement.

References

1. Grynkiewicz G, Gadzikowska M (2008) Tropane alkaloids as medicinally useful natural products and their synthetic derivatives as new drugs. Pharmacol Rep 60:439–463
2. Li F, Sun S, Wang J, Wang D (1998) Chromatography of medicinal plants and Chinese traditional medicine. Biomed Chromatogr 12:78–85
3. Cook EB, Dennis M, Ochillo RF (1981) Application of thin layer, ion exchange, and high performance liquid chromatography to separate pharmacologically active components of an African arrow poison of plant origin. J Chromatogr 4:549–557
4. Zhu L, Yang LM, Cui YY, Zheng PL, Niu YY, Wang H, Lu Y, Ren QS, Wei PJ, Chen HZ (2008) Stereoselectivity of satropane, a novel tropane analog, on iris muscarinic receptor activation and intraocular hypotension. Acta Pharmacol Sin 29:177–184
5. Chen H, Chen Y, Du P, Han F (2007) Structural elucidation of *in vivo* and *in vitro* metabolites of anisodine by liquid chromatography-mass spectrometry. J Pharm Biomed Anal 44:773–778
6. Chen Y, Du P, Han F, Chen H (2007) Characterization of *in vivo* and *in vitro* metabolic pathway of anisodamine by liquid chromatography–tandem mass spectrometry. J Liq Chromatogr Rel Technol 30:1933–1949
7. Chen H, Chen Y, Wang H, Du P, Han F, Zhang H (2005) Analysis of scopolamine and its eighteen metabolites in rat urine by liquid chromatography–tandem mass spectrometry. Talanta 67:984–991
8. Balunas MJ, Kinghorn AD (2005) Dug discovery from medicinal plants. Life Sci 78:431–441
9. Christen P (2000) Tropane alkaloids: old drugs used in modern medicine. In: Rahman A (ed) Studies in natural products chemistry, vol 22, bioactive natural products (part C). Elsevier, Amsterdam
10. Ali-Melkkilä T, Kanto J, Isalo E (1993) Pharmacokinetics and related pharmacodynamics of anticholinergic drugs. Acta Anaesthesiol Scand 37:633–642
11. Beyer J, Peters FT, Kraemer T, Maurer HH (2007) Detection and validated quantification of toxic alkaloids in human blood plasma – comparison of LC-APCI-MS with LC-ESI-MS/MS. J Mass Spectrom 42:621–633
12. Sauvage FL, Gaulier JM, Lachatre G, Marquet P (2008) Pitfalls and prevention strategies for liquid chromatography–tandem mass spectrometry in the selected reaction-monitoring mode for drug analysis. Clin Chem 54:1519–1527
13. Bjornstad K, Hulten P, Beck O, Helander A (2009) Bioanalytical and clinical evaluation of 103 suspected cases of intoxications with psychoactive plant materials. Clin Toxicol 47:566–572
14. Gerber R, Naude TW, de Kock SS (2006) Confirmed *Datura* poisoning in a horse most probably due to *D. ferox* in contaminated tef hay. J S Afr Vet Assoc 77:86–89
15. Steenkamp PA, Harding NM, van Heerden FR, van Wyk BE (2004) Fatal *Datura* poisoning: identification of atropine and scopolamine by high performance liquid chromatography/ photodiode array/mass spectrometry. Forensic Sci Int 145:31–39

16. Auriola S, Martinsen A, Oksman-Caldentey KM, Naaranlahti T (1991) Analysis of tropane alkaloids with thermospray high-performance liquid chromatography–mass spectrometry. J Chromatogr 562:737–744
17. Adams M, Wiedemann M, Tittel G, Bauer R (2006) HPLC-MS trace analysis of atropine in *Lycium barbarum* berries. Phytochem Anal 17:279–283
18. Jousse C, Vu TD, Tran TLM, Al Balkhi MH, Molinie R, Boitel-Conti M, Pilard S, Mathiron D, Hehn A, Bourgaud F, Gontier E (2010) Tropane alkaloid profiling of hydroponic *Datura innoxia* Mill. plants inoculated with *Agrobacterium rhizogenes*. Phytochem Anal 21:118–127
19. Banerjee S, Madhusudanan KP, Chattopadhyay SK, Rahman LU, Khanuja SPS (2008) Expression of tropane alkaloids in the hairy root culture of *Atropa acuminata* substantiated by DART mass spectrometric technique. Biomed Chromatogr 22:830–834
20. Humphrey AJ, O′Hagan D (2001) Tropane alkaloid biosynthesis. A century old problem unresolved. Nat Prod Rep 18:494–502
21. SPARC on-line calculator, release w4.6.1691–s4.6.1687, University of Georgia, USA http://archemcalc.com/sparc/smiles/smiles.cfm?CFID=22823&CFTOKEN=55164375. Accessed 11 Aug 2011
22. Sangster J (1994) LOGKOW databank. Sangster Research Laboratories, Montreal, Quebec, Canada
23. Kajbaf M, Jahanshahi M, Pattichis K, Gorrod JW, Naylor S (1992) Rapid and efficient purification of cimetropium bromide and mifentidine drug metabolite mixtures derived from microsomal incubates for analysis by mass spectrometry. J Chromatogr 575:75–85
24. Tang FPW, Leung GNW, Wan TSM (2001) Analyses of quaternary ammonium drugs in horse urine by capillary electrophoresis–mass spectrometry. Electrophoresis 22:2201–2209
25. Molinspiration Cheminformatics, http://www.molinspiration.com/cgi-bin/properties. Accessed 11 Aug 2011
26. U.S. Environmental Protection Agency, Mid-Continent Ecology Division. http://www.epa.gov/med/Prods_Pubs/smiles.htm. Accessed 11 Aug 2011
27. Hansch C, Leo A, Hoekman D (1995) Exploring QSAR – hydrophobic, electronic, and steric constants. American Chemical Society, Washington, DC
28. Ingelse BA, van Dam RCJ, Vreeken RJ, Mol HGJ, Steijger OM (2001) Determination of polar organophosphorus pesticides in aqueous samples by direct injection using liquid chromatography–tandem mass spectrometry. J Chromatogr A 918:67–78
29. Renner UD, Oertel R, Kirch W (2005) Pharmacokinetics and pharmacodynamics in clinical use of scopolamine. Ther Drug Monit 27:655–665
30. Tytgat GN (2007) Hyoscine butylbromide, a review of its use in the treatment of abdominal cramping and pain. Drugs 67:1343–1357
31. Flynn RA, Glynn DA, Kennedy MP (2009) Anticholinergic treatment in airways diseases. Adv Ther 26:908–919
32. Evangelista S (2004) Quaternary ammonium derivatives as spasmolytics of irritable bowel syndrome. Curr Pharm Design 10:3561–3568
33. Antimuscarinic Agents (1996) In: Reynolds JEF (ed.) Martindale the extra pharmacopoeia, 31st edn. Royal Pharmaceutical Society, London
34. Chapple CR, Khullar V, Gabriel Z, Muston D, Bitoun CE, Weinstein D (2008) The effects of antimuscarinic treatments in overactive bladder: an update of a systematic review and meta-analysis. Eur Urol 54:543–562
35. Schmeller T, Sporer F, Sauerwein M, Wink M (1995) Binding of tropane alkaloids to nicotinic and muscarinic acetylcholine receptors. Pharmazie 50:493–495
36. Freeman AJ, Bountra C, Dale TJ, Gardner CJ, Twissell DJ (1993) The vomiting reflex and the role of 5-HT_3 receptors. Anti-Cancer Drugs 4(Suppl 2):9–15
37. Gordon WP, Cheng H, Larsen DL, Ragner JA, Landmesser NG (1992) identification of urinary metabolites of 8-methyl-8-azabicyclo-[3,2,1] octan-3-yl 3,5-dichlorobenzoate (MDL 72,222) in the dog and monkey. Drug Metab Dispos 20:596–602

38. Poupko JM, Baskin SI, Moore E (2007) The pharmacological properties of anisodamine. J Appl Toxicol 27:116–121
39. Macor JE, Gurley D, Lanthorn T, Loch J, Mack RA, Mullen G, Tran O, Wright N, Gordon JC (2001) The 5-HT_3 antagonist tropisetron (ICS 205-930) is a potent and selective α7 nicotinic receptor partial agaonist. Bioorg Med Chem Lett 11:319–321
40. Chen H, Wang H, Chen Y, Zhang H (2005) Liquid chromatography–tandem mass spectrometry analysis of anisodamine and its phase I and II metabolites in rat urine. J Chromatogr B 824:21–29
41. World Health Organization 16th list (updated) March 2010. http://www.who.int/medicines/publications/essentialmedicines/en/index.html. Accessed 11 Aug 2011
42. Brown JH, Taylor P (1996) Muscarinic receptor agonists and antagonists. In: Hardman JG, Limbird LE (eds) The pharmacological basis of therapeutics, 9th edn. McGraw-Hill, New York
43. McDonough JH, Shih TM (2007) Atropine and other anticholinergic drugs. In: Marrs TC, Maynard RL, Sidell FR (eds) Chemical warfare agents: toxicology and treatment, 2nd edn. John Wiley and Sons, New York
44. Abbara C, Bardot I, Cailleux A, Lallement G, Le Bouil A, Turcant A, Clair P, Diquet B (2008) High-performance liquid chromatography coupled with electrospray tandem mass spectrometry (LC/MS/MS) method for the simultaneous determination of diazepam, atropine and pralidoxime in human plasma. J Chromatogr B 874:42–50
45. Aaltonen L, Kanto J, Iisalo E, Pihlajamäki K (1984) Comparison of radioreceptor assay and radioimmunoassay for atropine: pharmacokinetic application. Eur J Clin Pharmacol 26:613–617
46. Kentala E, Kaila T, Iisalo E, Kanto J (1990) Intramuscular atropine in healthy volunteers: a pharmacokinetic and pharmacodynamic study. Int J Clin Pharmacol Ther Toxicol 28:399–404
47. John H, Mikler J, Worek F, Thiermann H (2012) Application of an enantioselective LC-ESI MS/MS procedure to determine R- and S-hyoscyamine after intravenous administration of atropine to swine. 4:194–198
48. Siluk D, Mager DE, Gronich N, Abernethy D, Wainer IW (2007) HPLC-atmospheric pressure chemical ionization mass spectrometric method for enantioselective determination of R, S-propranolol and R, S-hyoscyamine in human plasma. J Chromatogr B 859:213–221
49. John H, Eyer F, Zilker T, Thiermann H (2010) High-performance liquid-chromatographic tandem-mass spectrometric methods for atropinesterase-mediated enantioselective and chiral determination of *R*- and *S*-hyoscyamine in plasma. Anal Chim Acta 680:32–40
50. John H, Binder T, Höchstetter H, Thiermann H (2010) LC-ESI MS/MS quantification of atropine and six other antimuscarinic tropane alkaloids in plasma. Anal Bioanal Chem 396:751–763
51. Chen H, Chen Y, Du P, Han F, Wang H, Zhang H (2006) Sensitive and specific liquid chromatographic-tandem mass spectrometric assay for atropine and its eleven metabolites in rat urine. J Pharm Biomed Anal 40:142–150
52. Chen HX, Chen Y, Du P, Han FM (2007) LC-MS for identification and elucidation of the structure of in-vivo and in-vitro metabolites of atropine. Chromatographia 65:413–418
53. Boermanns PAMM, Go HS, Wessels AMA, Uges DRA (2006) Quantification by HPLC-MS/MS of atropine in human serum and clinical presentation of six mild-to-moderate intoxicated atropine-adulterated-cocaine users. Ther Drug Monit 28:295–298
54. Ariffin MM, Anderson RA (2006) LC/MS/MS analysis of quaternary ammonium drugs and herbicides in whole blood. J Chromatogr B 842:91–97
55. Björnstad K, Beck O, Helander A (2009) A multi-component LC-MS/MS method for detection of ten plant-derived psychoactive substances in urine. J Chromatogr B 877:1162–1168
56. Kintz P, Villain M, Evans J, Pujol ML, Salquebre G, Cirimele V (2007) A case of abuse in which children were forced to take tablets containing scopolamine: segmental analysis of hair for scopolamine by ultra performance liquid chromatography–tandem mass spectrometry. Forensic Toxicol 25:49–52

57. Kintz P, Villian M, Barguil Y, Charlot JY, Cirimele V (2006) Testing of atropine and scopolamine in hair by LC-MS-MS after *Datura inoxia* abuse. J Anal Toxicol 30:454–457
58. Homesley HD, Hahne WF, McLees B, Heck K, Barrett RJ, Lentz SS, Woodlief L, Lovelace JV (1993) Randomized comparison of the antiemetic efficacy of a serotonin type 3 receptor anatagonist (MDL 72,222) with a high-dose metoclopramide regimen. Am J Clin Oncol 16:175–179
59. He H, McKay G, Midha KK (1995) Phase I and II metabolites of benztropine in rat urine and bile. Xenobiotica 25:857–872
60. Schmitt KC, Zhen J, Kharkar P, Mishra M, Chen N, Dutta AK, Reith MEA (2008) Interactions of cocaine-, benztropine-, and GBR12909-like compounds with wild-type and mutant human dopamine transporters: molecular features that differentially determine antagonist-binding properties. J Neurochem 107:928–940
61. Lee HW, Park WS, Kim YW, Cho SH, Kim SS, Seo JH, Lee KT (2006) A rapid and sensitive liquid chromatography/positive ion tandem mass spectrometry method for the determination of cimetropium in human plasma by liquid–liquid extraction. J Mass Spectrom 41:855–860
62. Kajbaf M, Jahanshahi M, Lamb JH, Gorrod JW, Naylor S (1992) Bioanalytical applications of tandem mass spectrometry in the in vitro metabolism of the anticholinergic drug cimetropium bromide to detect differences in species metabolism. Xenobiotica 22:641–655
63. Aehle E, Dräger B (2010) Tropane alkaloid analysis by chromatographic and electrophoretic techniques: an update. J Chromatogr B 878:1391–1406
64. Maurer HH (2005) Multi-analyte procedures for screening for and quantification of drugs in blood, plasma, or serum by liquid chromatography–single stage or tandem mass spectrometry (LC-MS or LC-MS/MS) relevant to clinical and forensic toxicology. Clin Biochem 38:310–318
65. Rook EJ, Hillebrand MJX, Rosing H, van Ree J, Beijnen JH (2005) The quantitative analysis of heroin, methadone, and their metabolites and the simultaneous detection of cocaine, acetylcodeine and their metabolites in human plasma by high-performance liquid chromatography coupled with tandem mass spectrometry. J Chromatogr B 824:213–221
66. Johansen SS, Bhatia HM (2007) Quantitative analysis of cocaine and its metabolites in whole blood and urine by high-performance liquid chromatography coupled with tandem mass spectrometry. J Chromatogr B 852:338–344
67. Bouzas NF, Dresen S, Munz B, Weinmann W (2009) Determination of basic drugs of abuse in human serum by online extraction and LC-MS/MS. Anal Bioanal Chem 395:2499–2507
68. Kraemer T, Paul LD (2007) Bioanalytical procedures for determination of drugs of abuse in blood. Anal Bioanal Chem 388:1415–1435
69. Barroso M, Gallardo E, Quelroz JA (2009) Bioanalytical methods for the determination of cocaine and metabolites in human biological samples. Bioanalysis 1:977–1000
70. Olver IN (2005) Update on anti-emetics for chemotherapy-induced emesis. Int Med J 35:478–481
71. Jiang Y, Lin M, Fan G, Chen Y, Li Z, Zhao W, Wu Y, Hu J (2006) Rapid determination of granisetron in human plasma by liquid chromatography coupled to tandem mass spectrometry and its application to bioequivalence study. J Pharm Biomed Anal 42:464–473
72. Woo JS (2006) Nasal absorption studies of granisetron in rats using a validated high-performance liquid chromatographic method with mass spectrometric detection. Arch Pharm Res 30:778–784
73. Nirogi RVS, Kandikere VN, Shukla M, Mudigonda K, Maurya S, Boosi R (2006) Quantification of granisetron in human plasma by liquid chromatography coupled to electrospray tandem mass spectrometry. Biomed Chromatogr 20:888–897
74. Van Zyl JM, Derendinger B, Seifart HI, Van der Bijl P (2008) Comparative diffusion of drugs through bronchial tissue. Int J Pharm 357:32–36
75. Nussdorf JD, Berman EL (2000) Anisocoria associated with the medical treatment of irritable bowel syndrome. J Neuro Ophthalmol 20:100–101
76. Leach K, Loiacono RE, Felder CC, McKinzie DL, Mogg A, Shaw DB, Sexton PM, Christopoulos A (2010) Molecular mechanisms of action and *in vivo* validation of an M_4 muscarinic acetylcholine receptor allosteric modulator with potential antipsychotic properties. Neuropsychol Pharmacol 35:855–869

77. Callegari E, Malhotra B, Bungay PJ, Webster R, Fenner KS, Kempshall S, LaPerle JL, Michel MC, Kay GG (2011) A comprehensive non-clinical evaluation of the CNS penetration potential of antimuscarinic agents for the treatment of overactive bladder. Br J Clin Pharmacol 72:235–246
78. Yiu KCH, Ho ENM, Wan TSM (2004) Detection of quaternary ammonium drugs in equine urine by liquid chromatography-mass spectrometry. Chromatographia 49:S45–S50
79. Manfio JL, Dos Santos MB, Favreto WAJ, Hoffmann FI, Mertin AC (2009) Validation of a liquid chromatographic/tandem mass spectrometric method for the determination of scopolamine butylbromide in human plasma: application of the method to a bioequivalence study. J AOAC Int 92:1366–1372
80. Wang J, Jiang Y, Wang Y, Li H, Fawcett JP, Gu J (2007) Highly sensitive assay for tiotropium, a quaternary ammonium, in human plasma by high-performance liquid chromatography/tandem mass spectrometry. Rapid Commun Mass Spectrom 21:1755–1758
81. Ding L, Tan W, Zhang Y, Shen J, Zhang Z (2008) Sensitive HPLC-ESI-MS method for the determination of tiotropium in human plasma. J Chromatogr Sci 46:445–449
82. Sanwald P, David M, Dow J (1996) Use of electrospray ionization liquid chromatography-mass spectrometry to study the role of CYP2D6 in the in vitro metabolism of 5-hydroxytryptamine receptor antagonist. J Chromatogr B 678:53–61
83. Deng P, Zhong D, Chen X (2009) Determination of tropisetron in human plasma by liquid chromatography–tandem mass spectrometry. J Pharm Biomed Anal 49:848–852
84. Hotha KK, Bharathi DV, Kumar SS, Reddy YN, Chatki PK, Ravindranath LK, Jayaveera KN (2010) Determination of the quaternary ammonium compound trospium in human plasma by LC-MS/MS: application to a pharmacokinetic study. J Chromatogr B 878:981–986
85. Fu J, Fang C, Cui YY, Yang LM, Zhu L, Feng XM, Zheng PL, Lu Y, Chen HZ (2008) Quantitative determination of a novel enantiomeric tropane analog, (−)-satropane, in biological fluids using liquid chromatography/tandem mass spectrometry. Biomed Chromatogr 23:1044–1050
86. Bonhage MR, Chilcoat CD, Li Q, Melendez V, Flournoy WS (2008) Evaluation of two scopolamine and physostigmine pre-treatment regimens against nerve agent poisoning in the dog. J Vet Pharmacol Ther 32:146–153
87. Chen H, Chen Y, Du P, Han F (2008) Liquid chromatography–electrospray ionization ion trap mass spectrometry for analysis of in vivo and in vitro metabolites of scopolamine in rats. J Chromatogr Sci 46:74–80
88. Ahmed S, Sileno AP, deMeireles JC, Dua R, Pimplaskar HK, Xia WJ, Marinaro J, Langenback E, Matos FJ, Putcha L, Romeo VD, Behl CR (2000) Effect of pH and dose on nasal absorption of scopolamine hydrobromide in human subjects. Pharm Res 17:974–977
89. Al-Ghananeem AM, Malkawi AH, Crooks PA (2007) Scopolamine sublingual spray: an alternative route of delivery for the treatment of motion sickness. Drug Dev Ind Pharm 33:577–582
90. Xu A, Havel J, Linderholm K, Hulse J (1995) Development and validation of an LC/MS/MS method for the determination of L-hyoscyamine in human plasma. J Pharm Biomed Anal 14:33–42
91. Snyder LR (1978) Classification of the solvent properties of common liquid solvents. J Chromatogr Sci 16:223–234
92. Becker C, Worek F, John H (2010) Chromatographic analysis of toxic phosphylated oximes (POX): a brief overview. Drug Test Anal 2:460–468
93. John H, Blum MM (2012) Review on UV-spectroscopic, chromatographic and electrophoretic methods for the cholinesterase reactivating antidote pralidoxime (2-PAM) 4:179–193
94. Musshoff F, Madea B (2007) Analytical pitfalls in hair testing. Anal Bioanal Chem 388:1475–1494
95. Wada M, Nakashima K (2006) Hair analysis: an excellent tool for confirmation of drug abuse. Anal Bioanal Chem 385:413–415
96. Kanazawa H, Nagat Y, Matsushima Y, Takai N, Uchiyama H, Nishimura R, Takeuchi A (1993) Liquid chromatography–mass spectrometry for the determination of medetomidine and other anaesthetics in plasma. J Chromatogr 631:215–220

97. Oertel R, Richter K, Ebert U, Kirch W (2001) Determination of scopolamine in human serum and microdialysis samples by liquid chromatography–tandem mass spectrometry. J Chromatogr B 750:121–128
98. Stetina PM, Madai B, Kulemann V, Kirch W, Joukhadar C (2005) Pharmacokinetics of scopolamine in serum and subcutaneous adipose tissue in healthy volunteers. Int J Clin Pharmacol Ther 43:134–139
99. Oertel R, Richter K, Fauler J, Kirch W (2002) Increasing sample throughput in pharmacological studies by using dual-column liquid chromatography with tandem mass spectrometry. J Chromatogr A 948:187–192
100. Boppana V, Miller-Stein C, Schaefer WH (1996) Direct plasma liquid chromatographic-tandem mass spectrometric analysis of granisetron and its 7-hydroxy metabolite utilizing internal surface reversed-phase guard columns and automated column switching devices. J Chromatogr B 678:227–236
101. Machtejevas E, John H, Wagner K, Ständker L, Marko-Varga G, Forssmann WG, Bischoff R, Unger KK (2004) Automated multi-dimensional liquid chromatography: sample preparation and identification of peptides from human blood filtrate. J Chromatogr B 803:121–130
102. John H, Huynh KD, Hedtmann C, Walden M, Schulz A, Anspach FB, Forssmann WG (2005) In vitro degradation of the antimicrobial human peptide HEM-γ 130–146 in plasma analyzed by a validated quantitative LC-MS/MS procedure. Anal Biochem 341:173–186
103. John H, Walden M, Schäfer S, Genz S, Forssmann WG (2004) Analytical procedures for peptide quantification in pharmaceutical research by liquid chromatography-mass spectrometry. Anal Bioanal Chem 378:883–897
104. John H, Worek F, Thiermann H (2008) LC-MS based procedures for monitoring of toxic organophosphorus compounds and the verification of pesticide and nerve agent poisoning. Anal Bioanal Chem 391:97–116
105. John H, Preissner KT, Forssmann WG, Ständker L (1999) Novel glycosylated forms of human plasma endostatin and endostatin-related fragments of collagen XV. Biochemistry 38:10217–10224
106. John H, Forssmann WG (2001) Determination of the disulfide bond pattern of the endogenous and recombinant angiogenesis inhibitor endostatin by mass spectrometry. Rapid Commun Mass Spectrom 15:1222–1228
107. Thevis M, Vollmer DA (2012) Recent instrumental progress in mass spectrometry: advancing resolution, accuracy and speed of drug detection 4:242–245
108. Würtinger P, Oberacher H (2012) Evaluation of the performance of a tandem mass spectral library with mass spectral data extracted from literature 4:235–241
109. Sauvage FL, Saint-Marcoux F, Duretz B, Deporte D, Lachatre G, Marquet P (2006) Screening of drugs and toxic compounds with liquid chromatography–linear ion trap tandem mass spectrometry. Clin Chem 52:1735–1742
110. Gray TR, Shakleya DM, Huestis MA (2009) A liquid chromatography tandem mass spectrometry method for the simultaneous quantification of 20 drugs of abuse and metabolites in human meconium. Anal Bioanal Chem 393:1977–1990
111. Kjaergaard Bjork M, Nielsen MKK, Markussen LO, Klinke HB, Linnet K (2010) Determination of 19 drugs of abuse and metabolites in whole blood by high-performance liquid chromatography–tandem mass spectrometry. Anal Bioanal Chem 396:2393–2401
112. Peters FT, Drummer OH, Musshoff F (2007) Validation of new methods. Forensic Sci Int 165:216–224
113. Drummer OH (2007) Requirements for bioanalytical procedures in post-mortem toxicology. Anal Bioanal Chem 388:1495–1503
114. Schroff CD (1852) Über belladonna, atropin und daturin. Z KK Gesell Arzte Wien 3:211–242
115. Van Zutphen LFM (1974) Serum esterase genetics in rabbits. I. Phenotypic variation of the prealbumin esterases and classification of atropinesterase and cocainesterase. Biochem Genetics 12:309–326

116. Maurer HH (2005) Advances in analytical toxicology: the current role of liquid chromatography-mass spectrometry in drug quantification in blood and oral fluid. Anal Bioanal Chem 381:110–118
117. Maurer HH (2007) Current role of liquid chromatography–mass spectrometry in clinical and forensic toxicology. Anal Bioanal Chem 388:1315–1325
118. Mueller CA, Weinmann W, Dresen S, Schreiber A, Gergov M (2005) Development of a multi-target screening analysis for 301 drugs using QTrap liquid chromatography/tandem mass spectrometry system and automated library searching. Rapid Commun Mass Spectrom 19:1332–1338
119. Ondra P, Zednikova K, Valka I (2006) Detection and determination of abused hallucinogens in biological material. Neuro Endocrinol Lett 27(Suppl 2):125–129
120. Pietsch J, Günther J, Henle T, Dreßler J (2008) Simultaneous determination of thirteen plant alkaloids in a human specimen by SPE and HPLC. J Sep Sci 31:2410–2416
121. Papoutsis I, Nikolaou P, Athanaselis S, Stefanidou M, Pistos C, Spiliopoulou C, Maravelias C (2010) Mass intoxication with *Datura innoxia* – case series and confirmation by analytical toxicology. Clin Toxicol 48:143–145
122. Mackenzie AL, Pigott JFG (1971) Atropine overdose in three children. Br J Anaesth 43:1088–1090
123. Pietsch J, Koch I, Hermanns-Clausen M, Hüller G, Wagner R, Dressler J (2008) Pediatric plant exposures in Germany, 1998–2004. Clin Toxicol 46:686–691
124. Frascht M, Schneider S, Schuman M, Wennig R (2007) Formation of scopolamine from *N*-butyl-scopolamine bromide in cigarettes. J Anal Toxicol 31:220–223
125. Sticht G, Kaferstein H, Staak M (1989) Results of toxicological investigations of poisonings with atropine and scopolamine. Acta Med Leg Soc (Liege) 39:441–447
126. Da Matta Chasin AA, Midio AF (2000) Validation of an ion-trap gas chromatographic-mass spectrometric method for the determination of cocaine and metabolites and cocaethylene in post mortem whole blood. Forensic Sci Int 109:1–13
127. Kaplan MM, Register DC, Bierman AH, Risacher RL (1974) A nonfatal case of intentional scopolamine poisoning. Clin Toxicol 7:509–512

Analysis of Illicit Drugs in Human Biological Samples by LC-MS^n

Manuel Sergi and Sabino Napoletano

Abstract There are several classes of illegal substances with different psychotropic effects. It is possible to select a biological matrix and a suitable analytical strategy depending on the aim of the analysis and/or on the availability of the sample or on which kind of information is needed. Liquid chromatography–mass spectrometry (LC-MS or LC-MS/MS) has provided a helpful tool in this field especially for hydrophilic, thermolabile, and nonvolatile analytes, which analysis is sometimes critical by gas chromatography–mass spectrometry (GC-MS). Specific guidelines or procedures have been adopted in order to assist the chemist and to direct him/her towards the practical applications, for which a new analytical method is being created, including measures of verification and external assessment.

1 Introduction

The consumption of illicit drugs is a growing social and health problem which involves people of different ages and social classes all over the world. This phenomenon has pushed the research in analytical and forensic toxicology toward a fast evolution: the development of even more sophisticated analytical techniques with better performances, reduced analysis time, and better knowledge of the biological matrices.

These drugs, which may be of synthetic or natural origin, produce psychic alterations and behavior modifications; often they induce the consumer to periodically

M. Sergi (✉)
Department of Food Science, University of Teramo, via C. Lerici 1, 64023 Mosciano S.A, TE, Italy
e-mail: msergi@unite.it

S. Napoletano
Department of Chemistry, Sapienza University of Rome, p.le A. Moro 5, 00185 Roma, Italy

Q. Alan Xu and Timothy L. Madden (eds.), *LC-MS in Drug Bioanalysis*,
DOI 10.1007/978-1-4614-3828-1_12,

repeat the administration. The continued use of some drugs drives the consumer towards consumption, despite being recognized as causing serious damage to health.

In 1993 the World Health Organization (WHO) defined "drug of abuse" as a substance capable of inducing modifications of perceptual, emotional, and cognitive abilities. As this definition is very vague, all the "illicit drugs" are included in specific tables, which comprise the substances possessing the following features (characteristics):

1. Induce a dependent (addictive) state.
2. Stimulate or depress the central nervous system (CNS).
3. Adverse effects similar to those listed above.
4. The presence of evidence showing that the selected susbtance may induce abuse or cause health and social problems.

1.1 Dependence

As previously reported, an important criterion to define an "illicit drug" is its capability to create a state of dependence, which can be psychological or physical and is typical of each substance. The user (referred to as an addict), who is periodically or chronically intoxicated, shows a compulsion to take the preferred substance (or substances), has great difficulty in voluntarily ceasing or modifying substance use, and exhibits a great determination to obtain the psychoactive substances. Typically, tolerance is prominent and a withdrawal syndrome frequently occurs when substance use is interrupted.

Psychological or psychic dependence refers to the experience of impaired control over drug use (craving, compulsion), while physiological or physical dependence refers to tolerance and abstinence symptoms.

The dependence syndrome may relate to a specific substance (e.g., tobacco, alcohol, or diazepam), a class of substances (e.g., opioids), or a wider range of pharmacologically different substances

The WHO define the *psychic dependence* as "a sensation of absolute need and the psychological tendency to periodical or continuous use." *Craving* is defined as a very strong desire for a psychoactive substance or for the intoxicating effects of that substance. *Pharmacological tolerance* is the need to progressively increase the dose to be administered to achieve the desired effects. *Addictive potential* is the ability of a substance to determine tolerance and dependence. This potential is used as a risk index for the consumption of a substance.

1.2 Action on the CNS

Illicit drugs work on the CNS by means of the mesolimbic dopamine system of the brain, which plays a key role in regulating mechanisms of reward [1]. Several scientific pieces of evidence showed the involvement of the dopamine neurotransmitter in the motivational properties of the active substances in the CNS [2]. These substances enhance mesolimbic dopaminergic transmission by increasing the concentration of extracellular dopamine in the nucleus accumbens, stimulating the release of the synaptic neurotransmitter, and/or blocking the neuronal reuptake [3–5].

2 Classification of Illicit Drugs

There are different means of classifying illicit drugs, depending on the highlighted characteristics of the substances. The main classification criteria are as follows: legal, danger, preparation, symptomatic, and pharmacological.

2.1 Legal Classification

(a) Licit drugs.
(b) Illicit drugs.

This classification strictly depends on the legislation of each State. Some substances may be legal in one State and illegal in another, and vice versa.

2.2 Danger Classification

(a) Hard drugs.
(b) Soft drugs.

This classification distinguishes between psychoactive drugs that are addictive and perceived as especially damaging and drugs that are believed to be nonaddictive (or minimally addictive) and with fewer dangers associated with their use.

2.3 Preparation Criteria

(a) Natural drugs.
(b) Semisynthetic drugs.
(c) Synthetic drugs.

2.4 Syntomatic Criteria

(a) Psycholeptics or depressants.
(b) Psychoanaleptics or stimulants.
(c) Psychodysleptics or hallucinogens.

It is a classification that highlights the various actions that a psychotropic substance may have on the CNS. The psycholeptics are substances that exhibit CNS depressant action by slowing the transmission of nerve impulses and body functions. The psychoanaleptic are substances that have a stimulating effect on the CNS by acting on the nerve impulse transmission and accelerating the body's functions. Finally the last group of psychodysleptic substances are those that alter the transmission of nerve impulses by altering the state of consciousness and causing the distortion of reality.

2.5 Pharmacological Criteria

(a) Opiates.
(b) Psychostimulants.
(c) Depressants.
(d) Ethyl alcohol.
(e) Nicotine and tobacco.
(f) Cannabinoids.
(g) Hallucinogens.
(h) Aryl-cyclohexan-amines.
(i) Inhalants.

This classification is based on the structural and pharmacological properties of the individual active substances.

2.6 Other Classifications

Reported below is a classification that gives particular prominence to the social aspects of the use of these drugs and the extent to which its consumption has increased [6]. Some commonly used terms are as follows:

(a) *Club drugs*: This definition combines substances while being dissimilar to each other, and shares prevailing modes of consumption that are favored in club scenes, raves and concerts.
(b) *Recreational drug*: Definition not different from the previous one while being less specific.
(c) *Disco drugs*: Similar to the first definition, but with more emphasis on the context of the nightclub as the environment of choice for the consumption of this group of drugs.

(d) *Smart drugs*: Substances with psychotropic effects similar to those of overt drugs of abuse, but which overcome the legal checks and can be usually legally sold in *smart shops*.
(e) *Eco drugs*: So called due to their natural origin.
(f) *Date rape drugs*: This class includes those substances whose consumption has been associated with violence or sexual crimes.
(g) *Designer drugs*: The term was first adopted in California by Henderson in the early 1980s [7]. This definition refers to a class of psychotropic substances which arise from the creativity of the chemist who, starting from basic structures, produces new ones. These compounds maintain the properties and effects of drugs banned by law but, because of their different chemical structure, are not included in tables and evade controls.

3 Opiates

The term opiate describes the class of molecules structurally and pharmacologically related to morphine, the main alkaloid of opium, which is a product of *Papaver somniferum*, a plant illegally cultivated in Asia, whose effects have been recognized since 4000 BC by the Sumers.

There are at least 25 alkaloids of opium belonging to different chemical classes: morphine (about 10 % of opium), noscapine or narcotine (about 5 %), papaverine (1 %), codeine (0.4 %), tebaine (0.4 %), narceine (0.2 %), but not all of these possess psychotropic effects. The illicit opiates may be natural molecules, such as morphine and codeine, and semisynthetic, such as heroin. Methadone and meperidine represent a group of totally synthetic molecules with opiate psychotropic effects.

3.1 Morphine

In 1803 Friedrich Serturner isolated morphine from opium, but the chemical structure (Fig. 1) was identified by Gulland and Robinson in 1925 and then confirmed by X-ray analysis in 1952 by Gates and Tschudi [8]. Only the *l* isomer is psychoactive, while the *d* form is totally inactive. It has low solubility in water and high solubility in organic solvents.

3.2 Codeine

Codeine is the methylic ether of morphine (3-methylmorphine) and can be isolated from opium during the extraction of morphine, but is usually prepared by the methylation of morphine. Codeine is used in medicine as an antitussive drug and furthermore it has analgesic properties. It may cause addiction, but less than morphine.

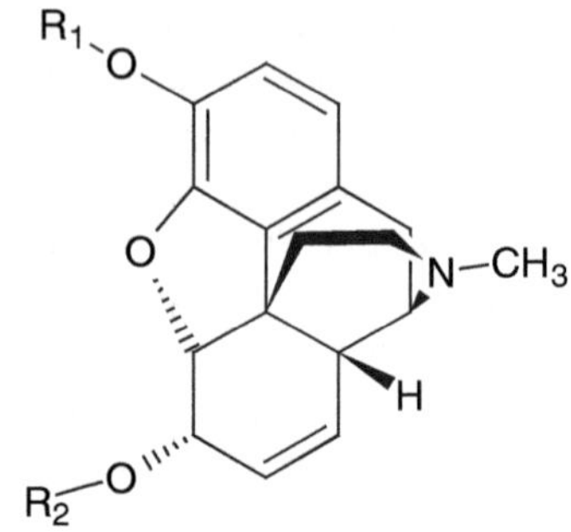

R_1	R_2	
-H	-H	Morphine
$-CH_3$	-H	Codeine
-H	$-COCH_3$	6-MAM
$-COCH_3$	$-COCH_3$	DAM

Fig. 1 Structures of the main opiates

3.3 Heroin

Heroin is a diester of morphine (3,6-diacetil morphine or DAM) obtained from the acetylation of both hydroxylic groups. It is a white, crystalline powder, with a bitter taste. It is unstable in air and light and typically produces an odor of acetic acid. Heroin, similarly to morphine, can be combined with acids producing water soluble salts, such as heroin hydrochloride. Similarly to morphine, it possesses narcotic and analgesic activity, but heroin causes strong addiction and intoxication. It is administered through inhalation, parenteral, subcutaneous, and above all intravenous.

3.3.1 Pharmacodynamics

The effects of heroin on the CNS are related to 6-monoacetylmorphine (6-MAM) and mainly morphine. The heroin crosses the blood–brain barrier due to the acetylic groups in position 3 and 6, that increase its lipophilicity. The morphinic compounds interact with the opioid receptors, mainly the μ receptors, acting as an antagonist by mimicking the endorphin effects [9]. Usually the first event occurs 30 s after its administration with a typical rush followed by a succession of euphoric sensations; after a few minutes the user becomes calm, satisfied, and indifferent to his surroundings; after 2–4 h typical restlessness accompanied with pain and an incessant demand for heroin sets in the craving.

The main symptoms of heroin use are miotic pupils, respiratory depression, decreased blood pressure, reduced body temperature, and decreased reflexes, as well as photophobia.

3.3.2 Pharmacokinetics

After intravenous administration heroin is very quickly metabolized (half-life is about 3 min) into 6-MAM and morphine. The 6-MAM reaches its maximum concentration in 6 min, then transforms itself into morphine, which reaches its peak in 20 min. In the liver and kidneys, deacetylation occurs, but in the liver other reactions also occur: N-demethylation that leads to normorphine, methylation that leads to codeine and conjugation with glucuronic acid.

The formation of glucuronide metabolites is the major catabolic route of morphine; in fact it is excreted at 85–90 % in the form of morphine 3-glucuronide (M3G) [10].

4 Psychostimulants

4.1 Cocaine

Cocaine is an alkaloid contained in great amounts in coca leaves and *Erythroxylon coca* and *Erythroxylon novogranatense*. *Erythroxylon coca* is a native shrub of Peru and Bolivia and is now widely cultivated in Central and South America and parts of Asia. The leaves of these plants contain between 0.4 and 2.5 % of alkaloids of which 50–60 % is represented by cocaine. This is a benzyl and methyl ester of ecgonine and is the active ingredient, which determines the psychotropic effects. It was isolated for the first time by the German chemist F. Gaedcke and was later characterized in 1860 by F. Wohler (Fig. 2); only the *l* form is pharmacologically active.

It is a white crystalline solid, slightly soluble in water, but soluble in acidic aqueous solutions and soluble in organic solvents, such as chloroform and diethyl ether. It is available in the illicit market as hydrochloride salt or free base (crack). Cocaine is usually mixed with other substances used as excipients.

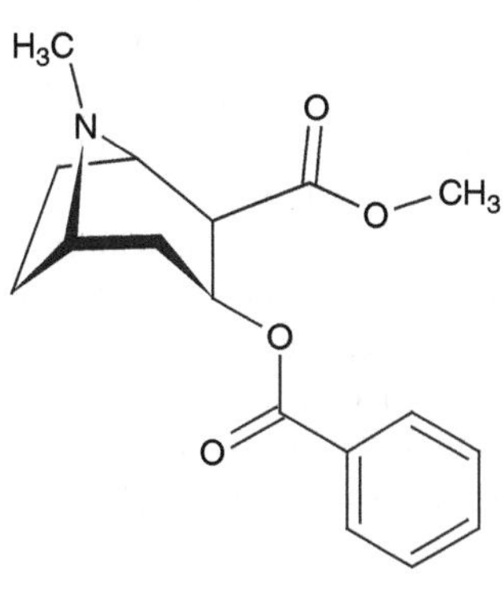

Fig. 2 Structure of cocaine

4.1.1 Pharmacodynamics

Cocaine acts as a potent local anesthetic and is a strong CNS stimulant: it extends and intensifies the effects of dopamine, norepinephrine, serotonin neurotransmitters [3]. The effects of cocaine can vary in relation to the individual characteristics, the administered dose, frequency of use, and route of administration. The intranasal administration causes plasma peak concentrations after 5–20 min, the euphoric effect in 15–20 min with a half-life of 40 min. The oral route involves a slow and low absorption with plasma peak concentrations after approximately 90 min and euphoric effect in 15–20 min. Intravenous plasma peak is immediate, euphoric effect occurs after 4–8 min with a half-life of about 40 min. Finally it may be administered through inhalation of combustion products or crack vapors, with great absorption speed.

4.1.2 Effects

A few seconds after the administration there is a subjective euphoric phase (rush), a feeling of extreme pleasure and comfort. Then a rapid rebound of the cortical depression (down) leads the subject to a condition of anxiety, depression, and irritability. When regularly used, the physical and psychological discomfort that accompanies the end of the effects is configured as a real abstinence syndrome (crash), whose main symptoms are depression, physical breakdown, irritability, and above all the compulsive and uncontrollable desire for cocaine.

The clinical symptoms of cocaine intake are the following: vasoconstriction, dilated and reactive pupils, hyperthermia, arrhythmias, increased blood pressure, dry mouth, increased sweating, tremors, dizziness, muscle spasms, hyperactivity, and insomnia. Cocaine-related deaths are generally caused by cardiac arrest or seizures followed by respiratory arrest [11].

4.1.3 Pharmacokinetics

Cocaine easily passes the blood brain barrier and the plasma levels are detectable for about 4–6 h after nasal ingestion; it is rapidly metabolized into inactive metabolites: benzoylecgonine (BEG), ecgonine methyl ester (EME), and ecgonine [12]. These molecules are biomarkers for the identification of cocaine abuse due to their higher biological half-life with respect to cocaine (about fivefold).

EME and BEG, which is mainly obtained by enzymatic hydrolysis, represents respectively the 32–49 % and 29–45 % of total urinary metabolites; cocaine can be converted, in small amounts, to norcocaine (NCOC) and psychoactive metabolite norbenzoylecgonine (NBE). The combined intake of alcohol and cocaine determines the formation of a pharmacologically active metabolite, the cocaethylene (ethyl ester of benzoylecgonine, CE), with a significant liver toxicity [13].

4.2 Amphetamine-Like Compounds

Amphetamines are a class of chiral molecules, synthesized from the precursor phenylethylamine (Fig. 3). Several formulations are usually found on the black market such as tablets, lozenges, drops, injections, powder, whose colors vary from white to pink to brown, depending on the impurities and adulterants.

They are usually taken orally and are the major class of CNS stimulants.

Amphetamine and methamphetamine possess an essentially pure psychostimulant effect; however, substituted derivatives in position 3 and 4 on benzene ring are defined as entactogene [14]. This class of substances includes methylenedioxymethamphetamine (MDMA), methylenedioxyamphetamine (MDA), methylenedioxyethyl-amphetamine (MDEA), and others such as the *N*-methyl-1-(3-4-methylenedioxyphenyl)-2-butanamine (MBDB), methoxymethylenedioxyamphetamine (MMDA) (Fig. 4). All of these substances are more active in the *d* form. There are also amphetamine-like substances which combine sympathomimetic (euphoric) and hallucinogen effects: they are primary amines, trisubstituted on the benzene ring, that produce effects similar to mescaline. Among these the 2,5-dimethoxy-4-methylamphetamine (DOM) is the most important.

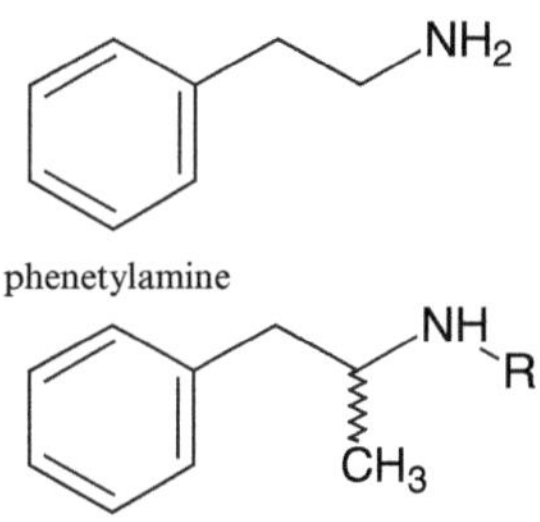

R
-H Amphetamine
$-CH_3$ Methamphetamine

Fig. 3 Structures of phenethylamine, amphetamine, and methamphetamine

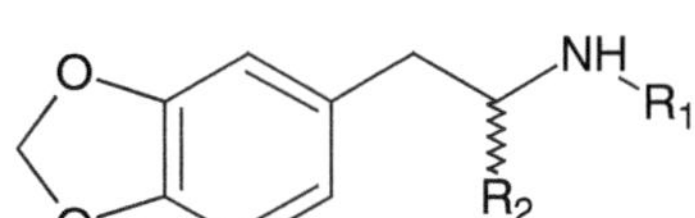

R_1	R_2	
-H	$-CH_3$	MDA
$-CH_3$	$-CH_3$	MDMA
$-C_2H_5$	$-CH_3$	MDE
$-CH_3$	$-C_2H_5$	MBDB

Fig. 4 Structures of methoxyamphetamines

4.2.1 Amphetamine

It is the parent compound of this class of substances, synthesized for the first time in 1887 and used at the beginning for treating asthma by inhalation. The stimulating effect was discovered later, in 1922.

4.2.2 Methamphetamine

Methamphetamine, known on the illicit market as *speed*, *ice*, or *crystal*, is a white crystalline powder, with a bitter taste and odor that dissolves easily in water or alcohol. Similarly to cocaine, it exerts a potent positive reinforcement and it is believed that the dextrorotatory form is responsible for the dopaminergic psychostimulant effect [15].

4.2.3 MDMA, MDA, MDEA

The 3-4-methylenedioxymethamphetamine (MDMA) is the prototype of many of the designer drugs and despite its well known neurotoxicity, is one of the most frequently used drug among young people around the world and especially among clubbers and ravers. Commonly known as ecstasy, MDMA was synthesized in 1898 by the German chemist Fritz Haber. Ecstasy is considered a stimulant with mild hallucinogenic effects [16].

MDA, known as the *Love Drug*, and MDEA, known as *Eve*, have very similar effects as MDMA, but the former has higher psychedelic power, while the latter is less effective.

Pharmacodynamics

Psychostimulant amphetamines induce the release of catecholamines, such as adrenaline and dopamine [15].

Entactogene and hallucinogenic amphetamines, furthermore, have an effect on the serotonin system by increasing the availability of serotonin, a neurotransmitter that controls sleep, mood, emotion, memory, perception, sexual behavior, and hunger [14].

For oral ingestion the effects start after about 20 min, while the effects of intranasal administration occur within 3–5 min without producing an intense rush.

Effects

The symptoms of amphetamine consumption are tremors, irritability, excessive loquacity, and anxiety. In the down phase symptoms of sleepiness, fatigue, heart disease, gastrointestinal, tremors, and convulsions may occur [11]. The effects of MDMA and related compounds are similar to those of amphetamine stimulants; they can also cause acute side effects and somatic neurotoxicity [17].

4.2.4 Pharmacokinetics

The main metabolic reaction is the deamination of amphetamine with the formation of phenylacetone, which is subsequently oxidized to benzoic acid, then conjugated with glycine to form hippuric acid. Side reactions include aromatic hydroxylation to form 4-hydroxyamphetamine (an active metabolite), the stereoselective β-hydroxylation for the isomer (+) of amphetamine leading to the formation of norephedrine (phenylpropanolamine) and finally the N-oxidation leading to the formation of a hydroxylamine derivative. The products of the hydroxyl and aromatic N-oxides can be conjugated with sulfate or glucuronic acid [18].

MDMA is effective in 20–30 min from intake and shows a peak plasma level after about 120 min with continued effects for up to 4–6 h; plasma half-life is 6–7 h. Metabolism proceeds by two routes. The principal one involves O-demethylation to 3,4-dihydroxymetamphetamine (HHMA) followed by O-methylation to 4-hydroxy-3-metoxymetamphetamine (HMMA) and 3,4-dihydroxyamphetamine (HHA) and subsequent O-glucuronide and sulfate conjugation. The second pathway involves an N-demethylation to MDA, followed by deamination and oxidation to the corresponding benzoic acid derivative, substantially conjugated with glycine. The MDA is a metabolite of both MDMA and MDEA [19].

5 Hallucinogens and Psychedelic Substances

These molecules cause significant changes in the perceptual system, producing sensory illusions and inducing a distortion of reality, the sense of space and time, to the point of hallucination. Natural, synthetic or semisynthetic substances can cause hallucinogenic effects: psilocin, psilocybin, mescaline, and LSD. Hallucinogenic effects are also created by phencyclidine and ketamine, but these molecules have a dissociative character.

5.1 Mescaline

Mescaline (3,4,5-trimethoxyphenethylamine) (Fig. 5) is mainly contained in *Lophophora williamsii peyote*, a cactus native to the deserts of Mexico. Peyote contains an average of about 1.5 % of mescaline. It is often synthetic and looks like a clear powder with colors ranging from white to brown, depending on purity.

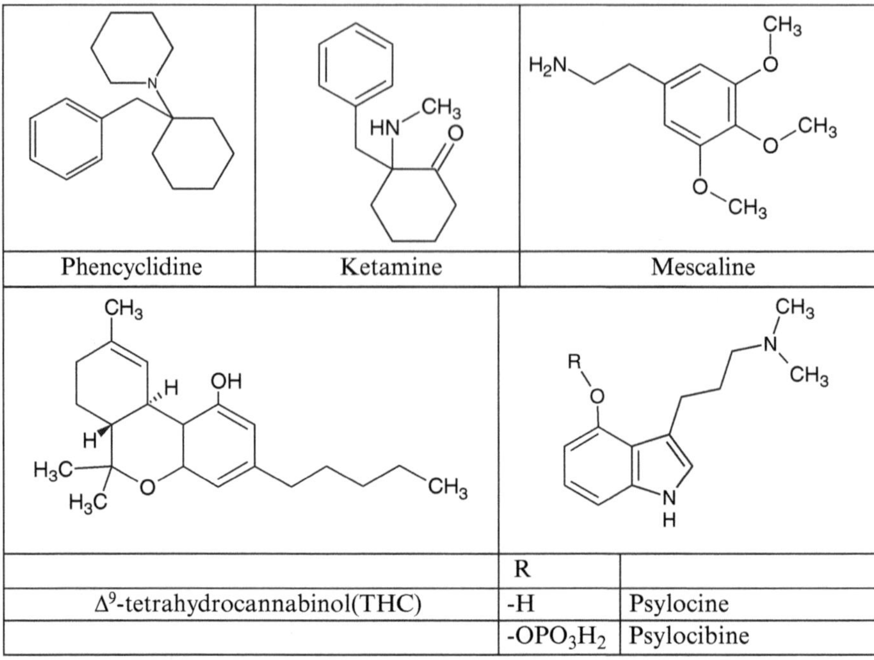

Fig. 5 Structures of the main hallucinogens

5.1.1 Pharmacodynamics

Mescaline acts similarly to other psychedelic agents. It activates the serotonin 5-HT2A receptor with a high affinity as a partial agonist. This receptor activation causes the release of glutamate in the prefrontal cortex and optic nerve correlated with cognitive and perceptual distortions.

5.1.2 Effects

The effects of the mescaline are strongly dose-dependent and occur 2 or 3 h after intake and last up to 12 h. Hallucinogenic effects are accompanied by states of excitement, insomnia, feelings of omnipotence, and logorrhea [20].

5.1.3 Pharmacokinetics

Mescaline is rapidly absorbed by the human organism, especially after oral intake. About 90 % of the dose is excreted in the urine within 24 h after intake. 39 % is found as an acid 3,4,5-trimethoxyphenethylacetic, inactive, and the remaining percentage mescaline is excreted as a glutamic conjugated acid after being subjected to hydrolysis in the liver. In smaller quantities the metabolic processes lead to the formation of *N*-acetyl-mescaline and *N*-acetyl-3,4-dimethoxy-5-hydroxyphenetylammine [21].

5.2 *Dissociative Drugs of Abuse*

Drugs such as phencyclidine and ketamine, the most important aryl-cyclohexane-amine on the illicit market, were initially used as a general anesthetic in surgery: these create distortions of sound and visual perception and feelings of dissociation from the environment and from themselves. These mental changes are not hallucinations, so phencyclidine and ketamine are more properly known as dissociative anesthetics. The dissociative drugs act by altering the distribution of glutamate in the brain.

5.2.1 Phencyclidine

Phencyclidine (PCP) was developed in the 1950s as an intravenous anesthetic used for surgical procedures. It has been used in veterinary but it then spread to the illicit drug market. PCP is currently known by more than thirty names such as *angel dust*, *zoom*, *crystal cyclone*, and *peace*. In its pure form PCP is a white crystalline powder that readily dissolves in water. It can be inhaled, smoked, injected, or chewed [22].

Pharmacodynamics

PCP, like ketamine, targets a particular receptor site, located in the channel for calcium, associated with excitatory amino receptors *N*-methyl-D-aspartate (NMDA) located in the brain [23].

Effects

PCP is able to alter the perceptual system up to the point of blocking its functioning, leading to a sort of trance. Alterations of perception and mental status, associated with the consumption of PCP, could also lead to difficulties in the articulation of language and motor coordination, drowsiness, nausea, vomiting, temporary amnesia, hallucinations, muscle spasms, convulsions, paralysis, and coma in severe forms.

Pharmacokinetics

PCP is metabolized in the liver and excreted through the urine. The main metabolic pathway is hydroxylation. In urine, two metabolites were identified in the form of glucuronic conjugates: the 4-phenyl-4-piperidine cyclohexanol and 1-(1-phenyl cyclohexyl)-4-hydroxy-piperidine (PCHP). About 30 % of the dose is excreted in the urine in 10 days: 16 % of this can be traced as a molecule as it is, while 30 % is in the form of conjugated hydroxy metabolite. The amount of substance, as it is, increases in acidic condition. The remaining dose is excreted in the feces [24].

5.2.2 Ketamine

Ketamine, 2-chlorophenyl-2-(methylamino)cyclohexan-1-one, a general anesthetic, is used for its ability to induce amnesia and loss of response to painful stimuli without loss of consciousness [25]. It has been used since 1967, but to date is still used only in veterinary. Even if taken in subanesthetic doses, psychotropic and hallucinogenic effects occur, so since the 1980s the drug has begun to be consumed in the world of clubs. Ketamine is produced as an injectable solution, but after evaporation remains a powder that can be used, as for other drugs of abuse, to produce tablets or to be sniffed. Some street names are used for identification: *K*, *Special K*, *Vitamin K*, *Superacid*, *Ket*, *Kit-Kat*, *Kitty*. The chemical structure, mechanism of action, and pharmacological effects are similar to those of PCP, although ketamine is less potent and produces only minor side effects; moreover, duration of the effect is reduced compared to the PCP.

Pharmacodynamics

Ketamine can be taken by intravenous, intramuscular, nasal or oral means and sometimes it can be smoked with tobacco. With nasal intake, plasmatic peak concentration is reached after about an hour, but results can be seen earlier within approximately 20 min if the intake is intramuscular [25]. Due to its high lipophilicity, ketamine is rapidly distributed in the brain and is then distributed to less vascularized organs and adipose tissue, where its accumulation is possible. Ketamine is also responsible for the increased adrenergic tone and cardiovascular symptoms activating [26].

Pharmacokinetics

Ketamine is metabolized primarily in the liver region leading to the formation of two main compounds: norketamine (NK) and dehydronorketamine (DHNK). NK, a compound much less powerful than the substance itself, is achieved through the demethylation, subsequent dehydrogenation generates DHNK [27, 28].

Both the ketamine as it is and the metabolites are hydroxylated and conjugated before elimination. About 90 % of the dose is excreted in the urine within 72 h, with about 2 % as it is, 2 % as NK, 16 % as DHNK, and 80 % as conjugates of hydroxylated metabolites [29].

5.3 *Cannabinoids*

The cannabinoids are a class of psychoactive substances contained both in *Cannabis* and *Echinacea* plants. The first reliable evidence of the effects of drugs derived from cannabis is a compendium of Chinese medicine (a herbarium), dated 2700 BC; today it has become very common in many parts of the world.

Phytocannabinoids are natural compounds present in *Cannabis sativa*, responsible for the pharmacological and toxicological properties of this plant. About 70 compounds belong to this class, the most important are Δ^9-tetrahydrocannabinol (THC), cannabinol (CBN), and cannabidiol (CBD).

THC is considered the most pharmacologically active cannabinoid of Cannabis [30] and is used as a reference substance for estimating the psychotropic power of the various preparations of this plant. THC is rapidly absorbed in large amounts and passes into the blood stream, and due to its lipophilicity it can spread throughout the body and penetrate the brain.

Derivatives of cannabis are hashish, which consists of a resin produced by the inflorescences (6–10 % THC), marijuana, which is in the air-dried leaves, flowers, and the stem (2–5 % THC), and hashish oil that looks like a viscous liquid, like tar, obtained by extraction with organic solvents (15–60 % THC). These derivatives are usually smoked.

5.3.1 Pharmacodynamics

There are specific binding sites for cannabinoids, whose endogenous antagonists belong to the class of endocannabinoids. The human organism produces endocannabinoids from long-chain polyunsaturated fatty acids, they bind to cannabinoid (CB) receptors and activate them. The CB receptors and endocannabinoids constitute the endocannabinoid system [31]. THC binds to both types of receptors: CB1 and CB2. The absence of cannabinoid receptors in the brain stem, the seat of the centers of respiration and other vital functions, explains the low toxicity of cannabinoids.

5.3.2 Effects

Activation of CB receptors by phytocannabinoids such as THC can cause a multitude of effects such as euphoria, anxiety, altered time perception, loss of concentration, and panic attacks. The most commonly researched ones are feeling of well-being, euphoria, and relaxation. THC produces an increase in heart rate, blood pressure, and body temperature (dose-dependent effects). It is also possible to experience dry mouth, increased hunger, and pain reduction. Very high doses of cannabis can cause anxiety, panic, or result in psychotic episodes.

5.3.3 Pharmacokinetics

THC undergoes metabolic degradation in the liver, where it is hydroxylated to 11-hydroxy tetrahydrocannabinol (THC-11OH). The latter, still with psychoactive activity, is oxidized to Δ^9-THC-COOH, an inactive metabolite which is conjugated as 11-nor-Δ^9-tetrahydrocannabinol-9-carboxy-glucuronide (Δ^9-THC-COOH-glu), more hydrophilic metabolite and therefore easily excreted in the urine [32].

6 Biological Matrices for the Determination of Illicit Drugs

It is possible to select a biological matrix and a suitable analytical strategy depending on the aim of the analysis (clinical, medicolegal, epidemiological) and/or on the availability of the sample or on which kind of information is needed.

The main biological matrices are the following:

- Plasma and urine, the most used and known for toxicological analysis.
- Hair and nails.
- Oral fluid (OF) and sweat.

Each one has a specific time window and often specific target analyte; when an analysis on more matrices is made, it is also possible to gather information on the means of administration.

Blood (plasma, serum, or whole blood) and, with some limits, OFs give detailed information about what is occuring at the time of testing; hair analysis demonstrates the history of abuse, that could backdate several months.

As for correlation among the different biological fluids, Toennes et al. have demonstrated that both OF and urine are able to predict the plasmatic positivity with 90 % of accuracy for most drugs, except for amphetamines, which have a detection time in OF greater than in plasma [33].

6.1 Blood, Plasma and Serum

Blood travels through the human body in more than 96,000 km of blood vessels and it is full of marker molecules [34]. The physiological pH is usually 7.4 with a complex buffer system (bicarbonate–carbonic acid, hemoglobinate–hemoglobin, phosphate buffer) [35].

Plasma is the liquid part of blood that is about 93 % water and 7 % of organic substances (proteins in suspension, nutrients such as glucose or lipids, hormones, metabolic products) and inorganic (electrolyte, carbon dioxide, oxygen).

Serum is deprived of fibrinogen, but all the other proteins are the same as plasma; the plasmatic proteins are the main problem for the analytical determination of drugs, because they bind many molecules, including the drugs of abuse.

The knowledge of the plasmatic concentration is very useful in legal testing, where it is necessary to correlate a person's actions with the effect of a substance (impairment).

Illicit drugs are usually quickly metabolized so it is necessary to perform testing a few hours after intake. In plasmatic analysis the possibility for adulteration is very low and so blood samples have always been considered for forensic analysis.

The problems related to the use of these matrices are the invasiveness of testing, the reduced volume of the sample, the need for specialized staff, and suitable storage until analysis [33].

6.2 Oral Fluids

6.2.1 Anatomy and Physiology

The glandular apparatus produces from 500 to 1,500 ml of saliva per day: this secretion, however, is not constant, being higher in the digestive phase (0.05 ml/min during sleep, at rest 0.3–0.6). There are several inorganic compounds, such as sodium, potassium, or bicarbonate, while organic molecules are mainly proteins, such as α-amylase and mucine, a glycoprotein which gives high viscosity to OF, even if the total protein concentration is 1 % than the plasmatic one (10–50 mg/ml). Most of the plasmatic substances are affected by a transcellular passive diffusion mechanism in the saliva in a concentration gradient [36]. The transfer between blood and OF depends on both their pH and on the p*K*a of the molecule: the OF pH is usually lower than the plasmatic one, so basic substances, which are unionized in plasma, can cross the membrane and remain in the salivary compartment at lower pH in ionized form [27].

The growing interest in the use of OF is due to the strict correlation between plasmatic concentration [28–30] and pharmacokinetics [31, 32]; furthermore, OF is easily available and its testing is absolutely noninvasive, fast, cheap and can be performed by untrained staff even on the road.

In OF, like in blood, the parent drugs are more concentrated than the metabolites: that is, cocaine is itself the target analyte and THC is the predominant cannabis product [37–39].

6.3 Urine

Urine is the matrix of choice in the analysis of drugs of abuse, as it allows noninvasive testing, the ability to analyze both the substance itself and the metabolites after several days, a large amount of sample, the concentration of analytes (up to 100 times than plasma) [40].

However, urine has no great relevance in quantitative analysis because the analyte concentrations vary depending on the dose, means of administration, physiological status (urinary pH, sex, age, weight, etc.), the time lag between intake and analysis, the addition of adulterants. So the analytical data of urine may only indicate the presence of a substance up to a defined cutoff point; the monitoring window (time interval in which a substance can be detected by ordinary analytical methods) varies from a few days for cocaine, amphetamine, methoxyamphetamine to 2–3 weeks for cannabinoids.

In order to prevent tampering, investigations are planned on the assessment of creatinine, specific gravity, pH, and the presence of oxidizing adulterants [41].

6.4 Hair

The hair matrix possesses a large monitoring window (months/years) that allows a retrospective analysis and the study of past history. Hair is also characterized by a high stability and a minimum possibility of adulteration [41, 42]. The structure of the hair can be considered as a repeated network of keratin fibers. There are also melanins, lipids, and all the compounds of cells that led to the formation of the stem. The growth of hair is in the range of 0.6–1.4 cm/month [43, 44].

6.4.1 Incorporation Mechanism

The most widely accepted model for the incorporation of substances into hair has been proposed by Henderson [45]: the incorporation of substances inside hair is primarily dependent on the diffusion of molecules from blood in the capillaries to the cells in growth. The drugs can also enter the hair fiber during its formation by spreading from the deep skin strata. Sebum and sweat, which themselves contain the substances, can carry them within the hair structure [46–51].

Moreover, substances from the environment can be absorbed in the hair surface and then sweat facilitates their spread within the hair.

7 LC-MS Analysis

The toxicological analysis is essential in the clinical and medicolegal process, aimed at seeking evidence of the presence of psychoactive substances in biological fluids and tissues. The fields of intervention of a clinical toxicology laboratory are mainly two:

- *Clinical toxicology*, where the analysis is made with a therapeutic purpose.
- *Forensic toxicology*, which performs analysis for administrative and forensic purposes.

Since errors in analytical forensic toxicology may lead to legal consequences for both the offender and the injured, obtaining meaningful and reliable analytical data is mandatory.

Liquid chromatography–mass spectrometry or tandem mass spectrometry (LC-MS or LC-MS/MS) has helped this field closing the gap with respect to hydrophilic, thermolabile, and nonvolatile analytes, which were not sufficiently covered by the established gold standard technique gas chromatography–mass spectrometry (GC-MS) [52]. Jenkins et al. compared with each other LC-MS, LC-UV, and GC-MS determination of the specific case of MDMA showing how the coupling LC-MS technique is very versatile, sensitive, and selective for the analysis of drugs of abuse and pretreatment is relatively simple [53]. LC-MS and LC-MS/MS are thus largely replacing GC-MS and GC-MS/MS in the analysis of drugs of abuse, because it does not require derivatization and is capable of simultaneous determination of both conjugated drugs of abuse and free forms in a single analysis without the intermediate step of hydrolysis pretreatment of the sample: these are important advantages in terms of analysis time and performance [54].

Its largest field of application, which will focus on this chapter, is the multiresidual analysis in biological samples. In so-called confirmatory analyses, LC-MS, or better LC-MS/MS, provides an unequivocal identification, due to the high selectivity of the MS detector. The results of Systematic Toxicological Screening Analysis (STA) or General Unknown Screening (GUS), other applications of LC-MS, needs of subsequent confirmatory analysis to be used in a court [52, 55].

7.1 *Development of an Analytical Method: Guidelines*

Specific guidelines or procedures have been adopted in order to assist the chemist and to direct him towards the practical applications, for which a new analytical method is being created, including measures of verification and external assessment.

In 1996 the American Board of Forensic Toxicology launched the Forensic Toxicology Accreditation Program based on the guidelines of the Society of Forensic Toxicologists in American Academy of Forensic Sciences (SOFT/AAFS): the final document was adopted and the latest version was published in 2006 [56]. In Europe there is a document relating to the performance of analytical methods and interpretation of results aimed to identification of drug residues in foodstuffs, which is sometimes used also for forensic purposes: the decision 2002/657/EC [57]. The German guidelines of the Society of Toxicological and Forensic Chemistry (GTFCh) are exquisitely forensic. In 2006 a document entitled "Drugs of Abuse Testing Guidelines" was published by AGSA (Swiss Working Group for Drugs of Abuse Testing Guidelines) and supported by the major public institutions in Switzerland. However, there are also those issued by the Food and Drugs Administration (FDA) relating bioanalytical method [58], sometimes supplemented with EMEA's homologue document [59] and SOFT-AAFS or 2002/657/EC [57]. SOFT-AAFS guidelines also assist and guide the laboratories on the procedures to be followed before chemical analysis: from sampling to acceptance of sample, rules to be followed to ensure the sample traceability.

All guidelines specify that in the case of analysis aimed at medicolegal purposes, the analytical steps envisaged are basically two: the *screening tests* and the *confirmatory tests*.

Screening tests allow to quickly analyze several samples, even by untrained personnel, in an economic, efficient and standardized way to prevent a negative samples. These tests identify samples in which the concentration is below or above a specific threshold value, defined cutoff (a reference value, expressed in concentration, above which the result of an analytical test is considered positive and below negative). These methods, therefore, should aim to limit the error of false negative below a certain threshold.

Confirmatory method means methods that provide full or complementary information enabling the substance to be unequivocally identified and if necessary quantified at the level of interest.

7.2 Validation

Validation involves documenting, through the use of specific laboratory investigations, that the performance characteristics of the method are suitable and reliable for the intended analytical applications. The acceptability of analytical data is related to the criteria used for the validation. Reliable analytical data are a prerequisite for correct interpretation of toxicological findings in the evaluation of scientific studies, as well as in daily routine work. Unreliable analytical data might not only be contested in court, but could also lead to unjustified legal consequences for the defendant or to wrong treatment of the patient. Therefore, new analytical methods to be used in forensic and/or clinical toxicology require careful method development and thorough validation of the final method.

The fundamental parameters for this validation include accuracy, precision, selectivity, sensitivity, reproducibility, and stability [58]. Other parameters are limits of detection (LODs) and quantification (LOQs), linear dynamic range (LDR) [56]. In the case of LC-MS/MS based procedures, appropriate steps should be taken to ensure the lack of matrix effects throughout the application of the method, as outlined by several authors [55, 60–64].

7.2.1 Accuracy

Accuracy means the closeness of agreement between a test result and the accepted reference value. It is determined by determining trueness and precision [57]. In a strict sense, the accuracy of a method is affected by systematic (bias) as well as random (precision) error components, but the term is often used to describe only the systematic error component, i.e., in the sense of bias. In this sense sometimes accuracy is simply termed trueness [60]. While this approach is still accepted in forensic toxicology, it is preferably to use a control material (certified reference material, CRM) with a specific target concentration, which will allow independent verification of calibration accuracy [56], as also assessed by EC Guidelines. If no CRM is available, instead of trueness, the recovery can be determined [57].

7.2.2 Recoveries

Recovery means the percentage of the true concentration of a substance recovered during the analytical procedure. It can be determined as follows: select 18 aliquots of a blank material and fortify six aliquots at each of 1, 1.5, and 2 times the minimum required performance limit or 0.5, 1, and 1.5 times the permitted limit, analyze the samples and calculate the concentration present in each sample, calculate the recovery for each sample, calculate the mean recovery and the coefficient of variation (CV) or the relative standard deviation (RSD) from the six results at each level [57]. Nevertheless in toxicological analysis, as assessed by Peters et al. [60], a high value for recovery is not mandatory, as long as the data for LLOQ, precision and accuracy (bias) are acceptable. It can be calculated as the percentage of the analyte response after sample workup compared to that of a solution containing the analyte at a concentration corresponding to 100 % recovery. In the validation of LC-MS-(MS) methods, it is therefore more appropriate to perform the recovery experiments together with ion suppression/enhancement experiments [60].

7.2.3 Precision

Precision should be measured using a minimum of five determinations per concentration. A minimum of three concentrations in the range of expected concentrations is recommended. The precision determined at each concentration level should not exceed 15 % of the coefficient of variation (CV) except for the LLOQ, where it should not exceed 20 % of the CV. Precision is further subdivided into within-run, intrabatch precision or repeatability, which assesses precision during a single analytical run, and between-run, interbatch precision or reproducibility, which measures precision with time, and may involve different analysts, equipment, reagents, and laboratories. Since it is not always easy to obtain data about the reproducibility in the strict sense it often makes sense to use intermediate precision: this expresses within-laboratories variations on different days, by different analysts, and on different equipment, etc. [60].

7.2.4 Selectivity

Proving the lack of response in blank matrix is the way to establish method selectivity. It needs to analyze blank samples and check for interferences (signals, peaks, ion traces) in the region of interest where the target analyte is expected to elute.

Due to the intrinsic variability of biological matrices, guidelines require to prove that in several independent sources of the same matrix: FDA requires a minimum of six, EC requires a minimum of 20 representative blank samples.

Another approach to verify selectivity is described (also reported in EC guidelines) as follows: once selected a range of chemically related compounds (metabolites, derivatives, etc.) or other substances likely to be encountered with the compound of interest that may be present in the sample, investigate after analysis whether the presence may lead to a false identification, the identification of the target analyte is hindered by the presence of one or more of the interferences, or the quantification is notably influenced.

7.2.5 Stability

Insufficient stability of the analyte or matrix constituents in the sample during storage or analysis may give rise to significant deviations in the outcome of the result of analysis. Furthermore, the stability of the calibration standard in solution should be checked. Usually analyte stability is well characterized under various storage conditions. Drug stability in a biological fluid is a function of the storage conditions, the chemical properties of the drug, the matrix, and the containing system. Stability procedures should evaluate the stability of the analytes during sample collection and handling, after long-term (frozen at the intended storage temperature) and short-term (bench top, room temperature) storage, and after going through freeze and thaw cycles. The procedure should also include an evaluation of analyte stability in stock solution.

7.2.6 Calibration

The relationship between the concentration of analyte in the sample and the corresponding response (in bioanalytical methods mostly the area ratio of analyte versus internal standard (IS)) must be investigated. There is general agreement that for bioanalytical methods, calibrators should be matrix-based, i.e., prepared by spiking the blank matrix with the standard. Calibrator concentrations must cover the whole calibration range and should be evenly spaced across the range [60]. An acceptable correlation coefficient is 0.99; however, in some cases 0.98 is minimally acceptable. In addition, it is good practice to evaluate the range of the calibration by calculating the value of each calibrator against the curve. Values of ±20 % are generally acceptable for most applications, although ±10 % is preferred. Single point calibrations are discouraged unless controls are used at or close to the upper and lower quantitative reporting limits [56].

Usually, linear models are preferable (linear ordinary, i.e., unweighted, least squares regression model is not appropriate in many cases, in which weighted least squares model should be applied), but, if necessary, nonlinear (e.g., second order) models can be used [60].

7.2.7 Limits

Limit of detection (LOD) is the lowest concentration of an analyte that the bioanalytical procedure can reliably differentiate from background noise [58]. SOFT/AAFS guidelines establish that in chromatographic assays, the LOD might be the smallest sample concentration of a drug needed to give a peak height three times the noise level of the background signal from a blank sample. Limit of quantification (LOQ) can be determined experimentally as the lowest concentration for which an acceptable coefficient of variation can be routinely achieved. LOQs may be administratively defined in terms of the concentration of the lowest calibrator [56]. FDA indeed defines lower limit of quantification (LLOQ) as the lowest amount of an analyte in a sample that can be quantitatively determined with suitable precision and accuracy (not more than 20 % in terms of CV). The upper limit of quantification (ULOQ) is the maximum analyte concentration of a sample that can be quantified with acceptable precision and accuracy (bias). In general, the ULOQ is identical with the concentration of the highest calibration standard.

7.2.8 Matrix Effects

A mandatory requirement when validating an LC-MS(-MS) based method according to the FDA guidelines is to evaluate and attempt to minimize the incidence of matrix effects [58]. Matrix effects refer to the direct or indirect alteration or interference in response due to the presence of interfering substances in the sample [52]. It can either reduce the analyte response (ion suppression) or enhance it (ion enhancement). Both can considerably compromise the accuracy of quantification and ion suppression may in the worst case even lead to decrease sensitivity and to false negative results.

It is acknowledged that ESI is more susceptible to matrix effects compared with APCI or APPI, and that ionization in the negative mode is more selective than the positive mode [55].

Two main methods exist for matrix effect evaluation and are widely cited. The method proposed by Bonfiglio et al. uses postcolumn infusion of target analytes to qualitate highlight regions of suppression/enhancement in a chromatogram [65], whilst that proposed by Matuszewski et al. [64] suggests a method to quantify the degree of matrix effects for particular analytes using peak area/height ratios from analyses carried out with and without the presence of matrix components. Another method to evaluate matrix effect has been used in literature: a comparison is performed between two calibration curves: one created with blank samples, fortified after the extraction process, and the other with standard solutions. The value of matrix effects, intended as ion suppression, is calculated by comparing the slopes of those curves, this method directly highlighted influence of matrix components on the sensitivity of the method [66, 67].

The use of adequate internal standard or better isotopic labeled internal standard can minimize the matrix effect.

7.2.9 Quality Management and Accreditation

Quality management and accreditation have become matters of increasing relevance in analytical toxicology in recent years. Forensic laboratory accreditation is based upon international standards (ISO/IEC 17025:2005 [68]) which include requirements for method validation and allows to control and monitor overtime the laboratory performance and its analysts by proficiency tests [69].

8 Mass Spectrometry

The quadrupole is the most common mass analyzers due to its cheapness and performance, given by high efficiency transmission. The first implementation of a two-dimensional mass spectrometry (MS^2) was obtained with the quadrupole (the triple quadrupole mass analyzer) that is now the most common. The first and third (Q1 and Q3) quadrupoles are separative, and the second one (Q2) is a collision cell, in which the ions can be fragmented. MS^2 enables a reduction of background noise and an increase in selectivity and also allows different types of experiments: Full Scan (FS), Selected Ion Monitoring (SIM), Product Ion Scan (PIS), Precursor Ion Scan, Selected Reaction Monitoring or Multi Reaction Monitoring (SRM or MRM). The last one is the most widely used mode for the confirmatory analysis on targeted molecules.

Also ion trap allows MS^2 experiments and since, separation occurs over time and not over space, MS^n experiments are possible; however, for increasing value of n there is a proportional loss in intensity.

Tandem mass spectrometry is considered as necessary to provide unequivocal identification and quantification of analytes, as pointed out by Maralikova et al. [70]. It is also clear that monitoring of a single transition could lead to false compound identifications [71]. This had been previously predicted in the 2002 document issued by the European Union Commission that contains detailed information about mass spectrometric identification of drug residues in foodstuffs [57], where a minimum of three so-called Identification Points (IPs) is required for compound identification.

8.1 An Odd Coupled

The interfacing HPLC with MS is problematic, as pointed out by the image proposed by Arpino of the coupling between HPLC-fish and MS-bird [72]:one provides for the separation of neutral molecules in solution with high operating pressures, while the other responds to ions in the gas phase and requires high vacuum. The most widely used technique is electrospray ionization (ESI). It is suitable for analysis of highly polar compounds which are ionized in solution, such as

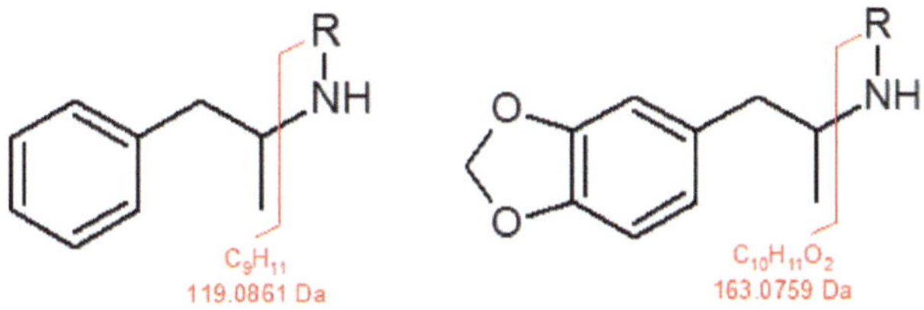

Fig. 6 Main MS cleavage for amphetamine and methylenedioxyamphetamine

urinary metabolites (glucuronide or sulfate conjugates). Signal is dependent not on the *absolute amount* of analyte entering the source, but the *concentration* of the analyte (i.e., the concentration of analyte in the injected volume) and the eluent flow-rate [73, 74].

Atmospheric pressure chemical ionization (APCI), is more efficient than ESI for nonpolar (hydrophobic) analytes, such as steroids, which do not readily form ions in solution [61]. Since APCI requires use of a heated nebulizer, thermally labile compounds may decompose in the ionization source.

Some modern instruments are supplied with dual-ionization sources, with the ability to rapidly switch between ESI and APCI during a chromatographic run or even carry out both processes simultaneously [75].

Since majority of illicit drugs are basic compounds, ESI in positive ionization mode is the most widely used source for these analysis, although negative ionization mode is also used in some cases.

8.2 MS Detection

Illicit drugs and their metabolites can have similar fragmentation pattern, so the use of multiple MRM transitions whenever possible, can also help to differentiate isobaric interferences.

In literature are reported fragmentation studies for most of known illicit drugs, here are two examples.

Derivatives of phenylethylamine: The chemical structure of phenylethylamine is shown in Fig. 3. Fragmentation can occur by cleaving of the bond in α or β amino nitrogen atom. The major route of fragmentation is in α to nitrogen, leading to the formation of a carbocation and a secondary amine, in fact this fragment appears to be the base peak for most of the CID spectra of these substances [76].

Molecules such as amphetamine and methamphetamine are different in their amino substitutions on nitrogen and, following the fragmentation pattern seen above, it is clear how these molecules have a common fragment of *m/z* 119 amu as can be seen in Fig. 6. The other fragment that appears in the CID spectra of these two substances falls at *m/z* of 91 amu, due to tropylium ion fragment, characteristic

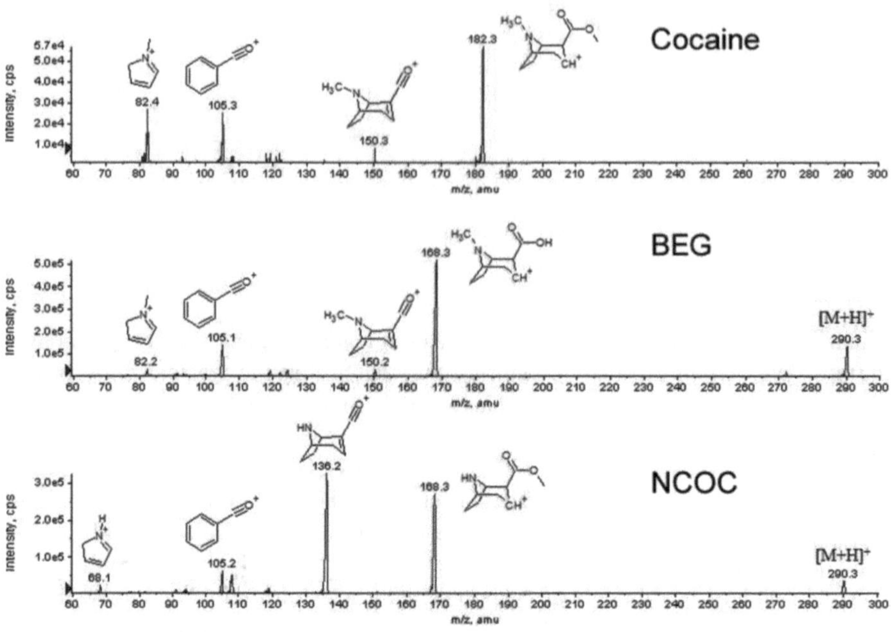

Fig. 7 Fragmentation pattern, registered in ESI, of Cocaine, BEG, and NCOC. Fragments at *m/z* 150 for BEG and 136 for NCOC are characteristic and permit to resolve these two metabolites

of all the molecules which have in their structure a benzene ring to which a methylene group is attached.

The MDA, MDMA, and MDEA only differ in the type of amino substitution on nitrogen: for these three molecules the most abundant fragment in the CID spectra has the same ratio *m/z* 163. For both spectra in addition to protonated adduct, it can be observed the mass peak of the fragment originated from the cleavage in α (Fig. 6) and the mass peak due to the formation of tropylium ion.

Cocaine and its metabolites, i.e., BEG and NCOC, follow a common pattern of fragmentation [77]. The ionization of cocaine, working in the ESI ion source in positive mode, leads mainly to the formation of charged species $[M+H]^+$ mass of 304 amu, while for BEG and NCOC $[M+H]^+$ mass of 209 amu. The metabolites of cocaine, here considered, have the same molecular weight as both are different from cocaine for fourteen mass units, corresponding at the replacement of a methyl with a hydrogen (cf. Sect. 4.1). This replacement occurs on the aminic nitrogen for the NCOC, while BEG originates by cocaine from the hydrolysis of the methyl ester. The use of different MRM transitions allows to discriminate these isobaric substances, and patterns of fragmentation of cocaine and metabolites are shown in Fig. 7. When this is not possible isobar masses must be resolved chromatographically.

9 Liquid Chromatography

High-performance liquid chromatography (HPLC) is a well-established separation technique; it is able to solve numerous analytical problems and there is the possibility of acting on the mobile phases with appropriate additives to improve the quality of the peak. Of course, any additive must be compatible with the MS detector: non-volatile buffer or eluent additives cannot be used; strong acids such as trifluoroacetic acid (TFA) may cause significant signal suppression in positive ionization. Different stationary phases are used as an alternative to the classical C18: Phenyl, HILIC, fluorinated, etc.

9.1 UHPLC and Fast Chromatography

Enhancing productivity and reducing costs could be a driving force for faster separations [78]. Several development have been done recently to improve the performances but the most important can be considered ultra high performance liquid chromatography (UHPLC) [79]. UHPLC systems are commercially available since 2004 and, regardless to major costs of instrumentation, are spreading in forensic laboratories [80]. Due to the cost of UHPLC hardware, chromatographic research has been addressed to improve HPLC performance in terms of faster separation and efficiency. So, when UHPLC is not available, alternative devices and technologies have been developed for *fast chromatography*: high-temperature liquid chromatography (HTLC), use of monolithic supports, use of column with superficially porous packing materials based on silica particles with nonporous cores [81].

HTLC operates chromatographic separation at elevated temperature (60–120 °C) resulting in a significant reduction of mobile phase viscosity, leading to higher diffusion coefficients for the compounds and improved mass transfer [82]. The authors also highlighted the practical limitations due to the instrumentation; moreover, limited number of stable stationary phases compatible with elevated temperature makes HTLC to be rarely used in routine analysis and only investigated in academic laboratories [81].

Monolithic supports consist of a single rod of porous material with several unique features in terms of permeability and efficiency. The low backpressure generated and good mass transfer enable use of elevated flow rates (3–10 times larger) [83]. First publication with the use of this support in LC was in 1989 by Hjerten et al. [84], but to date monoliths are not widely used.

The use of column with *superficially porous packing materials* based on silica particles with nonporous cores is the most recently reported strategy for improving chromatographic performance. This technology, originally developed by Kirkland in the 1990s to limit diffusion of macromolecules into the pores [85], became commercially available in 2007 [86]. In comparison with totally porous particles of similar diameters, the both A and C term of the Van Deemter curve are reduced [87, 88].

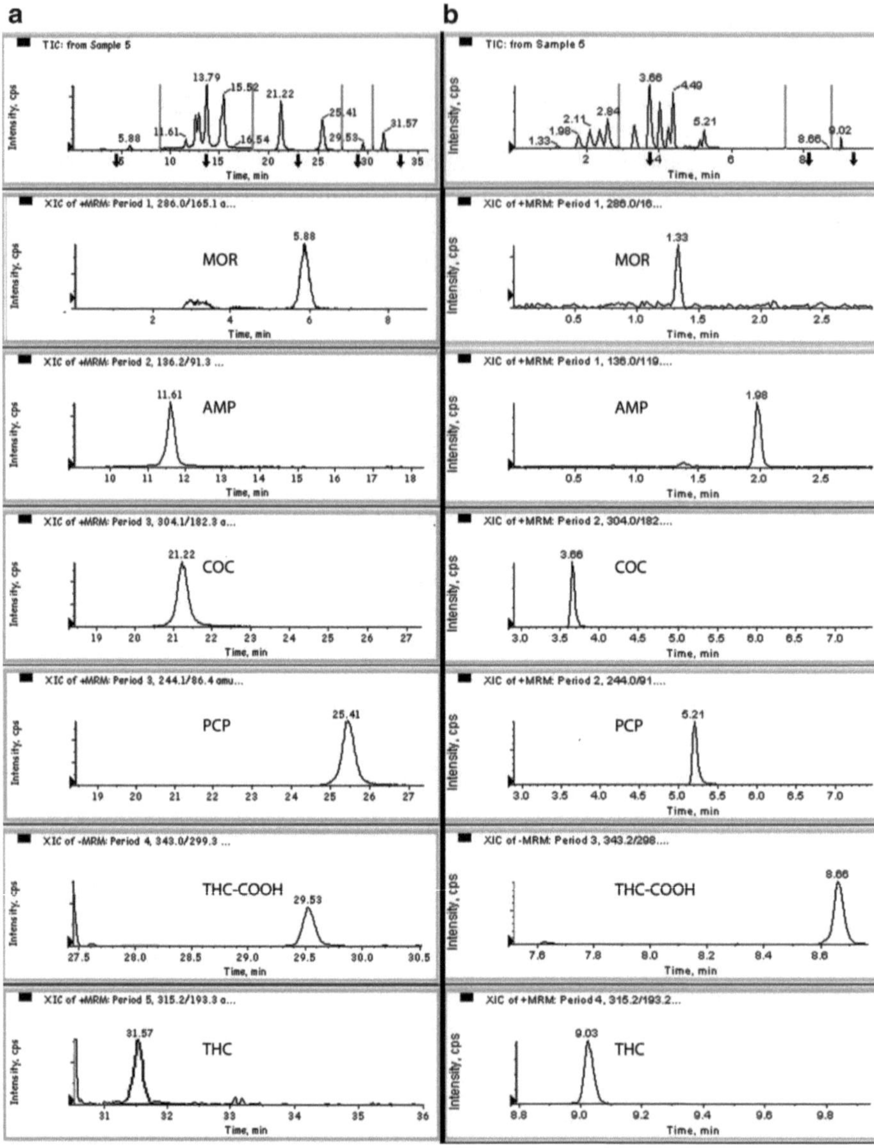

Fig. 8 HPLC-MS/MS chromatograms for multiclass analysis of illicit drugs obtained by normal bore (**a**) and fused core (**b**) columns

Good efficiency and shorter analysis time are keys of the usefulness of this new technology for the determination of illicit drugs in biological samples [66, 89]. In Fig. 8 we report chromatograms obtained with analytical column and chromatograms obtained with a core shell (or fused core) column.

10 Sample Preparation and Applications

Sample preparation prior to LC-MS analysis aims to reduce matrix effects by removing potential interferences and to get the analyte into an amenable form for the analysis. However, for drugs and low molecular mass compounds, coeluting components such as proteins, lipids, and salts may cause variability in the efficiency of analyte ionization [65]. It is common practice in the laboratory using appropriate internal standards, better stable isotope-labeled internal standards (ISTDs), which normalize signal excluding interferences. However, cases are reported in the literature in which a deuterated ISTD has different chromatographic properties [90, 91] or interacted in different matrices differently to the unlabeled analyte [92]. In their study Wang et al. [92] also suggested that ISTDs labeled with ^{13}C, ^{15}N, or ^{17}O may be better at compensating for matrix effects than deuterium labeled ISTDs (although they are often more expensive). When the cost of these labeled compounds is prohibitive, for example in multianalyte procedures, it was demonstrated a possibility to suitably select only some deuterium labeled ISTD, one by class of substance, without compromising data quality [66, 93].

10.1 Plasma, Serum, or Whole Blood

Serum, plasma, and even more whole blood are "dirty" matrices which need an effective pretreatment. Different strategies have been applied for this purpose.

10.1.1 Protein Precipitation

The simplest approach is to promote protein precipitation (PPT). Blanchard et al. in 1981 [94] evaluated different precipitating solvents. However, after a simple treatment such as PPT matrix effects may persist because of the endogenous compounds (mainly phospholipids [95]): in some cases the supernatant was diluted, but sensitivity got worse [96], or evaporated and reconstituted in mobile phase, with reproducibility getting worse in this case [55]. However, the use of suitable internal standards may be very helpful or even necessary, as demonstrated in a paper in 2009 by Sergi et al. [93]: the authors proposed an analytical procedure for the simultaneous determination in plasma and in OFs of 13 analytes belonging to different chemical and toxicological classes (amphetamine, methamphetamine, morphine, 6-MAM, MDA, MDE, MDMA, cocaine, BEG, THC, THC-COOH, ketamine, and PCP). The chromatographic run was also used as an online cleanup, allowing the selective elution of the analytes from the column and separating the whole injected solution into three major fractions, of which only the second was analyzed by LC-MS/MS. The use of a diverter valve enabled the maintenance of the source cleanliness for a long period of time and prevented sudden reductions of sensitivity during the analysis of real samples. Authors analyzed real samples with this method and good correlation was shown between plasma and OF results. Matrix effects were compensated by the use of ISTDs.

10.1.2 Liquid–Liquid Extraction

Many solvents are used, none of which meet the ideal criteria, and most of them require specialized storage, handling and disposal procedures. The most widely used organic solvents are ethyl acetate, diethyl acetate, hexane, chloroform, butyl acetate, dichloromethane. Liquid–liquid extraction (LLE) is simple, robust, and transferable; it shows good reductions in matrix effects for plasma and it may be more suited to urgent analyses than SPE [97]. In most cases, though, there is a need to evaporate the solvent extract before redissolving the residue in mobile phase. This extra step costs time and increases error. Gottardo et al. used LLE for the analyses of MA, MDA, MDEA, MDMA, methadone, cocaine, morphine, codeine, 6-MAM, BEG in blood. The sample was treated with ammonium sulfate and then extracted with chloroform–isopropanol (9:1) [98].

Couchman et al. developed a method for simultaneous analysis of amisulpride, metamphetamine, and amphetamine in plasma by HPLC-MS/MS. To the sample (200 μl), ISTDs solution and NaOH solution were added to the extraction solvent (butyl acetate–butanol, 9:1). HPLC separation was performed by strong cation-exchange. The use of methanol as eluent improved sensitivity for analytes due to more efficient in-source desolvation when compared with aqueous eluents [99].

10.1.3 Solid Phase Extraction

Many solid phase extraction (SPE) materials, when the sorbent becomes dry, risk inactivation of the bonded phase, leading to the poor recovery of analytes. However, some newer, polymeric sorbents are more resistant to this problem, can tolerate relatively *dirty* samples and are stable over a wider pH range compared to silica-based materials [100]. Mixed-mode materials, with a combination of interactions, allow efficient cleanup by using relatively harsh wash steps with minimal loss of analyte(s), and they are increasingly used for sample preparation [101]. Such methods typically involve the sequential elution of acidic, neutral, and basic compounds using solvents at appropriate pH, and this versatility has led to their increasing popularity. For comprehensive analyses, the elutes from different fractions are usually combined and evaporated before reconstitution in an LC-compatible solvent. Several SPE materials were evaluated for extraction of a range of drugs from serum/plasma, noting that a divinylbenzene (DVB) material offered the best combination of extraction capacity and desorption efficiency amongst those tested.

A series of SPE sorbents ranging from nonpolar, to mixed-mode, to polymeric, were tested for their performance in the STA of a diverse range of drugs (17 analytes) in whole blood [102]. In this experiment a C_8-modified silica material was found to offer the best overall recovery, followed closely by a mixed-mode polymeric sorbent. It was reported that the C_8 material offered the cleanest sample extracts. After SPE direct injection of eluate or its dilution prior to injection into LC-MS apparatus is possible.

Limitations of SPE include the coextraction of interfering compounds and poor extraction of some drugs, but an important advantage is the possibility of building automated systems and obtaining the analytical procedure online, as this is the recent trend.

Recently Bjock et al. [103] have developed an analytical method for the determination of 19 drugs of abuse and metabolites in whole blood. The following compounds were investigated: amphetamine, MDA, MDEA, MDMA, methamphetamine, cocaine, BEG, morphine, 6-MAM, codeine, methadone, buprenorphine, norbuprenorphine, ketobemidone, tramadol, *O*-desmethyltramadol, zaleplone, zolpidem, and zopiclone. Sample pretreatment consisted of PPT prior to automated SPE. The samples were mixed, sonicated, and centrifuged. SPE was performed by using mixed mode cartridge: the highest and most reproducible recoveries were obtained with the Isolute Confirm HCX (130 mg, 3 ml) SPE column with 80:20 toluene/ethyl acetate. An LC-MS system was used for separation of the target analytes and the total time of analysis was 35 min to ensure reequilibration between injections. The described method was successfully applied by the authors for screening and quantification of the 19 targeted drugs of abuse in 412 authentic forensic cases from October 2008 to February 2009 in DUIDs cases (Driving Under Influence of Drugs).

In the American Federal Bureau of Investigation (FBI) laboratories, Jagerdeo et al. analyzed cocaine and four metabolites: BEG, EME, ecgonine, and CE. Pretreatment of the samples to precipitate plasma proteins was carried out by adding zinc sulfate to blood sample. Filtered samples were transferred and were prepared for the online SPE system, with an anion exchange based cartridge. Detection was performed using a Qtrap Mass Spectrometry equipped with a turbo spray ion source operated in positive electrospray ionization mode. This method was validated and demonstrates excellent accuracy and precision, and an excellent lower limit of detection (also due to the high performing instrumentation) [104].

Bouzas et al. [95] developed a quantitation method for the determination of drugs of abuse (opiates, amphetamine and derivatives, cocaine, methadone and metabolites) in serum by using online extraction coupled to LC-MS/MS. The online extraction procedure described consists of an extraction column and an analytical column, which were coupled online. A PPT procedure was performed with zinc sulfate: an aliquot of 0.1 M zinc sulfate in methanol was added to the serum sample in a proportion of 2:1 (*v*/*v*) to serum. Analytes were extracted by a short pentafluorophenyl silica column and separated on a longer analytical column with the same stationary phase. Recoveries of all analytes were above 80 %. The proposed procedure by Matuszewski et al. [64] was used for the evaluation of matrix effect: $ME(\%) = B/A \times 100$, where ME is the matrix effect (suppression or enhancement) and B corresponds to peak areas for standards spiked after extraction into sample extracts and A to peak areas obtained in neat solution standards. Authors compared this method to offline SPE coupled with GC-MS and results showed that LC-MS/

MS is a valuable alternative. They have successfully applied the method to clinical and forensic serum samples of DUIDs cases and the results were in good correlation to those obtained by offline SPE-GC-MS analysis.

A different application of SPE which needs 100–200 μl of sample has recently been successfully applied to plasma. Micro SPE (μSPE) followed by HPLC-MS/MS was used by Napoletano et al. [67] to determinate amphetamine, methamphetamine, MDA, MDE, MDMA, cocaine, BEG, mescaline, ketamine, PCP, psilocybine in plasma. In this method 180 μl of plasma were submitted to slight PPT with 20 μl of methanol containing ISTD, 100 μl of supernatant were then collected and passed through C18 tips, adapted to automatic pipettes. Validation results showed μSPE allows to reduce matrix effect (≤10 %) keeping satisfactory recoveries. μSPE represent a simple, fast and reliable procedure with extremely reduced solvent consumption; it will be discussed in more detail below in its other applications.

10.2 *Oral Fluid as Alternative Matrix*

Mandel in 1990 wrote that "Saliva lacks the drama of blood, the sincerity of sweat and the emotional appeal of tears" [36]; however, the development of analytical procedures for the determination of illicit drugs in OF has dramatically increased over the last decade.

Sample collection is the first issue to study when developing analytical method: it is very important to uniform and control the sampling methods [105–107]. There are different procedures reported in literature: spitting in a tube [93, 108] or different sample collection devices [109–111].

10.2.1 PPT

A simple PPT was developed for plasma and OFs: recoveries of analytes from OF are often higher when compared to plasma, but sonication before precipitation has been proven necessary to help the analytes to release the protein binding, also in OF. Developing a multiclass method with PPT has proved to be simpler in respect to other extractive procedures, so it has been possible to obtain recoveries averaging 100 % and minimized matrix effects by the use of a deuterated ISTDs for each class of substances [93].

More recently Wang et al. have proposed analytical methods involving the use of UHPLC-MS/MS to determine four opiates and metabolites, five amphetamines, flunitrazepam and its two metabolites, cocaine and its four metabolites in OF. Samples were collected by spitting in a tube, then the fluids were diluted with twice the amount of distilled and deionized water and vortexed for 30 s. Liquid chromatography was performed with a gradient elution and the total run time was 7.5 min. The authors compared ESI, APCI, and APPI. The ion suppression ranged from 45 to 89 % and from 74 to 96 % on APCI and APPI, respectively. The authors compared these data with those related to other pretreatment such as LLE and SPE [93, 112, 113], proving that no great improvement was achieved [114].

10.2.2 LLE

Simões et al. have published an analytical method for the confirmation and quantitation of 24 substances, including opiates, amphetamines, cocaine, methadone, THC, benzodiazepines, and other medicines [109]. OF samples were collected in a Saliva Sampler collection device, followed by LLE. The samples were transferred into a tube, containing a buffer solution at pH 9 and a mixture of organic solvents, to extract organic bases and some neutral drugs from the biological matrix. Chromatographic separation was performed using gradient elution with acetonitrile/2 mM ammonium formate buffer and 2 mM ammonium formate buffer. This fully validated method has demonstrated a great feasibility to DUIDs real cases applications. However, LLE recoveries are not quantitative at lower concentrations.

10.2.3 Microwave Assisted Extraction

An interesting study was performed by Fernandez et al. to compare two extraction procedures for the determination of drugs of abuse [110]. Targeted analytes were morphine, 6-MAM, cocaine, CE, BEG, methadone, and 2-ethylene-1,5-dimethyl-3,3,-diphenylpyrrolidine.

OF samples were collected with the commercial device Salivette, which consists of a cotton swab that is inserted into the mouth for 2–3 min. Toxitubes A was compared to microwave assisted extraction (MAE). A volume of 1 ml of saliva was poured into each Toxitube and treated similarly as seen previously [109]. For MAE different solvents (chloroform, dichloromethane, hexane), temperatures (80, 90, and 100 °C), and time periods (5, 10, and 15 min) have been tested; finally 1 ml of saliva was mixed with 10 ml of chloroform and placed in the oven vessel for extraction at 100 °C for 10 min. Recoveries were found to range from 53 to 95 % with Toxitubes and from 83 to 100 % with MAE, so authors demonstrated that microwave-assisted extraction provides recoveries higher than those obtained for opiates with SPE.

10.2.4 SPE

Kala et al. proposed a similar multiclass method in 2008 for the simultaneous determination of amphetamines, opiates, PCP, and cocaine and BEG. In this paper the difficulty of establishing a common step of sample preparation for all of the substances is demonstrated. OF samples, collected by OF collection kits, were analyzed after screening. Amphetamines and PCP were extracted by LLE, while for opiates, cocaine and BEG extractions, authors used an SPE strategy. The separation of

AMPs was carried out using a C18 column, whereas opiates, PCP, cocaine, and BEG were separated on a PFP column. This is an example of a good analytical method that needs a time-consuming sample preparation and two chromatographic runs for the achievement of the best performances [115].

10.2.5 μSPE

Solid phase extraction usually needs at least 0.5 ml of sample, so it is not suitable when a small amount of sample is available. A method based on μ-SPE-HPLC-MS/MS for simultaneous determination of stimulants and hallucinogens was developed [66]. Amphetamine, methamphetamine, MDA, MDE, MDMA, cocaine, BEG, mescaline, ketamine, PCP, psilocybine were the target analytes. Extraction was carried out using C18 tips, adapted to automatic pipettes. These tips are made of a fiberglass, functionalized with apolar chains of octadecylsilane in a monolithic structure. Only 90 μl of OF is required: these were mixed with 10 μl of methanol containing deuterated ISTDs (one for each class of substances). The mixed solution was sonicated and centrifuged to precipitate proteins. The whole supernatant was then rescued and passed through the tip by loading and releasing five times with the pipette. The analytes elution was achieved using 100 μl of methanol. The analytes were separated using a reversed phase C18, packed with 3 μm average diameter particles. The complete separation of all substances occurs in 12 min.

Authors calculated the value of matrix effect comparing two calibration curves: one created with blank samples, fortified after the extraction process, and the other with standard solutions. Matrix effects, intended as ion suppression, calculated by comparing the slopes of those curves are lower than 6 % for each analyte.

10.3 Urine

Urine samples are often submitted to enzymatic cleavage of gluconic or sulfuric acid conjugates of drugs and/or their metabolites. This kind of treatment is a prerequisite for the sensitive GC-MS analysis of drugs excreted into urine in conjugated form, but it is not strictly required in LC-MS(/MS) analysis, because this technique allows the direct analysis of such conjugates, e.g., glucuronides of opiates [116–118]. Nevertheless, many of the recently published methods for LC-MS(/MS)-based urine analysis still include enzymatic conjugate cleavage.

Andersson et al. have recently proposed a direct injection LC-MS/MS method for the identification and quantification of amphetamine, methamphetamine, MDA, and MDMA in urine drug testing. The samples were prepared for analysis by fivefold dilution with ultrapure water [119]. A gradient elution was performed using two solvents: 25 mM formic acid containing 1 % acetonitrile and 25 mM formic acid containing 90 % acetonitrile. Authors observed matrix effects, in terms of ion suppression, about 25-fold. This method was used for real sample analysis and offers great potential for routine urine drug testing as the time for sample preparation can be shortened since no extraction and derivatization is needed. Same authors in 2007 had proposed direct injection of urine sample for confirmatory analysis of opiates and their glucuronide metabolites [118].

Peters noted in a recent review [52] that only a few works published have used simple dilution for the preparation of urine samples. The dilution factors ranged from 5- to 50-fold. Eichhorst et al. also used (tenfold) dilution in their procedure, but only after urine samples had been submitted to enzymatic conjugate cleavage [120].

10.3.1 LLE

Maquille et al. [121], due to the physicochemical properties (i.e., polarity and ionization state) of the investigated drugs (opiates, amphetamine, cocaine and metabolites), concluded that LLE should be selected. To automate the sample preparation procedure, this team proposed urine extraction by supported liquid–liquid extraction (SLE), a promising technique that appeared in 1997 [122], which can be easily automated in a 96-well plate format. It has been demonstrated that matrix effect is significantly minimized.

10.3.2 SPE

Fernandez et al. [123] proposed a high throughput analysis of THC-COOH. It is rare that one finds LC-MS/MS application to only one analyte, but sometimes there are specific requests. THC-COOH exists in both the free and glucuronide form, so authors, in order to improve sensitivity, performed sample enzymatic conjugate cleavage by basic hydrolysis. Similar hydrolysis procedures have been described by other authors [97, 124, 125]: these strong hydrolysis conditions can only be applied to molecules like THC or THC-COOH which are very resistant, thus, multiclass analysis address "softer" enzymatic hydrolysis. Elution of the analytes from the cartridge was achieved by the application of the LC mobile phases: 0.1 % formic acid and acetonitrile during the chromatographic run. Whilst the elution step was being performed, a new cartridge was conditioned, loaded, and washed in the left clamp. Although the analyte has a carboxylic group, the authors selected ESI in positive ionization mode to create their MRM experiments with $[M+H]^+$ as precursor ion.

The use of online SPE in the column-switching mode is particularly adapted to illicit drug analyses, on the other hand the offline SPE on multiwell plates is qualified for multianalyte procedures: since each sample is independently extracted, this format is compatible with different separation techniques working with different mechanisms (reversed or normal phase, ion exchange, etc.); reduces contamination risk; presents a great advantage in analyses where a legal aspect has to be taken into account; it does not require a particular technical skill [62].

Berg et al. [126] have developed a method for the determination of opiates and cocaine in urine. Sample preparation was performed by adding ISTDs and 0.5 ml of

0.67 M Sørensen phosphate buffer (pH 7.4) before mixed mode cation exchange SPE on Oasis MCX SPE cartridges. Chromatographic separation was performed at 60 °C on a UPLC BEH C18 column. Authors assessed that high pH mobile phase provided narrow and symmetrical peaks and repeatable retention times due to less silanol interactions, increased retention, and increased load capacity; this was possible due to the use of a column compatible with both low pH and high pH mobile phases. In literature there are several applications of MCX-SPE for urine sample pretreatment. For example Aturki et al. [127] used MCX-SPE for urine sample pretreatment, but after the centrifugation of samples and filtration of supernatant. Urine sample was loaded into the cartridge that was then dried under a stream of nitrogen for 5 min. The cartridge was washed with phosphoric acid (10 mM, 2 ml), acetic acid (100 mM, 1 ml), and methanol (2 ml). Retained analytes were eluted with 4 ml of 3 % ammonia solution in methanol.

10.3.3 μSPE

A *μ-SPE* approach was proposed for urine sample by Ellison et al. [128]. This group used Disposable Pipette Extraction (DPX). DPX incorporates pipette tips that contain solid-phase sorbent material loosely confined with a screen at the bottom narrow end of the tip and a barrier near the top. DPX tips were selected as containing sorbent with strong cation-exchange (CX) and reversed-phase characteristics. DPX-CX 1-ml tips were used to extract drugs of abuse from the prepared urine samples. No conditioning step was performed prior to extraction. The tips were loaded with the sample solution by slowly aspirating the solution using an attached syringe device (in place of a pipette) and vortex mixing for ~10 s (by holding the syringe with attached tip inside a small test tube) to allow the solution to thoroughly mix with sorbent and create a gel of high surface area. Extraneous material was washed from the tips by adding 500 μl of distilled water to the top of the sorbent and dispensing. Authors, in this way, developed a comprehensive extraction for 6-MAM, BEG, morphine, codeine, oxycodone, amphetamine, methamphetamine, MDMA, PCP, THC-COOH, and pentobarbital. Recoveries were measured between 80 and 105 %.

Another μSPE-based method has recently been proposed by Napoletano et al. [67]. Previously cited procedure applied at OF samples was exported with very slight modifications to urine and plasma samples. Recovery in urine were verified at SAMSHA cutoff concentration values [129], they ranged from 56 % for amphetamine and 90 % for PCP; RSD was always less or equal to 10 % and matrix effects, valuated as reported above, were less or equal to 10 %.

10.4 *Hair Analyses of Drugs of Abuse*

Although sample collection would seem easier than other matrices such as urine and plasma, this requires caution and has been encoded to be as significant as possible

[130, 131]. It is also possible to operate a segmentation of the hair to distinguish different temporalities; however, the variability of growth from person to person will never allow to assign the exact time of illicit drugs assumption.[42].

An analytical method for determining illicit drugs in the hair should include before analysis a proper washing procedure to eliminate possible external contamination, or traces of drugs that can be deposited on the outside of the hair, thus limiting the possibility of interference and false positives. The washes also remove residues of shampoo and cosmetics as well as sweat and sebum, which can cause problems of interference in the subsequent analytical procedures and increase the background noise recorded in instrumental analysis of the sample [42, 132].

10.4.1 Decontamination

The solvents can be divided between solvents that swell the keratin structure (swelling solvent), which penetrate the hair such as water and methanol, and solvents that do not swell the keratin structure (nonswelling solvent), failing to cross the cuticle and therefore not penetrating inside the hair. The washing procedure in most cases requires more than one step (2–3) [48, 133–146] generally processed through the subsequent use of different solvents or solutions. Only rarely is a washing procedure including a single step [51].

The used solvents, in subsequent steps or in mixtures, are as follows: ultrapure water [136, 138, 144], aqueous solutions of surfactants (dodecylsulphate [138], polysorbate [136, 140], generic surfactant [144]), and organic solvent (dichloromethane [51, 133, 141–143], acetone [137, 138, 143], methanol [134, 139, 143, 145], ethanol [146], hexane [137], isopropanol [135, 142, 147], petroleum ether [139]).

Aprotic solvents such as dichloromethane and acetone have the advantage of not possessing extractive capabilities [42], being unable to penetrate into the keratin structure. Extraction tests with conventional techniques, for dichloromethane, showed that the analytes were not found in the extracts [51].

10.4.2 Extraction and Digestion of the Keratin Matrix

The extraction methods are divided between those using the solvent and those providing for the dissolution of the keratinic matrix.

Solvent Extraction

This process allows to provide contact of hair, chopped or shredded to increase surface development, with an appropriate extracting solvent. Methanol is the most used as the extraction solvent and is often supported by the use of an ultrasonic bath to reduce the extraction time. The extraction times are usually between 16 and 20 h [51, 137, 148–150]; the use of ultrasonic bath shortens the extraction time up to 4 h

[139, 141] also combined with higher temperature [134, 139, 141, 143, 148]. It has been shown that higher temperature also increases the rate of degradation of some substances: above 65 °C a significant increase in the conversion of cocaine into benzoylecgonine occurs [151].

Furthermore, to increase the efficiency of extraction of basic molecules such as amphetamine, methamphetamine and cocaine, methanol can be acidified with HCl [151]. Methanol is a good extraction solvent that allows the simultaneous extraction of compounds sensitive to hydrolysis, such as cocaine and heroin, and very lipophilic compounds such as THC [148]. A disadvantage of using methanol is an extract with a high level of impurities, which must necessarily be subjected to a subsequent cleanup [42].

Another important factor is the pH of the extractive solution for not promoting the hydrolysis of certain substances such as cocaine and 6-MAM. In some methods, the extraction solvent is acetonitrile with addition of formic acid [135] or in mixtures with methanol, with extraction times of 18 h [145, 147].

The optimization of the extraction phase from hair with dichloromethane and methanol as modifier has recently been published by Fernandez et al. [140]. In this case MAE has been used, obtaining a very rapid extraction procedure and allowing the chlorinated solvent to permeate into the structure of the hair keratin.

The use of dilute aqueous solutions of HCl [142, 152, 153] or phosphate buffer (pH 6.5–5) [133] are very appropriate for the extraction of amphetamines and other compounds in basic character. In this case, the extraction times are about 20 h. When supported by increased temperature, the extraction lasts only for 2 h, but it is not effective for the extraction of lipophilic compounds such as THC. The use of aqueous solutions of HCl causes partial or complete hydrolysis of some substances, so phosphate buffer at pH 5 is preferred for the determination of cocaine and 6-MAM; in these conditions these two molecules appear to have increased stability [154]. In addition to acidic aqueous solutions, formic acid instead of hydrochloric acid solutions are also used in organic solvents [136].

Supercritical fluid extraction (SFE) has been used for the recovery of analytes from hair: this is a technique that offers several advantages due to the characteristics of low viscosity of supercritical fluid: the speed of extraction, high extraction efficiency, and ultimately the ability to easily remove the extraction solvent. Also there is the possibility of working with automated systems that also allow the recycling of the solvent. The major limitation of this technique is the cost of the instrumentation [42].

Matrix Dissolution

These processes provide a complete destruction of the structure of the hair keratin by the complete hydrolysis of proteins that compose it. Hydrolysis can occur enzymatically or chemically.

Enzymatic Digestion

The keratin structure is destroyed through the use of proteolytic enzymes such as pronase and proteinase K. They are often used with chemical agents such as urea and thioglycolic to cleave the disulfide bonds and increase the dissolution rate of enzyme activity. The extracting procedures, using enzymatic digestion, last about 4–6 h and must be conducted at constant temperature and pH for providing maximum enzyme activity [155].

Chemical Digestion

The hair treatment with concentrated solutions of sodium hydroxide allows the complete dissolution of the keratin structure by chemical hydrolysis of proteins in about 1 h. Under these conditions, some drugs such as amphetamines are volatile and thus there may be losses of the analyte. In more basic solutions occurs the complete hydrolysis of molecules such as cocaine, heroin, and 6-MAM [51]. Concentrated solutions of hydrochloric acid are also used: it eliminates the problem of volatile basic compounds, but increases the time of dissolution.

Matrix Solid Phase Dispersion

The matrix solid phase dispersion (MSPD) is a sample pretreatment technique used for extraction of analytes from solid samples. Typically applications are found in the treatment of complex biological matrices such as plant or animal tissues [156, 157]. The solid phase particles, in addition to disperse the sample, break the structure of the matrix, so that the contained analytes may be recovered. A cartridge is packed with the homogenized mixture and placed between two filters, and then the elution of the analytes is performed. This extraction technique allows for the use of small amounts of sample, extraction of high yields, and reduced extraction times. Recently the MSPD technique has been applied for the extraction of drugs of abuse from the hair by Míguez-Framil et al. [144]. Hair samples (approximately 0.050 g) were blended thoroughly with alumina in a glass mortar to obtain a homogeneous mixture. The mixture was quantitatively transferred by using a powder funnel to a 10 ml syringe. A previously conditioned Oasis HLB cartridge was then attached to the end of the MSPD syringe for on column analytes retention. In this way, the target drugs isolated from hair matrix were retained on column and adsorbed onto the solid support of the SPE cartridge.

Micropulverized Extraction

Miyaguchi and Inoue proposed micropulverized extraction for the determination of amphetamine, methamphetamine, dimethylamphetamine, MDA, MDMA, ketamine, norketamine, cocaine, and BEG in hair [158]. A washed hair specimen (0.2 mg) was precisely weighed in a polypropylene tube. A stainless steel bullet, 4 μl of 10 ng/ml methamphetamine-d5 (IS) and 50 ml of 0.1 M ammonium formate buffer (pH 3.5) were added. The tube was capped and shaken vertically for 5 min at 1,500 strokes per min with an automatic pulverizer. After a brief centrifugation, the bullet was removed with tweezers. The suspension was filtered using a centrifugal filter unit with a 0.45 mm polyvinylidene fluoride (PVDF) microporous membrane. The filtrate was transferred to a conical insert with polymer feet in a glass vial to be analyzed by LC-ESI-MS/MS.

A C18 column was used for separation at 40 °C. The mobile phase, delivered at a flow rate of 0.3 ml/min, was 5 mM ammonium formate (pH 3.5) in water–acetonitrile; proportion of acetonitrile was changed over a gradient. Full scan analysis with the Orbitrap analyzer was performed with a mass range of *m/z* 130–430 and resolving power of 30,000 in a profile mode. Method performances were investigated, specially accuracy was evaluated by analyzing standard reference material: the results obtained were in general agreement with the approved values. Sample pretreatment, consisting in 5 min extraction of a very small amount of hair, only 0.2 mg, validated with reference material, represented certainly a convenient approach.

10.4.3 Cleanup

In some cases the extract of hair is directly analyzed or after a solvent exchange, or after evaporation of the extracting phase [134–136, 139–141, 147]. Through direct injection strong signal suppression is recorded and an extract cleanup step prior to instrumental analysis is suggested [141]. The extract purification is important for increasing the selectivity and accuracy of the analysis by eliminating potential interference and restricting the matrix effects; it is important to make the sample compatible with the instrumental analysis as in this step may be a change of solvent or the concentration of analytes so as to fall within the detection limits of the instrument.

Furthermore, the sample purification is important for safeguarding the instrumentation, as the sample cleanup also removes all those substances that could damage it over time.

The procedures to purify the extracts appear to be mainly LLE (liquid–liquid extraction) and SPE (solid phase extraction).

An interesting application has been proposed by Bucelli et al. [133]: a LC-ESI-MS method for the analysis of 16 drugs (EME, cocaine, benzoylecgonine, morphine, 6-MAM, codeine, amphetamine, methamphetamine, MDA, MDEA, MDMA, MBDB, ephedrine, phentermine, phendimetrazine, buprenorphine) in human hair. The ion trap spectrometer afforded to work over the entire mass range in full scan mode, in MS/MS and MS^n mode. The mixture was incubated at 45 °C for 18 h and then centrifuged. Extraction was carried out with Bond Elut Certify cartridges. Analytes were eluted with a mixture of dichloromethane–isopropanol–ammonia. The eluent was removed under a gentle nitrogen stream at 45 °C. For chromatographic separation a C18 column was used and the elution solvents were ultrapure water–0.1 % formic acid and acetonitrile–0.1 % formic acid. When possible, the authors performed sectional analysis of hair samples [130]. Wang and coworkers used small volume LLE to perform cleanup after chemical hydrolysis of 20 mg of hair sample with NaOH [159]. Targeted analytes in this work were amphetamine, methamphetamine, MDA, MDMA. Digested hair were centrifuged, then 50 μl of chloroform and a small amount of solid KCl were added, finally the mixture was vortexed and centrifuged. Authors selected chloroform as extraction solvent because of its greater density than water, allowing it to remain at the bottom of a tapped vial, convenient when handling small volume. Authors tested volume of extraction solvent ranging from 20 to 200 μl: results showed that smaller the volume used in the extraction, the higher was the extraction of drugs. Although in this case the analysis is performed by GC, this LLE approach is interesting.

11 Conclusion

The versatility offered by LC, i.e., the possibility to vary mobile phase composition by adding suitable additives, as well as different types of commercially available stationary phase, makes LC particularly appropriate for multiclass analysis of drugs in complex matrices such as human biological matrices. Detection in MS and MS/MS has amplified the advantages, allowing the technique to reach unambiguous analytical data. This makes the *odd coupled* LC-MS an analytical technique applicable for forensic purposes, as the determination of illicit substances or substances subjected to government regulations. Drugs of abuse are certainly substances of interest in forensics and their determination in human biological matrices is a significant analytical problem. The most common are hallucinogenic substances such as THC, psychostimulants such as cocaine, amphetamines, and methoxyamphetamine or their derivatives, but also other psychotropic substances of emerging abuse such as ketamine and phencyclidine. Human biological matrices are different: from the most common plasma and urine to OFs, nails, sweat, and hair. In fact, Bush from SAMSHA wrote in 2008 about the fundamental understanding that the window of drug detection for urine, hair, OF, and sweat patch specimens are neither equal nor identical, but the results from each specimen can be used in a complementary manner [129].

A kind of guidance has been drawn for the chemist/toxicologist who approaches to develop a new analytical method (from sample pretreatment to chromatographic separation and MS detection) and to validate new methods. Different strategies are possible with MS detector, especially with high resolution instrumentation. This chapter focuses on analysis on targeted molecules (so-called confirmatory analysis), but there are many application of LC-MS in GUS/STA: in this case the aim is to identify unknown compounds that may be present in a selected matrix. However, in this case, it needs subsequent confirmatory analysis to be used in a court. The most cited international guidelines are compared with regard to the parameters to be monitored during method validation and how effectively to evaluate them; among these matrix effect plays a significant role in LC-MS methods.

Different developed analytical method are discussed in this chapter related to the determination of illicit substances in blood (either whole blood, plasma, or serum), OF, urine, and hair. These methods take into consideration the particular chemical and physical composition of the matrix and applies each time a suitable pretreatment to remove interfering and matrix effect, to maximize recoveries and to achieve a suitable enrichment if necessary. For liquid matrices the applications of the most common techniques are considered from simple PPT to SPE and LLE; the results of recent works from literature are reported and new trends as online SPE, μSPE, automated LLE (SLE) or MAE are examined. Several stationary phases have been shown to be suitable for determination of illicit drugs: C18, pentafluorophenyl, strong cation-exchange, and HILIC columns. The trend toward *fast chromatography* is investigated, both UHPLC and HPLC with appropriate arrangements; moreover, results obtained with different ion sources, ESI, APCI, and APPI are compared.

Without claiming to be exhaustive, these pages could be a good starting point to learn about the topic of analysis of illicit drugs by LC-MS^n; the aim is to direct the readers to specific aspects of analytical problem, which should be further focused depending on their needs.

References

1. Liberman JM, Cooper SJ (1989) The neuro-pharmacological basis of reward. Oxford University press, Oxford
2. Di Chiara G, Acquas E, Carboni E (1991) In: P. Willner and J. Scheel-Kruger (eds) Role of mesolimbic dopamine in the motivational effects of drug: brain dialysis and place preference studies. In: The mesolimbic dopamine system: from motivation to action. John Wiley and Sons Publ., Chichester, UK
3. Carboni E, Imperato A, Perezzani L, Di Chiara G (1989) Amphetamine, cocaine, phencyclidine and nomifensine increase extracellular dopamine concentrations preferentially in the nucleus accumbens of freely moving rats. Neuroscience 28(3):653–661
4. Mereu G, Yoon KW, Boi V, Gessa GL, Naes L, Westfall TC (1987) Preferential stimulation of ventral tegmental area dopaminergic neurons by nicotine. Eur J Pharmacol 141(3):395–399. doi:0014-2999(87)90556-5 [pii]
5. Grace AA (1988) The mesolimbic dopamine system. Ann N Y Acad Sci 537:51–76
6. Geninatti S, Bellavia F, Chieppa G (2001) Contributo alla definizione di nuove droghe. Boll Farmacodip e Alcoolis XXIV (2)
7. Henderson GL (1988) Designer drugs: past history and future prospects. J Forensic Sci 33(2):569–575
8. Gates M, Tschudi G (1952) The synthesis of morphine. J Am Chem Soc 74:1109–1110
9. Chang KJ, Cuatrecasas P (1979) Multiple opiate receptors enkephalins and morphine bind to receptors of different specificity. J Biol Chem 254(8):2610–2618
10. Osborne R, Joel S, Trew D, Slevin M (1990) Morphine and metabolite behavior after different routes of morphine administration: demonstration of the importance of the active metabolite morphine-6-glucuronide. Clin Pharmacol Ther 47:12–19
11. Olson KR (1999) Anfetamine, cocaine, oppioidi, cannabinoidi, allucinogeni. Intossicazioni acute, veleni; veleni, farmaci e droghe. Springer Ed., Milano
12. Warner A, Norman AB (2000) Mechanisms of cocaine hydrolysis and metabolism in vitro and in vivo: a clarification. Ther Drug Monit 22(3):266–270

13. Hearn WL, Rose S, Wagner J, Ciarleglio A, Mash DC (1991) Cocaethylene is more potent than cocaine in mediating lethality. Pharmacol Biochem Behav 32(2):531–533
14. Hardman JG, Limbird LE (1996) Parmacological basis of therapeutics. Ninth Edition edn, MCGrawHill
15. Logan BK (2002) Methamphetamine-effects on human performance and behavior. Forensic Sci Rev 14:133–151
16. Battaglia G, Brooks BP, Kulsakdinum C, De Souza EB (1988) Pharmacologic profile of MDMA (3,4-ethylenedioxymethamphetamine) at various brain recognition sites. Eur J Pharmacol 149:159–163
17. Lora-Tamayo C, Tena T, Rodriguez A (1997) Amphetamine derivative related deaths. Forensic Sci Int 85:149–157
18. Wright J, Cho AK, Gal J (1977) The metabolism of amphetamine in vitro by rabbit liver preparations: a comparison of R(−) and S(+) enantiomers. Xenobiotica 7(5):257–266
19. De la Torre R, Farré M, Roset PN, Pizarro N, Abanades S, Segura M, Segura J, Camí J (2004) Human pharmacology of MDMA: pharmacokinetics, metabolism, and disposition. Ther Drug Monit 26:137–144
20. Hoch PH, Cattell JP, Pennes HH (1952) Effects of mescaline and lysergic acid (d-lsd-25). Am J Psychiatry 108:579–584
21. Friedhoff AJ, Goldstein M (1962) New developments in metabolism of mescaline and related amines. N Y Acad Sci 96:5–13
22. Corelick DA, Balster RL (1995) Phencyclidine (PCP). In: Bloom EE, Kupfer DJ (eds) Psychopharmacology: the fourth generation of progress. Raven Presi, New York
23. Anis NA, Berry SC, Burton NR, Lodge D (1983) The dissociative anaesthetics, ketamine and phencyclidine, selectively reduce excitation of central mammalian neurones by *N*-methyl-aspartate. Br J Pharmacol 79(2):565–575
24. Uhl S, Schmid P, Schlatter C (1986) Pharmacokinetics of pentachlorophenol in man. Arch Toxicol 58(3):182–186
25. Curran HV, Monaghan L (2001) In and out of the K-hole: a comparison of the acute and residual effects of ketamine in frequent and infrequent ketamine users. Addiction 96:749–760
26. Koob GF (1992) Drugs of abuse: anatomy, pharmacology and function of reward pathways. Trends Pharmacol Sci 13:77–185
27. Aps JK, Martens LC (2005) Review: the physiology of saliva and transfer of drugs into saliva. Forensic Sci Int 150(2–3):119–131. doi:S0379-0738(05)00118-0 [pii] 10.1016/j.forsciint.2004.10.026
28. Wille SM, Raes E, Lillsunde P, Gunnar T, Laloup M, Samyn N, Christophersen AS, Moeller MR, Hammer KP, Verstraete AG (2009) Relationship between oral fluid and blood concentrations of drugs of abuse in drivers suspected of driving under the influence of drugs. Ther Drug Monit 31(4):511–519. doi:10.1097/FTD.0b013e3181ae46ea
29. Zuidema J, Hold KM, de Boer D, Maes RA (1996) Saliva as a specimen for therapeutic drug monitoring in pharmacies. Pharm World Sci 18(6):193–194
30. Zuidema J, van Ginneken CA (1983) Clearance concept in salivary drug excretion. Part I: theory. Pharm Acta Helv 58(3):88–93
31. Drummer OH (2005) Review: pharmacokinetics of illicit drugs in oral fluid. Forensic Sci Int 150(2–3):133–142. doi:S0379-0738(05)00119-2 [pii] 10.1016/j.forsciint.2004.11.022
32. Hold KM, de Boer D, Zuidema J, Maes RAA (1996) Saliva as an analytical tool in toxicology. Int J Drug Test 1:1–8
33. Toennes SW, Kauert GF, Steinmeyer S, Moeller MR (2005) Driving under the influence of drugs – evaluation of analytical data of drugs in oral fluid, serum and urine, and correlation with impairment symptoms. Forensic Sci Int 152(2–3):149–155. doi:S0379-0738(04)00443-8 [pii] 10.1016/j.forsciint.2004.08.002
34. Lim MD, Dickherber A, Compton CC (2011) Before you analyze a human specimen, think quality, variability, and bias. Anal Chem 83(1):8–13. doi:10.1021/ac1018974
35. Verstraete AG (2004) Detection times of drugs of abuse in blood, urine, and oral fluid. Ther Drug Monit 26(2):200–205

36. Mandel ID (1990) The diagnostic uses of saliva. J Oral Pathol Med 19(3):119–125
37. Samyn N, De Boeck G, Verstraete AG (2002) The use of oral fluid and sweat wipes for the detection of drugs of abuse in drivers. J Forensic Sci 47(6):1380–1387
38. Cone EJ (1993) Saliva testing for drugs of abuse. Ann N Y Acad Sci 694:91–127
39. Cone EJ, Kumor K, Thompson LK, Sherer M (1988) Correlation of saliva cocaine levels with plasma levels and with pharmacologic effects after intravenous cocaine administration in human subjects. J Anal Toxicol 12(4):200–206
40. Mancinelli R, Guiducci MS (2002) Procedural aspects and interpretation problems in the analysis of drugs of abuse. Ann Ist Super Sanita 38(3):305–313
41. Gallardo E, Queiroz JA (2008) The role of alternative specimens in toxicological analysis. Biomed Chromatogr 22(8):795–821. doi:10.1002/bmc.1009
42. Pragst F, Balikova MA (2006) State of the art in hair analysis for detection of drug and alcohol abuse. Clin Chim Acta 370(1–2):17–49. doi:S0009-8981(06)00122-7 [pii] 10.1016/j.cca.2006.02.019
43. Ribuffo A (1980) Manuale di dermatologia e venerorologia. Lombardo Editore, Roma
44. Harkey MR (1993) Anatomy and physiology of hair. Forensic Sci Int 63(1–3):9–18. doi:0379-0738(93)90255-9 [pii]
45. Henderson GL (1993) Mechanisms of drug incorporation into hair. Forensic Sci Int 63(1–3):19–29. doi:0379-0738(93)90256-A [pii]
46. Rothe M, Pragst F, Thor S, Hunger J (1997) Effect of pigmentation on the drug deposition in hair of grey-haired subjects. Forensic Sci Int 84(1–3):53–60
47. Rollins DE, Wilkins DG, Krueger GG, Augsburger MP, Mizuno A, O'Neal C, Borges CR, Slawson MH (2003) The effect of hair color on the incorporation of codeine into human hair. J Anal Toxicol 27(8):545–551
48. Kronstrand R, Forstberg-Peterson S, Kagedal B, Ahlner J, Larson G (1999) Codeine concentration in hair after oral administration is dependent on melanin content. Clin Chem 45(9):1485–1494
49. Potsch L, Skopp G, Moeller MR (1997) Influence of pigmentation on the codeine content of hair fibers in guinea pigs. J Forensic Sci 42(6):1095–1098
50. Claffey DJ, Stout PR, Ruth JA (2001) 3H-nicotine, 3H-flunitrazepam, and 3H-cocaine incorporation into melanin: a model for the examination of drug–melanin interactions. J Anal Toxicol 25(7):607–611
51. Wu YH, Lin KL, Chen SC, Chang YZ (2008) Simultaneous quantitative determination of amphetamines, ketamine, opiates and metabolites in human hair by gas chromatography/mass spectrometry. Rapid Commun Mass Spectrom 22(6):887–897
52. Peters FT (2011) Recent advances of liquid chromatography-(tandem) mass spectrometry in clinical and forensic toxicology. Clin Biochem 44(1):54–65. doi:S0009-9120(10)00348-6 [pii] 10.1016/j.clinbiochem.2010.08.008
53. Jenkins KM, Young MS, Mallet CR, Elian AA (2004) Mixed-mode solid-phase extraction procedures for the determination of MDMA and metabolites in urine using LC-MS, LC-UV, or GC-NPD. J Anal Toxicol 28(1):50–58
54. Stout PR, Bynum ND, Mitchell JM, Baylor MR, Ropero-Miller JD (2009) A comparison of the validity of gas chromatography–mass spectrometry and liquid chromatography–tandem mass spectrometry analysis of urine samples for morphine, codeine, 6-acetylmorphine, and benzoylecgonine. J Anal Toxicol 33(8):398–408
55. Couchman L, Morgan PE (2011) LC-MS in analytical toxicology: some practical considerations. Biomed Chromatogr 25(1–2):100–123. doi:10.1002/bmc.1566
56. SOFT, AAFS (2006) Forensic toxicology laboratory guidelines. Last access 10 May 2012 http://www.soft-tox.org/files/Guidelines_2006_Final.pdf
57. EC (2002) Consolidated guidelines for the implementation of decision 2002/657/EC. Last access 10 May 2012 http://ec.europa.eu/food/food/chemicalsafety/residues/cons_2004-2726rev4_en.pdf
58. FDA (2001) Guidance for industry: bioanalytical method validation. Last access 10 May 2012 http://www.fda.gov/downloads/Drugs/GuidanceComplianceRegulatoryInformation/Guidances/UCM070107.pdf

59. EMEA (1995) ICH Topic Q 2 (R1) Validation of analytical procedures: text and methodology. http://www.emea.eu.int
60. Peters FT, Drummer OH, Musshoff F (2007) Validation of new methods. Forensic Sci Int 165(2–3):216–224. doi:S0379-0738(06)00323-9 [pii] 10.1016/j.forsciint.2006.05.021
61. Maurer HH (2007) Current role of liquid chromatography–mass spectrometry in clinical and forensic toxicology. Anal Bioanal Chem 388(7):1315–1325. doi:10.1007/s00216-007-1248-5
62. Marchi I, Rudaz S, Veuthey JL (2009) Sample preparation development and matrix effects evaluation for multianalyte determination in urine. J Pharm Biomed Anal 49(2):459–467. doi:S0731-7085(08)00639-0 [pii] 10.1016/j.jpba.2008.11.040
63. Dams R, Huestis MA, Lambert WE, Murphy CM (2003) Matrix effect in bio-analysis of illicit drugs with LC-MS/MS: influence of ionization type, sample preparation, and biofluid. J Am Soc Mass Spectrom 14(11):1290–1294. doi:S1044030503005749 [pii]
64. Matuszewski BK, Constanzer ML, Chavez-Eng CM (2003) Strategies for the assessment of matrix effect in quantitative bioanalytical methods based on HPLC-MS/MS. Anal Chem 75(13):3019–3030
65. Bonfiglio R, King RC, Olah TV, Merkle K (1999) The effects of sample preparation methods on the variability of the electrospray ionization response for model drug compounds. Rapid Commun Mass Spectrom 13(12):1175–1185. doi:10.1002/(SICI)1097-0231(19990630)13:12<1175::AID-RCM639>3.0.CO;2-0 [pii] 10.1002/(SICI)1097-0231(19990630)13:12<1175::AID-RCM639> 3.0.CO;2-0
66. Sergi M, Compagnone D, Curini R, D'Ascenzo G, Del Carlo M, Napoletano S, Risoluti R (2010) Micro-solid phase extraction coupled with high-performance liquid chromatography–tandem mass spectrometry for the determination of stimulants, hallucinogens, ketamine and phencyclidine in oral fluids. Anal Chim Acta 675(2):132–137. doi:10.1016/j.aca.2010.07.011
67. Napoletano S, Montesano C, Compagnone D, Curini R, D'Ascenzo G, Roccia C, Sergi M (2011) Determination of illicit drugs in urine and plasma by micro-SPE followed by HPLC–MS/MS. Chromatographia. doi:10.1007/s10337-011-2156-6
68. ISO/IEC (2005) ISO/IEC 17025:2005. Incorporating Corrigendum no. 1. Published under the authority of the Standards Policy and Strategy Committee on 29 June 2005.
69. Drolc A, Cotman M, Ros M (2005) Integration of metrological principles in a proficiency-testing scheme in the field of water analysis. Anal Bioanal Chem 382(5):1311–1319. doi:10.1007/s00216-005-3264-7
70. Maralikova B, Weinmann W (2004) Confirmatory analysis for drugs of abuse in plasma and urine by high-performance liquid chromatography–tandem mass spectrometry with respect to criteria for compound identification. J Chromatogr B Analyt Technol Biomed Life Sci 811(1):21–30. doi:10.1016/j.jchromb.2004.04.039 S1570-0232(04)00642-7 [pii]
71. Sauvage FL, Gaulier JM, Lachatre G, Marquet P (2008) Pitfalls and prevention strategies for liquid chromatography–tandem mass spectrometry in the selected reaction-monitoring mode for drug analysis. Clin Chem 54(9):1519–1527. doi:clinchem.2008.105478 [pii] 10.1373/clinchem.2008.105478
72. Arpino PJ (1982) On-line liquid chromatography/mass spectrometry? An odd couple! Trends Anal Chem 1(7):154–158. doi:10.1016/0165-9936(82)80049-6
73. Polettini A (2006) Applications of LC-MS in toxicology. Pharmaceutical Press, London
74. Watson JT, Sparkman OD (2007) Introduction to mass spectrometry: instrumentation, applications and strategies for data interpretation, 4th edn. Wiley-Blackwell, Chichester
75. Gallagher RT, Balogh MP, Davey P, Jackson MR, Sinclair I, Southern LJ (2003) Combined electrospray ionization-atmospheric pressure chemical ionization source for use in high-throughput LC-MS applications. Anal Chem 75(4):973–977
76. Chena BH, Liu JT, Chena WX, Chena HM, CH L (2008) A general approach to the screening and confirmation of tryptamines and phenethylamines by mass spectral fragmentation. Talanta 74:512–517
77. Wang PP, Bartlett MG (1998) Collision-induced dissociation mass spectra of cocaine, and its metabolites and pyrolysis products. J Mass Spectrom 33:961–967
78. Al-Sayah MA, Rizos P, Antonucci V, Wu N (2008) High throughput screening of active pharmaceutical ingredients by UPLC. J Sep Sci 31(12):2167–2172. doi:10.1002/jssc.200700594

79. Wren SA, Tchelitcheff P (2006) Use of ultra-performance liquid chromatography in pharmaceuticaldevelopment.JChromatogrA1119(1–2):140–146.doi:S0021-9673(06)00385-2 [pii] 10.1016/j.chroma.2006.02.052
80. Nguyen DT, Guillarme D, Rudaz S, Veuthey JL (2006) Fast analysis in liquid chromatography using small particle size and high pressure. J Sep Sci 29(12):1836–1848
81. Guillarme D, Ruta J, Rudaz S, Veuthey JL (2010) New trends in fast and high-resolution liquid chromatography: a critical comparison of existing approaches. Anal Bioanal Chem 397(3):1069–1082. doi:10.1007/s00216-009-3305-8
82. Guillarme D, Heinisch S, Rocca JL (2004) Effect of temperature in reversed phase liquid chromatography. J Chromatogr A 1052(1–2):39–51
83. van Nederkassel AM, Aerts A, Dierick A, Massart DL, Vander Heyden Y (2003) Fast separations on monolithic silica columns: method transfer, robustness and column ageing for some case studies. J Pharm Biomed Anal 32(2):233–249. doi:S0731708503001316 [pii]
84. Hjerten S, Liao JL, Zhang R (1989) High-performance liquid-chromatography on continuous polymer beds. J Chromatogr 473(1):273–275
85. Kirkland JJ (1992) Superficially porous silica microspheres for the fast high-performance liquid-chromatography of macromolecules. Anal Chem 64(11):1239–1245
86. Fekete S, Fekete J, Ganzler K (2009) Characterization of new types of stationary phases for fast liquid chromatographic applications. J Pharm Biomed Anal 50(5):703–709. doi:S0731-7085(09)00351-3 [pii] 10.1016/j.jpba.2009.05.039
87. Cavazzini A, Gritti F, Kaczmarski K, Marchetti N, Guiochon G (2007) Mass-transfer kinetics in a shell packing material for chromatography. Anal Chem 79(15):5972–5979. doi:10.1021/ac070571a
88. Gritti F, Cavazzini A, Marchetti N, Guiochon G (2007) Comparison between the efficiencies of columns packed with fully and partially porous C18-bonded silica materials. J Chromatogr A 1157(1–2):289–303. doi:S0021-9673(07)00878-3 [pii] 10.1016/j.chroma.2007.05.030
89. Mallett DN, Ramirez-Molina C (2009) The use of partially porous particle columns for the routine, generic analysis of biological samples for pharmacokinetic studies in drug discovery by reversed-phase ultra-high performance liquid chromatography–tandem mass spectrometry. J Pharm Biomed Anal 49(1):100–107. doi:S0731-7085(08)00548-7 [pii] 10.1016/j.jpba.2008.09.041
90. Stokvis E, Rosing H, Beijnen JH (2005) Stable isotopically labeled internal standards in quantitative bioanalysis using liquid chromatography/mass spectrometry: necessity or not? Rapid Commun Mass Spectrom 19(3):401–407. doi:10.1002/Rcm.1790
91. Lindegardh N, Annerberg A, White NJ, Day NP (2008) Development and validation of a liquid chromatographic-tandem mass spectrometric method for determination of piperaquine in plasma stable isotope labeled internal standard does not always compensate for matrix effects. J Chromatogr B Analyt Technol Biomed Life Sci 862(1–2):227–236. doi:S1570-0232-(07)00874-4 [pii] 10.1016/j.jchromb.2007.12.011
92. Wang S, Cyronak M, Yang E (2007) Does a stable isotopically labeled internal standard always correct analyte response? A matrix effect study on a LC/MS/MS method for the determination of carvedilol enantiomers in human plasma. J Pharm Biomed 43(2):701–707. doi:10.1016/j.jpba.2006.08.010
93. Sergi M, Bafile E, Compagnone D, Curini R, D'Ascenzo G, Romolo FS (2009) Multiclass analysis of illicit drugs in plasma and oral fluids by LC-MS/MS. Anal Bioanal Chem 393(2):709–718. doi:10.1007/s00216-008-2456-3
94. Blanchard J (1981) Evaluation of the relative efficacy of various techniques for deproteinizing plasma samples prior to high-performance liquid chromatographic analysis. J Chromatogr 226(2):455–460
95. Bouzas NF, Dresen S, Munz B, Weinmann W (2009) Determination of basic drugs of abuse in human serum by online extraction and LC-MS/MS. Anal Bioanal Chem 395(8):2499–2507. doi:10.1007/s00216-009-3036-x
96. Kirchherr H, Kuhn-Velten WN (2006) Quantitative determination of forty-eight antidepressants and antipsychotics in human serum by HPLC tandem mass spectrometry: a multi-level, single-sample approach. J Chromatogr B Analyt Technol Biomed Life Sci 843(1):100–113. doi:S1570-0232(06)00445-4 [pii] 10.1016/j.jchromb.2006.05.031

97. Wille SM, Lambert WE (2007) Recent developments in extraction procedures relevant to analytical toxicology. Anal Bioanal Chem 388(7):1381–1391. doi:10.1007/s00216-007-1294-z
98. Gottardo R, Polettini A, Sorio D, Pascali JP, Bortolotti F, Liotta E, Tagliaro F (2008) Capillary zone electrophoresis (CZE) coupled to time-of-flight mass spectrometry (TOF-MS) applied to the analysis of illicit and controlled drugs in blood. Electrophoresis 29(19):4078–4087. doi:10.1002/elps.200800087
99. Couchman L, Morgan PE, Flanagan RJ (2010) Basic drug analysis by strong cation-exchange liquid chromatography–tandem mass spectrometry: simultaneous analysis of amisulpride, and of metamfetamine and amfetamine in serum/plasma. Biomed Chromatogr. doi:10.1002/bmc.1530
100. Schellen A, Ooms B, van de Lagemaat D, Vreeken R, van Dongen WD (2003) Generic solid phase extraction-liquid chromatography–tandem mass spectrometry method for fast determination of drugs in biological fluids. J Chromatogr B Analyt Technol Biomed Life Sci 788(2):251–259. doi:S1570023202010139 [pii]
101. Yawney J, Treacy S, Hindmarsh KW, Burczynski FJ (2002) A general screening method for acidic, neutral, and basic drugs in whole blood using the Oasis MCX column. J Anal Toxicol 26(6):325–332
102. Decaestecker TN, Coopman EM, Van Peteghem CH, Van Bocxlaer JF (2003) Suitability testing of commercial solid-phase extraction sorbents for sample clean-up in systematic toxicological analysis using liquid chromatography-(tandem) mass spectrometry. J Chromatogr B Analyt Technol Biomed Life Sci 789(1):19–25. doi:S1570023203002083 [pii]
103. Bjork MK, Nielsen MK, Markussen LO, Klinke HB, Linnet K (2010) Determination of 19 drugs of abuse and metabolites in whole blood by high-performance liquid chromatography–tandem mass spectrometry. Anal Bioanal Chem 396(7):2393–2401. doi:10.1007/s00216-009-3268-9
104. Jagerdeo E, Montgomery MA, Sibum M, Sasaki TA, LeBeau MA (2008) Rapid analysis of cocaine and metabolites in urine using a completely automated solid-phase extraction-high-performance liquid chromatography–tandem mass spectrometry method. J Anal Toxicol 32(8):570–576
105. Pujadas M, Pichini S, Civit E, Santamarina E, Perez K, de la Torre R (2007) A simple and reliable procedure for the determination of psychoactive drugs in oral fluid by gas chromatography–mass spectrometry. J Pharm Biomed Anal 44(2):594–601. doi:10.1016/j.jpba.2007.02.022
106. Spiehler V (2004) Drugs in saliva. In: Anthony C. Moffat, M. David Osselton, Widdop B (eds) Clarke's analysis of drugs and poisons, 3rd edn. Pharmaceutical Press, London
107. Langel K, Engblom C, Pehrsson A, Gunnar T, Ariniemi K, Lillsunde P (2008) Drug testing in oral fluid-evaluation of sample collection devices. J Anal Toxicol 32(6):393–401
108. Maurer HH (2005) Advances in analytical toxicology: the current role of liquid chromatography–mass spectrometry in drug quantification in blood and oral fluid. Anal Bioanal Chem 381(1):110–118. doi:10.1007/s00216-004-2774-z
109. Simoes SS, Ajenjo AC, Franco JM, Vieira DN, Dias MJ (2009) Liquid chromatography/tandem mass spectrometry for the qualitative and quantitative analysis of illicit drugs and medicines in preserved oral fluid. Rapid Commun Mass Spectrom 23(10):1451–1460. doi:10.1002/rcm.4020
110. Fernandez P, Morales L, Vazquez C, Lago M, Bermejo AM (2008) Comparison of two extraction procedures for determination of drugs of abuse in human saliva by high-performance liquid chromatography. J Appl Toxicol 28(8):998–1003. doi:10.1002/jat.1365
111. Bosker WM, Huestis MA (2009) Oral fluid testing for drugs of abuse. Clin Chem 55(11):1910–1931
112. Wood M, Laloup M, Ramirez Fernandez Mdel M, Jenkins KM, Young MS, Ramaekers JG, De Boeck G, Samyn N (2005) Quantitative analysis of multiple illicit drugs in preserved oral fluid by solid-phase extraction and liquid chromatography–tandem mass spectrometry. Forensic Sci Int 150(2–3):227–238. doi:S0379-0738(05)00131-3 [pii] 10.1016/j.forsciint.2004.11.027
113. Badawi N, Simonsen KW, Steentoft A, Bernhoft IM, Linnet K (2009) Simultaneous screening and quantification of 29 drugs of abuse in oral fluid by solid-phase extraction and

ultraperformance LC-MS/MS. Clin Chem 55(11):2004–2018. doi:clinchem.2008.122341 [pii] 10.1373/clinchem.2008.122341

114. Wang IT, Feng YT, Chen CY (2010) Determination of 17 illicit drugs in oral fluid using isotope dilution ultra-high performance liquid chromatography/tandem mass spectrometry with three atmospheric pressure ionizations. J Chromatogr B Analyt Technol Biomed Life Sci 878(30):3095–3105. doi:10.1016/j.jchromb.2010.09.014
115. Kala SV, Harris SE, Freijo TD, Gerlich S (2008) Validation of analysis of amphetamines, opiates, phencyclidine, cocaine, and benzoylecgonine in oral fluids by liquid chromatography–tandem mass spectrometry. J Anal Toxicol 32(8):605–611
116. Taylor K, Elliott S (2009) A validated hybrid quadrupole linear ion-trap LC-MS method for the analysis of morphine and morphine glucuronides applied to opiate deaths. Forensic Sci Int 187(1–3):34–41. doi:S0379-0738(09)00085-1 [pii] 10.1016/j.forsciint.2009.02.011
117. Svensson JO, Andersson M, Gustavsson E, Beck O (2007) Electrospray LC-MS method with solid-phase extraction for accurate determination of morphine-, codeine-, and ethylmorphine-glucuronides and 6-acetylmorphine in urine. J Anal Toxicol 31(2):81–86
118. Gustavsson E, Andersson M, Stephanson N, Beck O (2007) Validation of direct injection electrospray LC-MS/MS for confirmation of opiates in urine drug testing. J Mass Spectrom 42(7):881–889. doi:10.1002/jms.1219
119. Andersson M, Gustavsson E, Stephanson N, Beck O (2008) Direct injection LC–MS/MS method for identification and quantification of amphetamine, methamphetamine, 3,4-methylenedioxyamphetamine and 3,4-methylenedioxymethamphetamine in urine drug testing. J Chromatogr B Analyt Technol Biomed Life Sci 861:22–28. 10.1016/j.jchromb.2007.11.025
120. Eichhorst JC, Etter ML, Rousseaux N, Lehotay DC (2009) Drugs of abuse testing by tandem mass spectrometry: a rapid, simple method to replace immunoassays. Clin Biochem 42(15):1531–1542. doi:S0009-9120(09)00322-1 [pii] 10.1016/j.clinbiochem.2009.07.019
121. Maquille A, Guillarme D, Rudaz S, Veuthey JL (2009) High-throughput screening of drugs of abuse in urine by supported liquid–liquid extraction and UHPLC coupled to tandem MS. Chromatographia 70(9–10):1373–1380. doi:10.1365/s10337-009-1337-z
122. Breitenbucher JG, Arienti KL, McClure KJ (2001) Scope and limitations of solid-supported liquid-liquid extraction for the high-throughput purification of compound libraries. J Comb Chem 3(6):528–533. doi:cc010039f [pii]
123. Fernandez MDR, Wille SMR, Samyn N, Wood M, Lopez-Rivadulla M, De Boeck G (2009) On-line solid-phase extraction combined with liquid chromatography–tandem mass spectrometry for high throughput analysis of 11-nor-Delta(9)-tetrahydrocannabinol-9-carboxylic acid in urine. J Chromatogr B 877(22):2153–2157. doi:10.1016/j.jchromb.2009.04.047
124. Mercolini L, Musenga A, Comin I, Baccini C, Conti M, Raggi MA (2008) Determination of plasma and urine levels of Delta9-tetrahydrocannabinol and its main metabolite by liquid chromatography after solid-phase extraction. J Pharm Biomed Anal 47(1):156–163. doi:S0731-7085(07)00779-0 [pii] 10.1016/j.jpba.2007.12.023
125. Robandt PV, Klette KL, Sibum M (2009) Automated solid-phase extraction-liquid chromatography–tandem mass spectrometry analysis of 11-nor-Delta9-tetrahydrocannabinol-9-carboxylic acid in human urine specimens: application to a high-throughput urine analysis laboratory. J Anal Toxicol 33(8):456–460
126. Berg T, Lundanes E, Christophersen AS, Strand DH (2009) Determination of opiates and cocaine in urine by high pH mobile phase reversed phase UPLC-MS/MS. J Chromatogr B Analyt Technol Biomed Life Sci 877(4):421–432. doi:S1570-0232(08)00951-3 [pii] 10.1016/j.jchromb.2008.12.052
127. Aturki Z, D'Orazio G, Fanali S, Rocco A, Bortolotti F, Gottardo R, Tagliaro F (2009) Capillary electrochromatographic separation of illicit drugs employing a cyano stationary phase. J Chromatogr A 1216(17):3652–3659. doi:S0021-9673(08)02202-4 [pii] 10.1016/j.chroma.2008.12.041
128. Ellison ST, Brewer WE, Morgan SL (2009) Comprehensive analysis of drugs of abuse in urine using disposable pipette extraction. J Anal Toxicol 33(7):356–365

129. Bush DM (2008) The U.S. mandatory guidelines for federal workplace drug testing programs: current status and future considerations. Forensic Sci Int 174(2–3):111–119. doi:10.1016/j.forsciint.2007.03.008
130. SOHT (2004) Recommendations for hair testing in forensic cases. Forensic Sci Int 145:83–84
131. UNDPC (2001) Guidelines, hair, sweat and saliva. UN, New York
132. Musshoff F, Madea B (2007) Analytical pitfalls in hair testing. Anal Bioanal Chem 388(7):1475–1494. doi:10.1007/s00216-007-1288-x
133. Bucelli F, Fratini A, Bavazzano P, Comodo N (2009) Quantification of drugs of abuse and some stimulants in hair samples by liquid chromatography–electrospray ionization ion trap mass spectrometry. J Chromatogr B Analyt Technol Biomed Life Sci 877(31):3931–3936. doi:S1570-0232(09)00671-0 [pii] 10.1016/j.jchromb.2009.09.026
134. Dwivedi P, Hill HH Jr (2008) A rapid analytical method for hair analysis using ambient pressure ion mobility mass spectrometry with electrospray ionization (ESI-IMMS). Int J Ion Mobil Spec 11:61–69
135. Hegstad S, Khiabani HZ, Kristoffersen L, Kunoe N, Lobmaier PP, Christophersen AS (2008) Drug screening of hair by liquid chromatography–tandem mass spectrometry. J Anal Toxicol 32(5):364–372
136. Tabernero MJ, Felli ML, Bermejo AM, Chiarotti M (2009) Determination of ketamine and amphetamines in hair by LC/MS/MS. Anal Bioanal Chem 395(8):2547–2557. doi:10.1007/s00216-009-3163-4
137. Klys M, Rojek S, Kulikowska J, Bozek E, Scislowski M (2007) Usefulness of multi-parameter opiates-amphetamines-cocainics analysis in hair of drug users for the evaluation of an abuse profile by means of LC-APCI-MS-MS. J Chromatogr B Analyt Technol Biomed Life Sci 854(1–2):299–307. doi:S1570-0232(07)00337-6 [pii] 10.1016/j.jchromb.2007.04.040
138. Xiang P, Shen M, Zhuo X (2006) Hair analysis for ketamine and its metabolites. Forensic Sci Int 162(1–3):131–134. doi:S0379-0738(06)00389-6 [pii] 10.1016/j.forsciint.2006.05.046
139. Hoelzle C, Scheufler F, Uhl M, Sachs H, Thieme D (2008) Application of discriminant analysis to differentiate between incorporation of cocaine and its congeners into hair and contamination. Forensic Sci Int 176(1):13–18. doi:S0379-0738(07)00750-5 [pii] 10.1016/j.forsciint.2007.07.020
140. Fernandez P, Lago M, Lorenzo RA, Carro AM, Bermejo AM, Tabernero MJ (2009) Optimization of a rapid microwave-assisted extraction method for the simultaneous determination of opiates, cocaine and their metabolites in human hair. J Chromatogr B Analyt Technol Biomed Life Sci 877(18–19):1743–1750. doi:S1570-0232(09)00305-5 [pii].1016/j.jchromb.2009.04.035
141. Huang DK, Liu C, Huang MK, Chien CS (2009) Simultaneous determination of morphine, codeine, 6-acetylmorphine, cocaine and benzoylecgonine in hair by liquid chromatography/electrospray ionization tandem mass spectrometry. Rapid Commun Mass Spectrom 23(7):957–962. doi:10.1002/rcm.3955
142. Quintela O, Lendoiro E, Cruz A, de Castro A, Quevedo A, Jurado C, Lopez-Rivadulla M (2010) Hydrophilic interaction liquid chromatography–tandem mass spectrometry (HILIC-MS/MS) determination of cocaine and its metabolites benzoylecgonine, ecgonine methyl ester, and cocaethylene in hair samples. Anal Bioanal Chem 396(5):1703–1712. doi:10.1007/s00216-009-3393-5
143. Yazdi AS, Es'haghi Z (2005) Surfactant enhanced liquid-phase microextraction of basic drugs of abuse in hair combined with high performance liquid chromatography. J Chromatogr A 1094:1–8
144. Miguez-Framil M, Moreda-Pineiro A, Bermejo-Barrera P, Alvarez-Freire I, Tabernero MJ, Bermejo AM (2010) Matrix solid-phase dispersion on column clean-up/pre-concentration as a novel approach for fast isolation of abuse drugs from human hair. J Chromatogr A 1217(41):6342–6349. doi:S0021-9673(10)01105-2 [pii] 10.1016/j.chroma.2010.08.034
145. Nielsen MK, Johansen SS, Dalsgaard PW, Linnet K (2010) Simultaneous screening and quantification of 52 common pharmaceuticals and drugs of abuse in hair using UPLC-TOF-MS. Forensic Sci Int 196(1–3):85–92. doi:S0379-0738(09)00528-3 [pii] 10.1016/j.forsciint.2009.12.027

146. Nishida M, Yashiki M, Namera A, Kimura K (2006) Single hair analysis of methamphetamine and amphetamine by solid phase microextraction coupled with in matrix derivatization. J Chromatogr B Analyt Technol Biomed Life Sci 842(2):106–110. doi:S1570-0232(06)00580-0 [pii] 10.1016/j.jchromb.2006.07.039
147. Kronstrand R, Nystrom I, Strandberg J, Druid H (2004) Screening for drugs of abuse in hair with ion spray LC-MS-MS. Forensic Sci Int 145(2–3):183–190. doi:10.1016/j.forsciint.2004.04.034 S037907380400249X [pii]
148. Pujol ML, Cirimele V, Tritsch PJ, Villain M, Kintz P (2007) Evaluation of the IDS One-Step ELISA kits for the detection of illicit drugs in hair. Forensic Sci Int 170(2–3):189–192. doi:S0379-0738(07)00549-X [pii] 10.1016/j.forsciint.2007.02.032
149. Lee S, Park Y, Yang W, Han E, Choe S, Lim M, Chung H (2009) Estimation of the measurement uncertainty of methamphetamine and amphetamine in hair analysis. Forensic Sci Int 185(1–3):59–66. doi:S0379-0738(08)00502-1 [pii] 10.1016/j.forsciint.2008.12.012
150. Lee S, Han E, Park Y, Choi H, Chung H (2009) Distribution of methamphetamine and amphetamine in drug abusers' head hair. Forensic Sci Int 190(1–3):16–18. doi:S0379-0738-(09)00208-4 [pii] 10.1016/j.forsciint.2009.05.004
151. Barroso M, Dias M, Vieira DN, Queiroz JA, Lopez-Rivadulla M (2008) Development and validation of an analytical method for the simultaneous determination of cocaine and its main metabolite, benzoylecgonine, in human hair by gas chromatography/mass spectrometry. Rapid Commun Mass Spectrom 22(20):3320–3326. doi:10.1002/rcm.3738
152. Gottardo R, Bortolotti F, De Paoli G, Pascali JP, Miksik I, Tagliaro F (2007) Hair analysis for illicit drugs by using capillary zone electrophoresis–electrospray ionization–ion trap mass spectrometry. J Chromatogr A 1159(1–2):185–189. doi:S0021-9673(07)00039-8 [pii] 10.1016/j.chroma.2007.01.011
153. Lin YH, Lee MR, Lee RJ, Ko WK, Wu SM (2007) Hair analysis for methamphetamine, ketamine, morphine and codeine by cation-selective exhaustive injection and sweeping micellar electrokinetic chromatography. J Chromatogr A 1145(1–2):234–240. doi:S0021-9673-(07)00094-5 [pii] 10.1016/j.chroma.2007.01.054
154. Romolo FS, Rotolo MC, Palmi I, Pacifici R, Lopez A (2003) Optimized conditions for simultaneous determination of opiates, cocaine and benzoylecgonine in hair samples by GC-MS. Forensic Sci Int 138(1–3):17–26. doi:S0379073803003657 [pii]
155. Hill V, Cairns T, Schaffer M (2008) Hair analysis for cocaine: factors in laboratory contamination studies and their relevance to proficiency sample preparation and hair testing practices. Forensic Sci Int 176(1):23–33. doi:S0379-0738(07)00752-9 [pii] 10.1016/j.forsciint.2007.08.011
156. Sergi M, Gentili A, Perret D, Marchese S, Materazzi S, Curini R (2007) MSPD extraction of sulphonamides from meat followed by LC tandem MS determination. Chromatographia 65(11–12):757–761. doi:10.1365/s10337-007-0245-3
157. Capriotti AL, Cavaliere C, Giansanti P, Gubbiotti R, Samperi R, Lagana A (2010) Recent developments in matrix solid-phase dispersion extraction. J Chromatogr A 1217(16):2521–2532. doi:10.1016/j.chroma.2010.01.030
158. Miyaguchi H, Inoue H (2011) Determination of amphetamine-type stimulants, cocaine and ketamine in human hair by liquid chromatography/linear ion trap-Orbitrap hybrid mass spectrometry. Analyst 136(17):3503–3511. doi:10.1039/c0an00850h
159. Meng P, Zhu D, He H, Wang Y, Guo F, Zhang L (2009) Determination of amphetamines in hair by GC/MS after small-volume liquid extraction and microwave derivatization. Anal Sci 25(9):1115–1118. doi:JST.JSTAGE/analsci/25.1115 [pii]

Use of Matrix Assisted Laser Desorption/Ionization Imaging Mass Spectrometry (MALDI-IMS) in the Development of Novel Small Molecule Drugs

Jihai Pang and Timothy L. Madden

Abstract Modern drug research and development includes the study of drug distribution and its metabolites within the biological organ and tissues. The high performance liquid chromatography coupled to a tandem mass spectrometer (HPLC-MS/MS) is currently widely employed in the study of drug metabolism and pharmacokinetic characterization of new drug molecules.

Today, the newest frontier, which adds the mass spectrometry to the arsenal of tools for the direct analysis of tissue biopsies and molecular diagnosis, is typically called Matrix Assisted Laser Desorption/Ionization Imaging Mass Spectrometry (MALDI-IMS). This powerful analytical tool is a new molecular imaging technique which in essence takes a mass spectral "snapshots" of intact tissue sections, revealing how the macromolecules, such as peptides and protein, and small molecular drugs and their metabolites are spatially distributed within a given biological tissue sample. In this chapter, MALDI-IMS technology is presented as well as applications of such technology in drug development and in clinical pharmacology. It provides direct, unambiguous evidence for the presence of drug substance in tissue sections.

J. Pang (✉) • T.L. Madden
Department of Experimental Therapeutics, Pharmaceutical Development Center,
The University of Texas MD Anderson Cancer Center, 8000 El Rio Street,
Houston, TX 77054, USA
e-mail: jipang@mdanderson.org

Q. Alan Xu and Timothy L. Madden (eds.), *LC-MS in Drug Bioanalysis*,
DOI 10.1007/978-1-4614-3828-1_13,

Abbreviation

3D	Three-dimensional
AgNPs	Silver nanoparticles
APCI	Atmospheric pressure chemical ionization
AP-IR-MALDI-MS	Atmospheric pressure infrared MALDI mass spectrometry
CHCA	α-Cyano-4-hydroxycinnamic acid
DESI	Desorption electrospray ionization
DHB	2,5-Dihydroxybenzoic acid
ESI	Electrospray ionization
HPLC-MS/MS	High performance liquid chromatography - mass spectrometer
LAESI	Laser ablation electrospray ionization mass spectrometry
LA-ICP-MS	Laser ablation-inductively coupled plasma-mass spectrometry
LDI	Laser desorption ionization
LDPI	Laser desorption postionization
LEV	Levofloxacin
MALDI-IMS	Matrix assisted laser desorption/ionization imaging mass spectrometry
MALDI-FT-ICR-MS	MALDI-Fourier transform ion cyclotron resonance mass spectrometry
MALDI-Q-IM-TOFMS	MALDI-Q-ion mobility-TOFMS
MALDI-QIT-TOF-MS	MALDI-Quadrupole-ion-trap-TOF mass spectrometry
MALDI-TOF-IMS	MALDI-TOF-imaging mass spectrometry
MALDI-TOF-MS	MALDI Time-of –flight mass spectrometry
MRI	Magnetic resonance imaging
MSF	Matrix solution fixation
MXF	Moxifloxacin
Nano-PALDI-IMS	Nanoparticle-assisted laser desorption/ionization imaging mass spectrometry
NIMS	Nanostructure initiator mass spectrometry
OCN	Oscillating capillary nebulizer
OLZ	Olanzapine
PCa	Prostate cancers
SA	Sinapinic acid
SALDI-MS	Surface-assisted laser desorption/ionization mass spectrometry
SIMS	Secondary ion mass spectrometry
TB	Tuberculosis
TCA	Taurocholic acid
TCDCA	Taurochenodeoxycholic acid

TFA	Trifluoroacetatic acid
TLC	Thin layer chromatography
UPLC	Ultraperformance liquid chromatography

1 Introduction

Mass spectrometry has provided a wealth of information on chemical structural determination, identification, and a trace level chemical analysis. It has many applications in the biochemistry, pharmaceutical, and clinical diagnostic research field.

Electrospray ionization (ESI), a very powerful MS ionization mode, has been coupled with liquid chromatography becoming a popular tool in biomolecular analysis and drug analysis applications [1]. Use of HPLC or ultra performance liquid chromatography (UPLC) combined with a mass spectrometry through the interface—ESI has become a powerful and routinely used analytical tool for many fields in scientific research, pharma industry, healthcare, and clinical diagnostic applications.

A fast-growing application of the MALDI technique is use in combination with Time-of-Flight mass spectrometry (MALDI-TOF-MS) and has been applied to the determination of molecular weight for biological macromolecules [2–4]. Currently, analytes that have molecular weights of <1,000 Da can be analyzed by MALDI-TOF-MS [5–7].

A direct molecular imaging detection method that does not require the use of specific labels for visualization of the biodistribution of target biomolecules in tissues is based on MALDI-TOF-MS. The MALDI-TOF-Imaging mass spectrometry (IMS) is rapidly developing into a powerful technique that enables us to identify and localize biological compounds directly on tissue sections within a user-defined area and without the need for radioactive, fluorescence labels, or immunochemical reagents [8–14]. It has enjoyed broad use for studying macromolecular entities such as proteins and peptides [15–26], lipids [27–37], carbohydrates [38], oligosaccharides and oligonucleotides [39, 40], biomarkers [41–48], and small pharmaceutical compounds and their metabolites [49–58] as well as others. More recently MALDI-TOF-MS imaging has been developed to directly determine the distribution of drugs and drug metabolites in tissue sections benefiting the drug development process.

MALDI-TOF-IMS direct analysis and imaging of tissue section show continuously increased performances in term of detected molecules, sensitivity, and applications. MALDI-TOF-IMS analysis of protein and peptides in three-dimensional (3D) volume reconstruction explores the proteome of complex tissue such as the brain [59].

The MALDI-Fourier Transform Ion Cyclotron Resonance Mass Spectrometry (MALDI-FT-ICR-MS) and MALDI-Quadrupole-Ion-trap-TOF Mass Spectrometry (MALDI-QIT-TOF-MS) can be used on IMS. The FT-ICR-MS could provide the high resolution, expansive mass range and high sensitivity imaging MS data and good for determining the elemental composition of small molecules [60].

This chapter describes and discusses the mostly used IMS technology—MALDI-TOF-IMS, including sample preparation, data interpretation/handling, and its applications.

2 Principles of MALDI Mass Spectrometry

MALDI was introduced in the late 1980s and is one of the most successfully developed MS soft ionization techniques that uses the matrix assists laser ablation of sample-coated target to vaporize gas-phase ions for injection into a mass spectrometer. The advantage of MALDI is its gentleness compared with ESI and Atmospheric Pressure Chemical Ionization (APCI) and its ability to analyze the polar, nonvolatile, and large molecules. It has been very successfully used for the analysis of both biopolymers compounds and small molecular organic compounds (<1,500 Da).

2.1 The Principle of MALDI Ionization Method

The Laser Desorption Ionization (LDI) was investigated by Franz Hillenkamp and Michael Karas [2, 3]. LDI involved sample bombardment with short and intense pulses from a laser light to effect both desorption and ionization of the analyte molecules. It has become a soft desorption ionization method for mass spectrometric analyses of biological macromolecules and small molecular compounds. The MALDI technique was improved by Koichi Tanaka.

A laser beam serves as the source of desorption and ionization. Many different types of laser light have been studied for MALDI-TOF-MS. The most used lasers include pulsed N_2 laser with a wavelength of 337 nm and Nd-YAG solid-state laser with a wavelength of 355 nm. The ideal laser should deliver an efficient and controllable quantity of energy to the samples, and in order to avoid thermal decomposition this energy must be transferred quickly. The samples have absorption of the laser energy radiation and ionization.

2.2 Choice of Matrix

MALDI-MS employs a matrix and the use of a matrix with the samples has several purposes: extraction of analyte from the cocrystallization surface, formation of analyte-dropped crystals, and absorption of the laser energy for soft-ionization of analyte molecules into MS analyzer. The commonly used MALDI matrices include α-cyano-4-hydroxycinnamic acid (CHCA), 2,5-Dihydroxybenzoic acid (DHB), Sinapinic acid (SA), et al. [61]. A typical matrix solution would comprise the matrix at a concentration of 10–20 mg/mL in a solvent that is compatible with the target compounds to be analyzed. The solvents used for MALDI-MS can be water, ethanol, methanol, acetonitrile, acetone, chloroform, tetrahydrofuran, and a mixture of solvents. Generally, the MALDI matrix solutions should be prepared freshly as they are susceptible to light-induced decomposition. Often a matrix solution is saturated and allowed to settle. Trifluoroacetic acid at 0.1–0.2 % could be added to the matrix solution to enhance ionization.

2.3 MALDI Sample Preparation

To prepare a sample, an analyte of interest is dissolved in an appropriate solvent at a concentration of about 20 μM. This solution and a matrix solution (50:50, vol/vol) are mixed with an appropriate matrix–analyte ratio at ≥10:1 in a small vial. If an internal standard is used, the internal standard is dissolved in the sample solution before combining with the matrix. A tiny drop (0.5–2.0 μl) of the final solution is transferred and applied to the MADLI sample target plate which can be a stainless steel or a Gold-coated MALDI sample plate. The sample–matrix solution is allowed to dry at room temperature and under normal pressure for several minutes until all of solvents are completely evaporated. The precipitate is the cocrystallization of matrix and analyte, ready for MALDI analysis.

There are several ways to coat a MALDI plate. One is two-layer overlayer [62, 63]: In this method, the first layer on the MALDI target plate comprises the densely packed matrix microcrystals formed by the quick solvent evaporation of a matrix solution in acetone or a mixture of acetonitrile and water at room temperature. A mixture solution containing both matrix and sample is then deposited onto the first layer of small crystals to form uniform analyte–matrix microcrystals. The overlayer method is reported to provide improved results, particularly for analyses of proteins and peptides [64–66].

The second method is sandwich method which was developed from the two-layer method and used first for the analysis of the single mammalian cell lysates [67]. The first thin layer is formed by the matrix-only solution. It is followed by deposition of the analyte solution and then deposition of second layer of matrix on the top of analyte layer. The sample is basically sandwiched between the two matrix layers and preparation results in a matrix–sample–matrix sandwich. This method is specifically used for detecting protein and peptides [67–69]. The three-layer matrix–sample preparation method was also used as matrix–matrix–sample mode [70].

Han and Schey [71] used a special matrix coating method in processing the bovine lenses sections (30–40 μm) for MS imaging. The tissue sections were first sprayed with an acetonitrile–water (50:50, vol/vol) solution resulting in a tightly bound section. After drying, the tissue sections were coated with a thin layer matrix of SA at 15 mg/mL in ethanol–water (50:50, vol/vol). After it was dried, the tissue sections were finally sprayed with several cycles of SA matrix solution at 15 mg/mL in ethanol–water–formic acid (44:44:12, vol/vol/vol).

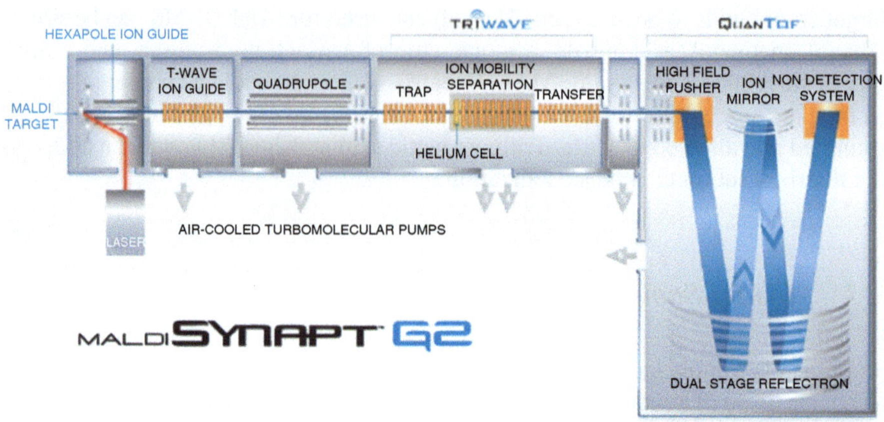

Fig. 1 The scheme of Waters MALDI SYNAPT G2 HDMS system with TOF W-mode (Reproduced with permission from Waters Corporation)

2.4 *MALDI-TOF Mass Spectrometry*

A MALDI-TOF mass spectrometer is a modern TOF mass spectrometer interfaced through a pulse laser desorption ionization technique possessing a theoretically unlimited mass scan range, high ion-transmission efficiency, multiplex detection capability, and simplicity in instrument design and maintenance. It can record simultaneously all the ions produced from each single laser pulse. The basic concept of TOF Mass Analyzer is that the ions are separated based on the time it takes for the ion to drift down the flight tube to the detector with different speeds which are inversely proportional to their mass over charge values. The lighter ions have higher velocities than heavier ions and reach the detector first. A tandem mass spectrometry, such as MALDI-TOF-MS/MS, could reduce the background noise and increase the sensitivity of the drug and its metabolites.

The MALDI-TOF has advantages on the high ion transmission, highest practical mass range of all MS analyzers, very low detection limits. It also has some disadvantages, such as the fast digitizers used in TOF can have limited dynamic range, and has limited mass measurement accuracy and limited high resolution. This was overcome by using orthogonal acceleration technique (oa-TOF) and combining a quadrupole mass analyzer with a TOF-MS (Q-TOF). The oa-TOF and Q-TOF systems revolutionize the application of TOF-MS system for the structural analysis with tandem mass spectrometry. It provides high mass accuracy and resolving power. The ion mobility separation was introduced to Q-TOF instrument that can separate the ion with the same mass, but have different spatial shape [72]. Figure 1 shows an example of MALDI-TOF instrument.

3 Principles of MALDI Imaging Mass Spectrometry

MALDI-TOF-IMS allows for the analysis of large numbers of compounds ranging from small molecular drug and its metabolite molecules ($m/z < 1{,}000$) [73, 74] to large proteins with a molecular weight of 10^4 Da [75].

3.1 *Ionization Methodologies and Instrumentation for Imaging MS*

The ionization methods reported for IMS included MALDI [41, 76–80], Secondary Ion Mass Spectrometry (SIMS) [19, 81–86], Matrix-enhanced (ME)-SIMS [87, 88], Desorption Electrospray Ionization (DESI) [89–99], Nanostructure Initiator Mass Spectrometry (NIMS) [100–102], Atmospheric Pressure Infrared MALDI Mass Spectrometry (AP-IR-MALDI-MS) [103], Laser Ablation-inductively Coupled Plasma-Mass Spectrometry (LA-ICP-MS) [104–106], Laser Desorption Postionization (LDPI) [107], Laser Ablation Electrospray Ionization Mass Spectrometry (LAESI) [108, 109], and Surface-assisted Laser Desorption/ionization Mass Spectrometry (SALDI) [110–112]. Another method was called probe electrospray ionization (PESI) that was used for both liquid solution and the direct sampling on wet samples.

The MALDI-MS instrumentation used for IMS included MALDI-TOF, MALDI-Q-TOF, MALDI-TOF-TOF, MALDI-Q-Ion Mobility-TOFMS (MALDI-Q-IM-TOFMS) [113–117], MALDI-FT-ICR-MS, MALDI-Ion Trap [73, 118–120], and MALDI-QIT-TOF [121].

3.2 *Tissue Sample Preparation*

The sample preparation plays a very important role for the analysis of a drug and its metabolites in biological tissue sections using mass spectrometric imaging. Several factors in IMS sample preparation must be considered, from sample collection to surface treatment prior to analysis in order to produce high quality, reliable, and reproducible results.

The sample preparation procedures for the direct analysis of small molecules in tissue have been described by several papers [120–124]. Tissues (brain, heart, lung, kidney, liver, etc.), were immediately frozen and stored at −80 °C after harvest. The frozen tissues were subsequently cut into serial 10–20 μm thick section which was typically prepared by cryosectioning on a microtome at a temperature of −20 °C. The adjacent sections were gently mounted onto a conductive surface, MALDI imaging target plate or glass slides. These plates were desiccated under low vacuum for a short period of time until dry, then robotically or manually coated with the matrix solution using a MALDI matrix sprayer or a spotter, and dried at room temperature inside of a vacuum desiccator, and analyzed via direct ionization of analyte from the tissue by MALDI-TOF mass spectrometer. Any contamination of tissue sections should be prevented during tissue sample preparation.

3.3 MALDI Matrix Selection and Application

The matrix selection and deposition for MALDI-IMS is another crucial step in the sample preparation protocol. The matrix, solvent, concentration of matrix, and the protocol of coating matrix have to be optimized first and then used for the real samples detection.

Matrix selection: The successful detection of analyte molecules depends on the correct choice of the MALDI matrix. The matrix must not react with the analyte in the tissue sections. The MALDI matrices generally used for MALDI-TOF-MS are CHCA, SA, and DHB. For small molecule drugs, CHCA and DHB are the most widely used matrices [125].

Solvent: Different organic solvents and ratios of organic solvent–water have different performances in MALDI-TOF-MS assay. Acetonitrile, methanol, ethanol, and acetone are usually used as solvents. Five to twenty percent water containing 0.1–0.2 % TFA is usually used.

Matrix concentration: MALDI-MS signal and the formation of homogeneous matrix cocrystal layer are influenced by the matrix concentration. The formation of matrix cocrystal layer can be evaluated using a microscope. The tighter the crystals are, the better the signal level is.

Matrix coating application: A good protocol of coating matrix on the biological tissue section should meet the following purposes: (1) it does not relocate the analyte; (2) the matrix and the analyte form a homogeneous crystal layer; (3) the size of cocrystal is small enough for the mass spectrometry resolution and sufficient sensitivity; (4) it is easy to apply and reproducible. Once the tissue section is mounted onto the sample plate and dried completely, the matrix is deposited over tissue section to extract analytes onto the surface of the tissue and produce crystals. [52, 126].

Agar et al. [127] developed a fast, inexpensive, histology-compatible tissue section preparation method in which the tissue sections were prepared via simultaneous fixation and matrix deposition. This method was called the matrix solution fixation (MSF). It was accomplished by incorporating the MALDI matrix into solvents that preserved tissue integrity when applied according to standard histology procedures. The sinapinic acid solution was in ethanol–methanol–ACN–0.1%TFA (in water) in a 2:2:1:1 ratio and treated the samples at −20 °C for 5 min. The MSF method improved upon time and significantly increased the resolution of the MS imaging. Leinweber et al. [128] reported a new matrix deposition method of five layers sandwich for increasing protein signal at high molecular mass. A 14 μm slice of frozen tissue was coated with four layers of SA solution. It enhanced signals of proteins above 25 kDa with the potential to go above 100 kDa. The Gold Deposition on top of matrix-coated rat brain tissue sections strongly enhanced imaging quality and signal intensity for both SIMS and MALDI Imaging [85, 129, 130].

3.4 Matrix Coating Devices (Sprayer and Spotter)

Matrix desorption is one of most important factors that governs the quality of MALDI imaging of mass resolution, detection sensitivity, spatial resolution, and reproducibility. Therefore, the MALDI matrix deposition should be carefully carried out. It is currently achieved by two major methods: spray and spotting. For a large quantity of samples, the matrix deposition can be automated. The commercially available automated matrix application systems are based on either spray or microspotting deposition of the matrix onto the biological tissue sections. For example, a Portrait 630TM reagent multi-spotter (Labcyte Inc., Sunnyvale, CA) is a fully automated device using acoustic droplet ejection system optimized to deposit matrix onto tissue section. The droplet volume could be low to 170 picoliter (pL) droplets. A TM-Sprayer Imaging mass spectrometry sample preparation system (LEAP Technologies, Carrboro, NC) uses Microsoft Windows to operate interface and graphical selection of the desired deposition area. The spray deposition can be carried out either in linear or in serpentine modes and its track width can vary from 1 to 20 mm. An ImagePrep Station (Bruker Daltonics GmbH; Bremen, Germany) is a piezoelectric nebulizer which creates spray. The matrix layer thickness and wetness are sensor-controlled. The 20 μm droplet size is used for getting high resolution MALDI imaging. Baluya et al. [131] reported custom made automation for matrix deposition, employing an Epson desktop inkjet printer with 5,760 × 1,440 dpi resolution and 3 pL of minimum droplet size for MALDI matrix deposition. Chen Y. et al. [132] described an oscillating capillary nebulizer (OCN) to spray small droplets of matrix aerosol onto the sample surface to improve matrix homogeneity, reduce crystal size, and control solvent effects.

For a few tissue samples, the matrix deposition can be performed manually. A matrix solution is sprayed onto the tissue section with a hand-held thin layer chromatography (TLC) sprayer or an artist airbrush. The reproducibility of manual matrix deposition is an issue. When the manual sprayer is used, the MALDI target plate with the tissue section is held vertically about 15–25 cm from the sprayer nozzle. It is recommended to spray multiple coats of matrix across the tissue section and each coating cycle consists of passing the sprayer two to five times across the tissue section and allowing the tissue to dry for about 1–5 min. This process is usually repeated between 10 and 20 cycles.

A TLC reagent sprayer/or a glass spray nebulizer was used for the matrix coating onto tissue section by several researchers [36, 44, 76, 122, 133]. Puolitaival et al. [134] reported a solvent free matrix Dry-Coating method. The matrix (DHB) was ground for 15 min using a mortar and pestle into crystals between 1 and 40 μm in diameter. A mouse brain section was cut and mounted on a glass slide and stained with hematoxylin and eosin for anatomical visualization. A layer of the finely ground DHB was disbursed over the tissue section through 20-μm stainless steel sieve. Goodwin et al. [135] also deposited the finely ground CHCA over a rat brain section for the quantitation of a small molecular drug. Hankin et al. [136] described a matrix sublimation method for imaging MS. Sublimation of matrix (DHB, CHCA, and SA) produced an even layer of small crystals across the sample plate and the deposition was readily controlled with time, temperature, pressure setting and was highly reproducible from one sample to another.

3.5 IMS Data Acquisition

After a tissue section is coated with a MALDI matrix, an optical imaging is first scanned and restored. Then, the MALDI target plate with the coated tissue section is loaded into the MALDI source of the mass spectrometer. The tissue section is moved in two dimensions (*X*, *Y*) to define the exact region of interest while the laser beam remains fixed to the inlet of MALDI-MS and the mass spectra are acquired. The mass spectrometer records the spatial distribution of molecular species such as peptides, protein, and/or small molecules. The nanoparticle-assisted laser desorption/ionization (nano-PALDI) imaging MS [137] was used to visualize lipids and peptides at a resolution of 15 μm in mammalian tissue. Bouschen et al. [138] introduced the scanning microprobe matrix-assisted laser desorption/ionization mass spectrometry (SMALDIMS) and reported that mass spectrometric images of lateral distribution of sample components and impurities were obtained with a lateral resolution of 1 μm. The image resolution is affected by many factors, including crystal size and laser beam spot diameter. Once the crystal size is optimized, the laser spot diameter remains the critical parameter. Generally, most commercially available MALDI-TOF-MS spectrometers equipped with N_2 (337 nm) laser or a frequency tripled Nd:YAG (355 nm) laser provide a laser spot size at 100 μm. The smaller the diameter of a laser spot is, the higher the resolution of a MS image is.

3.6 IMS Data Processing and Evaluation

The size of data generated by an IMS experiment is dependent on the special resolution and MS scan area. It can vary from 100 megabytes up to a few gigabytes per sample. After acquisition, mass spectral data are converted into image data. A suitable image processing software package can be used to import data from the mass spectrometer, to generate imaging from position correlated mass spectra and to allow visualization and comparison with the optical imaging of the biological sample. The BioMap software package [139–141] provides visualization and a storage plate form. It allows displaying the mass spectrum from selected single point with a cursor on the generated imaging. Selected ions for different compounds are displayed as pseudocolor imaging in which the color intensity correspond to ion signal abundance. The different compounds are separated based on mass scan results (MS and/or MS/MS).

Klerk et al. [142] reported a new method to deal with extremely large image hyperspectral datasets. Authors introduced the use of a data format capable of efficiently storing sparse datasets for multivariate analysis. Recent work has also demonstrated the feasibility to generate three-dimensional (3D) molecular images composed of two spatial dimension using MALDI-MS [59, 143] and MALDI-MS/MS imaging technology [144] and coregistration of these image volumes to other imaging modalities such as magnetic resonance imaging (MRI) [145]. Norris et al. [146] presented a data processing workflow that they had used to successfully increase the amount of useful information that could be derived from mass spectrometry profiling and imaging of biological specimens.

4 Applications to Small Molecular Drug Discovery and Development

The MALDI-TOF-IMS has been applied to tissue imaging of pharmaceuticals and their metabolites for supporting drug discovery and development. Drug distribution is commonly studied by using radiolabeled compounds and autoradiography. MALDI-TOF-IMS allows for the study of drug distribution using nonradiolabeled compounds and offers the potential to simultaneously study drug and metabolites.

The IMS assay has been applied to the analysis of animal organ tissue, skin, whole body, human and animal cancer tissue, and drug formulation. It was also used for MS imaging study on mammalian cell [147], single neurons [148, 149], bacteria [150], and MS imaging of features smaller than the size of laser beam [151].

4.1 Application to Animal Tissue

4.1.1 Brain

Astemizole [152] examined the spatial distribution of astemizole and its metabolites in rat brain slices with and without perfusion with saline solution. The Sprague-Dawley rats were treated orally with the drug at 100 mg/kg in 0.4 % methylcellulose. Matrix solution (DHB, 10 ml) coated by 15–20 coats over the entire surface of tissue sections by a glass reagent sprayer. MALDI-MS/MS images showed the distribution of astemizole and its metabolite (M-14) in rat brain slice. Astemizole appeared to be the major drug-related component in rat brain (Fig. 2).

Clozapine and norclozapine [52] were used as model compounds to investigate fundamental parameters such as matrix and solvent effects and irradiance dependence

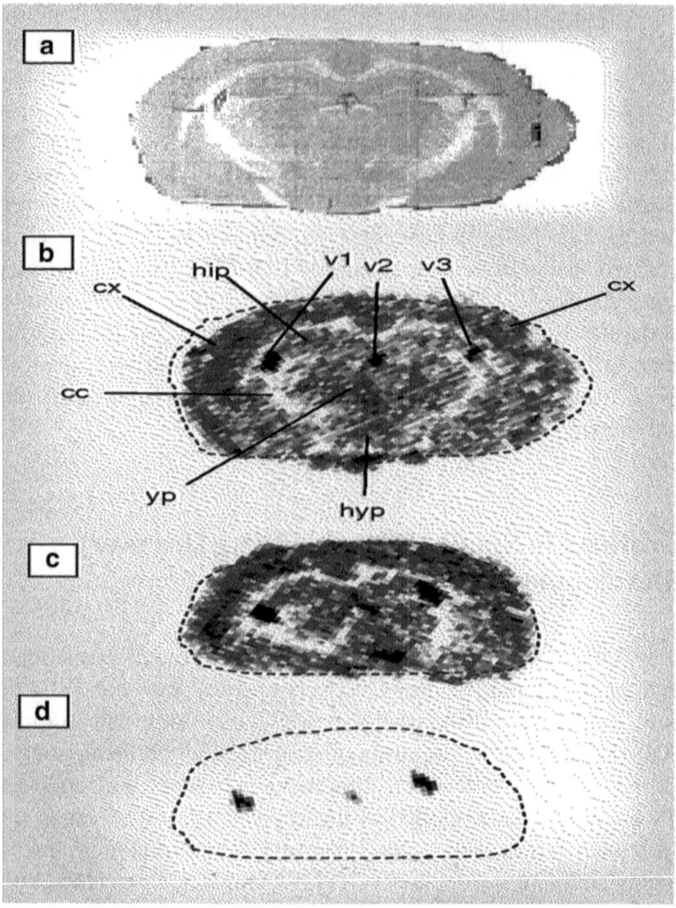

Fig. 2 (**a**) The optical image of a rat brain from a coronal section. (**b**) Matrix-assisted laser desorption/ionization (MALDI)–mass spectrometry (MS)/MS images of astemizole in the rat brain slice without perfusion and (**c**) with perfusion; cortex, hippocampus, corpus callosum, hypothalamic region, thalamus region, choroid plexus, dorsal third ventricle, and lateral ventricle are indicated by arrows. (**d**) MALDI–MS/MS images of M-14 metabolite of astemizole in the rat brain slice (Li et al. [152], Reproduced with permission from Future Science Ltd)

on MALDI intensity. This study also addressed issues such as (1) uniform coating by the matrix, (2) linearity of MALDI signals, and (3) redistribution of surface analytes. Matrices tested were SA, CHCA, DHB, 3-hydroxypicolinic acid, ferulic acid, and caffeic acid. Matrix solutions were prepared at around 25 mg/mL in acetonitrile–water or methanol–water (80:20). Among these matrices, SA yielded the highest sensitivity.

4.1.2 Spleen-Liver

Signor [54] reported qualitative and quantitative analysis of erlotinib and its metabolites in rat liver and spleen following an oral administration at a dose of

5 mg/kg. Sinapinic acid (SA) matrix was used. The tissue section samples were analyzed by MALDI-TOF-MS/MS. The parent compound (*m/z* 394, $[M+H]^+$) was detected in all tissues analyzed. The metabolites (*m/z* 380, $[M+H]^+$) following drug *O*-dealkylation could also be detected in liver sections. MS imaging experiments using MALDI in MS/MS mode allowed visualizing the distribution of the parent compound in liver and spleen tissues. By calculating the ratio between the total ion intensities of MS images for liver and spleen sections, a value of 6:1 was found, which was in good agreement with the quantitative data obtained by LC-MS/MS analysis.

4.1.3 Kidney–Liver–Brain

Cornett [60] reported images for olanzapine (OLZ) in kidney and liver and Imatinib in glioma/mouse brain by MSI. OLZ was administered p.o. at dose of 8 mg/kg to 10 week-old male Fischer 344 rats. Animals were euthanized at 2 h postdose. Three milligrams of Imatinib in 100 mL of PBS was administered by esophageal gavage to a mouse with 14-day old tumor. Tissue sections (12 μm thick) were cut on a cryostat, thaw mounted onto gold-plated stainless-steel plates. A matrix solution of HCCA at 10 mg/mL was prepared in water/acetonitrile (1:1). Arrays of matrix microdroplets were applied onto tissue sections using a Portrait 630 (Labcyte Inc., Sunnyvale CA). Four major ions observed at the nominal *m/z* of the 2-hydroxymethyl metabolite of OLZ were detected within a range of 200 mDa in each tissue.

Bouslimani [153] studied oxaliplatin in rat kidney. Frozen kidney tissue sections of 15 μm thickness were coated using 10 mg/mL CHCA in ACN–0.1%TFA (50:50, vol/vol) either manually or using an Airbrush sprayer. The MALDI imaging was performed using a MALDI-TOF/TOF analyzer. Oxaliplatin in kidney tissues was detected after incubation under HIPEC conditions. This experiment confirmed that oxaliplatin and its metabolites, monocysteine and monomethionine complexes were located at the cortex with weak penetration inside kidney (0.2–0.5 mm).

4.1.4 Skin

Bunch et al. [154] first reported using MALDI-MS to examine the absorption and image the distribution of antifungal agent into skin. A porcine epidermal tissue was treated with a medicated shampoo containing ketoconazole as active ingredient. Following incubation for 1 h at 37 °C all excess formulation was washed from the surface. Due to poor matrix coverage from application methods, a cross section of the drug-treated tissue was blotted onto a cellulose membrane that was precoated with matrix CHCA by airspray deposition. The MS image by MALDI showed that the ketoconazole penetrated into the skin only as far as the dermis with no absorption into the underlying connective tissue

4.1.5 Whole-Body

Stoeckli et al. [155] reported the whole-body imaging by MALDI. Three HLA-B27 transgenic mice were dosed intravenously with 0.5 mL of a saline solution of β-peptide at 2 mg/mL. A fourth mouse was dosed with 0.5 mL of a saline solution of the Rheuma α-peptide at 2 mg/mL as a control. Mice treated with β-peptide were sacrificed at 5 min, 1 h, and 24 h postdose (one mouse per time point). Matrix coating was obtained by spraying 4 mL per plate of a solution of 10 mg/mL CHCA in acetonitrile–water (50:50) using a TLC sprayer.

MS imaging was carried out on whole-body sections of mice. No α-peptide control was detected at 1 h postdose, while retention of the β-peptide was observed for longer than 24 h postdose.

4.1.6 Cancer Tissue

Reyzer et al. [156] reported a study of the distribution of antitumor drug, SCH226374, in mouse tumor tissue. The matrix SA was prepared at a 20 mg/mL in ACN/0.2%TFA (50:50). The matrix solution was applied onto the tissues using the spotting and spray. The MALDI images were created using the selected reaction monitoring (SRM) technique to monitor the transition of $[M+H]^+$ *m/z* 695 > *m/z* 228. The MS image indicated that the drug was present over most of the tumor but was concentrated in the outer periphery.

4.1.7 Retina

Hayasaka et al. [157] reported the determination of the fatty acid distribution in mouse retina by using AgNPs in nano-PALDI-IMS. The sections were sliced to a thickness of 10 μm and sprayed with AgNPs or DHB matrix solution at 50 mg/mL in 70 % methanol/0.1%TFA. The mouse retinal sections were analyzed at a high spatial resolution with a scan pitch of 10 μm. The MS images showed the distribution of palmitic acid, linoleic acid, oleic acid, stearic acid, eicosapentaenoic acid (EPA), arachidonic acid, and docosahexaenoic acid (DHA).

4.2 *Application to Human Cancer Tissue*

Atkinson [158] reported a study on the distribution of banoxatrone (AQ4N) in human cancer tissue by MALDI MS. In hypoxic cells AQ4N is reduced to the topoisomerase II inhibitor AQ4. CHCA at 25 mg/mL in ethanol with 1.0 % TFA was used as matrix. The distribution of AQ4N and AQ4 in treated H460 human tumor xenografts was examined by MALDI MS imaging in positive ion acquisition mode. Images of the distribution of AQ4N and AQ4 showed little overlap.

The distribution of ATP (negative ion acquisition) in the tumor xenografts was also studied and indicated that the cytotoxic metabolite AQ4 was confined to hypoxic regions of the tumor as intended.

4.3 Application to Pharmaceutical Drug Formulation

Earnshaw et al. [159] recently used MALDI-MS to directly analyze a range of commercially available and prescription tablet formulations, including aspirin, paracetamol, sildenafil citrate (Viagra), and a batch of tablets in development. The tablet sections approximately 1 mm in thickness were created by using a tablet cutter to fracture the tablet if it is necessary. The exposed surface of the tablet was coated with 25 mg/mL CHCA in EtOH containing 0.1 % TFA by an airspray. MALDI-MS acquired data with MS full scan function. Drugs were clearly observed both in the MS spectra and in the images as protonated molecules, or as sodium adducts. MALDI-MS imaging also provided semiquantitative information.

5 Conclusion

Since its inception about 15 year ago, MALDI-IMS has been developed into a powerful and versatile tool for biomedical research. It allows for the investigation of the spatial distribution of molecules at complex surfaces. The combination of molecular speciation with local analysis makes a chemical microscope that can be used for the direct biomolecular characterization of histological tissue section surface. However, successful detection of the analytes of interest at the desired spatial resolution requires careful attention to several steps in the IMS protocol: matrix selection, matrix coating, data acquisition, and data processing. MALDI-IMS is increasingly playing an important role in the drug discovery and development and disease treatment.

References

1. Fenn J, Mann M, Meng C, Wong S, Whitehouse C (1989) Electrospray ionization for mass spectrometry of large biomolecules. Science 246:64–71. doi:10.1126/science.2675315
2. Karas M, Bachmann D, Hillenkamp F (1985) Influence of the wavelength in high-irradiance ultraviolet laser desorption mass spectrometry of organic molecules. Anal Chem 57:2935–2939. doi:10.1021/ac00291a042
3. Karas M, Hillenkamp F (1988) Laser desorption ionization of protein with molecular masses exceeding 10000 Daltons. Anal Chem 60:2299–2301. doi:10.1021/ac00171a028
4. Tanaka K, Waki H, Ido Y, Akita S, Yoshida Y, Yoshida T (1988) Protein and polymer analysis up to m/z 100000 by laser ionization time-of-flight mass spectrometry. Rapid Commun Mass Spectrom 2:151–153. doi:10.1002/rcm.1290020802

5. Kiyohara M, Hama Y, Yamaguchi K, Ito M (2006) Structure of β-1,3-xylooligosaccharides generated from *caulerpa racemosa var. laete-virens* β1,3-xylan by the action of β1,3-xylanase. J Biochem 140:369–373. doi:10.1093/jb/mvj173
6. Mechref Y, Kang P, Novotny M (2006) Differentiating structural isomers of sialylated glycans by matrix-assisted laser desorption/ionization time-of-flight/time-of-flight tandem mass spectrometry. Rapid Commun Mass Spectrom 20:1381–1389. doi:10.1002/rcm.2445
7. Troendle F, Reddick C, Yost R (1999) Detection of pharmaceutical compounds in tissue by matrix-assisted laser desorption/ionization and laser desorption/chemical ionization tandem mass spectrometry with a quadrupole ion trap. J Am Soc Mass Spectrom 10:1315–1321
8. Caprioli R, Farmer T, Gile J (1997) Molecular imaging of biological samples: localization of peptides and proteins using MALDI-TOF MS. Anal Chem 69:4751–4760
9. Koomen J, Stoeckli M, Caprioli R (2000) Mapping of surrogate markers of cellular components and structures using laser desorption/ionization mass spectrometry. J Mass Spectrom 35:258–264
10. Chaurand P, Schwartz S, Caprioli R (2002) Imaging mass spectrometry: a new tool to investigate the spatial organization of peptides and protein in mammalian tissue sections. Curr Opin Chem Biol 6:676–681
11. Schwartz S, Weil R, Johnson M, Toms S, Caprioli R (2004) Protein profiling in brain tumors using mass spectrometry: feasibility of a new technique for the analysis of protein expression. Clin Cancer Res 10:981–987
12. Cornett D, Reyzer M, Chaurand P, Caprioli R (2007) MALDI imaging mass spectrometry: molecular snapshots of biochemical systems. Nat Methods 10:828–833. doi:10.1038/nmeth1094
13. Grey A, Chaurand P, Caprioli R, Schey K (2009) MALDI imaging mass spectrometry of integral membrane protein from ocular lens and retinal tissue. J Proteome Res 8:3278–3283. doi:10.1021/pr800956y
14. Oppenheimer S, Mi D, Sanders M, Caprioli R (2010) Molecular analysis of tumor margins by MALDI mass spectrometry in renal carcinoma. J Proteome Res 9:2182–2190. doi:10.1021/pr900936z
15. Chaurand P, Rahman M, Hunt T, Mobley J, Gu G, Latham J, Caprioli R, Kasper S (2008) Monitoring mouse prostate development by profiling and imaging mass spectrometry. Mol Cell Proteomics 7:411–423. doi:10.1074/mcp.M700190-MCP200
16. Dickinson D, La Duc M, Haskins W, Gornushkin I, Winefordner J, Powell D, Venkateswaran K (2004) Species differentiation of a diverse suite of *Bacillus* spores by mass spectrometry-based protein profiling. Appl Environ Microbiol 70:475–482. doi:10.1128/AEM.70.1.475-482.2004
17. Gustafsson J, Oehler M, McColl S, Hoffmann P (2010) Citric acid antigen retrieval (CAAR) for tryptic peptide imaging directly on archived formalin-fixed paraffin-embedded tissue. J Proteome Res. doi:10.1021/pr9011766
18. Liu N, Liu F, Xu B, Gao Y, Li X, Wei K, Zhang X, Yang S (2008) Establishment of imaging mass spectrometry for biological tissue and its application on the proteome analysis of microwave radiated hippocampus. Chin J Anal Chem 36(4):421–425
19. Djidja M, Claude E, Snel M, Francese S, Scriven P, Carolan V, Clench M (2010) Novel molecular tumor classification using MALDI-mass spectrometry imaging of tissue micro-assay. Anal Bioanal Chem 397:587–601. doi:10.1007/s00216-3554-6
20. Monroe E, Annangudi S, Hatcher N, Gutstein H, Rubakhin S, Sweedler J (2008) SIMS and MALDI MS imaging of the spinal cord. Proteomics 8:3746–3754. doi:10.1002/pmic.200800127
21. Nakanishi T, Ando E, Furuta M, Tsunasawa S, Nishimura O (2007) Direct on-membrane peptide mass fingerprinting with MALDI-MS of tyrosine-phosphorylated protein detected by immunostaining. J Chromatogr B 847:24–29. doi:10.1016/j.jchromb.2006.08.024
22. Thiery G, Shchepinov M, Southern E, Audebourg A, Audard V, Terries B, Gut I (2007) Multiplex target protein imaging in tissue sections by mass spectrometry – TAMSIM. Rapid Commun Mass Spectrom 21:823–829. doi:10.1002/rcm.2895

23. Uys J, Grey A, Wiggins A, Schwacke J, Schey K, Kalivas P (2010) Matrix-assisted laser desorption/ionization tissue profiling of secretoneurin in the nucleus accumbens shell from cocaine-sensitized rats. J Mass Spectrom 45:97–103. doi:10.1002/jms.1697
24. Rauser S, Marquardt C, Balluff B, Deininger S, Albers C, Belau E, Hartmer R, Suckau D, Specht K, Ebert M, Schmitt M, Aubele M, Höfler H, Walch A (2010) Classification of HER2 receptor status in breast cancer tissue by MALDI imaging mass spectrometry. J Proteome Res 9:1854–1863. doi:10.1021/pr901008d
25. Remoortere A, Zeijl R, Oever N, Franck J, Longuespée R, Wisztorski M, Salzet M, Deelder A, Fournier I, McDonnell L (2010) MALDI imaging and profiling MS of higher mass proteins from tissue. J Am Soc Mass Spectrom 21:1922–1929. doi:0.1016/j.jasms.2010.07.011
26. Verhaert P, Conaway M, Pekar T, Miller K (2007) Neuropeptide imaging on an LTQ with vMALDI source: the complete 'all-in-one' peptidome analysis. Int J Mass Spectrom 260: 177–184. doi:10.1016/j.ijms2006.11.008
27. Chaurand P, Cornett D, Angel P, Caprioli R (2011) From whole-body sections down to cellular level, multiscale imaging of phospholipids by MALDI mass spectrometry. Mol Cell Proteomics 10(2):O110.004259. doi:10.1074/mcp.O110.004259
28. Colsch B, Woods A (2010) Localization and imaging of sialylated glycosphingolipids in brain tissue sections by MALDI mass spectrometry. Glycobiology 20:661–667. doi:10.1093/glycob/cwq031
29. Dill A, Ifa D, Manicke N, Zheng O, Cooks R (2009) Mass spectrometric imaging of lipids using desorption electrospray ionization. J Chromatogr B 877:2883–2889. doi:10.1016/j.jchromb.2008.12.058
30. Hankin J, Murphy R (2010) Relationship between MALDI IMS intensity and measured quantity of selected phospholipids in rat brain sections. Anal Chem 82:8476–8484. doi:10.1021/ac101079v
31. Jackson S, Wang H, Woods A (2007) In situ structure characterization of glycerophospholipids and sulfatides in brain tissue using MALDI-MS/MS. J Am Soc Mass Spectrom 18:17–26. doi:10.1016/j.jasms.2006.08.015
32. Koizumi S, Yamamoto S, Hayasaka T, Konishi Y, Yamaguchi-Okada M, Goto-Inoue N, Sugiura Y, Setou M, Namba H (2010) Imaging mass spectrometry revealed the production of lyso-phosphatidylcholine in the injured ischemic rat brain. Neuroscience 168:219–225. doi:10.1016/j.neuroscience.2010.03.056
33. Meriaux C, Franck J, Wisztorski M, Salzet M, Fournier I (2010) Liquid ionic matrixes for MALDI mass spectrometry imaging of lipids. J Proteomics 73:1204–1218. doi:10.1016/j.jprot.2010.02.010
34. Mikawa S, Suzuki M, Fujimoto C, Sato K (2009) Imaging of phosphatidylcholines in the adult rat brain using MALDI-TOF MS. Neurosci Lett 451:45–49. doi:10.1016/j.neulet.2008.12.035
35. Sjövall P, Lausmaa J, Johansson B (2004) Mass spectrometric imaging of lipids in brain tissue. Anal Chem 76:4271–4278. doi:10.1021/ac049389p
36. Wang H, Post S, Woods A (2008) A minimalist approach to MALDI imaging of glycerophospholipids and sphingolipids in rat brain sections. Int J Mass Spectrom 278:143–149. doi:10.1016/j.ijms.2008.04.005
37. Veloso A, Astigarraga E, Barreda-Gómez G, Manuel I, Ferrer I, Giralt M, Ochoa B, Fresnedo O, Rodriguez-Puertas R, Fernandez J (2011) Anatomical distribution of lipids in human brain cortex by imaging mass spectrometry. J Am Soc Mass Spectrom 22:329–338. doi:10.1007/s13361-010-0024-5
38. Naven T, Harvey D, Brown J, Critchley G (1997) Fragmentation of complex carbohydrates following ionization by matrix-assisted laser desorption with an instrument fitted with time-lag focusing. Rapid Commun Mass Spectrom 11:1681–1686
39. Kong Y, Zhu Y, Zhang J (2001) Ionization mechanism of oligonucleotides in matrix-assisted laser desorption/ionization time-of-flight mass spectrometry. Rapid Commun Mass Spectrom 15:57–64
40. Molin L, Cristoni S, Crotti S, Bernardi L, Seraglia R, Traldi P (2008) Sieve-based device for MALDI sample preparation. I. Influence of sample deposition conditions in oligonucleotide

analysis to achieve significant increases in both sensitivity and resolution. J Mass Spectrom 43:1512–1520. doi:10.1002/jms.1428
41. Anker J, Hall W, Lambert M, Velasco P, Mrksich M, Klein W, Van Duyne R (2009) Detection and identification of bioanalytes with high resolution LSPR spectroscopy and MALDI mass spectrometry. J Phys Chem 113:5891–5894. doi:10.1021/jp900266k
42. Bauer J, Chakravarthy A, Rosenbluth J, Mi D, Seeley E, Granja-Ingram N, Olivares M, Kelley M, Mayer I, Meszoely I, Means-Powell J, Johnson K, Tsai C, Ayers G, Sanders M, Schneider R, Formenti S, Caprioli R, Pietenpol J (2010) Proteomic and genomic analyses of breast tumors from patients receiving neoadjuvant paclitaxel and radiation. Clin Cancer Res 16(2):681–690. doi:10.1158/1078-0432.CCR-09-1091
43. Benabdellah F, Touboul D, Brunelle A, Laprévote O (2009) In situ primary metabolites localization on a rat brain section by chemical mass spectrometry imaging. Anal Chem 81:5557–5560. doi:10.1021/ac9005364
44. Goodwin R, Dungworth J, Cobb S, Pitt A (2008) Time-dependent evolution of tissue markers by MALDI-MS imaging. Proteomics 8:3801–3808. doi:10. 1002/pmic.2008002001
45. Lemaire R, Menguellet S, Stauber J, Marchaudon V, Lucot J, Collinet P, Farine M, Vinatier D, Day R, Ducoroy P, Salzet M, Fournier I (2007) Specific MALDI imaging and profiling for biomarker hunting and validation: fragment of the 11 S proteasome activator complex, rag alpha fragment, is a new potential ovary cancer biomarker. J Proteome Res 6:4127–4134. doi:10.1021/pr0702722
46. Lemaire R, Stauber J, Wisztorski M, Van Camp C, Desmons A, Deschamps M, Proess G, Rudlof I, Woods A, Day R, Salzet M, Fournier I (2007) Tag-mass: specific molecular imaging of transcriptome and proteome by mass spectrometry based on photocleavable tag. J Proteome Res 6:2057–2067. doi:10.1021/rp0700044
47. Meistermann H, Norris J, Aerni H, Cornett D, Friedlein A, Erskine A, Augustin A, Mudry M, Ruepp S, Suter L, Langen H, Caprioli R, Ducret A (2006) Biomarker discovery by imaging mass spectrometry, transthyretin is a biomarker for gentamicin-induced nephrotoxicity in rat. Mol Cell Proteomics 5:1876–1886. doi:10.1074/mcp.M500399-MCP200
48. Stauber J, Lemaire R, Franck J, Bonnel D, Croix D, Day R, Wisztorski M, Fournier I, Salzet M (2008) MALDI imaging of formalin-fixed paraffin-embedded tissues: application to model animals of Parkinson disease for biomarker hunting. J Proteome Res 7:969–978. doi:10.1021/pt070464x
49. Chen J, Hsieh Y, Knemeyer I, Crossman L, Korfmacher W (2008) Visualization of first-pass drug metabolism of terfenadine by MALDI-imaging mass spectrometry. Drug Metab Lett 2:1–4
50. Cohen L, Gusev A (2001) Small molecule analysis by MALDI mass spectrometry. Anal Bioanal Chem 373:571–586. doi:10.1007/s00216-002-1321-z
51. Cristoni S, Brioschi M, Rizzi A, Sironi L, Gelosa P, Tremoli E, Bernardi L, Banfi C (2006) Analysis of rosuvastatin by imaging mass spectrometry. Rapid Commun Mass Spectrom 20:3483–3487. doi:10.1002/rcm.2744
52. Hsieh Y, Casale R, Fukuda E, Chen J, Knemeyer I, Wingate J, Morrison R, Korfmacher W (2006) Matrix-assisted laser desorption/ionization imaging mass spectrometry for direct measurement of clozapine in rat brain tissue. Rapid Commun Mass Spectrom 20:965–972. doi:10.1002/rcm.2397
53. Mangé A, Chaurand P, Perrochia H, Roger P, Caprioli R, Solassol J (2009) Liquid chromatography-tandem and MALDI imaging mass spectrometry analyses of RCL2/CS100-fixed, paraffin-embedded tissue: proteomics evaluation of an alternate fixative for biomarker discovery. J Proteome Res 8:5619–5628. doi:10.1021/pr9007128
54. Signor J, Varesio E, Staack R, Starke V, Richter W, Hopfgartner G (2007) Analysis of erlotinib and its metabolites in rat tissue sections by MALDI quadrupole time-of-flight mass spectrometry. J Mass Spectrom 42:900–909. doi:10.1002/jms.1225
55. Su A, Liu J, Lin C (2005) Rapid drug-screening of clandestine tablets by MALDI-TOF mass spectrometry. Talanta 67:718–724. doi:10.1016/j.talanta.2005.03.013

56. Wang H, Jackson S, McEuen J, Woods A (2005) Localization and analyses of small drug molecules in rat brain tissue sections. Anal Chem 77(20):6682–6686. doi:10.1021/ac050868d
57. Woods A, Jackson S (2006) Brian tissue lipidomics: direct probing using matrix-assisted laser desorption/ionization mass spectrometry. AAPS J 8(2):E391–E395
58. Woods A, Wang H, Jackson S (2007) A snapshot of tissue glycerolipids. Curr Pharm Des 11:2577–2591
59. Andersson M, Groseclose N, Deutch A, Caprioli R (2008) Imaging mass spectrometry of proteins and peptides: 3D volume reconstruction. Nat Methods 5(1):101–109. doi:10.1038/nmeth1145
60. Cornett D, Frappier S, Caprioli R (2008) MALDI-FTICR imaging mass spectrometry of drugs and metabolites in tissue. Anal Chem 80:5648–5653. doi:10.1021/ac800617s
61. Hillenkamp F, Peter-Katalinić J (2007) MALDI MS: a practical guide to instrumentation, methods and applications. Wiley-VCHGmbH & Co. KGaA, Weinheim, pp 19
62. Vorm O, Roepstorff P, Mann M (1994) Improved resolution and very high sensitivity in MALDI TOF of matrix surfaces made by fast evaporation. Anal Chem 66:3218–3287
63. Dai Y, Whittal R, Li L (1996) Confocal fluorescence microscopic imaging for investigating the analyte distribution in MALDI matrices. Anal Chem 68:2494–2500
64. Dai Y, Whittal R, Li L (1999) Two-layer sample preparation: a method for MALDI-MS analysis of complex peptide and protein mixtures. Anal Chem 71:1087–1091. doi:10.1021/ac980684h
65. Dong H, Kemptner J, Marchetti-Deschmann M, Kubicek C, Allmaier G (2009) Development of a MALDI two-layer volume sample preparation technique for analysis of colored conidia spores of *Fusarium* by MALDI linear TOF mass spectrometry. Anal Bioanal Chem 395:1373–1383. doi:10.1007/s00216-009-3067-3
66. Schriemer D, Li L (1996) Detection of high molecular weight narrow polydisperse polymers up to 1.5 million daltons by MALDI mass spectrometry. Anal Chem 68:2721 2725
67. Li L, Golding R, Whittal R (1996) Analysis of single mammalian cell lysates by mass spectrometry. J Am Chem Soc 18:11662–11663
68. Kemptner J (2009) Evaluation of matrix-assisted laser desorption/ionization (MALDI) preparation techniques for surface characterization of intact *Fusarium* spores by MALDI linear time-of-flight mass spectrometry. Rapid Commun Mass Spectrom 23:877–884. doi:10.1002/rcm.3949
69. Kussmann M, Nordhoff E, Rahbek-Nielsen H, Haebel S, Rossel-Larsen M, Jakobsen L, Gobom J, Mirgorodskaya E, Kroll-Kristensen A, Palm L, Roepstorff P (1997) Matrix-assisted laser desorption/ionization mass spectrometry sample preparation techniques designed for various peptides and protein analytes. J Mass Spectrom 32:593–601
70. Keller B, Li L (2006) Three-layer matrix/sample preparation method for MALDI MS analysis of low nanomolar protein samples. J Am Soc Mass Spectrom 17:780–785. doi:10.1016/j.jasms.2006.02.012
71. Han J, Schey K (2006) MALDI tissue imaging of ocular lens α-crystalline. Invest Ophthalmol Vis Sci 47:2990–2996
72. Kanu A, Dwivedi P, Tam M, Matz L, Hill H Jr (2008) Ion mobility-mass spectrometry. J Mass Spectrom 443:1–22. doi:10.1002/JMS.1383
73. Garrett T, Prieto-Conaway M, Kovtoun V, Bui H, Izgarian N, Stafford G, Yost R (2007) Imaging of small molecules in tissue sections with a new intermediate-pressure MALDI linear ion trap mass spectrometer. Int J Mass Spectrom 260:166–176. doi:10.1016/j.ijms.2006.09.019
74. Khatib-Shahidi S, Andersson M, Herman J, Gillespie T, Caprioli R (2006) Direct molecular analysis of whole-body animal tissue sections by imaging MALDI mass spectrometry. Anal Chem 78(18):6448–6456. doi:10.1021/ac060788p
75. Chaurand P, Norris J, Cornett D, Mobley J, Caprioli R (2006) New developments in profiling and imaging of protein from tissue sections by MALDI mass spectrometry. J Proteome Res 5:2889–2900. doi:10.1021/pr060346u

76. Acquadro E, Cabella C, Ghiani S, Miragoli L, Bucci E, Corpillo D (2009) Matrix-assisted laser desorption ionization imaging mass spectrometry detection of a magnetic resonance imaging contrast agent in mouse liver. Anal Chem 81:2779–2784. doi:10.1021/ac900038y
77. Burrell M, Earnshaw C, Clench M (2007) Imaging matrix assisted laser desorption ionization mass spectrometry: a technique to map plant metabolites within tissue at high spatial resolution. J Exp Bot 58(4):757–763. doi:10.1093/jxb/erl139
78. Cazares L, Troyer D, Mendrinos S, Lance R, Nyalwidhe J, Beydoun H, Clements M, Drake R, Semmes O (2009) Imaging mass spectrometry of a specific fragment of mitogen-activated protein kinase/extracellular signal-regulated kinase kinase kinase 2 discriminates cancer from uninvolved prostate tissue. Clin Cancer Res 15(17):5541–5551. doi:10.1158/1078-4032.CCR-08-2892
79. Garden R, Sweedler J (2000) Heterogeneity within MALDI samples as revealed by mass spectrometric imaging. Anal Chem 72:30–36. doi:10.1021/ac9908997
80. McCombie G, Staab D, Stoeckli M, Knochenmuss R (2005) Spatial and spectral correlations in MALDI mass spectrometry imaging by clustering and multivariate analysis. Anal Chem 77:6118–6124. doi:10.1021/ac051081q
81. Broersen A, Liere R, Altelaar A, Heeren R, MacDonnell L (2008) Automated, feature-based imaging alignment for high-resolution imaging mass spectrometry of large biological samples. J Am Soc Mass Spectrom 19:823–832. doi:10.1016/j.jasms.2008.02.015
82. Brunelle A, Touboul D, Laprévote O (2005) Biological tissue imaging with time-of flight secondary ion mass spectrometry and cluster ion sources. J Mass Spectrom 40:985–999. doi:10.1002/jms.902
83. Lorey D II, Morrison G, Chandra S (2001) Dynamic secondary ion mass spectrometry analysis of boron from boron neutron capture therapy drug in co-cultures: single-cell imaging of two different cell types within the same ion microscopy field of imaging. Anal Chem 73: 3947–3953. doi:10.1021/ac0103266
84. Seyer A, Riu A, Debrauwer L, Bourgès-Abella N, Brunelle A, Laprévote O, Zalko D (2010) Time-of-flight secondary ion mass spectrometry imaging demonstrates the specific localization of deca-bromo-diphenyl-ether residues in the ovaries and adrenal glands of exposed rats. J Am Soc Mass Spectrom 21:1836–1845. doi:10.1016/j.jasms.2010.06.019
85. Touboul D, Halgand F, Brunelle A, Kersting R, Tallarek E, Hagenhoff B, Laprévote O (2004) Tissue molecular ion imaging by gold cluster ion bombardment. Anal Chem 76:1550–1559. doi:10.1021/ac
86. Zheng L, McQuaw C, Ewing A, Winograd N (2007) Sphingomyelin/phosphatidylcholine and cholesterol interactions studied by imaging mass spectrometry. J Am Chem Soc 129(51):15730–15731. doi:10.1021/ja0741675
87. Luxembourg S, McDonnell L, Duursma M, Guo X, Heeren R (2003) Effect of local matrix crystal variations in matrix-assisted ionization techniques for mass spectrometry. Anal Chem 75:2333–2341. doi:10.1021/ac026434p
88. McDonnell L, Piersma S, Altelaar A, Mize T, Luxembourg S, Verhaert P, van Minnen J, Heeren R (2005) Subcellular imaging mass spectrometry of brain tissue. J Mass Spectrom 40: 160–168. doi:10.1002/jms.735
89. Ellis S, Wu C, Deeley J, Zhu X, Truscott R, Panhuis J, Cooks R, Mitchell T, Blanksby S (2010) Imaging of human lens lipids by desorption electrospray ionization mass spectrometry. J Am Soc Mass Spectrom 21:2095–2104. doi:10.1016/j.jasms.2010.09.003
90. Girod M, Shi Y, Cheng J, Cooks G (2010) Desorption electrospray ionization imaging mass spectrometry of lipids in rat spinal cord. J Am Soc Mass Spectrom 21:1177–1189. doi:10.1016/j.jasms.2010.03.028
91. Ifa D, Wiseman J, Song Q, Cooks G (2007) Development of capabilities for imaging mass spectrometry under ambient conditions with desorption electrospray ionization (DESI). Int J Mass Spectrom 259:8–15. doi:10.106/j.ijms.2006.08.003
92. Kertesz V, van Berkel G (2008) Improved imaging resolution in desorption electrospray ionization mass spectrometry. Rapid Commun Mass Spectrom 22:2639–2644. doi:10.1002/rcm.3662

93. Kertesz V, van Berkel G, Vavrek M, Koeplinger K, Schneider B, Covey T (2008) Comparison of drug distribution images from whole-body thin tissue sections obtained using desorption electrospray ionization tandem mass spectrometry and autoradiography. Anal Chem 80: 5168–5177. doi:10.1021/ac800546a
94. Manicke N, Dill A, Ifa D, Cooks R (2010) High-resolution tissue imaging on an orbitrap mass spectrometry by desorption electrospray ionization mass spectrometry. J Mass Spectrom 45:223–226. doi:10.1002/jms.1707
95. Takáts Z, Wiseman J, Gologan B, Cooks G (2004) Mass spectrometry sampling under ambient conditions with desorption electrospray ionization. Science 306:471–473. doi:10.1126/science.1104404
96. Takáts Z, Wiseman J, Cooks G (2005) Ambient mass spectrometry using desorption electrospray ionization (DESI): instrumentation, mechanisms and application in forensics, chemistry, and biology. J Mass Spectrom 40:1261–1275. doi:10.1002/jms.922
97. Takats Z, Kobliha V, Sevcik K, Novak P, Kruppa G, Lemr K, Havlicek V (2008) Characterization of DESI-FTICR mass spectrometry-from ECD to accurate mass tissue analysis. J Mass Spectrom 43:196–203. doi:10.1002/jms.1285
98. Watrous J, Hendricks N, Meehan M, Dorrestein P (2010) Capturing bacterial metabolic exchange using thin film desorption electrospray ionization-imaging mass spectrometry. Anal Chem 82:1598–1600. doi:10/1021/ac9027388
99. Wiseman J, Ifa D, Zhu Y, Kissinger C, Manicke N, Kissinger P, Cooks R (2008) Desorption electrospray ionization mass spectrometry: imaging drug and metabolites in tissues. Proc Natl Acad Sci USA 105(47):18120–18125. doi:10.1073/pnas.0801066105
100. Northen T, Yanes S, Northen M, Marrinucci D, Uritboonthai W, Apon J, Golledge S, Nordström A, Siuzdak G (2007) Clathrate nanostructures for mass spectrometry. Nature 449:1033–1037. doi:10.1038/nature06195
101. Patti G, Woo H, Yanes O, Shriver L, Thomas D, Uritboonthai W, Apon J, Steenwyk R, Manchester M, Siuzdak G (2010) Detection of carbohydrates and steroids by cation-enhanced nanostructure-initiator mass spectrometry (NIMS) for biofluid analysis and tissue imaging. Anal Chem 82:121–128. doi:10.1021/ac9014353
102. Yanes O, Woo H, Northen T, Oppenheimer S, Shriver L, Apon J, Estrada M, Potchoiba M, Steenwyk R, Manchester M, Siuzdak G (2009) Nanostructure initiator mass spectrometry: tissue imaging and direct biofluid analysis. Anal Chem 81:2969–2975. doi:10.1021/ac802576q
103. Li Y, Shrestha B, Vertes A (2007) Atmospheric pressure molecular imaging by infrared MALDI mass spectrometry. Anal Chem 79:523–532. doi:10.1021/ac061577n
104. Gholap D, Izmer A, Samber B, van Elteren J, Šelih V, Evens R, Schamphelaere K, Janssen C, Balcaen L, Lindemann I, Vincze L, Vanhaecke F (2010) Comparison of laser ablation-inductively coupled plasma-mass spectrometry and micro-X-ray fluorescence spectrometry for elemental imaging. Anal Chim Acta 664:19–26. doi:10.1016/j.aca.2010.01.052
105. Dobrowolska J, Dehnhardt M, Matusch A, Zoriy M, Palomero-Gallagher N, Koscielniak P, Zilles K, Becker J (2008) Quantitative imaging of zinc, copper and lead in three distinct regions of the human brain by laser ablation inductively coupled plasma mass spectrometry. Talanta 74:717–723. doi:10.1016/j.talanta.2007.06.051
106. Santos M, Wagner M, Wu B, Scheider J, Oehlmann J, Cadore S, Becker J (2009) Biomonitoring of metal contamination in marine prosobranch snail (nassarius reticulates) by imaging laser ablation inductively coupled plasmas mass spectrometry (LA-ICP-MS). Talanta 80:428–433. doi:10.1016/j.talanta.2009.06.080
107. Akhmetov A, Moore J, Gasper G, Koin P, Hanley L (2010) Laser desorption postionization for imaging MS of biological material. J Mass Spectrom 45:137–145. doi:10.1002/jms.1716
108. Nemes P, Barton A, Li Y, Vertes A (2008) Ambient molecular imaging and depth profiling of live tissue by infrared laser ablation electrospray ionization mass spectrometry. Anal Chem 80:4575–4582. doi:10.1021/ac8004082
109. Nemes P, Woods A, Vertes A (2010) Simultaneous imaging of small metabolites and lipids in rat brain tissue at atmospheric pressure by laser ablation electrospray ionization mass spectrometry. Anal Chem 82:982–988. doi:10.1021/ac902245p

110. Liu Q, Xiao Y, Pagan-Miranda C, Chiu Y, He L (2009) Metabolite imaging using matrix-enhanced surface-assisted laser desorption/ionization mass spectrometry (ME-SALDI-MS). J Am Soc Mass Spectrom 20:80–88. doi:10.1016/j.jasms.2008.09.011
111. Liu Q, He L (2009) Ionic matrix for matrix-enhanced surface-assisted laser desorption/ionization mass spectrometry (ME-SALDI-MS). J Am Soc Mass Spectrom 20:2229–2237. doi:10.1016/j.jasms.2009.08.011
112. Watanabe T, Kawasaki H, Yonezawa T, Arakawa R (2008) Surface-assisted laser desorption/ionization mass spectrometry (SALDI-MS) of low molecular weight organic compounds and synthetic polymers using zinc oxide (ZnO) nanoparticles. J Mass Spectrom 43:1063–1071. doi:10.1002/jms.1385
113. Jackson S, Wang H, Woods A, Ugarov M, Egan T, Schultz A (2005) Direct tissue analysis of phospholipids in rat brain using MALDI-TOFMS and MALDI-Ion mobility-TOFMS. J Am Soc Mass Spectrom 16:133–138. doi:10.1016/j.jasms.2004.10.002
114. Jackson S, Ugarov M, Egan T, Post J, Langlais D, Schultz J, Woods A (2007) MALDI-ion mobility-TOFMS imaging of lipids in rat brain tissue. J Mass Spectrom 42:1093–1098. doi:10.1002/jms.1245
115. McLean J, Ridenour W, Caprioli R (2007) Profiling and imaging of tissues by imaging ion mobility-mass spectrometry. J Mass Spectrom 42:1099–1105. doi:10.1002/jms.1254
116. Stauber J, MacAleese L, Franck J, Claude E, Snel M, Kaletas B, Wiel I, Wisztorski M, Fournier I, Fournier I (2010) On-tissue protein identification and imaging by MALDI-Ion mobility mass spectrometry. J Am Soc Mass Spectrom 20:338–347. doi:10.1016/j.jasms.2009.09.016
117. Trim P, Henson C, Avery J, McEwen A, Snel M, Claude E, Marshall P, West A, Princivalle A, Clench M (2008) Matrix-assisted laser desorption/ionization-ion mobility separation-mass spectrometry imaging of vinblastine in whole body tissue sections. Anal Chem 80:8628–8634. doi:10.1021/ac8015467
118. Drexler D, Garrett T, Cantone J, Diters R, Mitroka J, Conaway M, Adams S, Yost R, Sanders M (2007) Utility of imaging mass spectrometry (IMS) by matrix-assisted laser desorption ionization (MALDI) on an ion trap mass spectrometer in the analysis of drugs and metabolites in biological tissue. J Pharmacol Toxicol Methods 55:279–288. doi:10.1016/j.vascn.2006.11.004
119. Landgraf R, Conaway M, Garrett T, Stacpoole P, Yost R (2009) Imaging of lipids in spinal cord using intermediate pressure matrix-assisted laser desorption-liner ion trap/orbitrap MS. Anal Chem 81:8488–8496. doi:10.1021/ac901387u
120. Végvári Á, Fehniger T, Gustavsson L, Nilsson A, Andrén P, Kenne K, Nilsson J, Laurell T, Marko-Varga G (2010) Essential tactics of tissue preparation and matrix nano-spotting for successful compound imaging mass spectrometry. J Proteomics 73:1270–1278. doi:10.1016/j.jprot.2010.02.021
121. Shimma S, Sugiura Y, Hayasaka T, Zaima N, Matsumoto M, Setou M (2008) Mass imaging and identification of biomolecules with MALDI-QIT-TOF based system. Anal Chem 80:878–885. doi:10.1021/ac071301v
122. Nilsson A, Fehniger T, Gustavsson L, Andersson M, Kenne K, Marko-Varga G, Andrén P (2010) Fine mapping the spatial distribution and concentration of unlabeled drugs within tissue micro-compartments using imaging mass spectrometry. PLoS One 5(7):1–8
123. Snel M, Fuller M (2010) High-spatial resolution matrix-assisted laser desorption ionization imaging analysis of glucosylceramide in spleen sections from a mouse model of gaucher disease. Anal Chem 82(9):3664–3670. doi:10.1021/ac902939k
124. Liu Q, Guo Z, He L (2007) Mass spectrometry imaging of small molecules using desorption/ionization on silicon. Anal Chem 79:3535–3541. doi:10.1021/ac0611465
125. Francese S, Dani F, Traldi P, Mastrobuoni G, Pieraccini G, Moneti G (2009) MALDI mass spectrometry imaging, from its origins up to today: the state of the art. Comb Chem High Throughput Screen 12:156–174
126. Schwartz S, Reyzer M, Caprioli R (2003) Direct tissue analysis using matrix-assisted laser desorption/ionization mass spectrometry: practical aspects of sample preparation. J Mass Spectrom 38:699–708

127. Agar N, Yang H, Carroll R, Lack P, Agar J (2007) Matrix solution fixation: histology-compatible tissue preparation for MALDI mass spectrometry imaging. Anal Chem 79(19):7416–7423. doi:10.1021/ac071460e
128. Leinweber B, Tsaprailis G, Monks T, Lau S (2009) Improved MALDI-TOF imaging yields increased protein signals at high molecular mass. J Am Soc Mass Spectrom 20:89–95. doi:10.1016/j.jasms.2008.09.008
129. Altelaar A, Klinkert I, Jalink K, de Lange R, Adan R, Heeren R, Piersma S (2006) Gold-enhanced biomolecular surface imaging of cells and tissue by SIMS and MALDI mass spectrometry. Anal Chem 78:734–742. doi:10.1021/ac0513111
130. Goto-Inoue N, Hayasaka T, Zaima N, Kashiwagi Y, Yamamoto M, Nakamoto M, Setou M (2010) The detection of glycosphingolipids in brain tissue sections by imaging mass spectrometry using gold nanoparticles. J Am Soc Mass Spectrom 21:1940–1943. doi:10.1016/j.jasms.2010.08.002
131. Baluya D, Garrett T, Yost R (2007) Automated MALDI matrix deposition method with inkjet printing for imaging mass spectrometry. Anal Chem 79:6862–6867. doi:10.1021/ac070958d
132. Chen Y, Allegood J, Liu Y, Wang E, Cachón-González B, Cox T, Merrill A, Sullards C (2008) Imaging MALDI mass spectrometry using an oscillating capillary nebulizer matrix coating system and its application to analysis of lipids in brain from a mouse model of Tay-sachs/sandhoff disease. Anal Chem 80:2780–2788
133. Schwamborn K, Krieg R, Reska M, Jakse G, Knuechel R, Wellmann A (2007) Identifying prostate carcinoma by MALDI-imaging. Int J Mol Med 20:155–159
134. Puolitaival S, Burnum K, Cornett D, Caprioli R (2008) Solvent-free matrix dry-coating for MALDI imaging of phospholipids. J Am Soc Mass Spectrom 19:882–886. doi:10.1016/j.jasms.2008.02.013
135. Goodwin R, Scullion P, MacIntyre L, Watson D, Pitt A (2010) Use of a solvent-free dry matrix coating for quantitative matrix-assisted laser desorption ionization imaging of 4-bromophenyl-1,4- diazabicyclo(3,2,2)nonanne-4-carbonxylate in rat brain and quantitative analysis of the drug from laser microdissected tissue regions. Anal Chem 82:3868–3873. doi:10.1021/ac100398y
136. Hankin J, Barkley R, Murphy R (2007) Sublimation as a method of matrix application for mass spectrometric imaging. J Am Soc Mass Spectrom 18:1646–1652. doi:10.1016/j.lasms.2007.06.010
137. Taira S, Sugiura Y, Moritake S, Shimma S, Ichiyanagi Y, Setou M (2008) Nanoparticle-assisted laser desorption/ionization based mass imaging with cellular resolution. Anal Chem 80(12):4761–4766. doi:10.1021.ac800081z
138. Bouschen W, Spengler B (2007) Artifacts of MALDI sample preparation investigated by high-resolution scanning microprobe matrix-assisted laser desorption/ionization (SMALDI) imaging mass spectrometry. Int J Mass Spectrom 266:129–137. doi:10.1016/j.ijms.2007.07.017
139. Aerni H, Cornett D, Caprioli R (2006) Automated acoustic matrix deposition for MALDI sample preparation. Anal Chem 78:827–834. doi:10.1021/ac051534r
140. Rohner T, Staab D, Stoeckli M (2005) MALDI mass spectrometric imaging of biological tissue sections. Mech Ageing Dev 126:177–185. doi:10.1016/j.mad.2004.09.032
141. Stoeckli M, Staab D, Staufenbiel M, Wiederhold K, Signor L (2002) Molecular imaging of amyloid β peptides in mouse brain sections using mass spectrometry. Anal Biochem 311: 33–39. doi:10.1016/S0003-2697(02)00386-X
142. Klerk L, Broersen A, Fletcher I, Liere R, Heeren R (2007) Extended data analysis strategies for high resolution imaging MS: new method to deal with extremely large image hyperspectral datasets. Int J Mass Spectrom 260:222–236. doi:10.1016/j.ijms.2006.11.014
143. Nemes P, Barton A, Vertes A (2009) Three-dimensional imaging of metabolites in tissue under ambient conditions by laser ablation electrospray ionization mass spectrometry. Anal Chem 81:6668–6675. doi:10.1021/ac900745e
144. Chen R, Hui L, Sturm R, Li L (2009) Three dimensional mapping of neuropeptides and lipids in crustacean brain by mass spectral imaging. J Am Soc Mass Spectrom 20:1068–1077. doi:10.1016/j.jasms.2009.01.017

145. Sinha T, Khatib-Shahidi S, Yankeelov T, Mapara K, Ehtesham M, Cornett D, Dawant B, Caprioli R, Gore J (2008) Integration spatially resolved three-dimensional MALDI IMS with in vivo magnetic resonance imaging. Nat Methods 5(1):57–59. doi:10.1038/NMETH1147
146. Norris J, Cornett D, Mobley J, Andersson M, Seeley E, Chaurand P, Caprioli R (2007) Processing MALDI mass spectra to improve mass spectral direct tissue analysis. Int J Mass Spectrom 260:212–221. doi:10.1016/j.ijms.2006.10.005
147. Berman E, Fortson S, Checchi K, Wu L, Felton J, Wu K, Kulp K (2008) Preparation of single cells for imaging/profiling mass spectrometry. J Am Soc Mass Spectrom 19:1230–1236. doi:10.1016/j.jasms.2008.05.006
148. Rubakhin S, Greenough W, Sweedler J (2003) Spatial profiling with MALDI MS: distribution of neuropeptides within single neurons. Anal Chem 75:5374–5380. doi:10.1021/ac034498+
149. Zimmerman T, Rubakhin S, Sweedler J (2011) MALDI mass spectrometry imaging of neuronal cell cultures. J Am Soc Mass Spectrom 22:828–836. doi:10.1007/s13361-011-0111-2
150. Yang Y, Xu Y, Straight P, Dorrestein P (2009) Translating metabolic exchange with imaging mass spectrometry. Nat Chem Biol 5(12):885–887. doi:10.1038/nchembio.252
151. Jurchen J, Rubakhin S, Sweedler J (2005) MALDI-MS imaging of features smaller than the size of the laser beam. J Am Soc Mass Spectrom 6:1654–1659. doi:10.1016/j.jasms.2005.06.006
152. Li F, Hsieh Y, Kang L, Sondey C, Lachowicz J, Korfmacher W (2009) MALDI-tandem mass spectrometry imaging of astemizole and its primary metabolites in rat brain sections. Bioanalysis 1(2):299–307. doi:10.4155/BIO.09.27
153. Bouslimani A, Bec N, Glueckmann M, Hirtz C, Larroque C (2010) Matrix-assisted laser desorption/ionization imaging mass spectrometry of oxaliplatin derivatives in heated intraoperative chemotherapy (HIPEC)-like treated rat kidney. Rapid Commun Mass Spectrom 24:415–421. doi:10.1002/rcm.4408
154. Bunch J, Clench M, Richards D (2004) Determination of pharmaceutical compounds in skin by imaging matrix-assisted laser desorption/ionization mass spectrometry. Rapid Commun Mass Spectrom 18:3051–3060. doi:10.1002/rcm.1725
155. Stoeckli M, Staab D, Schweitzer A, Gardiner J, Seebach D (2007) Imaging of a β-peptide distribution in whole-body mice sections by MALDI mass spectrometry. J Am Soc Mass Spectrom 18:1921–1924. doi:10.1016/j.jasms.2007.08.005
156. Reyzer M, Hsieh Y, Ng K, Korfmacher W, Caprioli R (2003) Direct analysis of drug candidates in tissue by matrix-assisted laser desorption/ionization mass spectrometry. J Mass Spectrom 38:1081–1092. doi:10.1002/jms.525
157. Hayasaka T, Goto-Inoue N, Zaima N, Shrivas K, Kashiwagi Y, Yamamoto M, Nakamoto M, Setou M (2010) Imaging mass spectrometry with silver nanoparticles reveals the distribution of fatty acids in mouse retinal sections. J Am Soc Mass Spectrom 21:1446–1454. doi:10.1016/j.jasms.2010.04.005
158. Atkinson S, Loadman P, Sutton C, Patterson L, Clench M (2007) Examination of the distribution of the bioreductive drug AQ4N and its active metabolite AQ4 in solid tumors by imaging matrix-assisted laser desorption/ionization mass spectrometry. Rapid Commun Mass Spectrom 21:1271–1276. doi:1002/rcm.2952
159. Earnshaw C, Carolan V, Richards D, Clench M (2010) Direct analysis of pharmaceutical tablet formulations using matrix-assisted laser desorption/ionization mass spectrometry imaging. Rapid Commun Mass Spectrom 24:1665–1672. doi:10.1002/rcm.4525

A Planar Integrated Micro-mass Spectrometer

Jörg Müller, Grigoriy Quiring, Maria Reinhardt-Szyba, Régulo Miguel Ramírez Wong, and Henning Wehrs

Abstract A planar fully integrated micro-mass spectrometer fabricated in a full wafer based state-of-the-art MEMS technology in a glass–silicon–glass sandwich is presented. Within a volume of $7 \times 10 \times 1.3$ mm³ it contains all components of a mass spectrometer, i.e., a microwave plasma electron source for ionization, an ionization chamber, the electron and ion extraction, acceleration and focusing electrodes, a new type of mass separator, a Faraday detector as well as structures for the pressure management within the system for analyte, plasma gas, optics, and mass separation. Also a spring arrangement to insert a self-aligning and contacting microchannel plate (MCP) is included. The complete system is transferred from one single photolithographic mask layer into a 2½ dimensional structure in a silicon substrate by ICP-etching. The designs of the subsystems, especially that of a new type of separation principle, are presented and the layout of the injection system and the batch processing of the device are outlined. A completely newly developed hardware and software of the electronics to drive the system is presented including its physical layout and operational scheme. Actual spectra obtained with the system demonstrate a mass resolution of 43 in a mass range of 0.5–200 and a sensitivity of <100 ppm. Means to adapt the size of the periphery like vacuum pumps, inlet pressure stages, and handling of liquid analytes, which would allow for a really handheld device, conclude the contribution.

J. Müller (✉) • G. Quiring • M. Reinhardt-Szyba • R.M. Ramírez Wong • H. Wehrs
Institute for Microsystem Technology, Hamburg University of Technology, Hamburg, Germany
e-mail: j.mueller@tu-harburg.de; gregor.quiring@tu-harburg.de; reinhardt@tu-harburg.de; regulo.ramirez@tu-harburg.de; henning.wehrs@tu-harburg.de

Q. Alan Xu and Timothy L. Madden (eds.), *LC-MS in Drug Bioanalysis*,
DOI 10.1007/978-1-4614-3828-1_14,

1 Introduction

Mass spectroscopy is an extremely powerful technique to analyze gases and vaporized liquids or solids both with respect to mass as well as concentration. In combination with gas chromatography such systems are the core of the most selective and sensitive microanalytical systems. Therefore, their application ranges from sophisticated laboratory experiments with extreme resolution in mass and concentration to rather robust applications, e.g., in process monitoring. Furthermore, they could be powerful analytical systems in patient monitoring, healthcare, or biotechnology, and harsh environments like under water, space, and environmental analysis. In spite of this wide range of possible useful and valuable applications in reality their use is still rather restricted due to the fact that present systems need extended resources with respect to volume, power, infrastructure, and specialist knowledge and are rather sensitive in handling. Standard systems due to their mostly extended dimensions require low pressures in the 10^{-3} Pa range and a correspondingly extensive vacuum equipment. Thus, the application of mass spectrometers is presently restricted to a laboratory environment and the few portable devices are still rather clumsy.

In recent years a variety of attempts were undertaken to generate miniaturized microanalytical systems like gas chromatographs [1, 2], flame ionization detectors [3, 4], chemical and gas sensors [5], and the like. Using the very powerful technologies of microfabrication, which offer a number of well-established, reliable, low cost, and mass production proven processes, very sophisticated, complex, and highly diverse structures can be fabricated in large volume. Typical examples of this technology are pressure and acceleration sensors, positioning sensors, or electronic compasses, e.g., for automotive and mobile phone applications. In this context also several approaches have been undertaken to generate mass spectrometers as a microsystem. This research mostly, however, is concentrated on the generation and optimization of one or the other of the various components of a standard mass spectrometer, i.e., chambers to ionize the sample, ionization sources—commonly accelerated electrons emitted from an electron source—ion focusing and acceleration sections, mass separators, and lastly detectors, in many cases containing a multiplier stage and a Faraday cup. Different types of mass separators have been miniaturized, e.g., ion traps [6], time of flight separators [7, 8], and several designs of quadrupole mass separators [9, 10] and also miniaturized electron sources like thermal emitters [11] and carbon nanotube cold emitters [12] have been proposed. Mostly these devices are just miniaturized versions of their macro sized counterparts. Such isolated microcomponents are then usually introduced into a standard mass spectrometer environment, i.e., a real micro-mass spectrometer is not realized this way.

A closer look into the origin of this limited success of microtechnology in this field is, that either technologies like LIGA (German acronym for Lithographie, Galvanoformung, Abformung—Lithography, Electroplating, and Molding [13]) or rather simple silicon bulk micromachining technologies are used, which significantly restrict available geometries, features, and also electrical functions—e.g., highly conductive metal or silicon structures, which may cause large RF-losses (Radiofrequency) due to capacitive coupling, and limitations in applying appropriate electrical potentials. In addition complex and mostly individualized 3D-assembly procedures are finally necessary to fix the systems. This opposes the principle advantage of micromachining on fine machining processes. On the other hand, however, mostly positive results with respect to system performance of such not real microsystem mass spectrometer designs can be demonstrated rather quickly, also because most of the periphery including the electronics is standard. Unfortunately such designs are far off the performance a full microsystem approach would allow, since interfaces between micro- and macrodesigns are problematic with respect to size accommodation, dead volumes, and adjustment, the vacuum as well as electric and electronic. Furthermore, volume requirements are hardly improved and the cost of a complete mass spectrometer keeps the same or is even higher due to the microcomponent, since the effort for its fabrication, assembly, and integration is comparable to or even larger than that of fine machined parts and lot size is low.

The intention of this contribution is to demonstrate that such a micro-mass spectrometer can be realized with features which in part resemble that of state-of-the-art spectrometers for standard applications closely and that the state of the art of this system is at the verge to make it a commercial, easily to handle analytical system.

To allow the reader to understand the philosophy behind this development and the physical and technological challenges and already obtained results, the text is divided into the following subsections: Sect. 2 shortly overviews the basic system design. In Sect. 3 the simulation and derived design of the different subsystems of the micro-mass spectrometer are presented and how they entwine to form the complete mass spectrometer. The subsystem to manage the sample and plasma gas pressure and flow is discussed in Sect. 4. After a description of the applied basic microsystem technologies and the process flow to fabricate the system in Sect. 5, Sect. 6 summarizes the main features of the electronics in hardware and software to drive and control the system automatically, extract the raw data, and evaluate them to generate calibrated spectra. In Sect. 7 experimental results for operating parameters, mass resolution and spectral width, selectivity, sensitivity, and characteristic time constants are given and typical spectra are presented and compared to the simulations. The contribution ends with a summary and outlook for work presently in progress and still to do in Sect. 8 and finishes with an acknowledgment.

2 The Planar Integrated Micro-mass Spectrometer

In order to avoid the limitations of macro scale mass spectrometers and those where just single subsystems of such standard systems are exchanged for microdesigns, as mentioned in the introduction, a fully integrated mass spectrometer containing all the necessary subsystems was designed. It can be fabricated "in one step," i.e., all the different components of a mass spectrometer are aligned and confined in one mask layer. For its fabrication well established micromachining processes are applied, which allow to transfer the fabrication directly into an industrial environment without substantial engineering efforts and time delay. This approach also allows to get around the cited shortcomings of a hybrid assembly and adjustment to a "macro"-periphery.

To follow this approach subsystem layout and functions as well as microfabrication processes have to be selected, which are apt to generate and integrate the various subsystems of a mass spectrometer into one substrate in one batch process, i.e., all the components are defined in their geometry and orientation among each on one photo mask, which is then transferred into features in the substrate. Thus, submicron accuracy of device features and their orientation and adjustment can reproducibly be guaranteed, i.e., many identical devices can be fabricated simultaneously in one run. This, however, also means that all the subsystems are fabricated in the same material within a narrow range of the necessarily different dimensions.

Such an approach is evidently inherently rather challenging, since mostly due to the restriction to a 2½-D geometry in a single material at least part of the subsystems will need a complete redesign as compared to established mass spectrometer geometries or even the introduction of alternative physical principles with completely new structures. Furthermore, all subsystems must be compatible with respect to fabrication, size, function, and pressure regime as well as electrical interfaces. Finally such a micro-mass spectrometer will need a modified and adapted hardware and software of the electronics—which actually means a completely new one.

Another difficulty is that all these subsystems have to function at once, i.e., the development, characterization, and optimization of the individual subsystems is limited, and there is no way of a mechanical fine tuning. This means that the total system has to be designed on very reliable simulations and pattern transfer, which comprise not only the geometrical features for potentials and trajectories of electrons and ions, but also the behavior of the mass separator or pressure regimes in different areas of the system. Such an approach may appear rather adventurous and risky; at least it may and will take much more time to generate a working system than the demonstration of a single subsystem implemented into a standard mass spectrometer environment.

When successful, however, the full advantage of microsystem design can be utilized, i.e., low power consumption, small volume and weight, low cost, cheap, and reproducible mass production, as well as low vacuum requirements, portability, and even long-term application on battery or even energy harvesting power supply can be envisioned.

Figure 1 shows the actual planar integrated micro-mass spectrometer (PIMMS). It is fabricated as a three wafer glass–silicon–glass sandwich with all the relevant structures realized in a highly doped silicon wafer via a deep reactive ion etch (DRIE) process, which generates 2½-D structures, i.e., deliberate structures in the plane and vertical walls with depths of several 100 μm. All features of the mass spectrometer are contained in one photo mask layer. The thickness of the Pyrex glass wafers is 500 μm, that of the double side polished silicon wafer

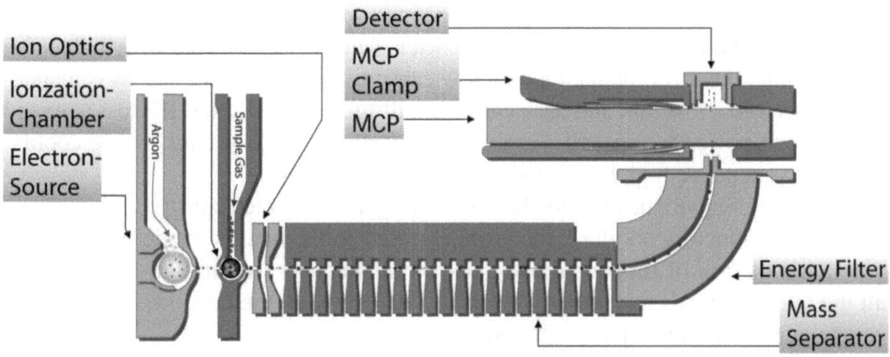

Fig. 1 The planar integrated micro-mass spectrometer (PIMMS)

300 µm. The final chip size presently is $7 \times 10 \times 1.3$ mm³ ($w \times l \times h$), i.e., about 0.1 cm³. Typically about 100 complete mass spectrometers are fabricated in parallel on a 100 mm diameter wafer with a high yield even in a laboratory fabrication environment, which would be 250 on a 150 mm wafer, the present standard of the "microsystem" state of the art.

The silicon wafer is attached to two Pyrex glass plates on both sides via anodic bonding. This encapsulation allows sealing silicon to glass hermetically without an additional interface layer. The bottom glass wafer also carries the electrical conductors to apply the potentials and electric signals to the silicon electrodes and together with the top glass wafer forms vacuum tight seals to install channels and chambers of different pressure regimes as well as guiding structures for the gas capillaries to supply the sample and process gases.

As already mentioned the planar integration forces all the subsystems needed for the spectrometer to be in one plane and the available structures are restricted to a 2½-D geometry, i.e., only subsystem designs are allowed, which can be realized in such a geometry. Furthermore, the devices are to be fabricated by standard microfabrication processes on wafer level, such as masking by photolithography, deep reactive ion etching, wet chemical etching, thin film deposition, and full wafer anodic bonding. In contrast structures typically used in common mass spectrometer designs are mostly cylindrical in shape.

The basic functional blocks, which a mass spectrometer has to comprise, are an ionizer for the sample gas, a source for the ionization, an ion extractor, optic, and accelerator, a mass separator and ion detector mostly following a multiplier to amplify the ion current and improve the signal to noise ratio (SNR). All surfaces and components exposed to the sample gas should be inert and chemically stable, not to limit the type of chemistry to be analyzed. As these subsystems partly function in very different pressure regimes "pressure shields," i.e., pressure stages have to be incorporated.

Then such a system can be operated at one base pressure, which due to the small volume of the device can be generated by a small vacuum pump. Since a wide range of different, especially also very high voltages and very high frequencies, respectively, have to be applied to the different subsystems, isolated electric wiring with high isolation and shielding in combination with low capacitive loading and low loss is mandatory. Otherwise cross talk and dark currents will limit the SNR and RF-losses would require high power drivers and—especially critical for microsystems, where cooling is limited—generate high losses and cause intolerable heating.

As will be shown, the chosen glass–silicon–glass sandwich construction with the electrical conductors on glass, the areas exposed to the analyte made of silicon or glass, and the use of functional elements simultaneously for different tasks fulfills all the cited demands and, due to their special design, all the necessary subsystems can be generated in one patterning process.

Before the design, simulation, fabrication, pressures, potentials, electronics, and software are discussed in detail in the sections to follow, an overview on the basic functional principles of the system is given. According to Fig. 1 starting from left to right the system comprises electron generation and acceleration including an electron optic, analyte ionization and its supply, ion extractor, optic, and accelerator, mass separator, energy filter, and the detector with an optional amplifier.

Electrons for the ionization of the sample gas are generated in a microwave (2.45 GHz) argon plasma burning in a chamber with a diameter of 150 μm at a pressure of about 100 Pa. Argon is supplied from a small bottle at a pressure of 6 bar via a capillary which, in conjunction with the electron extraction gap in the chamber, reduces the pressure to the desired 100 Pa. The plasma is ignited by a spark generated by a high voltage pulse between two integrated platinum electrodes on the bottom glass.

The electrons are extracted through a slit in the chamber wall, the anode, that simultaneously acts as electron optic and a pressure stage between the 100 Pa in the plasma chamber and <1 Pa outside, to which the mass spectrometer chip is evacuated in an appropriate housing.

The electrons are accelerated across this low pressure region between the plasma chamber and the adjoining ionization chamber to typically 100 eV, the optimum for the ionization of most gases. They are focused to the input slit into ionization chamber, into which the sample gas is introduced via another capillary, alternatively via a microchannel chip, which reduces the pressure from atmospheric again to about 15 Pa, the pressure for maximum ionization efficiency for the chamber diameter of also 150 μm. Analyte ions are extracted from the ionization chamber through a second slit on the opposite side into an ion optic and are then accelerated into the mass separator to an energy of typically 100 eV.

This mass separator is a new type, especially accommodated to be compatible with the 2½-D geometries available in this fabrication process. The "synchronous ion shield (SIS)" analyzer selects the mass, for which the filter is transparent, by a traveling high frequency rectangular pulse stream (rise and fall time <1 ns) of a swept frequency (0–270 MHz), which is supplied to a comb shaped electrode arrangement.

According to Fig. 2, which depicts the electric field between the electrodes (top) perpendicular to the ion trajectory, ions which have the same speed as the pulse

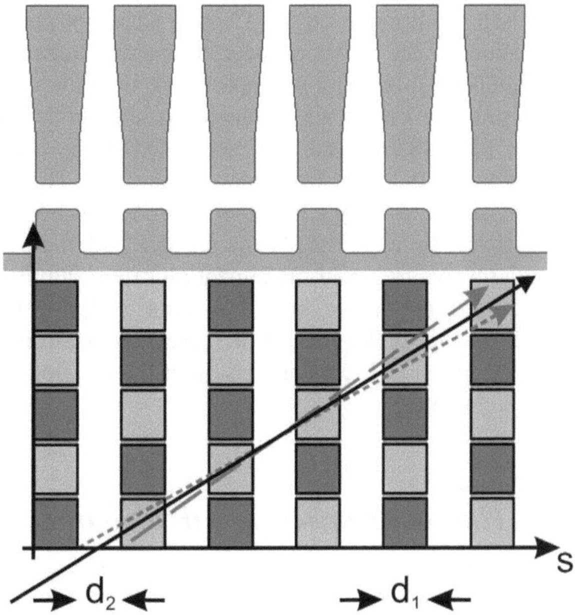

Fig. 2 Principle of mass separation in the SIS separator

stream will travel ballistically through the filter, as they pass a dynamically field free volume, ions which are faster (lower mass) or slower (higher mass) are subjected to a field perpendicular to the trajectory and are deflected towards the electrodes. By changing the pulse stream frequency ions of the same energy are selected according to their velocity, i.e., mass.

The separator is followed by a 90 ° energy filter, which selects the right energy-mass combination via an output slit. The selected ions finally hit the detector, which is a Faraday cup. Alternatively between the detector and the output of the energy filter a MCP is hybridly integrated by inserting it into a self-adjusting and contacting integrated spring arrangement, which is also fabricated in this process.

The total system is evacuated to a pressure of <1 Pa through the sideways open structures in the glass–silicon–glass sandwich. For electrical contact the chip is presently bonded to a PCB, which also serves as vacuum seal.

3 Simulation and Layout

3.1 Overview

For the layout of the PIMMS-structures a variety of boundary conditions and limitations have to be considered with respect to both the physical principles as well as technological and size restrictions. Since the functions of the different PIMMS-elements are based on different physical effects, it is not possible to use one all-embracing simulation for the layout of these structures. Consequentially the PIMMS-elements are simulated in parallel and optimized, with the interfaces adapted to each other. The results are verified via the characterization of the complete system.

The PIMMS-elements can be divided in the fields of fluidic, plasma physics, electrostatic, and high frequency technology. The fluidic part covers the supply with plasma and sample gas, their distribution inside, and their evacuation out of the system. By means of plasma physics both the ionization of the plasma gas by the microwave field and the ionization of the sample gas by impact ionization are described. Electrostatic theory governs to design the elements for acceleration, focusing, and deflection of the plasma electrons and the sample gas ions. Electromagnetic wave theory describes the creation of a microwave-field inside the plasma chamber.

3.2 *Plasma Chamber: Electron Source*

The electron source of the PIMMS is an argon plasma. Inside the plasma chamber the gas is ionized by a 2.45 GHz microwave field, ignited by an electric spark. In the plasma chamber free electrons are created, that are accelerated by a static electric field for impact ionization of the sample gas atoms. The layout of the plasma chamber has to incorporate both the fluidic and the electrostatic requirements. On the one hand the gas apertures of the chamber must have the appropriate dimensions to assure that the gas flow out of the chamber is low. On the other hand the geometry must be such that most of the electrons are generated close to the outlet of the chamber and can be extracted through this small aperture. Electrons should be generated close to the acceleration field, which intrudes the chamber only to a small depth.

Furthermore, the geometry must be such that two electrodes can be placed at a distance that allows a spark generation in the available pressure range inside the chamber to ignite the plasma [14].

The layout of the plasma chamber is optimized for a high electron yield at a low gas flow. Special attention was paid to obtain a low reflected power under fluctuating pressure conditions. This is important to protect the RF-generator and to ensure a stable plasma, which is crucial for a reliable ionization rate of the sample gas.

Figure 3 shows the resulting plasma chamber. Its diameter is 600 μm and its outlet aperture is 40 μm wide.

Besides the silicon microwave electrodes two metal electrodes on the bottom substrate for spark ignition of the plasma are depicted.

As previously mentioned the electrode geometry in particular affects the electron beam, since the extracting field intrudes the chamber only to a limited depth. During operation of the RF-plasma the electrodes are biased due to the different mobility of ions and electrons. This biasing depletes electrons close to the electrodes, i.e., the chamber walls. In this “dark space” or “sheath” no secondary electrons are created.

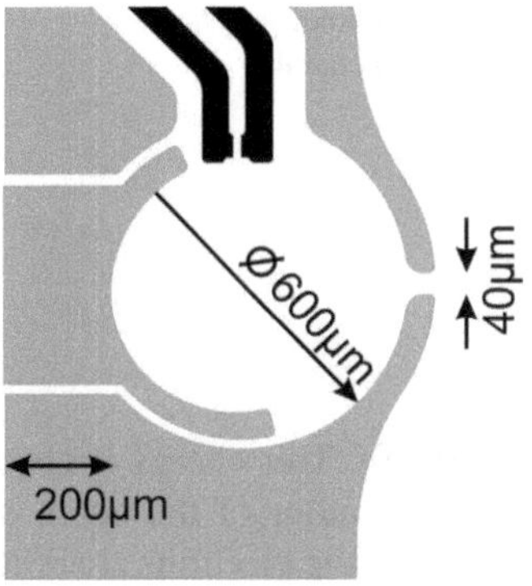

Fig. 3 Optimized plasma chamber design with reduced side gaps, optimized electrode surface ratio and without bottom metallization

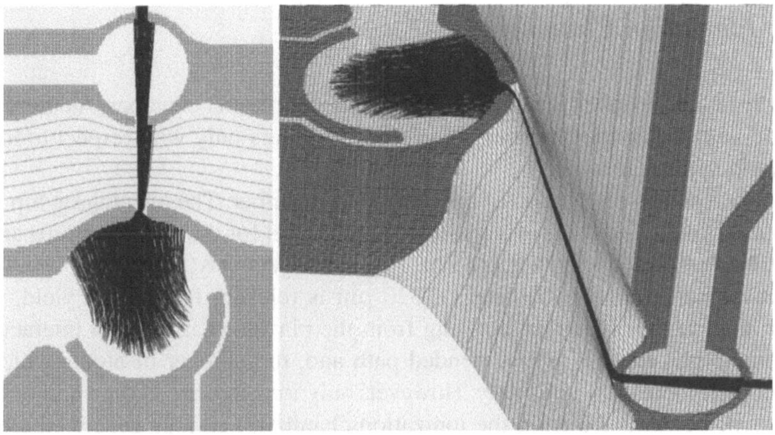

Fig. 4 Simulated electron trajectories between plasma and ionization chamber

The potential difference in this region and therewith the extension of the sheath for not too different electrode areas can be approximated by [15–17]:

$$\frac{U_1}{U_2} = \left(\frac{A_2}{A_1}\right)^4.$$

The sheath potential is minimized, when the electrode area ratio equals 1. The design of Fig. 3 fulfills this requirement. The lowest possible pressure, for which a stable plasma can be established, is 30 Pa.

For this layout of the plasma chamber the trajectories of the electrons as well as the ions in an applied electrical field were simulated with the SIMION 8.0 software [18]. Figure 4 depicts the extraction of electrons out of the plasma chamber for a potential difference of 100 V between the plasma and the ionization chamber [19] at vacuum conditions, i.e., without electron–gas interaction. The corresponding electron energy

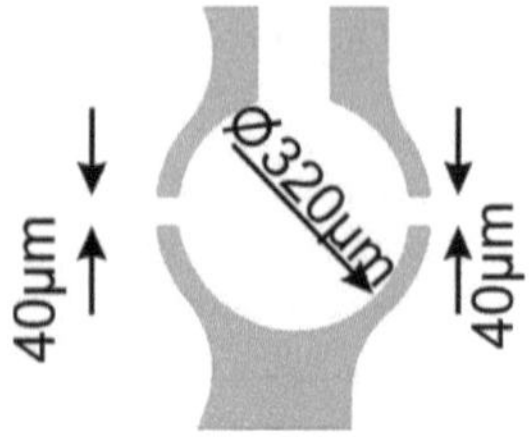

Fig. 5 Geometry of the ionization chamber

is 100 eV. This energy is about the optimum for a maximum ionization yield. Due to the contour of the outer walls of both the plasma and the adjacent ionization chamber, the electrons are efficiently focused into the ionization chamber orifice.

3.3 Ionization Chamber

The ionization chamber is of similar design as the plasma chamber. As there is no electric field inside the chamber, its geometry matters only with respect to ionization yield and gas flow.

Figure 5 shows the outline of the ionization chamber. The opening on top is the sample gas inlet. The other two apertures are the inlet for the electrons (left) and the outlet for the ionized sample gas atoms (right), respectively.

The chosen chamber diameter of 320 μm is relevant for the ion yield. For a larger diameter the electrons entering from the plasma chamber can interact with the sample gas along a more extended path and, the number of atoms which are ionized will increase accordingly. However, only ions that are in the vicinity of the outlet will be extracted from the ionization chamber, i.e., ions generated close to the electron entrance have to travel to the exit aperture to be exposed to the extracting field. A small chamber diameter therefore is advantageous. A compromise with respect to the probability of ionization and ion extraction efficiency will generate the maximum ion current [20].

Besides the chamber diameter, the width of the apertures is an important structural parameter. It affects the density (pressure) of the sample gas and the overall gas flow into the system. Ion trajectory simulations identified an optimum ion yield for entrance and exit orifices of 40 μm and a chamber diameter of 320 μm [19], as experimentally validated in the real systems.

Figure 6 demonstrates generation of ions due to the incoming electrons (left) and the extraction of the ions out of the chamber biased at +100 V by the first pair of electrodes of the ion optics (right).

3.4 Ion Extraction and Ion Focus Electrodes

The ion extraction and focus electrode arrangement comprises two pairs of electrodes located between the ionization chamber and the mass separator. Their function

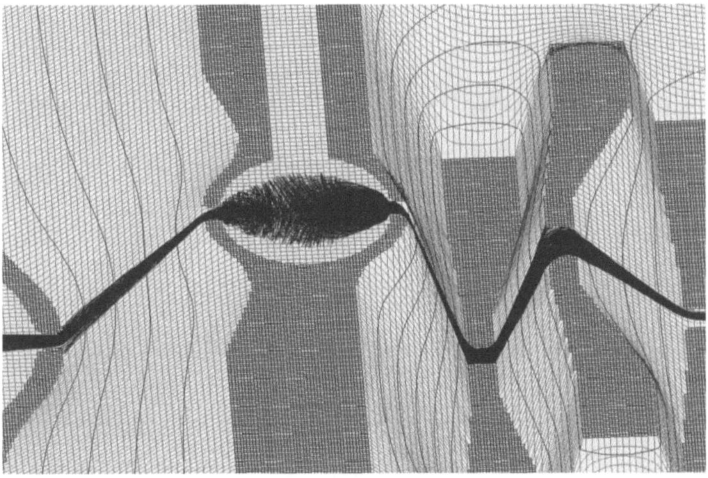

Fig. 6 Simulated potential of the ionization chamber and ion extraction

Fig. 7 Ion extraction and focus optic

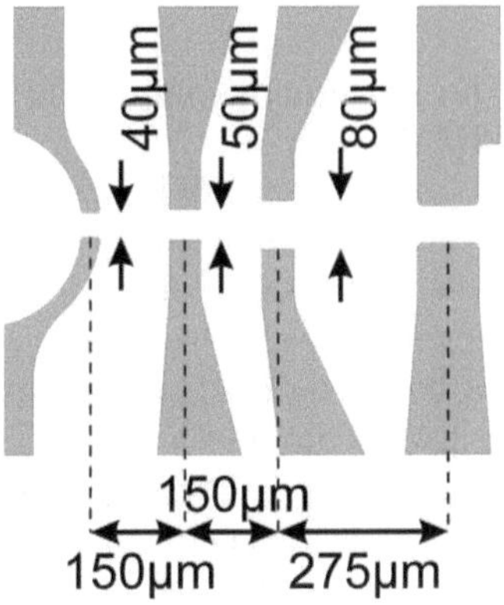

is to extract and generate a concentrated parallel ion beam of defined energy. Accordingly, the first pair of electrodes (IE) extracts and accelerates the ions (see above), whereas the second pair (IFO) decelerates and thereby focuses them [21]. The design is similar to conventional mass spectrometers [22].

Figure 7 shows the structures of the ion optics with the ionization chamber on the left and the entrance of the mass-analyzer on the right, which are described in the next section. The arrangement and shape of these structures were also computed via ion trajectory simulations. A result is shown in Fig. 8. At an extraction voltage (IE) of −25 V and separator input voltage (WFP) of 0 V an ion focus voltage (IFO) of 67 V generates the highest ion yield for the separator.

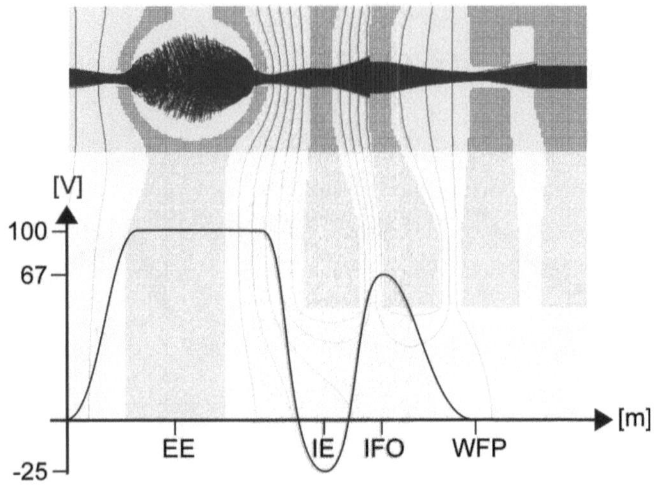

Fig. 8 Simulation of the ion optics (*top*) potential in the ion optics (*bottom*)

3.5 Synchronous Ion Shield (SIS) Mass Analyzer

The synchronous ion shield (SIS) mass analyzer is the core component of the PIMMS [23].

The upper part of Fig. 9 shows a schematic of this analyzer. The working principle is related to time-of-flight filters, yet the SIS analyzer is operated in a high frequency, rectangular pulse voltage mode. This voltage is applied perpendicular to the flight direction between the finger electrodes and the comb structure. The voltage is alternating both in terms of time and space, which means that neighbored finger electrodes are actuated at inverted voltages. This results in an electrical field which is traveling in the same direction as the ion beam. When its velocity matches the ion velocity, a number of ions can travel through the analyzer inside the momentarily field-free regions.

The lower part of Fig. 9 shows the time–space diagram of the SIS analyzer, which resembles a chess pattern. The dark gray colored regions depict the channel area between the comb electrode and the "active" electrodes, where the ions are deflected sideways by the electric field. The light gray and white regions depict inactive fingers and gaps in between, respectively.

The mass analyzer works primarily as a velocity filter because the slope in the time–space diagram corresponds to the ion velocity. An ion moving on a line within the light gray and white areas will not be deflected and pass the analyzer, ions with other velocities will be subjected to an orthogonal electrical field and be deflected. Ions of different mass but equal energy are selected according to:

$$E = \frac{1}{2}mv^2,$$

due to their different velocity, i.e., slope in the diagram, by varying the frequency of the traveling pulse stream accordingly.

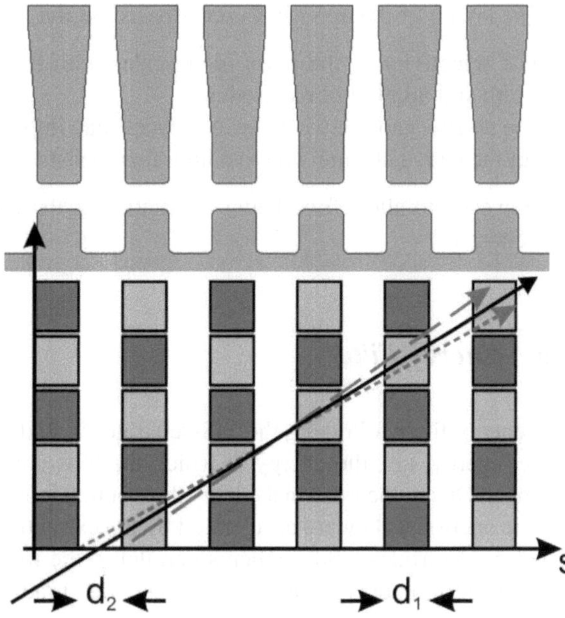

Fig. 9 Functional Principle of the Synchronous Ion Shield Mass Analyzer

Both the time and the space domain show field free (white) areas with respect to the actual potentials. In the time domain these effects emerge during switching. In the space domain the electric field in the gaps between the finger electrodes cannot be controlled fully. This is the main reason for some blurr in the filter process.

The relative mass resolution can be approximated as:

$$\frac{m}{\Delta m} = \frac{\left(A^2 - d_1^{\,2}\right)^2}{4d_2A\left(A - d_1^{\,2} - d_1d_2\right)},$$

A stands for the overall length of the filter. This formula is optimistic as it assumes that an ion that crosses a dark region is safely deflected. Experiments show that the actual selectivity is slightly lower than calculated.

This formula suggests a preferably long filter with small gaps between the electrodes, whereas the width of the fingers hardly affects the selectivity. According to theory there is no absolute optimum for the resolution, i.e., the more electrodes and the longer the separator, the higher will be the resolution. However, to stay within microsystem dimensions, reduce the effort for evacuation (dimensions < ion mean free path), and limit the loss of "straying" ions, an overall length of the filter of 4 mm was chosen. The minimum width for the gaps between the fingers is limited to 20 μm due to fabrication restrictions.

The layout of the mass analyzer is based on the following geometrical impacts:

1. The selectivity of the mass filter scales with its length and is affected by the width and number of electrodes.
2. Too narrow gaps between the electrodes may impair the removal of captured ions out the active channel due to a high flow resistance.

However, to allow for a better evacuation of the ion channel a width of 100 μm was chosen.

3.6 Energy Filter

The energy filter following the SIS-separator is to deflect ions that do not have the preset energy, i.e., the energy at which the ions are extracted from the ionization chamber. Due to their thermal energy distribution ions will start with a non-negligible kinetic energy as they are accelerated by the extraction field. Furthermore, between adjacent electrodes field gradients parallel to the ion movement will occur, which contribute to ion energy variations. These field gradients appear at the edges of the active finger electrodes and can either increase or decrease the ion energy.

An ion subjected to the electric field inside the semicircular shaped energy filter will be deflected towards the inner radius r_i of the filter. This deflecting force along with the initial kinetic energy of the ion results in a circular trajectory. For the cylindrical layout, the applied voltage U defines the ion energy that can pass the filter as [24]:

$$E_{kin}(U) = \frac{qU}{2\ln\left(\frac{r_i}{r_o}\right)},$$

r_o is the outer diameter of the filter channel. Ions that do not have the selected energy are deflected to the sidewalls and thus neutralized.

Two further important parameters have to be defined for the energy filter layout: the channel length and the width of the outlet aperture. The channel length is important because the ion beam enters the channel with a certain width. Since ions entering the channel close to the inner or outer radius are deflected differently, ions of the desired energy are focused at 63 °.

A good configuration therefore is when the focal point is in the center of the outlet aperture.

Figure 10 shows simulations of different energy filter configurations. In Fig. 10a the filter ends at the focal point which is exactly in the center of the outlet aperture. Hence a small outlet aperture is sufficient and a high selectivity is expected. If the filter is extended to 90 °, as it is usually the case in macroscopic systems (Fig. 10b)

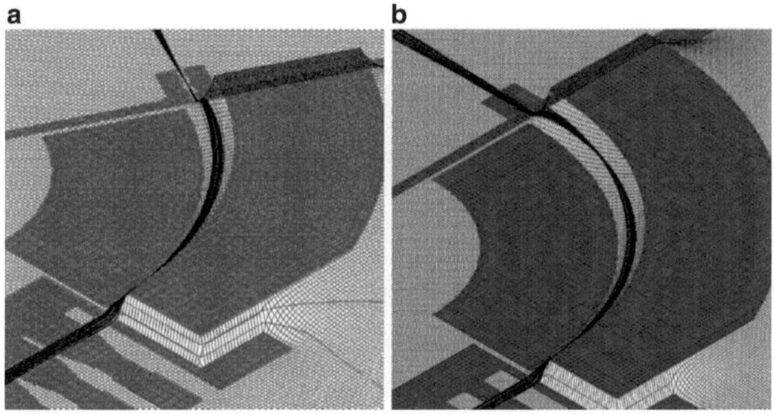

Fig. 10 Trajectory simulations of the ions inside the energy filter with a focal point in the center of the outlet aperture (**a**) and with a focal point before the outlet aperture (**b**)

the ion beam expands again and a wider outlet aperture is necessary and the selectivity is inferior. Nevertheless this 90 ° layout was chosen because it allows for a better utilization of the total system area.

3.7 Microchannel Plate Amplifier

To intensify the ion current, an amplification via electron multiplication is advantageous. For this purpose a MCP is introduced into the PIMMS. An MCP rectangular cut in shape from a standard MCP [25, 26] works as a secondary electron multiplier (SEM) as shown in Fig. 11. The MCP is hybrid-integrated into the spectrometer by clamping with flexible silicon springs 20 μm in thickness. They serve both as mechanical fixture and electrical contact. The MCP is operated between 400 and 1,300 V and amplifies the current by more than 1,000, which allows for measurements into the <100 ppm region.

3.8 Detector

The detector of the PIMMS is a U-shaped Faraday cup surrounded by guard electrodes.

Figure 12 shows the detector structure and the surrounding guard electrodes. The dark regions depict metalized structures on the bottom glass substrate; the light gray regions are silicon structures. The guard assures that no electrical field can build up in front of the detector, which would deteriorate the ion measurement. The detectable currents range from 0.5 to 200 pA.

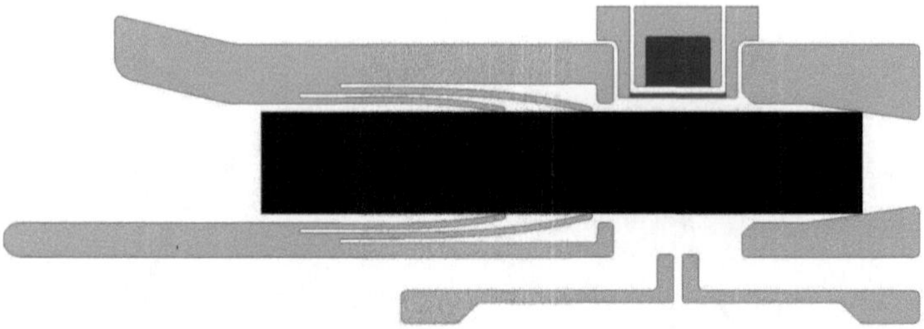

Fig. 11 Microchannel plate (MCP) hybrid-integrated in the PIMMS

Fig. 12 Faraday detector

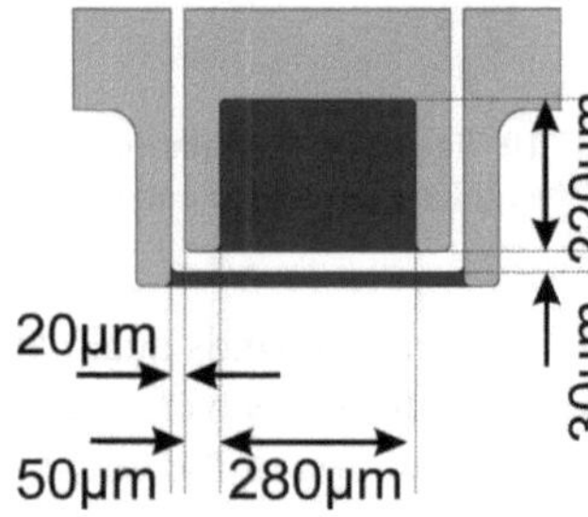

4 Injection System

For an adequate performance of the PIMMS it is necessary to install appropriate pressures and gas flows in the system, as well as a minimum time delay for the sample to reach the ionization chamber. The flow time of the sample gas governs the temporal resolution of the PIMMS measurements. Since the pressure of the plasma and sample gases are usually much higher at their origin than allowed in the plasma and the ionization chamber, respectively, there is a need for pressure reduction which will inevitably introduce a time delay.

Both plasma and sample gases are fed into the PIMMS system via glass capillaries as shown in Fig. 13. They not only physically connect the gas supplies to the PIMMS but also reduce the gas pressure via their flow resistance to the adequate operation range. Typically 10 cm long, 100 µm diameter capillaries are sufficient to induce the desired pressure reduction in the chambers. These capillaries are glued to the PIMMS.

As already mentioned the plasma is ignited by a high voltage discharge in the plasma chamber at a relatively high pressure. In continuous operation, a stable RF plasma will burn at much lower chamber pressures [13]. This requires switching between two different flow resistances, i.e., capillary lengths, which is accomplished by a four-way valve operated as shown in Fig. 14.

Due to the intricate layout of the PIMMS, the pressure distribution throughout the system is difficult to predict theoretically. Even though an integration of a pressure

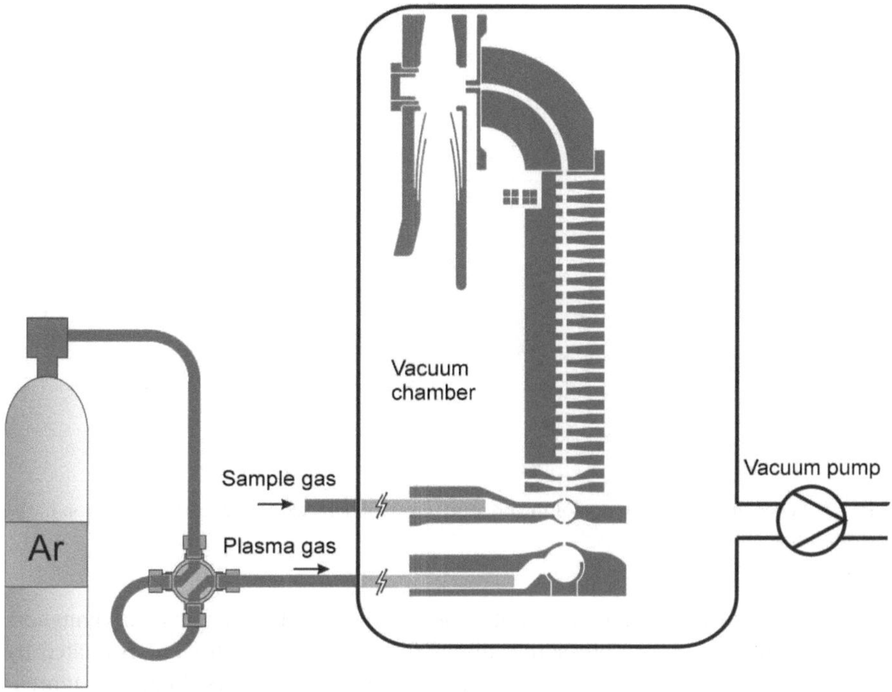

Fig. 13 Schematic representation of the injection system of the PIMMS

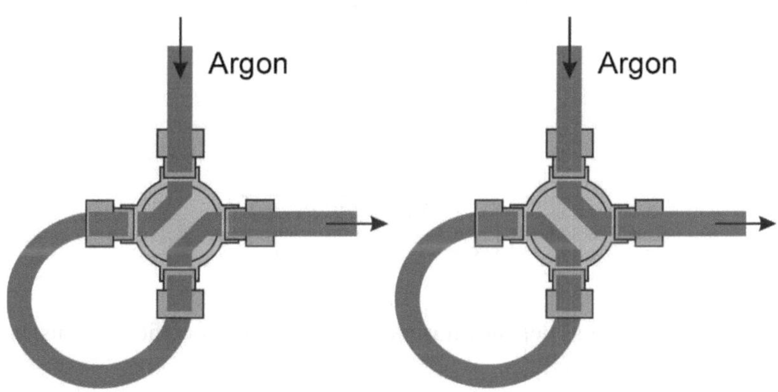

Fig. 14 Four-way valve for switching between ignition and operation pressure. The setting for the ignition pressure (*left side*) bypasses the capillary loop

transducer directly into the plasma and ionization chambers is possible, it would further complicate the layout and operation of the system. Therefore, in order to introduce an appropriate pressure stage, a simplified pressure distribution model is used (Fig. 15). This is based on a chamber (plasma or ionization chamber, respectively) with an aperture of cross-sectional area A through which gas is pumped out.

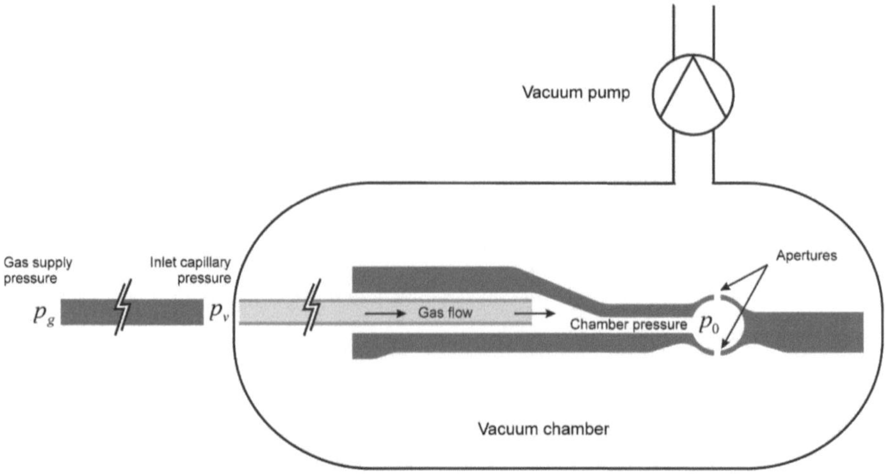

Fig. 15 Schematic representation of the pressure drop model used for pressure calculation in the plasma and ionization chambers

Three pressure stages are relevant to design the capillary lengths and diameters as shown in Fig. 15 The pressure regimes for stable operation were evaluated by measurements using pressure sensors and mass flow controllers. They are summarized in Table 1:

Assuming a constant pressure p_0 in the chamber, the throughput, which is the flow out of the chamber into vacuum, or gas flow pressure product q_{pV} at steady state is [27]:

$$q_{pV} = A \cdot K_g \cdot p_0;$$

where

$$K_g = \left(\frac{2}{\kappa+1}\right)^{\frac{1}{(\kappa-1)}} \sqrt{\frac{2 \cdot \kappa}{\kappa+1} \frac{R \cdot T}{M_{molar}}},$$

κ, R and M_{molar} are the gas dependent isentropic exponent, specific gas constant, and molar mass, respectively.

The conversion of the throughput q_{pV} into volume flow yields the values listed in Table 1. Since in the capillaries laminar flow prevails (Reynolds number RE << 2,300), the pressure drop along the capillaries follows *Hagen–Poiseuille´s equation*:

$$q_{pV} = \frac{\pi \cdot d^4}{128 \cdot \eta \cdot l} \frac{\left(p_v^2 - p_0^2\right)}{2},$$

where η, l, and d are the dynamic viscosity of the gas, length of the capillary, and the capillary diameter, respectively.

Table 1 Pressure regimes for stable operation of the PIMMS

Chamber	Mode	Gas flow (μL/min)	Gas supply pressure, p_g (Pa)	Capillary pressure, p_V (Pa)	Chamber pressure, p_0 (Pa)
Plasma	Ignition	1,000	1.5×10^5	6×10^4	900
	Operation	100	1.5×10^5	2×10^4	100
Ionization	Operation	45	1×10^5	1×10^4	15

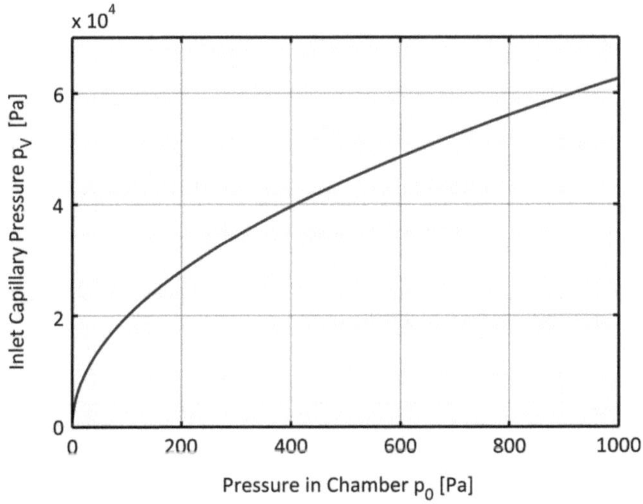

Fig. 16 Pressure in the chamber vs. inlet pressure; capillary: length, 10 cm; diameter, 100 μm; aperture size A, $2\times40\times300$ μm²

By defining:

$$K_c = \frac{\pi\cdot d^4}{128\cdot\eta\cdot l}$$

the pressure p_0 results:

$$p_0 = -\frac{A\cdot K_g}{K_c} + \sqrt{\left(\frac{A\cdot K_g}{K_c}\right)^2 - p_v^2}.$$

In Fig. 16 the result of the pressure drop model for the chosen geometry is shown. In this curve only the 10 cm long capillary is considered, since that is the point where p_v is defined at according to Fig. 15.

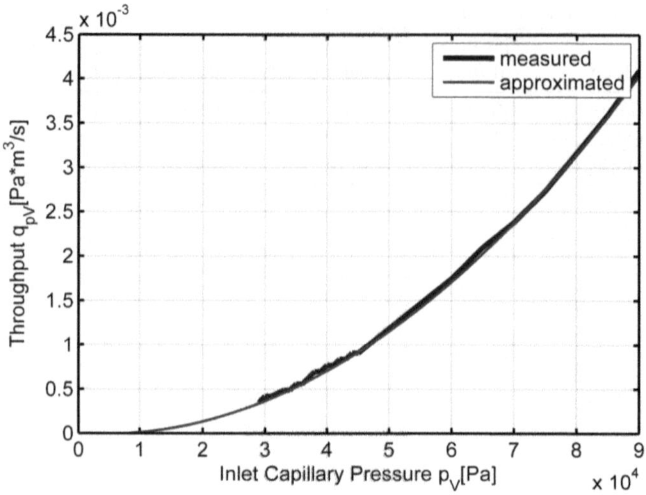

Fig. 17 This figure depicts the computed and measured throughput of the plasma chamber vs. the inlet capillary pressure extrapolated to low flows

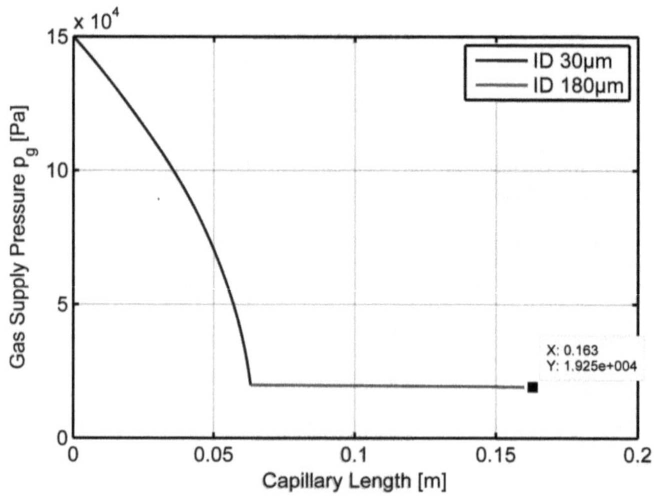

Fig. 18 Pressure drop diagram for the plasma chamber in operation mode. The inlet capillary pressure is 2e4 Pa

Finally, the gas supply pressure p_g is transformed to the desired inlet capillary pressures p_v listed in Table 1, according to the *Hagen–Poiseuille based equation*:

$$p_v = \sqrt{p_g^2 - \frac{2 \cdot q_{pV}}{K_C}}$$

with q_{pV} as determined above (Fig. 17).

The current gas supply consists of two capillaries of different diameters that satisfy the previously mentioned pressure values (30 and 180 μm). Thus, a rapid pressure drop across a short length is accomplished with a short delay time, which is about 350 ms in this arrangement. The pressure drop of the sample gas for the plasma chamber in operation mode is shown in Fig. 18.

5 Fabrication

MEMS (microelectromechanical systems) are systems with small device sizes of 1–100 μm. They are typically driven by electrical signals. To fabricate such systems materials like semiconductors, metals, and polymers are commonly used. MEMS technology fabrication is very cost-efficient. The structures are transferred by processes, which are applied to many systems on one substrate or even many of them simultaneously. The most important fabrication processes are: physical vapor deposition (PVD), chemical vapor deposition (CVD), lithography, wet chemical etching, and dry etching. Typical examples for MEMS are pressure, acceleration, and gyro sensors [28, 29], DLPs [30], ink jets [31], compasses [32], and also (bio)medical devices.

The fabrication of the PIMMS-chips is based only on MEMS standard processes.

Figure 19 shows a cross sectional draft of all the MEMS processes which are necessary to generate the PIMMS. All parts of the mass spectrometer are planar integrated on one chip, i.e., no individual adjustments of the single components are needed. The mechanical composition is a glass–silicon–glass"sandwich." For electrical contacts the metals titanium, nickel and gold are used.

The fabrication process follows the sequence:

1. *Bottom glass wafer*

   - Deposition of Ni and Au on a glass wafer substrate via sputtering
   - Patterning of Au layer by photolithography and wet chemical etching
   - Patterning of Ni layer by photolithography and wet chemical etching

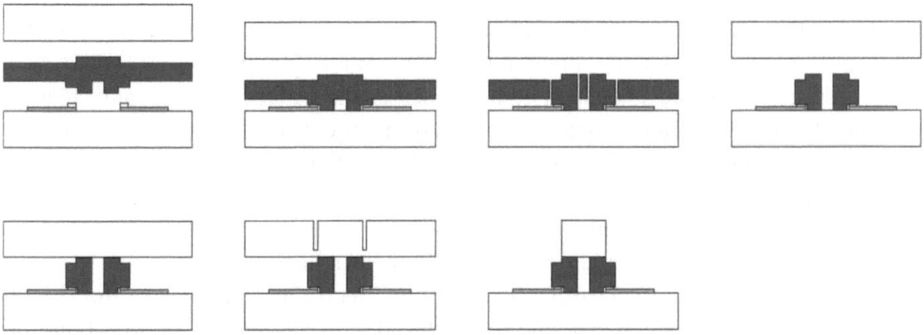

Fig. 19 MEMS process steps

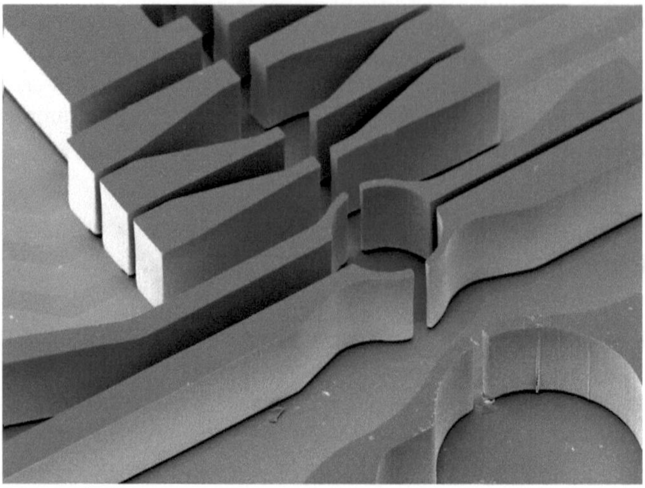

Fig. 20 REM picture of the silicon structures

2. *Silicon wafer*
   - Photolithography and DRIE etching of the 500 nm step
   - Photolithography and DRIE etching of the 5 μm step
3. *Assembly of the sandwich*
   - Anodic bonding of the bottom glass wafer and the silicon wafer
   - Photolithography and DRIE etching through the silicon wafer
   - Removing of released structures (Puzzle-Process)
   - Anodic bonding of the top glass wafer

All mechanical structures of the PIMMS are fabricated in a one mask anisotropic deep reactive-ion etching process (DRIE, Bosch- or ASE(Advanced Silicon-Etch)-Process. This process allows the production of silicon structures of high aspect ratio, because it etches highly anisotropically with approximately vertical side walls. The characteristic aspect ratio is achieved by combining isotropic plasma-etching and passivation in alternating process steps. In order to achieve separated highly conductive structures, such as plasma chamber walls and the finger electrodes of the separator, the DRIE process is applied to a low resistive n-doped silicon wafer (300 μm), which is anodically bonded to a glass wafer (500 μm).

In order to guarantee a constant etch rate across the wafer independent of feature size the etching is restricted to narrow gaps around the desired patterns. This allows to remove narrow as well as extended areas of the silicon. The areas to be removed are not bonded to the bottom glass substrate. The released areas are removed afterwards just by turning the sandwich upside down.

Figure 20 shows a REM picture of the silicon structures.

Metal conductors are necessary for applying electrical potentials to the silicon structures. Due to the small size and the material a direct welding or wire bonding would be tedious. Therefore, the bottom glass of the sandwich is furnished with nickel tracks with gold pads on top. The gold supplies the actual contact between nickel conductors and silicon.

These metal structures are deposed via sputtering and patterned by wet chemical etching. During anodic bonding the gold pads alloy with the silicon and form an eutectic contact.

To minimize mechanical stress between the nickel conductors and the silicon, a 500 nm step, which corresponds to the height of the metal tracks, is etched into the silicon before the wafers are bonded. Another 5 μm deep step is etched at areas, which are to be removed, i.e., anodic bonding is to be prevented.

After the DRIE-process the second glass wafer is bonded on top of the silicon structures to close the plasma and the ionization chamber. It also serves to mechanically stabilize the system. To install the electrical contacts and the gas supply the PIMMS chip is glued to a PCB and the gold pads of the chip are connected via wire bonding to the PCB. After that, as described in Sect. 4, capillaries are glued to the plasma-/ionization-chamber inlet to connect the PIMMS to plasma and sample gas supplies.

6 Hardware and Software

From the numerous and partly new functional principles as discussed in previous sections the number and complexity of the peripheral components needed to operate the PIMMS can be estimated. Besides the PIMMS-chip itself and the environment providing an appropriate vacuum the electronic hardware is another very important parameter, which determines the cost and size of the device. As the many subsystems of the spectrometer and their interactions are not standard, in many respects completely new electronic has to be generated to drive the system and read and evaluate the measurements, respectively. To allow real-time applications, such as online control, several hardware components have to work independently from a central controller, which is another challenge for the firmware and software implementation. This section presents an overview on present state of the hardware, firmware, and software infrastructure for the PIMMS and will outline the further steps for industrialization.

6.1 Electronic Components

As a result of miniaturization several electrical extremes are concentrated on the chip area of ca. 1 cm²: There are DC-voltage-sources in range of −2,000 to 200 V driving electrodes and the MCP, the micro-plasma electron source is powered by a RF-generator at 2.45 GHz, which is ignited by an arc discharge. For the ion separation steep pulse trains of several volts up to 270 MHz with a bandwidth of

Table 2 PIMMS structures and the assembly of electronic components

	DC-Sources	HV-Source	RF-Generator	Plasma-DC-bias-detection	HV-pulse-generator	pA-current measurement	Traveling wave Gen.	Valve control	Pump and pressure control
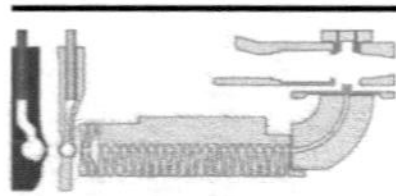 Plasma Chamber			×	×	×			×	×
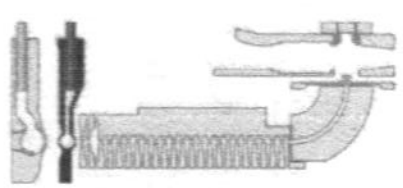 Ionization chamber	×							×	×
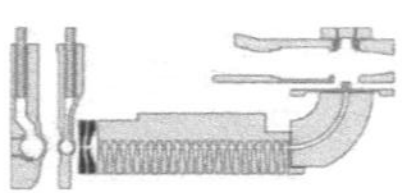 Ion optic	×								×
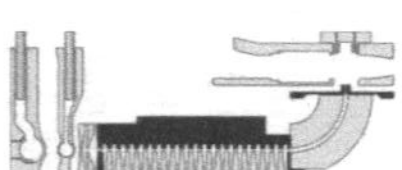 Separator (DC)	×								×

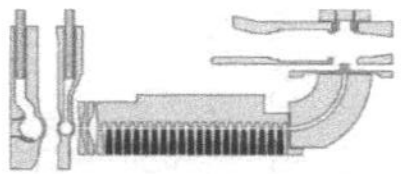 Separator (AC)	×			×	×
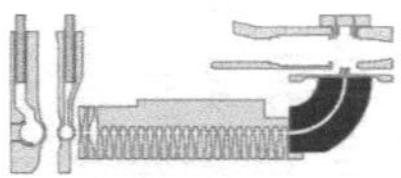 Energy filter	×				×
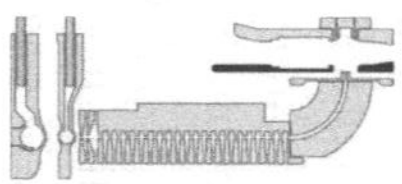 Microchannel plate (MCP)		×			×
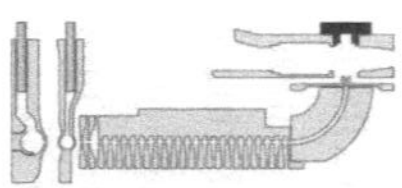 Detector	×		×		×

approximately 2 GHz are needed. Ion currents typically in the range of femto- to a few picoamperes are to be measured. Some of these extremes at these small dimensions can only coexist under vacuum conditions, i.e., will damage the chip at atmospheric pressure. Table 2 summarizes the assembly of electronic components and the PIMMS structures which they are to drive.

6.2 *Pumps, Valves, and Pressure Monitoring*

The amount of electrons extracted from the microplasma, the ionization rate of the sample gas, and the mean free path length depend on the pressure at the different locations of PIMMS-chip. Therefore, exact knowledge and control of pressure and gas flow rates will be necessary for quantitative analysis. Pressure sensors, valves for gas inlets, and vacuum pumps are the components, which have to be read out and controlled, respectively, by electronic and software to install and stabilize appropriate pressure regimes.

Presently a commercially available two stage vacuum system comprising a membrane (Pfeiffer MVP 006-4) and a turbo pump (Pfeiffer HiPace™ 10) in combination with a pressure sensor (Leybold Vacuum Ionivac ITR 90) establish a pressure of about 0.1 Pa in the system. Three electric valves are used to control the gas flow into the capillary system and for the bypasses. The use of macro devices simplifies the handling of the experimental setup and also the electronic control. Pressure drops for plasma and sample gases are accomplished by an appropriate combination of capillaries with different diameters and lengths as described in Sect. 4.

Presently the electronics controlling the pumps and valves predominantly protect the turbo pump and the PIMMS from damage. Since the high voltages driving the system could initiate detrimental arc discharges between electrodes and metal structures, they are only to be applied at low pressure. Therefore, this control of high priority is supplied by an independent unit.

6.3 *DC-Sources*

Extraction and acceleration of electrons and ions, focusing, and energy filtering of the ion beam rely on stable but tunable electric fields. Different potentials have to be applied to the electrode structures to generate these fields.

Figure 21 and Table 3 summarize the necessary DC-sources applied to the structures of the chip. Presently all DC-sources are programmable and variable in the range of −230 to +230 V. This flexibility is essential to determine the optimal parameter setting during system optimization. Once it is known many of these sources can be replaced by fixed ones. Still, due to production tolerances and long term drift effects the sources for ion extraction and focus (U_{IFO}), energy filter (U_{SI}, U_{SO}), and MCP (U_{HV}) will have to remain variable for optimum device performance.

Instability or noise of any of these sources will lead to loss of resolution and/or sensitivity in the mass-spectra. For example, noise on the electron extraction voltage

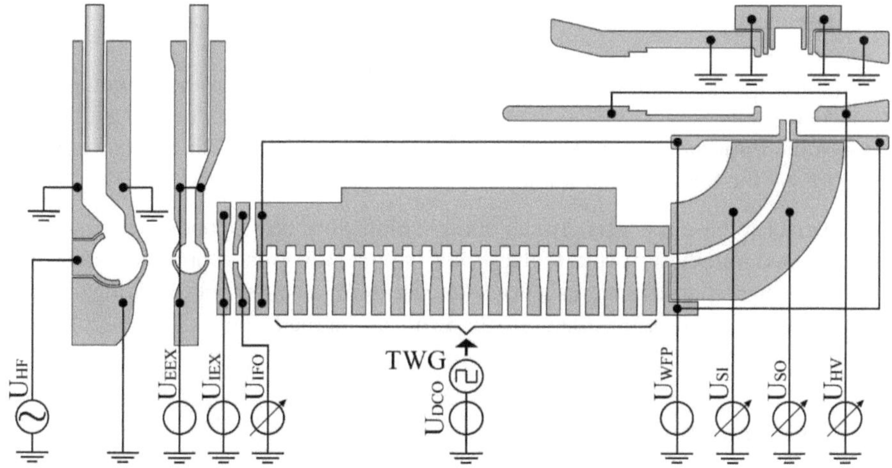

Fig. 21 Configuration of DC-sources applied to electrodes of the PIMMS

Table 3 Voltages applied on electrodes of PIMMS [19]

DC-Source	Signal	Possible range (V)	Typical values (V)
U_{WFP}	Traveling field potential	–230 ... 230	0
U_{EEX}	Electron extraction	–230 ... 230	100
U_{IEX}	Ion extraction	–230 ... 230	–25
U_{IFO}	U_{IFO} ion focus	–230 ... 230	30 ...75
U_{SI}	Inner sector	–230 ... 230	0 ... –25
U_{SO}	Outer sector	–230 ... 230	0 ... 25
U_{DCO}	DC-offset	–230 ... 230	2.5
U_{HV}	High Voltage	–2,000 ... 0	–1,300 ... –400

U_{EEX} will cause varying ionization rates in the ionization chamber as well as a broadening of the ion energy distribution focused in the separator:

$$E_{ion} = q \cdot \left(\underbrace{\left(U_{EEXstable} + U_{noise}\right)}_{U_{EEX}} - U_{WFP} \right)$$

with q the unit charge. Noise of U_{IFO}, U_{SI}, and U_{SO} is even more critical, since these voltages strongly affect the shape of the ion beam.

As already mentioned U_{IFO}, U_{SI}, and U_{SO} are variable DC-sources. Their settings are optimized for each chip individually with respect to maximum ion throughput at deactivated separator. This follows the algorithm:

1. Energy filter optimization by varying U_{SI} and U_{SO}: The voltages are swept in the ranges [U_{WFP}, U_{WFP} − 25 V] for U_{SI} and [U_{WFP}, U_{WFP} + 25 V] for U_{SO} simultaneously, while the ion current is measured at the detector and plotted versus the

difference $U_{SO} - U_{SI}$. The settings corresponding to the maximum in Fig. 22a are optimum.
2. By varying U_{IFO} the focus point of ion beam is varied. Again the maximum ion current against U_{IFO} is the optimum. The voltage is varied between 10 and 120 V with the optimum typically between 30 and 75 V. Figure 22b shows a typical result for the ion focus variation.

To perform these not time critical optimization steps corresponding routines are implemented into the software of the present setup. However, in order to reduce the power-on and response times, as it is advantageous for a number of operating modes, this procedure will be embedded into the firmware in the future.

6.4 HV-Source

A MCP is used in the PIMMS as a secondary electron multiplier (see Sect. 3.7). The electron current measured after MCP compared to the initial ion current is amplified by a factor of 10–1,000. The secondary electron emission coefficient is an averaged number of secondary electrons emitted after each impact. This number depends on the initial energies of the electrons or ions and so on the voltage applied to the MCP. The amplification factor of a MCP configuration is expressed as:

$$F = \left(c_{se}\left(U_{HV}\right)\right)^{n}$$

$c_{se}\left(U_{HV}\right)$ is the voltage and material dependent secondary electron emission coefficient and n the average number of impact stages. A noise level of 1 $V_{p\text{-}p}$ for U_{HV}=2,000 V (0.05 %) generates a noise on the amplified signal in order of the signal magnitude. Thus, the HV-source should be as noiseless as possible. Higher MCP-voltages (>3 kV), as commonly used in standard mass spectrometers, require chamber pressures <0.1 Pa, higher pressures, however, restrict allowable voltages to <2,000 V.

Currently, a commercial HV-Source is used (Spellman MPS3N 10/24). A miniaturized, integrated, and noise reduced source is presently developed. It is based on the EMCO GPMT module, a commercially available HV-source designed to drive photo multiplier tubes.

6.5 Micro-Plasma Support

To ignite the microplasma the pressure of the inert gas in the plasma chamber is temporally increased to allow a controlled arc discharge according to the Paschen law. With the RF-power applied between the center electrode and the chamber walls

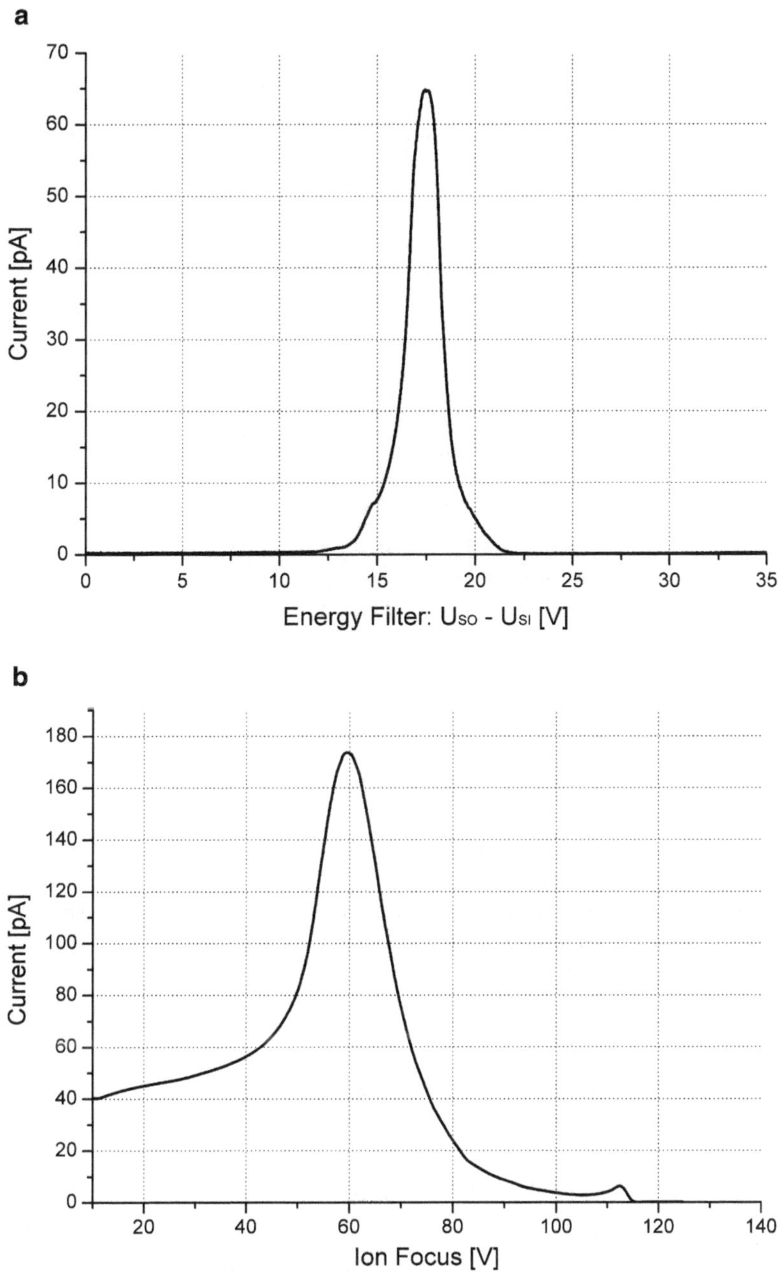

Fig. 22 (**a**) Energy filter variation; (**b**) ion focus variation

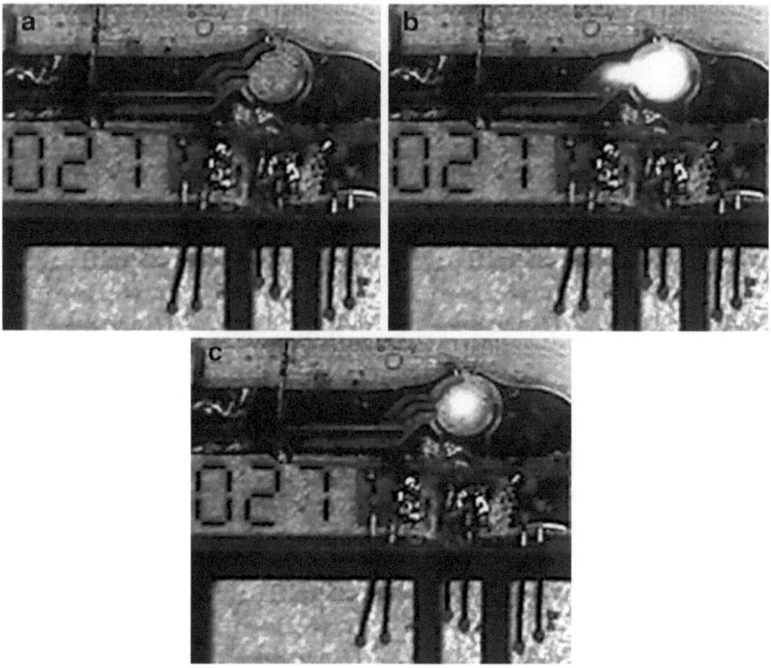

Fig. 23 Plasma chamber (**a**) before ignition; (**b**) at ignition pressure after arc discharge; (**c**) at operation pressure

the plasma is sustained and the gas pressure can be reduced. Figure 23 demonstrates the ignition of plasma in a PIMMS-chip:

As known from plasma physics the potential of the electrode with smaller area, which is the powered electrode in PIMMS, will turn negative. This DC-bias voltage is used for the detection that the plasma is on.

Thus, to ignite, sustain, and control the microplasma the following electronic components are necessary:

- pressure and valve controller, as described in Sect. 6.2,
- high voltage pulse generator to generate the arc discharges,
- RF-Generator,
- DC-Bias-Voltage measurement of the powered electrode.

The high voltage pulses are generated by discharging of a capacitor via the secondary winding of a transformer. As a consequence the primary winding delivers a high voltage pulse. Since the igniting electrodes in the plasma chamber are very closely spaced, an arc discharge is easily generated, when the local gas pressure is high (Fig. 3). The arc current has to be limited to avoid evaporation of the electrodes. With the present solution >50,000 ignition cycles were performed on one chip without observing any structural change of the electrodes.

The RF-generator (designed by Krohne Messtechnik GmbH) delivers about 1 W at a frequency of 2.45 GHz (IMS-band). Due to the very low impedance of the plasma presumably most of this power is dissipated between oscillator an plasma chamber. Once the chip will be packaged in a small housing the RF-power can be significantly reduced to <100 mW and the generator be further integrated. Obviously, the impedance of the plasma changes dramatically from ignition to stable operation, which complicates impedance matching and presently also makes an oversized RF-source necessary.

The DC-Bias-Voltage on the RF-electrode of the plasma chamber is used during the plasma ignition procedure to monitor a successful ignition and stop the pulse train by software. The DC-Bias-Voltage is also a measure for the intensity of the plasma. As the plasma state changes abruptly during this process as shown in Fig. 23b, c, the DC-Bias-Voltage drops abruptly.

6.6 Traveling Wave Generator

Complementary to the physical design of the chip the signal generator driving SIS-separator governs the measurable mass range and resolution of the system. As described in Sect. 3.5, the traveling wave generator (TWG) generates rectangular signals. These are applied to the finger electrodes to generate the local electric fields in the separator which deflect all ions, which are not meant to be detected.

The TWG is one of the most challenging components of the electronics. In order to analyze a wide spectrum of masses within a few seconds the time to change from one selected mass to the next, i.e., from one stable TWG-frequency to the next is in the order of a few milliseconds. The mass-to-charge ratio depends on frequency according [33]:

$$qU = \frac{1}{2}mv^2 \Rightarrow \frac{m}{q} = \frac{2U}{(df)^2},$$

U is the ion acceleration voltage and d the distance of the finger electrodes driven by the same signal of the generator. For a linear sweep of the mass spectrum, the frequency sweep has to be nonlinear in time, i.e., an independent controller for a fast computation of the sweep is necessary.

The minimum requirements for the TWG are:

- Two 180 ° phase shifted rectangular signals with 50 % duty cycle.
- Programmable signal amplitude from 2 $V_{p\text{-}p}$ to 5 $V_{p\text{-}p}$, to adapt different electrode configurations. (In future fixed amplitudes may be sufficient).
- Rise and fall time of the signals have to be very short, since a fast switching of the electric field within the separator determines its mass resolution. The actual generation of TWG achieves rise and fall times less than 0.5 ns at 5 $V_{p\text{-}p}$ amplitude.

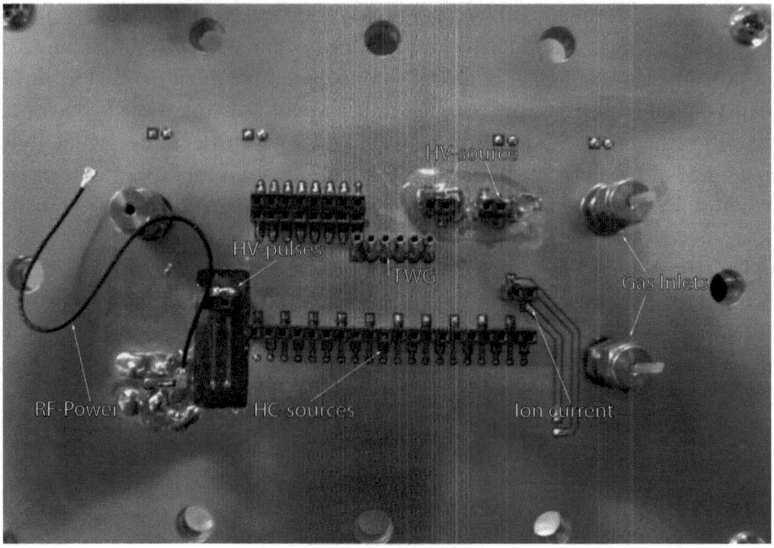

Fig. 24 The vacuum seal and electrical contact vias through the PCB

- For a mass range of 1 *m/z* to ca. 1,200 *m/z* a frequency range for the pulse train of 250–5 MHz is needed for the actual chip design.
- Ramp time of the staircase shaped frequency change has to be <1 ms.
- For the desired mass resolution the frequency resolution has to be better than 100 Hz.
- For the optimization steps for the ion extraction, as described in Sect. 6.3, all generator signals have to be set to 0 V.

As the broadband rectangular signals cannot be matched to the impedance of the finger electrodes, the TWG has to be placed physically close to the evacuated PIMMS-chip and the signals tracks to the chip have to be as short as possible. However, the generator cannot be placed into the vacuum chamber, e.g., because of insufficient heat dissipation. In the present concept a PCB, as shown in Fig. 24, serves as both the vacuum seal and via for the electronics within and outside the vacuum chamber. Thus, the critical signal tracks are reduced to few millimeters. In future an appropriate vacuum tight housing of the chip will meet this demand even more effectively, as the driving stages of TWG can be placed directly next to the chip housing on the PCB.

In an early version a more complex TWG with four variable phase shifted signals with controllable duty cycle [34] was investigated, primarily to determine the optimum signal shape driving the SIS-Separator. From these results the actual version was derived, which due to the concentration of the electronics and signals to the inevitable functions can be minimized with respect to size, cost, and power consumption even further. Still a generator-PCB circuit of 10×5 cm² (Fig. 25a) appears rather large as compared to the PIMMS-Chip.

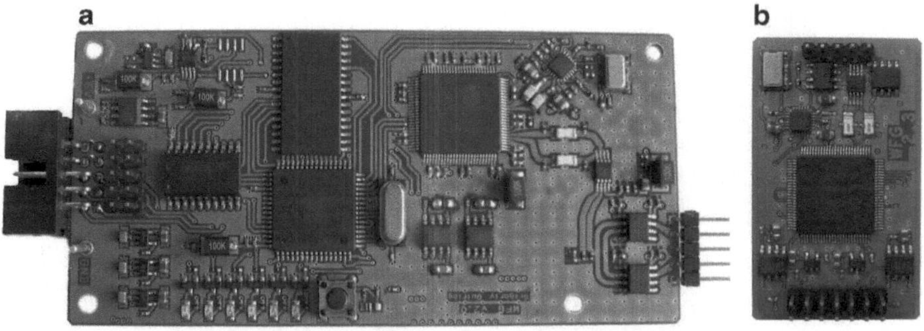

Fig. 25 (**a**) TWG2.0 with PCB size of 10×5 cm² and (**b**) TWG2.3 with PCB

The next TWG-redesign on a PCB of 5×3 cm² as shown in Fig. 25b will be used for the next PIMMS-generation.

6.7 Current Measurement

The maximum ion currents to be detected, when amplified by an MCP, are in the range of few hundred picoamperes. The current is negative, since the MCP generates electrons, and positive, if the ion current is directly measured.

The present setup, therefore, allows capturing currents between −250 and 250 pA. It incorporates an 18-bit high speed successive approximation analog to digital converter (ADC). Since 1 bit determines the sign, the signal resolution is 17 bits, i.e., 131,072 distinct values. Two neighbored values differ by 1.9 fA. In order to detect gas concentrations of 1 ppm and less another order of magnitude is necessary corresponding to a >20 bit resolution. This is accomplished by the high speed sampling capability of the ADC. By oversampling in hardware typically a thousand values are sampled within 1 ms, which increases the number of interpretable bits to 22.

For currents in the pico- and femtoampere range the signal to noise ratio (SNR) becomes low. Without post processing of the measured signal absolute noise levels of about 140 fA were achieved. This would make a ppm resolution impossible. The described oversampling reduces noise to <25 fA. Further noise reduction to <100 aA is achieved by averaging in firmware and software (see below). It should be noted that this way to reduce noise will correspondingly increase the time to capture the spectra. The analog to digital conversion hardware is shown in Fig. 26. The size of the PCB is 5×3 mm²; it contains the low noise power supplies, the ADC, and the averaging stage,

For lower noise also the input stage of the converter has to be placed as close as possible to the detector. This input stage is the only electronic component placed within the vacuum chamber. The assembly of the PIMMS on the PCB with the metal box containing the shielded input stage of the pA-electronics is shown in Fig. 27.

Fig. 26 ADC with hardware averaging on a PCB 5×3 cm²

Fig. 27 Shielded input stage of current measurement placed near to PIMMS in the vacuum chamber

6.8 Firmware and Software

As already discussed, most of the components or groups of components of the PIMMS periphery are controlled by their own microcontroller. Their independence enables secure and fast operation. It is also important to keep the functional groups separate at present stage of development, in order to improve and exchange them individually. In a final system a higher level of integration can be achieved. Using a central unit with enough processing power, e.g., an ASIC (Application-Specific Integrated Circuit) comprising all digital and most analog components will guarantee a very small and cost-effective design, once mass production of PIMMS-devices is intended. Such an ASIC, e.g., could implement the DDS-core of the TWG, the

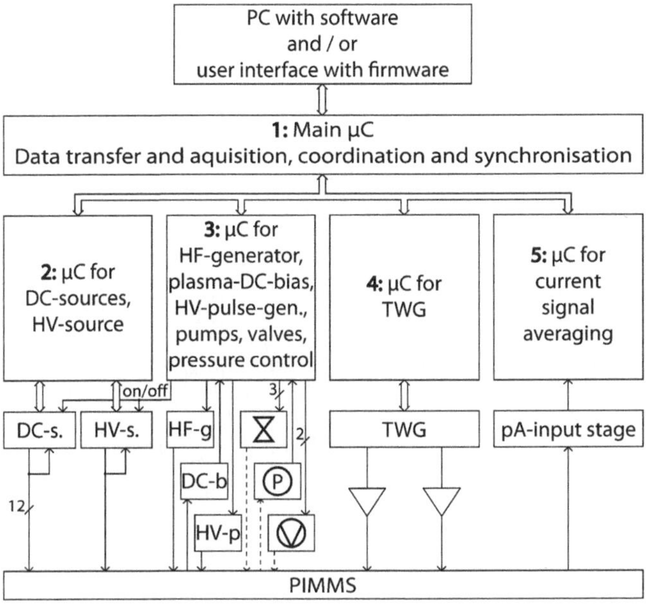

Fig. 28 Present topology of electronic components

ADCs for the DC-sources, the averaging stage for the current measurement, and the user interface.

The realized topology is shown in Fig. 28. Five firmware components are necessary in this configuration. The tasks of the components 2–5 were described in previous sections. With the exception of averaging microcontroller (µC) for the current measurement, all firmware components are implemented in C on standard general purpose microcontrollers. The averaging µC is a CPLD (Complex Programmable Logic Device) and is programmed in Verilog. The main µC serves as a coordinating unit. It responds to user requests from a PC and/or user interface, assembles states of other components, processes data, and transfers them between components and PC. During the measurements it also synchronizes the involved components. For communication between the microcontrollers an SPI-Bus (Serial Peripheral Interface) is used. PC and main µC communicate via USB. The software presently implemented on the PC is of experimental character. It was designed to evaluate the performance of the different chip designs and to determine the optimum operational conditions and their settings. It comprises a graphical user interface, which enables fast access to all settings of the setup and the measurement of variations and spectra. Furthermore, a script language is integrated into the software, which allows to create customized sequences, protocols, and automatic benchmarking of the results, as well as mathematical post processing of measured spectra.

Fig. 29 Design of a hand portable device

Presently a portable prototype is under construction, which includes a fully equipped PIMMS within a volume of 4,000 cm^3 at a weight of <7 kg (Fig. 29).

7 Results

The aspects considered for a reliable evaluation of the performance of the miniaturized mass spectrometer are repeatability, resolution, and detection limit. Over the last years, these parameters could be improved by theoretical analyses and validated with successive chip generations [34, 35].

The first spectra obtained with the mass spectrometer date back to 2007, where gas mixtures of Neon, Air and Argon were measured [23]. This PIMMS-device was the first in which the separation principle relied on the new Synchronous Ion Shield (SIS) Separator, and the first proof-of-principle for the total integration of a mass spectrometer using batch processes of MEMS technology. Also a quantitative correlation could be stated from the peak heights for different gas concentrations.

Figure 30 displays a spectrum of lab air, neon, and argon. The Ar peak represents the leakage from the plasma chamber and thus can be used for mass calibration. The measured resolution $m/\Delta m$ at full width at half maximum (FWHM) is 10.8, and 4.3

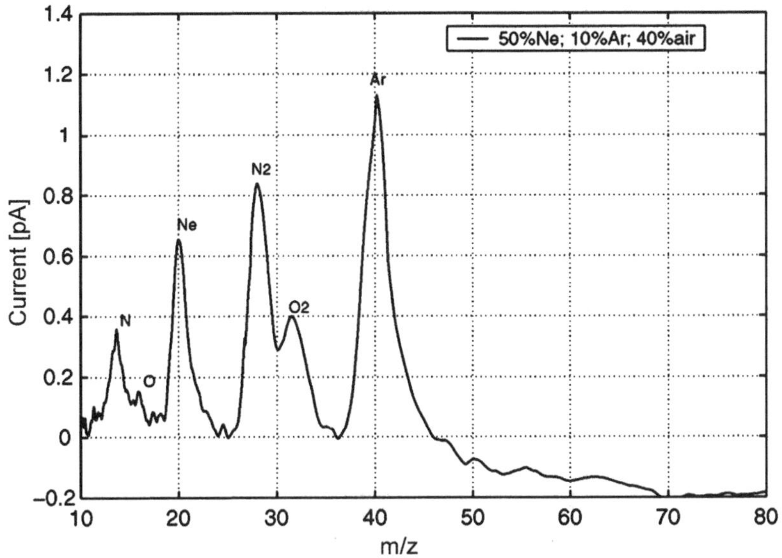

Fig. 30 Spectrum of Ne, Ar, and air 2007 [19]

at the base, respectively, in good agreement with theory. Due to the maximum available signal frequency of 35 MHz the lower mass detection limit at the chosen ion energy of 50 eV the theoretical detection limit *m/z* was about 6, as confirmed by the carbon peak at $m/z=12$.

Based on the simulations discussed in Sect. 3 for the succeeding chip generation the total length of the mass separator was extended from 2.3 to 3.9 mm. This extension, for identical finger width d_1 and the gap d_2, should increase the base resolution of the mass separator R according to the theory from Sect. 3.5 to 9.5. Additionally the maximum operating frequency of the mass separator and the ion energy were doubled (70 MHz and 100 eV) allowing for a detection limit of $m/z=0.5$.

According to Sect. 3.5, this increases the detection limit by a factor of 4. In this measurement methane was fed into the system. The resolution at the base is approximately *m*, m=8.5, which is smaller than what was theoretically expected (9.5). Resolution measured at FWHM is 32. Hence, the improvement in comparison to the previous chips generation is evident. In order to push the detection limit down to $m/z=0.5$, the finger electrodes were wired such that the spacing d_1 was artificially doubled.

Figure 31a, b, show a measurement with the extended mass separator. The methane spectrum in Fig. 31a is shown with a hydrogen peak with $m/z=1$. The spectrum in Fig. 31b was measured with a Xenon plasma, this allows to detect the 400 ppm argon in air. The current performance of the PIMMS is summarized in Table 4.

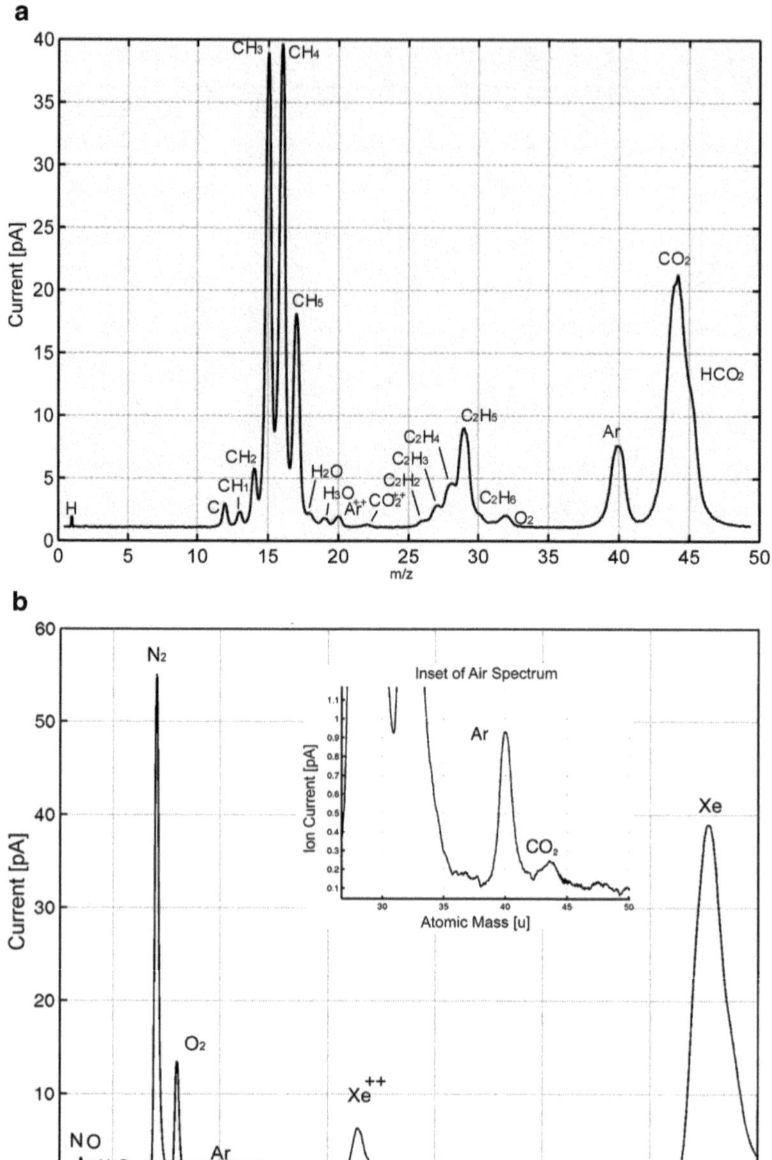

Fig. 31 (**a**) Spectrum of methane and (**b**) spectrum of air and xenon

Table 4 Performance of the PIMMS

Mass range	Resolution	Sensitivity	Scan time
0.5–200 m/z	43	100 ppm	150 m/z in 4.5 s

8 Conclusions and Outlook

The PIMMS is presently able to detect molecules in a mass range from 1 to about several 100 amu with a resolution of 43.

Due to the batch oriented fabrication by standard MEMS processes the PIMMS has a high potential for a cost-effective fabrication in large volumes. Because of its small size it allows to access application areas, where conventional mass spectrometers can or will not be applied, e.g., mobile applications or in line real-time monitoring in the chemical industry, environmental surveillance, patient or air condition monitoring in buildings or vehicles.

However, to address these applications a size reduction of the periphery will be essential. This holds for the electronics as well as for the vacuum system and sample supply. Since there is presently not much research in this field, these problems have been also addressed in course of the PIMMS development.

Microsubsystems presently under investigation are:

- an integrated ion/electron amplifier
- a sample and plasma pressure stage with low dead volume
- a liquid sample vaporizer and
- vacuum pumps

8.1 Integration of Ion/Electron Multiplier

The presently used amplifier in the system is a hybrid-integrated MCP. This hybid integration is disadvantageous for a mass production, since it makes an individual handling of each system necessary to insert the MCP. To allow for an integrated amplification, a MCP-type device is not the optimum choice. Instead a "classical" dynode arrangement appears to be appropriate, since it can be fabricated in the same way and processes as the rest of the PIMMS and is less sensitive to higher pressures than an MCP, as they are intended in the PIMMS. Such a secondary electron amplifier is presently under investigation.

8.2 Pressure Stage in Silicon

Despite the miniaturized size of the PIMMS, the way gases are fed into the system is not optimum for total system integration. Capillaries have to be glued to each system individually, which in addition to tedious handling decreases the manufacturing yield due to defects such as blocked capillaries and/or chambers. Furthermore, according to pressure analysis, the quadratic pressure drop dependence versus length is critical when the vertex of the parabola is approached. A slight variation in the capillary length (in the sub-mm range) causes a pronounced variation in the

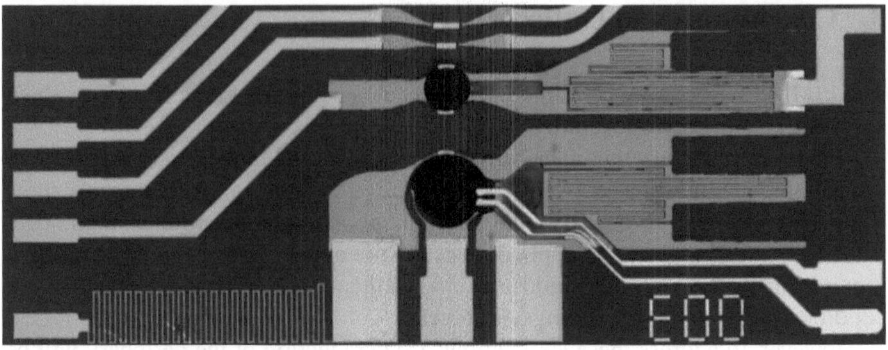

Fig. 32 Microchannels in the PIMMS for gas intake

actual pressure in the chamber, which requires being very accurate for adequate function. Since rather long the capillaries have to be used, small tolerance may have a strong impact.

Microsystem technology with its high precision can be used instead to fabricate the required flow resistances as etched channels in silicon. The advantages of this approach are multiple: the pressure stage length can be easily controlled and by using different channel geometries (width, depth) connected in series, the area necessary to fix the capillaries is avoided, the injection system can be manufactured along with the chip, and gluing of the capillaries is avoided. As a result the injection system is directly integrated into the PIMMS.

In first experiments for this integration very narrow channels with a well-defined length were etched into silicon. Thus, the desired pressure drop was achieved in a process-integrated solution. Figure 32 shows a PIMMS prototype with integrated pressure stages. In this example, capillaries are still glued to the system for connecting the gas supply. The robustness and batch manufacturing capability of this injection system makes it very attractive for mobile and harsh environment applications.

8.3 Microvaporizer

The variety of applications of the PIMMS can be significantly extended, if liquids can be analyzed in an appropriately sized vaporizer.

Consequently such a microvaporizer uses a glass–silicon–glass stack with a microstructured silicon evaporator. This evaporator is heated up by a pulsed NIR-laser diode. Thus, the liquids which are in contact with the evaporator within a very short period of time are heated beyond their evaporation temperature. This arrangement not only requires a low energy to evaporate in comparison to the commonly used resistive heating, but the very rapid heating of the silicon vaporizer also allows for simultaneous evaporation of sample mixtures with components of different evaporation temperatures and partial pressures, respectively. Furthermore, due to the small sample intake, a fast switching of samples, selective selection of samples in a gas flow (e.g., plug flows), and lock-in type process for sample generation and detection is possible.

A combination of the pressure stage and the microvaporizer will also ensure that the vaporized samples will not condense on their way to the ionization chamber, even if the total system is at room temperature.

8.4 Miniaturized Vacuum Pumps

Another part of the periphery that is not yet in accordance with the above-mentioned applications is the vacuum system of the PIMMS. As mentioned in Sect. 4, the gas input to the system is very low. In continuous operation barely 150 μL/min are fed into the PIMMS. All available vacuum pumps on the market are oversized in pump power by several orders of magnitude and are correspondingly large, heavy, and energy consuming. These currently commercial vacuum pumps (Pfeiffer MVP 006-4 and Pfeiffer HiPace™ 10) are used with a total volume of about 2.5 L and a weight of 3.6 kg. For a real mobile field device a miniaturized vacuum pump system is advantageous. Efforts have been made to develop a miniaturized vacuum pump using microsystems technology. At the moment three pumping principles and their transfer to the MEMS domain are investigated also at the TUHH.

As in high vacuum generation in macroscopic scale, the micropump is divided into two parts. As a backing pump for the pressure range from a few 10^2 Pa to atmospheric pressure a micro-sorption pump is under investigation, which relies on surface adsorption effects, as well as a scroll pump, which ranks with its displacement principle among the classic backing pumps.

The high precision with which structures can be etched using DRIE allows for a precise alignment of the structures with minimal clearance between scrolls. Wear problems can be avoided by deposition of polymer films. First experiments have been done with a pump with 10 mL/min pump capacity.

Sorption effects in the macroscopic scale are usually used for non-continuous pumping. They contain a highly porous sorption material like activated carbon or zeolites with a huge inner surface. The sorption material is usually cooled down by liquid nitrogen. They are regenerated by heating the sorption material to temperatures of several hundred degrees Celsius after a disconnection of pump and vacuum-chamber.

This pumping principle can be transferred to the microsystems. In particular the low thermal masses allow for fast heating and cooling. Hence with two or more pumping units, which work parallel in a push–pull cycle, a quasi-continuous pumping can be obtained. Due to the short cycle time freshly outgassed sorption material is available most of the time and because the very low temperatures used in macroscopic pumps are not necessary for the pumping of most gases, a cost-effective cooling with Peltier-elements is sufficient.

For the lower pressures the principle of the diffusion pump is investigated, that uses the momentum transfer between colliding molecules [36].

The diffusion pumps principle has been known for decades and these pumps were widely used till the 1970s. It relies on momentum transfer of a condensable vaporized fluid to the molecules to be pumped. Macroscopic diffusion pumps can generate vacuum down to 10^{-6} Pa.

The transfer of this principle also benefits from the characteristic conditions that count for microfluidic systems. By using MEMS technologies the geometry of the steam nozzles can be reduced drastically without losing relative accuracy. Thus, the overall dimensions of the pump and also the amount of steam that is necessary for operation is reduced. Another advantage of a micro-diffusion pump is that the capillary forces overbalance gravity forces, which are decisive in the macroworld. Hence it is possible to construct a pump that can be operated orientation-independent.

All the above-mentioned pumping principles have been proofed and are momentarily adjusted for the operation to supply the necessary pressure and pump power regimes for the PIMMS.

Acknowledgment This work was funded and supported by several organizations and scientific as well as industrial partners for more than a decade. A number of PhD students, and many Diploma and Master students worked on this subject. It was funded by the City of Hamburg, the German Research Council (DFG), the European Union as well as the German Ministry of Research and Development; scientific partners were LETI and industrial partners Leda Mass, now part of MKS, Bayer Technology Services, and Krohne Messtechnik. PhD students besides the coauthors were Volker Relling, Ralph Siebert, Gerald Petzold, Jan-Peter Hauschild, and Eric Wapelhorst. Neither the PhD students nor the funders ever lost their confidence that finally we would succeed in realizing such a complex and fully integrated mass spectrometer.

Finally we appreciate the critical reading of the text by Winfred Kuipers.

References

1. Lehmann U, Krusemark O, Müller J, Vogel A, Binz D (2000) Micro machined gas chromatograph based on a plasma polymerized stationary phase. Kluwer Academic Publishers, Enschede
2. μGC Technology, Elster Instromet, [Online]. Available www.elster-instromet.com/en/micro_GC_technology.html. Accessed 17 Nov 2011
3. Zimmermann S, Krippner P, Müller J (2002) Miniaturized flame ionisation detector for gas chromatography. Sensors and Actuators B: Chemical 83(1–3):285–289
4. Kuipers W, Müller J (2010) Sensitivity of a planar micro-flame ionization detector. Talanta 82(5):1674–1679
5. Helbling T, Pohle R, Durrer L (2008) Sensing NO2 with individual suspended single-walled carbon nanotubes. Sensors and Actuators B: Chemical 132(2):491–497
6. Brucker GA, Rathbone GJ (2010) Autoresonant Trap Mass Spectrometry (ART MS) for remote sensing applications. International Journal of Mass Spectrometry 295:133–137
7. Tassetti C-M, Duraffoung L, Danel J-S, Lagutére T, Progent F (2011) Poster, Grenoble, France; Arpajon, France: HEMS 2011
8. Wapelhorst E, Hauschild J-P, Müller J (2007) Complex MEMS: a fully integrated TOF micro mass spectrometer. Sensors and Actuators A 138:22–27
9. Brkic B, France N, Clare AT, Sutcliffe CJ (2009) Development of Quadrupole Mass Spectrometers Using Rapid Prototyping Technology. American Society for Mass Spectrometry 20:1359–1365

10. Hogan T, Taylor S, Cheung K, Velasquez-Garcia L, Akinwande A, Pedder R (2010) Performance Characteristics of a MEMS Quadrupole Mass Filter With Square Electrodes: Experimental and Simulated Results. IEEE Transactions on Instrumentation and Measurement 59(9):2458–2467
11. Mastrangelo C, Yeh J-J, Muller R (1992) Electrical and optical characteristics of vacuum-sealed polysilicon microlamps. IEEE Transactions on Electron Devices 39(6):1363–1375
12. Han K, Lee Y, Jun D, Lee S, Jung KW, Yang SS (2011) Field Emission Ion Source Using a Carbon Nanotube Array for Micro Time-of-Flight Mass Spectrometer. Japanese Journal of Applied Physics 50:06GM04
13. Madou M (2002) Fundamentals of microfabrication: the science of miniaturization. CRC, Boca Raton, FL
14. Ramírez Wong RM, Hauschild J-P, Wapelhorst E, Müller J (2009) Optimization of Microplasma for the Application in a Micro Mass Spectrometer. VDE, Berlin
15. Vossen JL, Kern W (1979) Thin Film Processes. Academic, New York, NY
16. Chapnam B (1980) Glow Discharge Processes: Sputtering and Plasma Etching. Wiley, New York, NY
17. Lieberman MA (2004) Principles of Plasma Discharges and Materials Processing. Wiley, New York, NY
18. Scientific Instrument Services, [Online]. Available http://www.sisweb.com. Accessed 18 Nov 2011
19. Hauschild J-P, Wapelhorst E, Müller J (2007) Mass spectra measured by a fully integrated MEMS mass spectrometer. Int J Mass Spectrom 264(2007):53–60
20. Ewald H, Hintenberg H (1953) Methoden und Anwendungen der Massenspectroskopie. Verlag Chemie-GmbH, Weinheim
21. Orloff J (1997) Handbook of Charged Particel Optics. CRC, New York, NY
22. Blaum K, Geppert C, Müller P, Nörtershäuser W, Otten EW, Schmitt A, Trautmann N, Wendt K, Bushaw BA (1998) Properties and Performance of a Quadrupole Mass Filter used for Resonance Ionization Mass Spectrometry. International Journal of Mass Spectrometry 181:67–87
23. Hauschild J-P, Wapelhorst E, Müller J (2009) The novel synchronous ion shield mass analyzer. International Journal of Mass Spectrometry 44:1330–1337
24. Herzog R (1934) Ionen- und Elektronenoptische Zylinderlinsen und Prismen I. Z f Physik 89:447–473
25. PHOTONIS France S.A.S., [Online]. Available http://www.photonis.com. Accessed 18 Nov 2011
26. Reinhardt-Szyba M, Hauschild J-P, Wapelhorst E, Müller J (2009) Ein Mikromassenspektrometer mit integrierter Mikrokanalplatte, Proceedings mikrosystemtechnik kongress 2009, no. VDE VERLAG
27. Jousten K (2008) Handbook of Vacuum Technology. Wiley-Blackwell, Weinheim
28. Qu H, Fang D, Sadat A, Yuan P, Xie H (2004) High-resolution integrated micro-gyroscope for space applications. 41st Space Congress
29. B.S. GmbH, Bosch Sensortec GmbH, [Online]. Zugriff am 18 Nov 2011
30. DLP & MEMS, Texas Instruments, [Online]. Available www.ti.com/analog/docs/dlptoplevel.tsp?sectionId=622&tabId=2441&familyId=1743. Accessed 18 Nov 2011
31. KYOCERA Inc Jet Printhead, Kyocera, [Online]. Available http://global.kyocera.com/prdct/tfc/pdf/07_08.pdf. Accessed 18 Nov 2011
32. Sensitec GmbH, [Online]. Available http://www.sensitec.com/. Accessed 18 Nov 2011
33. Quiring G, Hauschild J-P, Wapelhorst E, Müller J (2009) Optimierung der ansteuerung des SIS-massenseparators im planar integrierten mikro-massenspektrometer. Proceedings mikrosystemtechnik kongress 2009, no. VDE VERLAG
34. Reinhardt M, Quiring G, Ramírez Wong RM, Wehrs H, Müller J (2010) Helium detection using a planar integrated micro-mass spectrometer. International Journal of Mass Spectrometry 295:145–148
35. Quiring G, Reinhardt-Szyba M, Müller J (2011) PIMMS, ein universell einsetzbares Mikromassenspektrometer. Proceedings mikrosystemtechnik kongress 2011, no. VDE VERLAG
36. Doms M, Müller J (2007) A micromachined vapor-jet vacuum pump. Transducers 2007 - international solid-state sensors, actuators and microsystems conference

Index

Q. Alan Xu and Timothy L. Madden (eds.), *LC-MS in Drug Bioanalysis*,
DOI 10.1007/978-1-4614-3828-1, © Springer Science+Business Media, LLC 2012

MIX
Papier aus verantwortungsvollen Quellen
Paper from responsible sources
FSC® C105338

If you have any concerns about our products,
you can contact us on
ProductSafety@springernature.com

In case Publisher is established outside the EU,
the EU authorized representative is:
Springer Nature Customer Service Center GmbH
Europaplatz 3, 69115 Heidelberg, Germany

Printed by Libri Plureos GmbH
in Hamburg, Germany